Flow Control

PASSIVE, ACTIVE, AND REACTIVE FLOW MANAGEMENT

The ability to actively or passively manipulate a flow field to bring about a desired change is of immense technological importance. The potential benefits of improving flow control systems range from saving billions of dollars in fuel costs for land, air, and sea vehicles to achieving more economically competitive and environmentally sound industrial processes involving fluid flows.

This book provides a thorough, up-to-date treatment of the basics of flow control and control practices that can be used to produce desired effects. Among topics covered are transition delay, separation prevention, drag reduction, lift augmentation, turbulence suppression, noise abatement, and heat and mass transfer enhancement. The final chapter explores the frontiers of flow control strategies, especially as applied to turbulent flows.

Intended for engineering and physics students, researchers, and practitioners, *Flow Control* brings together in a single source a wealth of information on current practices and state-of-the-art developments in this very active field.

Mohamed Gad-el-Hak received his Ph.D. in fluid mechanics from the Johns Hopkins University in 1973. He has since taught and conducted research at the University of Southern California, University of Virginia, Institut National Polytechnique de Grenoble and Université de Poitiers, has lectured extensively at seminars in the United States and overseas, and is currently Professor of Aerospace and Mechanical Engineering at the University of Notre Dame.

Professor Gad-el-Hak is world renowned for advancing several novel diagnostic tools for turbulent flows, including the laser-induced fluorescence (LIF) technique for flow visualization; for discovering the efficient mechanism via which a turbulent region rapidly grows by destabilizing a surrounding laminar flow; for introducing the concept of targeted control to achieve drag reduction, lift enhancement and mixing augmentation in boundary layer flows; and for developing a novel viscous pump suited for microelectromechanical systems (MEMS) applications.

The author is a fellow of the American Society of Mechanical Engineers, a life member of the American Physical Society, an associate fellow of the American Institute of Aeronautics and Astronautics and a research fellow of the American Biographical Institute.

In 1998, Professor Gad-el-Hak was named the Fourteenth ASME Freeman Scholar. In 1999, he was awarded the prestigious Alexander von Humboldt Prize—Germany's highest research award for senior U.S. scientists and scholars in all disciplines—as well as the Japanese Government Research Award for Foreign Scholars.

Flow Control
PASSIVE, ACTIVE, AND REACTIVE FLOW MANAGEMENT

MOHAMED GAD-EL-HAK

University of Notre Dame

CAMBRIDGE
UNIVERSITY PRESS

CAMBRIDGE UNIVERSITY PRESS
Cambridge, New York, Melbourne, Madrid, Cape Town, Singapore, São Paulo

Cambridge University Press
The Edinburgh Building, Cambridge CB2 2RU, UK

Published in the United States of America by Cambridge University Press, New York

www.cambridge.org
Information on this title: www.cambridge.org/9780521770064

First published 2000
This digitally printed first paperback version 2006

A catalogue record for this publication is available from the British Library

Library of Congress Cataloguing in Publication data
Gad-el-Hak, M.
    Flow control : passive, active, and reactive flow management / Mohamed Gad-el-Hak.
        p.   cm.
    Includes bibliographical references and index.
    ISBN 0-521-77006-8 (hb)
    1. Fluid dynamics.   2. Hydraulic control.   I. Title.
    QC151.G32   2000
    629.8′042 – dc21

                                                                99-052941

ISBN-13 978-0-521-77006-4 hardback
ISBN-10 0-521-77006-8 hardback

ISBN-13 978-0-521-03671-9 paperback
ISBN-10 0-521-03671-2 paperback

To Dilek, Kamal, and Yasemin, *sine qua non*

Contents

Preface

Words are food for thought; utter the right ones and you will have created a delicacy to be savored. And so will begin every chapter in this book with the rich words spoken through the ages by the wisest of all men.

The ability to actively or passively manipulate a flow field to effect a desired change is of immense technological importance, and this undoubtedly accounts for the subject's being hotly pursued at present by more scientists and engineers than any other topic in fluid mechanics. The art of flow control is as old as prehistoric man, whose sheer perseverance resulted in the invention of streamlined spears, sickle-shaped boomerangs, and fin-stabilized arrows. The German engineer Ludwig Prandtl pioneered the science of flow control at the beginning of the twentieth century.

The potential benefits of realizing efficient flow-control systems range from saving billions of dollars in fuel cost for land, air, and sea vehicles to achieving economically and environmentally more competitive industrial processes involving fluid flows. The purpose of this book is to provide an up-to-date view of the fundamentals of some basic flows and control practices that can be employed to produce needed effects. Understanding of some basic mechanisms in free and wall-bounded turbulence has increased substantially in the last few years. This understanding suggests that taming of turbulence—the quintessential challenge in the field of flow control—is possible so as to eliminate some of its deleterious effects while enhancing its useful traits.

The present monograph covers current ideas and methods for effecting such control to achieve transition delay, separation prevention, drag reduction, lift augmentation, turbulence suppression, noise abatement, and heat and mass transfer (including combustion) enhancement in both free-shear and wall-bounded flows. Futuristic flow-control strategies will be explored as well. Sophisticated reactive control systems consisting of a colossal number of microsensors and microactuators communicating with each other via massively parallel computers will in the future improve, by leaps and bounds, the fluids engineer's ability to control the all too common turbulent flows.

Speaking of turbulence, some readers may rightly sense that this particular topic is getting more than its fair share of coverage. The argument is often made that laminar flows are easier to control and that every effort should be made to control the flow at this stage *before it is too late*. I could not agree more. However, there are no known strategies that could delay transition (provided that is what one wants to accomplish) beyond a Reynolds number, based on freestream velocity and distance from leading edge, of $Re_x = \mathcal{O}[10^7]$. The boundary layer flow around the fuselage of a large commercial airplane reaches that Reynolds number after only 1 m, which leaves more than 50 m to go. So turbulence for better or worse is there to stay and must be dealt with whatever the difficulties.

The present book is intended for engineers, physicists, educators, and graduate students at universities and research laboratories in government and industry who have an interest in the physics and control of laminar, transitional, and turbulent flows. The different chapters here are written in a pedagogic style and are designed to attract newcomers to the field. A knowledge of basic fluid dynamics is assumed, and the book should help the reader in navigating through the colossal literature available on flow-control fundamentals and practices.

The book is organized into fourteen chapters as follows: The subject is introduced and the important governing equations are derived in the first two chapters. Chapter 3 provides the unifying principles for the broad field of flow control. Coherent structures and Reynolds number effects are reviewed in the next two chapters. Chapter 6 introduces aspects of transition control, and compliant coatings are discussed in Chapter 7. The all-important separation control and low-Reynolds-number aerodynamics are covered in Chapters 8 and 9. The technologically very important drag reduction is reviewed in Chapter 10. Mixing enhancement and noise reduction are briefly discussed in Chapters 11 and 12. Microelectromechanical systems (MEMS) is an emerging field that offers unique possibilities for fabricating inexpensive, communicative, dense arrays of microsensors and microactuators for use in reactive flow control. Both flow physics and typical applications of MEMS are covered in Chapter 13. The final chapter explores the frontiers of flow control, including futuristic ideas for achieving a leap in the performance of control systems.

Hans Wolfgang Liepmann, my academic grandfather, wrote in a recent retrospective of the early days of research on boundary-layer transition (*Applied Mechanics Reviews*, vol. 50, no. 2, pp. R1–R4, 1997): "The importance of asking the right question is as obvious in principle as it is nebulous in research. Always present is the temptation to look for the key under the street light and not in the dark corner where it was lost. Indeed, in my opinion, much of the extensive turbulence research today follows this line and fails to address the real problems." How true those words of wisdom are to the field of flow control and especially to the difficult task of taming turbulence. The author of the present book aspires for the right questions to be asked and for the answers to be sought even in the darkest corners.

Mohamed Gad-el-Hak
Notre Dame, Indiana
1 January 2000

Nomenclature

Few of the symbols below are used to indicate more than one variable or parameter. This is done to preserve long-held conventions. For example, c symbolizes both chord length and wave speed. The proper usage is clearly indicated when the symbol appears in the text.

Non-Greek Symbols

$\mathcal{A}$	control surface
$\lvert A \rvert$	wave amplitude
a	channel half-width; also pipe radius; also rotor radius
a_{o}	speed of sound
B	second constant in the log-law
Br	Brinkman number ($\equiv Ec\, Pr$)
C	Clauser's outer-layer shape parameter
C_D	drag coefficient ($\equiv 2D/\rho_\infty U_\infty^2\, c$; per unit span)
C_f	local skin-friction coefficient ($\equiv 2\tau_{\mathrm{w}}/\rho_o U_o^2$)
C_L	lift coefficient ($\equiv 2L/\rho_\infty U_\infty^2\, c$; per unit span)
C_q	suction coefficient ($\equiv \lvert v_{\mathrm{w}}\rvert/U_o$)
c	airfoil's chord; also complex wave speed
c_{i}	imaginary part of c (determines temporal growth of disturbance)
c_ℓ	propagation velocity for longitudinal waves
c_{p}	specific heat at constant pressure
c_{r}	disturbance phase velocity (real part of c)
c_{t}	propagation velocity for transverse waves
c_{v}	specific heat at constant volume
c_x^+	streamwise phase velocity in wall units
D	drag; also damping coefficient
De	Dean number ($\equiv (Re/2)(\sqrt{a/R})$)
$\mathcal{D}_{\mathrm{K.E.}}$	dissipation of turbulence kinetic energy
d	depth
dE	energy flux
E	elastic modulus
Ec	Eckert number ($\equiv v_o^2/c_{\mathrm{p}}\,\Delta T$)
e	internal energy per unit mass
F	external forcing term

$\mathcal{F}$	flexural rigidity of a compliant surface; also probability distribution
$\mathbf{F}$	external force vector
F_{ij}	force field between two molecules i and j
$F_v(T)$	dimensionless function that characterizes viscosity variation with temperature
f	frequency; also normalized velocity distribution
$f^{(0)}$	Maxwellian velocity distribution
G	first elastic constant (shear modulus of rigidity)
Gr	Grashof number ($\equiv g \beta \Delta T L^3/v^2$)
g	acceleration of gravity
g_i	component of body force per unit mass
H	shape factor (ratio of displacement thickness to momentum thickness)
h	riblet height; also channel depth; also convective heat transfer coefficient
i	imaginary number ($i = \sqrt{-1}$); also used as a suffix in tensor notations
$J(f, f^*)$	collision integral
Kn	Knudsen number ($\equiv \mathcal{L}/L$)
k	spring constant; also Boltzmann constant
k_ℓ	longitudinal wave number
L	lift; also characteristic flow dimension
$\mathcal{L}$	mean free path of gas molecules
L_s	slip length; also characteristic dimension of solid
L_s^o	limiting value of slip length
ℓ	probe length; also mixing length
ℓ_v	viscous length scale ($=v/u_\tau$)
ℓ^+	probe length in wall units ($=\ell/\ell_v$)
$M(\zeta)$	dispersion relation
Ma	Mach number ($\equiv v_o/a_o$)
m	mass; also molecular mass
N	number of molecules
Nu	Nusselt number (dimensionless temperature gradient at the surface)
n	modal amplification; also exponent in velocity power-law; also number density
$\mathbf{n}$	outward unit normal vector
P	bursting period
$\mathcal{P}$	sensor–actuator power consumption
$\overline{P}$	mean pressure
$\mathcal{P}_{K.E.}$	production of turbulence kinetic energy
Pe	Péclet number ($\equiv v_o L/v_T = Pr\, Re$)
Pr	Prandtl number ($\equiv v/v_T$)
p	thermodynamic pressure
p_i	inlet pressure
p_o	outlet pressure
p'	pressure fluctuations
q^2	twice turbulence kinetic energy
q_k	heat flux vector ($k = 1, 2, 3$)
q_x	tangential heat flux
q_y	normal heat flux
R	radius of curvature

$\mathcal{R}$	gas constant
Ra	Rayleigh number $(=Gr\,Pr)$
Re	Reynolds number $(\equiv v_{\mathrm{o}}\,L/\nu)$
Re_a	Reynolds number based on centerline velocity and pipe radius
Re_{c}	Reynolds number based on chord and freestream velocity
Re_{crit}	critical Reynolds number
Re_x	Reynolds number based on distance from leading edge and freestream velocity
Re^*	ratio of boundary-layer thickness (or pipe radius) to viscous length scale
r_{c}	cut-off distance
S	slip factor $(\equiv \frac{2-\sigma_v}{\sigma_v}\,Kn)$
St	Strouhal number $(\equiv f\,L/v_{\mathrm{o}})$
s	riblet spanwise spacing; also entropy
T	temperature
$\mathcal{T}$	period of forcing signal; also time constant for temperature change
T_{gas}	gas temperature at wall
T_ℓ	longitudinal tension of thin plate
T_{o}	reference temperature
T_{w}	wall temperature
T_∞	ambient temperature
t	time
t_ν	viscous time-scale (given by the inverse of mean wall vorticity)
U	upper wall speed in a Couette cell
$\overline{U}$	time-averaged streamwise velocity; also bulk velocity
$U(y)$	unidirectional laminar flow
U_{c}	centerline velocity
$\overline{U}_i$	mean velocity component
$U_{\mathrm{o}}(x,t)$	velocity outside boundary layer
U_∞	freestream velocity
$\mathbf{u}$	velocity vector
u_{fluid}	fluid velocity at wall
u_k	instantaneous velocity component (u,v,w)
u'_k	velocity fluctuations
u_{wall}	wall velocity
u_τ	viscous velocity scale; also called friction velocity $(=\ell_\nu/t_\nu = \sqrt{\tau_{\mathrm{w}}/\rho}\,)$
u_*	another symbol for friction velocity
$\overline{u_i u_k}$	(kinematic) Reynolds stress
$\overline{uv}$	tangential Reynolds stress
$\mathcal{V}$	control volume
$\overline{V}$	time-averaged normal velocity
$V(r)$	potential energy
$V_{ij}(r)$	potential between two molecules i and j
$\overline{v}_{\mathrm{m}}$	mean molecular speed
v_{w}	injection–suction velocity; also upward–downward wall velocity
v_{o}	characteristic velocity
$W(\frac{y}{\delta})$	universal wake function
$\mathbf{x}$	position vector
x_k	Cartesian coordinates $(k=1,2,3)$

x_i	spatial coordinate
y^+	distance normal to surface in wall units
y_p^+	location of peak Reynolds stress (expressed in wall units)

Greek Symbols

α	angle of attack; also complex wave number; also isothermal compressibility coefficient
$\dot{\alpha}$	rate of change of angle of attack $(= \frac{d\alpha}{dt})$
α_i	imaginary part of α (determines spatial growth of disturbance)
α_r	wave number (real part of α)
β	coefficient of thermal expansion (also called bulk expansion coefficient)
Γ	circulation
γ	specific heat ratio (isentropic exponent)
$\gamma(x)$	local circulation in a vortex sheet per unit length
$\dot{\gamma}$	shear rate
$\dot{\gamma}_c$	critical value of shear rate
Δp^*	pressure drop normalized using a viscous pressure scale $(\equiv \Delta p \, 4a^2 / \rho \, \nu^2)$
ΔT	characteristic temperature difference
$\Delta u|_w$	tangential velocity slip
ΔU^+	maximum deviation of mean velocity profile from log-law (in wall units)
Δx	center-to-center distance between discrete vortices
δ	boundary-layer thickness; also distance between molecules
δ_{ki}	Kronecker delta (unit second-order tensor)
δ_θ	momentum thickness
δ^*	displacement thickness
ϵ	characteristic energy scale
ϵ^{wf}	energy scale for wall–fluid interaction
ε	normalized rotor eccentricity
ζ	ratio of surface wave speed to transverse wave speed
η	compliant coating displacement in y-direction; also Kolmogorov length scale
Θ	second elastic constant
κ	von Kármán constant
κ_f	thermal conductivity of fluid
κ_s	thermal conductivity of solid
Λ	Pohlhausen parameter (ratio of pressure forces to viscous forces)
λ	second coefficient of viscosity
λ_z^+	spanwise mean-streak-spacing (in wall units)
μ	first coefficient of viscosity
ν	kinematic viscosity $(= \mu / \rho)$
ν_T	thermal diffusivity $(= \kappa_f / \rho c_p)$
ξ	compliant coating displacement in x-direction
$\boldsymbol{\xi}$	molecular velocity vector
ξ_r	relative speed between two molecules
Π	profile parameter
ρ	fluid density

ρ_s	solid density
Σ_{ki}	stress tensor (surface force per unit area)
σ	molecular diameter (molecular length scale)
σ_T	thermal accommodation coefficient
σ_ν	tangential-momentum-accommodation coefficient
σ^{wf}	length scale for wall-fluid interaction
τ	shear stress; also molecular time scale; also tangential momentum flux
τ_w	wall shear stress
Υ	characteristic roughness height
ϕ	dissipation function
ϕ^*	dissipation function for incompressible flows
$\phi(y)$	perturbation amplitude-function in the Orr–Sommerfeld equation
ψ	stream function
$\overline{\Omega}$	mean vorticity
$\overline{\Omega}_z$	mean spanwise vorticity
ω	radian frequency $(=2\pi f)$; also rotational speed
ω_i	instantaneous vorticity component

Subscripts

0	conditions at $y = 0$
i	imaginary part; also incident variable
ℓ	longitudinal direction
o	conditions outside boundary layer; also reference quantity
r	real part; also reflected variable
rms	root mean square
s	solid variable
t	transverse direction
w	conditions at wall
∞	far-field or ambient conditions

Superscripts

$\prime$	turbulence fluctuations; also derivative w.r.t. (y/δ)
$+$	variable expressed in wall units
$*$	dimensionless variable; also postcollision value
$\wedge$	dimensionless variable

Abbreviations

AFOSR	(U.S.) Air Force Office of Scientific Research
AGARD	(NATO) Advisory Group for Aerospace Research and Development
AIAA	American Institute of Aeronautics and Astronautics
AICHE	American Institute of Chemical Engineers
AIP	American Institute of Physics

ARC	(British) Aeronautical Research Council
ARO	(U.S.) Army Research Office
ASME	American Society of Mechanical Engineers
CFD	computational fluid dynamics
DNS	direct numerical simulations
DOD	(U.S.) Department of Defense
DOE	(U.S.) Department of Energy
FISI	flow-induced surface instabilities
ICAS	International Council of the Aeronautical Sciences
IEEE	Institute of Electrical and Electronics Engineers
IOP	(British) Institute of Physics
LEBU	large-eddy breakup device
LES	large-eddy simulations
LFC	laminar flow control
MEMS	microelectromechanical systems
MRS	Moore–Rott–Sears (criterion)
NACA	(U.S.) National Advisory Committee for Aeronautics
NASA	(U.S.) National Aeronautics and Space Administration
NATO	North Atlantic Treaty Organization
NLF	natural laminar flow
NSF	(U.S.) National Science Foundation
OGY	Ott–Grebogi–Yorke (strategy for controlling chaos)
OLD	outer-layer device
ONR	(U.S.) Office of Naval Research
SAE	(U.S.) Society of Automotive Engineers
TSI	Tollmien–Schlichting instabilities
TWF	traveling-wave flutter

CHAPTER ONE

Introduction

Thinking is one of the greatest joys of humankind.
    (Galileo Galilei, 1564–1642)

The farther backward you can look, the farther forward you are likely to see.
    (Sir Winston Leonard Spencer Churchill, 1874–1965)

PROLOGUE

The subject of flow control is broadly introduced in this first chapter, leaving much of the details to the subsequent chapters of the book. The ability to manipulate a flowfield actively or passively to effect a desired change is of immense technological importance, and this undoubtedly accounts for the subject's being more hotly pursued by scientists and engineers than any other topic in fluid mechanics. The potential benefits of realizing efficient flow-control systems range from saving billions of dollars in annual fuel costs for land, air, and sea vehicles to achieving economically and environmentally more competitive industrial processes involving fluid flows. In this monograph both the classical tools and the more modern strategies of flow control are covered. Methods of control to achieve transition delay, separation postponement, lift enhancement, drag reduction, turbulence augmentation, and noise suppression are considered. The treatment is tutorial at times, which makes the material accessible to the graduate student in the field of fluid mechanics. Emphasis is placed on external boundary-layer flows, although applicability of some of the methods discussed for internal flows as well as free-shear flows will be mentioned. An attempt is made to present a unified view of the means by which different methods of control achieve a variety of end results. Performance penalties associated with a particular method such as cost, complexity, and trade-off will be elaborated upon throughout the book.

1.1   What Is Flow Control?

Before we get started it might be a good idea to explain what is meant by flow control. The topic, as discussed in this monograph, is not related to flow-rate control via manual or automatic valves. Rather it is the attempt to alter the character or disposition of a flowfield favorably that is of concern to us here. Many definitions for the topic exist, and some differentiate between (active) flow control and (passive) flow management. My favorite definition was offered by Flatt[1] in 1961 as applied to wall-bounded flows but could easily be extended to free-shear flows: "Boundary layer control includes any mechanism or process through which the boundary layer of a fluid flow is caused to behave differently than it normally would were the flow developing naturally along a smooth straight surface." Prandtl (1904) pioneered the modern use of flow control in his epoch-making presentation to the Third International Congress of Mathematicians held at Heidelberg, Germany. In just eight pages (as required for acceptance by the congress) of a paper titled *Über Flüssigkeitsbewegung bei sehr kleiner Reibung* (*On Fluid Motion with Very Small Friction*), Prandtl introduced the boundary-layer theory, explained the mechanics of steady separation, opened the way for understanding the motion of real fluids, and described several experiments in which the boundary layer was controlled.

Prandtl (1904) used active control of the boundary layer to show the great influence such a control exerted on the flow pattern. He used suction to delay boundary-layer separation from the surface of a cylinder. Notwithstanding Prandtl's success, aircraft designers in the three decades following his convincing demonstration were accepting lift and drag of airfoils as predestined characteristics with which no man could or should tamper (Lachmann 1961). This predicament changed mostly owing to the German research in boundary-layer control pursued vigorously shortly before and during the Second World War. In the two decades following the war, extensive research on laminar flow control, in which the boundary layer formed along the external surfaces of an aircraft is kept in the low-drag laminar state, was conducted in Europe and the United States, culminating in the successful flight-test program of the X-21 in which suction was used to delay transition on a swept wing up to a chord Reynolds number of 4.7×10^7. The oil crisis of the early 1970s brought renewed interest in novel methods of flow control to reduce skin-friction drag even in turbulent boundary layers. In the 1990s, the need to reduce the emissions of greenhouse gases and to construct supermaneuverable fighter planes, faster and quieter underwater vehicles, and hypersonic transport aircraft (e.g., the U.S. National Aerospace Plane) provided new challenges for researchers in the field of flow control.

The major goal of this monograph is to introduce the subject broadly, to present a unified view of the different control methods to achieve a variety of end results, and to provide an up-to-date view of the fundamentals of some basic flows and control practices. The number of journal articles and proceedings available on flow control is daunting. For example, a recent bibliography on the rather narrow sub-subject of skin-friction reduction by polymers and other additives cites over 4900 references (Nadolink and Haigh 1995). Therefore, citations in the present book will be rather selective—indeed very limited. To avoid having all 441 pages of this book as a gigantic list of references, numerous good articles will not be cited. At times, the treatment of the subject matter herein is pedagogical and is designed to attract newcomers to the field and to help the reader in navigating through the colossal literature available on flow-control fundamentals and practices. Emphasis is placed on the technologically important external boundary-layer flows, although applicability of some of

[1] All references are listed alphabetically in the bibliography at the end of the book.

the methods discussed for internal flows as well as free-shear flows will be covered as appropriate. The same vorticity considerations brilliantly employed by Lighthill (1963a) to place the boundary layer correctly in the flow as a whole are used to explain many of the flow-control techniques discussed in this book. The history of the subject is briefly and broadly recalled in the following section.

1.2 Five Eras of Flow Control

Flow control involves passive or active devices to effect a beneficial change in wall-bounded or free-shear flows. Whether the task is to delay or advance transition, to suppress or enhance turbulence, or to prevent or provoke separation, useful end results include drag reduction, lift enhancement, mixing augmentation, and flow-induced noise suppression. Broadly, there are perhaps five distinct eras in the development of the art and science of this challenging albeit very useful field of research and technology: the empirical era (prior to 1900), the scientific era (1900–1940), the World War II era (1940–1970), the energy crisis era (1970–1990), and the 1990s and beyond.

The art of flow control probably has its roots in prehistoric times when streamlined spears, sickle-shaped boomerangs, and fin-stabilized arrows evolved empirically (Williams 1987) by the sheer perseverance of archaic Homo sapiens who knew nothing about air resistance or aerodynamic principles. Three *aerodynamically correct* wooden spears were recently excavated in an open-pit coal mine near Hannover, Germany (Dennell 1997). Archaeologists dated the carving of those complete spears to about 400,000 years ago (Thieme 1997), which strongly suggests early Stone Age ancestors possessing resourcefulness and skills once thought to be characteristics that came only with fully modern Homo sapiens.

Modern man also artfully applied flow-control methods to achieve certain technological goals. Relatively soon after the dawn of civilization and the establishment of an agricultural way of life 8000 years ago, complex systems of irrigation were built along inhabited river valleys to control the waterflow, thus freeing man from the vagaries of the weather. Some resourceful albeit mischievous citizens of the Roman Empire discovered that adding the right kind of diffuser to the calibrated convergent nozzle ordinarily installed at home outlets of the public water main significantly increased the charge of potable water over that granted by the emperor. For centuries, farmers knew the value of windbreaks to keep topsoil in place and to protect fragile crops.

The science of flow control originated with Prandtl (1904), who introduced the boundary-layer theory, explained the physics of the separation phenomena, and described several experiments in which a boundary layer was controlled. Thus, the scientific method to control a flowfield, or the second era of flow control, was born. Slowly but surely, the choice of flow-control devices was no longer a trial and error feat, but physical reasoning and even first principles were more often than not used for rational design of such artifacts.

Stimulated by the Second World War and the subsequent Cold War, that trend accelerated significantly during the third era (1940–1970). Military needs of the superpowers dictated the development of fast, highly maneuverable, efficient aircraft, missiles, ships, submarines, and torpedoes, and flow control played a major role in achieving these goals. Natural laminar flow (in which shaping is used to delay transition), laminar flow control (in which nonpassive strategies are employed to keep the flow laminar), and polymer drag reduction were notable achievements during this era. Partial summaries of flow-control research during this period are presented in the books edited by Lachmann (1961) and Wells (1969).

The energy crises exemplified by the 1973 Arab oil embargo brought about a noticeable shift of interest from the military sector to the civilian one. During the period 1970–1990, government agencies and private corporations around the world, but particularly in the industrialized countries, invested valuable resources in the search for methods to conserve energy, and hence drag reduction for civilian air, sea, and land vehicles; for pipelines; and for other industrial devices was emphasized. The availability of fast, inexpensive computers made it possible to simulate numerically complex flow situations that had not been approachable analytically. Some control strategies, for example, transition-delaying compliant coatings (Gad-el-Hak 1996a), were rationally optimized using computational fluid dynamics. Large-eddy breakup devices (LEBUs) and riblets are examples of control methods developed during this period to reduce skin-friction drag in turbulent boundary layers. Good sources of information on these and other devices introduced during the fourth era are the books edited by Hough (1980); Bushnell and Hefner (1990); and Barnwell and Hussaini (1992). Numerous meetings devoted to flow control, particularly drag reduction, were held during this period. Plentiful fuel supplies during the 1990s and the typical short memory of the long gas lines during 1973 have, unfortunately, somewhat dulled the urgency and enthusiasm for energy conservation research as well as practice. Witness—at least in the United States—the awakening of the long-hibernated gas-guzzler automobile and the recent run on house-size sport utility vehicles, a.k.a. land barges.

For the 1990s and beyond, more complex reactive control devices geared specifically toward manipulating the omnipresent coherent structures in transitional and turbulent shear flows (Cantwell 1981; Robinson 1991) are being pursued by several researchers. Theoretical advances in chaos control and the development of micro-electromechanical systems (MEMS) and neural networks should help such efforts. Articles specifically addressing reactive control strategies include those by Wilkinson (1990); Moin and Bewley (1994); and Gad-el-Hak (1994, 1996b). The MEMS are discussed in Chapter 13, and reactive flow control is emphasized in the last chapter of this book.

All five eras of flow control are seen from the perspective of the history of the universe timeline shown in Figure 1.1.

1.3 Has the Field Crested?

There are those who consider fluid mechanics to be a mature subject that led to very useful technological breakthroughs in the past but that the pace of improvements is fast reaching the point at which returns on investment in research are not sufficiently impressive. As we approach the twenty-first century, the skeptics claim, little new scientific or engineering breakthroughs are to be expected from the aging field of study. It may be worth remembering that much the same was said about physics toward the end of the nineteenth century. Self-satisfied that almost all experimental observations of the time could be fitted into either Newton's theory of mechanics

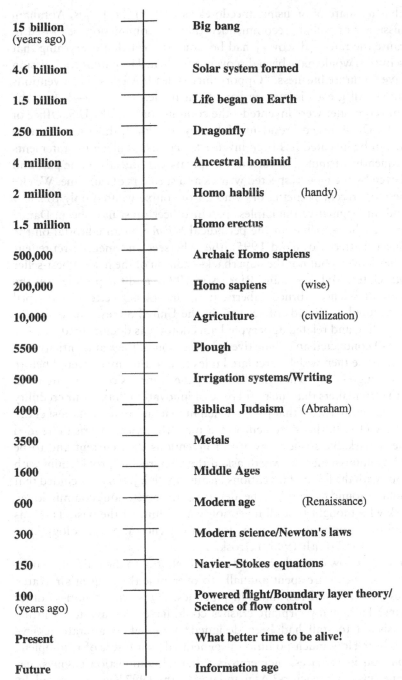

15 billion (years ago)	Big bang
4.6 billion	Solar system formed
1.5 billion	Life began on Earth
250 million	Dragonfly
4 million	Ancestral hominid
2 million	Homo habilis (handy)
1.5 million	Homo erectus
500,000	Archaic Homo sapiens
200,000	Homo sapiens (wise)
10,000	Agriculture (civilization)
5500	Plough
5000	Irrigation systems/Writing
4000	Biblical Judaism (Abraham)
3500	Metals
1600	Middle Ages
600	Modern age (Renaissance)
300	Modern science/Newton's laws
150	Navier–Stokes equations
100 (years ago)	Powered flight/Boundary layer theory/ Science of flow control
Present	What better time to be alive!
Future	Information age

Figure 1.1: History of the universe timeline. All dates are approximate, and the timescale is highly nonlinear.

or Maxwell's theory of electromagnetics, the majority of physicists believed that the work of their successors *would be merely to make measurements to the next decimal place.* That was just before the theory of relativity and quantum mechanics were discovered!

Technology has its share of amusing anecdotes as well. In the 1860s, Abraham Lincoln's commissioner of patents recommended that the commission be closed in a few years because the rate of discovery had become so great that everything that needed to be discovered would have been discovered by then. The patent commission would simply have no future business. "Opportunity is dead! All possible inventions have been invented. All great discoveries have been made." In 1899—before the airplane, laser, and computer were invented—the commissioner of the U.S. Office of Patents, Charles H. Duell, urged President McKinley to abolish this office because "everything that can be invented has been invented." Foolish, fallacious statements like those are frequently attributed to various myopic patent officials of the past and are perpetuated even by the most respected writers and speakers of our time. We cite here three of the most recent perpetuators who all of course wanted only to show how ignorant and unimaginative the hapless patent officer must have been: Daniel E. Koshland, Jr., the editor-in-chief of the periodical *Science*, in an editorial on the future of its subject matter (Koshland 1995); the cyberseer and megaentrepreneur Bill Gates in the hardcover—but not the paperback—edition of the instant best-seller *The Road Ahead* (Gates, Myhrvold, and Rinearson 1995); and the president of the National Academy of Sciences, Bruce Alberts, in a fund-raising letter dated April 1997, widely distributed to 'friends of science' in the United States. The definitive history of the preceding and related apocryphal anecdotes was documented by Eber Jeffery, who in 1940 conducted an exhaustive investigation of their authenticity and origin. Jeffery traced the then widely circulated tales to a testimony delivered before the United States Congress by Henry L. Ellsworth, the commissioner of patents in 1843, who told the lawmakers that the rapid pace of innovation "taxes our credulity and seems to presage the arrival of that period when human improvement must end." According to Jeffery (1940), this statement was a mere rhetorical flourish intended to emphasize the remarkable strides forward in inventions then current and to be expected. Indeed, Commissioner Ellsworth asked the Congress to provide him with extra funds to cope with the flood of inventions he anticipated. Jeffery concluded that no document could be found to establish the identity of the mysterious commissioner, examiner, or clerk who thought that all invention was a thing of the past. This was not true then and is certainly not true now, for both science and technology have indeed an 'endless frontier' (Bush 1945; Petroski 1997).

Our own subfield of flow control provides a good rebuttal to the fluid mechanics critics. How much resources are spent annually to overcome the drag of air, water, and land vehicles and the fluid resistance in pipelines, ducts, and countless other man-made systems? How many airplane crashes could have been averted if lifting surfaces more resistant to stall had been designed? Can fighter aircraft achieve supermaneuverability? How much pollution is generated as a result of incomplete, inefficient combustion in furnaces, incinerators, internal combustion engines, gas turbines, and numerous other reactors? As mandated by the 1997 Kyoto Protocol, by the year 2012 the industrialized countries must reduce their emissions of greenhouse gases 6–8% *below* those produced in 1990. How can they achieve this target? How much does it cost to fly across continents at hypersonic or even supersonic speeds? Can we economically utilize ultrafast submarines, instead of snail-paced ships, for hauling people and commerce across the seas? For transport through urban areas whose already congested roads have no more room for growth, can we envision a personal helicopter (skycar) that, aided by a sophisticated network of computers, is as easy to navigate and safer to drive than the automobile? And how about medical applications such as artificial organs, microdosage delivery systems, and diagnostic tools? All these challenges are areas that future advances in flow control in particular and fluid mechanics in general, unforeseen to us today, could help overcome. The book edited by Lumley, et al. (1996) provides an excellent outline of the near-future prospects of 27 different fluid dynamics research areas, whereas the one edited by Gad-el-Hak, Pollard, and Bonnet (1998) focuses exclusively on flow control.

True, fluid mechanics will be a very different discipline during the third millennium. The entire field, together with much of the rest of the physical sciences, may be set for dramatic changes. In no small part, rapidly advancing computer technology will be responsible for those changes. As argued in the editorial by Gad-el-Hak and Sen (1996) and the paper by Gad-el-Hak (1998a), gigantic computers combined with appropriate software may be available during the twenty-first century to routinely integrate the full, instantaneous Navier–Stokes equations. The black box would prompt its operator for the geometry and flow conditions and would then present a first-principles numerical solution to the specific engineering problem, high-Reynolds-number turbulent flows included.

In a turbulent flow, the ratio of the large eddies (at which the energy maintaining the flow is input) to the Kolmogorov microscale (the flow smallest length scale) is proportional to $Re^{3/4}$ (Tennekes and Lumley 1972). Each excited eddy requires at least one grid point to describe it. Therefore, to adequately resolve, via direct numerical simulations (DNS), a three-dimensional flow, the required number of modes would be proportional to $(Re^{3/4})^3$. To describe the motion of small eddies as they are swept around by large ones, the time step must not be larger than the ratio of the Kolmogorov length scale to the characteristic rms velocity. The large eddies, on the other hand, evolve on a time scale proportional to their size divided by their rms velocity. Thus, the number of time steps required is again proportional to $Re^{3/4}$. Finally, the computational work requirement is the number of modes × the number of time steps, which scales with Re^3, that is, an order of magnitude increase in computer power is needed as the Reynolds number is doubled (Karniadakis and Orszag 1993). Because the computational resource required varies as the cube of the Reynolds number, it may not be possible to simulate very high-Reynolds-number turbulent flows any time soon (Karniadakis 1999).

Despite the bleak assessment above, one wonders whether gigantic computers combined with appropriate software will be available during the twenty-first century to routinely solve, using DNS, practical turbulent flow problems? The black box would prompt its operator for the geometry and flow conditions and would then generate a numerical solution to the specific engineering problem. Nobody, except the software developers, needs to know the details of what is going on inside the black box, not even which equations are being solved. This situation is not unlike using a present-day word processor or hand calculator. A generation of users of the Navier–Stokes computers would quickly lose the aptitude, and the desire, to perform simple analysis based on physical considerations, much the same as the inability of some of today's users of hand calculators to manually carry out long divisions. The need for rational approximations, so prevalent today in fluid mechanics teaching and practice, would gradually wither.

During the late 1990s, the supercomputer power approached the teraflop (i.e., 10^{12} floating-point operations per second). This is about right to compute a flow with a characteristic Reynolds number of 10^8, sufficient to simulate the flow around an airfoil via DNS, around a wing via large-eddy simulation, or around an entire commercial aircraft via Reynolds-averaged calculation. It is anticipated that the 10-teraflop computer will be available by mid-2000, and the 100-teraflop machine by 2004, bettering the Moore's law by tripling—instead of doubling—the chip speed every 18 months. An exaflop (10^{18} flops) computer is needed to carry out direct numerical simulation of the complete airplane (Karniadakis and Orszag 1993), and thus we are still a long way from that.

Silicon-based computer powers have manifested spectacular recent advances—something like a factor of 10,000 improvement in speed and capacity during the past 20 years.[2] If one is to extrapolate those recent advances to the next 50 years or so, using direct numerical simulations to solve the turbulence problem for realistic geometries and field Reynolds numbers may begin to approach feasibility. Unfortunately, however, silicon microchips are rapidly approaching their physical limits with little room for further growth. The smallest feature in a microchip is defined by its smallest linewidth, which in turn is related to the wavelength of light employed in the basic lithographic process that is used to create the chip. The linewidths of chips manufactured in 1999 is 180 nm, and most experts doubt that chips using current complementary metal oxide semiconductor (CMOS) technology can shrink below 50-nm linewidths—a size that could be reached around 2012 (Lerner 1999). There is even a natural—in contrast to a technological—barrier to extreme miniaturization (Schulz 1999). A recent report by a team of researchers from Lucent Technologies (Muller et al. 1999) has uncovered a troubling natural brick wall. The thin layer of silicon dioxide insulation that separates the various conducting or semiconducting components on a chip is rendered useless if its thickness falls below 0.7 nm or the equivalent of four silicon atoms across. At the current rate of progress, such a fundamental limit on the thinnest usable silicon dioxide gate dielectric will also be reached around 2012.

Fortunately, this kind of linear thinking may be misleading. Revolutionary computing machines that bear little resemblance to today's chip-based computers may be developed in the future. A recent article in *Science* (Glanz 1995) discusses five such futuristic computing concepts: quantum dots, quantum computers, holographic association, optical computers, and deoxyribonucleic acid (DNA) computers. Very recently, the prospect for molecular computing was unveiled (Collier et al. 1999; Service 1999; Markoff 1999). We elaborate below on the latter two computers as examples.

The DNA computer is a novel concept introduced and actually demonstrated only late in 1994 by Adleman, who in turn was inspired by Feynman's (1961) original vision of building even smaller submicroscopic computers. The idea for the massively parallel DNA computing system is to use the four basic chemical units of DNA (adenine, thymine, guanine, and cytosine, designated A, T, G, and C, respectively) as computing symbols and to utilize the genetic material for information storage and computations. Computer theorists argue that a problem could be set up by synthesizing DNA molecules with a particular sequence that represents numerical

---

[2] Although loosely related, this is consistent with the law named after the cofounder of Intel Corporation, Gordon Moore, who in 1965 predicted that the transistor density on a semiconductor chip would double and its price would halve roughly every 18 months. Incidentally, Moore's law has been bettered in 1997.

information and by letting the molecules react in a test tube, producing new molecules whose sequence is the answer to the problem. Thus, the same genetic machinery that generates living organisms could be used to solve previously unapproachable mathematical puzzles. Crude estimates indicate that a mere 500 g of DNA molecules (a human body contains about 300 g of deoxyribonucleic acid) suspended in 1000 l of fluid would have a memory equivalent to all the electronic computers ever made! In such primordial reacting soup, 4 months of manipulating the DNA molecules would yield an answer to a problem that would have required more operations than all those ever performed on all the conventional computers ever built—at least up to the mid-1990s.

Just as this book was going to the press, a research team from Hewlett–Packard company and the University of California at Los Angeles (Collier et al. 1999) announced that they have fashioned simple computing components no thicker than a single rotaxane[3] molecule, thus plunging deeply into a Lilliputian world that promises computers 10^{11} times as fast as today's most powerful personal computers. This infant field of research is termed *moletronics* or molecular electronics and uses chemical processes rather than light to form molecular-thin on–off switches. Future developments in moletronics will open new windows onto a once speculative but now increasingly probable vista of molecular-scale computers, sensors, and actuators.

To close this section on a positive note, neither fluid mechanics nor flow control appears to be approaching a pinnacle. The best is yet to come. Future computers will allow solutions to the most daunting problems in both fields, and rational designs based on first principles will be routine for the most complex fluid systems. Sophisticated reactive control systems consisting of a colossal number of microsensors and microactuators communicating with each other via massively parallel computers will in the future improve, by leaps and bounds, the fluids engineer's ability to reduce drag, increase lift, suppress flow-induced noise, enhance mixing, and achieve any other desired flow-control goal.

---

[3] Rotaxane is a synthetic organic compound created by chemists at the University of California at Los Angeles.

CHAPTER TWO

Governing Equations

No knowledge can be certain if it is not based upon mathematics.
(Leonardo da Vinci, 1452–1519)

You are not educated until you know the Second Law of Thermodynamics.
(Charles Percy (Baron) Snow, 1905–1980)

PROLOGUE

There is no doubt that rational design (i.e., based on first principles) of flow-control devices is always preferable to a trial and error approach. Rational design of course is not always possible owing to the extreme complexity of the equations involved, but one tries either analytically or, more commonly to date, numerically. The search for useful compliant coatings, discussed in Chapter 7, is a case in point. The window of opportunity for a successful coating is so narrow that the probability of finding the right one by experimenting is near nil. Fortunately, the analytical and numerical tools to guide the initial choice for a transition-delaying compliant surface are currently available. On the other hand, the flowfield associated with a typical, deceivingly simple vortex generator for airplane wings is so complex that its design is still done to date more or less empirically.

The proper first principles for flow control are those for fluid mechanics itself. The principles of conservation of mass, momentum, and energy govern all fluid motions. Additionally, all processes are constrained by the second law of thermodynamics. In general, a set of partial, nonlinear differential equations expresses those principles, and, together with appropriate boundary and initial conditions, constitute a well-posed problem. In the following we provide a simple derivation of the general form of the three conservation relations. Readers desirous of more details could consult any advanced textbook in fluid dynamics (e.g., Batchelor 1967; Hinze 1975; Landau and Lifshitz 1987; Tritton 1988; Sherman 1990; Kundu 1990; Currie 1993; Panton 1996).

2.1 The Fundamental Equations

Each of the fundamental laws of fluid mechanics, conservation of mass, momentum, and energy, will be expressed first as an integral relation and then as a differential

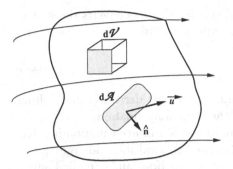

Figure 2.1: Arbitrary control volume.

equation. The fluid motions under consideration are nonrelativistic (i.e., their characteristic velocities are far below the speed of light). Thus, mass and energy are not interchangeable, and each is separately conserved. The only additional assumption we will make initially is that the fluid is a continuum, implying that the derivatives of all the dependent variables exist in some reasonable sense. In other words, local properties such as density and velocity are defined as averages over elements large compared with the microscopic structure of the fluid but small enough in comparison with the scale of the macroscopic phenomena to permit the use of differential calculus to describe them. The continuum approximation is almost always met. The resulting equations therefore cover a very broad range of situations, the exception being flows with spatial scales that are not much larger than the mean distance between the fluid molecules, as for example in the case of rarefied gas dynamics; shock waves that are thin relative to the mean free path; or flows in micro and nano-devices (see Chapter 13). The present general approach will become particularly useful when considering high-speed flows, variable-viscosity flows (e.g., in the case of surface heating and cooling), non-Newtonian fluids (e.g., concentrated solutions of drag-reducing polymers), and other unconventional situations.

Consider the arbitrary, simply connected control volume $\mathcal{V}$ fixed in an inertial, Eulerian frame and depicted in Figure 2.1. Fluid can pass in or out through the control surface $\mathcal{A}$, which is also fixed in space and time. The outward unit normal vector field is $\hat{n}$. To derive an integral form of any of the three conservation relations, we simply state that the rate of change of any physical property contained within the control volume must be equal to the net rate of influx of that property through the control surface plus the net rate of gain from surface as well as volume sources. The three properties we consider in turn in the following subsections are mass, momentum, and energy.

2.1.1 Conservation of Mass

For a chemically nonreacting fluid, that mass cannot be destroyed or created translates into the following statement: Rate of change of mass inside the control volume = −(mass convected out through the control surface). Thus, the integral form of the continuity equation reads in Cartesian tensor notations

$$\frac{\partial}{\partial t} \iiint_{\mathcal{V}} \rho \, d\mathcal{V} = -\iint_{A} \rho u_k \, d\mathcal{A}_k \tag{2.1}$$

where ρ is the fluid density and u_k is an instantaneous velocity component (u, v, w). The independent variables are time t and the three spatial coordinates x_1, x_2, and x_3 or (x, y, z). The infinitesimal volume $d\mathcal{V} = dx \, dy \, dz$. Finally, the Einstein's summation convention applies to all repeated indices.

Applying the Gauss divergence theorem of vector calculus converts the surface integral in Eq. (2.1) into a volume one, and the equation reads

$$\iiint_{\mathcal{V}} \frac{\partial \rho}{\partial t} \, d\mathcal{V} = -\iiint_{\mathcal{V}} \frac{\partial}{\partial x_k}(\rho u_k) \, d\mathcal{V} \tag{2.2}$$

Note that because time and space are independent, the time derivative and volume integral are commuted in the left-hand side of the preceding equation.

The last step to derive the differential form of the continuity equation involves simply moving the right-hand side of Eq. (2.2) to the left-hand side and arguing that because the control volume is arbitrary, the resulting equation can be fulfilled only if the integrand is zero at each point in space. Thus, at every point in space–time

$$\frac{\partial \rho}{\partial t} + \frac{\partial}{\partial x_k}(\rho u_k) = 0 \tag{2.3}$$

2.1.2 Newton's Second Law

If the physical property under consideration is momentum, then according to Newton's second law surface and body forces constitute, respectively, the surface and volume sources of momentum. Conservation of momentum thus reads: Time rate of increase of momentum inside the control volume = −(rate of convective outflux of momentum across the control surface) + surfaces forces + body forces. The integral form of the momentum equation can therefore be written as[1]

$$\frac{\partial}{\partial t} \iiint_{\mathcal{V}} \rho u_i \, d\mathcal{V} = -\iint_{\mathcal{A}} \rho \, (u_k \, n_k) \, u_i \, d\mathcal{A}$$
$$+ \iint_{\mathcal{A}} \Sigma_{ki} \, n_k \, d\mathcal{A} + \iiint_{\mathcal{V}} \rho g_i \, d\mathcal{V} \tag{2.4}$$

where Σ_{ki} is the second-order stress tensor (surface force per unit area), and g_i is the body force per unit mass. Body forces apply to the entire mass of a fluid element and are due to external fields such as gravity and electromagnetic potential. A discussion of the latter kind of body force will be included in Chapter 14. Equation (2.4) is of course a vector equation valid for $i = 1, 2,$ and 3.

Once again all the surface integrals in Eq. (2.4) are converted to volume integrals using the divergence theorem, and thus the momentum integral equation reads

$$\iiint_{\mathcal{V}} \frac{\partial}{\partial t}(\rho u_i) \, d\mathcal{V} = -\iiint_{\mathcal{V}} \frac{\partial}{\partial x_k}(\rho u_k u_i) \, d\mathcal{V}$$
$$+ \iiint_{\mathcal{V}} \frac{\partial \Sigma_{ki}}{\partial x_k} d\mathcal{V} + \iiint_{\mathcal{V}} \rho g_i \, d\mathcal{V} \tag{2.5}$$

The differential form of the momentum equation follows immediately from equating the integrand in Eq. (2.5) to zero, thus

$$\frac{\partial}{\partial t}(\rho u_i) + \frac{\partial}{\partial x_k}(\rho u_k u_i) = \frac{\partial \Sigma_{ki}}{\partial x_k} + \rho g_i \tag{2.6}$$

---

[1] If the frame of reference is noninertial (i.e., if it accelerates linearly, rotates relative to an inertial frame, or both), additional acceleration terms are needed in the momentum balance.

This can readily be simplified by invoking the continuity equation (2.3) to read finally

$$\rho\left(\frac{\partial u_i}{\partial t} + u_k \frac{\partial u_i}{\partial x_k}\right) = \frac{\partial \Sigma_{ki}}{\partial x_k} + \rho g_i \tag{2.7}$$

Each term in the Cauchy's equation of motion (2.7) has units of force per unit volume. An equation for the angular momentum can be derived in a way similar to the derivation of the linear momentum equation above. Also, taking the curl of Eq. (2.7) leads to the equation for vorticity.

2.1.3 Conservation of Energy

Here the conserved physical property under consideration is the intrinsic energy of a thermodynamic system. For convenience, we choose the intrinsic energy per unit mass to be[2] the sum of internal energy e and kinetic energy $\frac{1}{2} u_i u_i$. For nonrelativistic motions, energy cannot be created or destroyed. The first law of thermodynamics states: Time rate of change of intrinsic energy inside the control volume = −(rate of convective outflux of energy across the control surface) − (rate of heat transfer due to conduction and radiation out of the control surface) + rate of work done by surface and body forces. The integral form of the energy equation can therefore be written as

$$\frac{\partial}{\partial t} \iiint_\mathcal{V} \rho\left(e + \frac{1}{2} u_i^2\right) d\mathcal{V} = -\iint_\mathcal{A} \rho u_k \left(e + \frac{1}{2} u_i^2\right) d\mathcal{A}_k$$

$$- \iint_\mathcal{A} q_k \, d\mathcal{A}_k + \iint_\mathcal{A} u_i \Sigma_{kl} \, d\mathcal{A}_k + \iiint_\mathcal{V} \rho u_i \, g_i \, d\mathcal{V} \tag{2.8}$$

where q_k is the sum of heat flux vectors due to conduction and radiation, and a summation over i is implied in u_i^2. All the surface integrals in Eq. (2.8) are converted to volume integrals using the divergence theorem, and thus the integral form of the conservation of energy equation now reads

$$\iiint_\mathcal{V} \frac{\partial}{\partial t} \left[\rho\left(e + \frac{1}{2} u_i^2\right)\right] d\mathcal{V}$$

$$= -\iiint_\mathcal{V} \frac{\partial}{\partial x_k} \left[\rho u_k \left(e + \frac{1}{2} u_i^2\right)\right] d\mathcal{V}$$

$$- \iiint_\mathcal{V} \frac{\partial q_k}{\partial x_k} d\mathcal{V} + \iiint_\mathcal{V} \frac{\partial (u_i \Sigma_{ki})}{\partial x_k} d\mathcal{V} + \iiint_\mathcal{V} \rho u_i g_i \, d\mathcal{V} \tag{2.9}$$

The differential form of the energy equation follows from equating the integrand in Eq. (2.9) to zero and invoking the continuity equation (2.3), thus

$$\rho\left[\frac{\partial (e + \frac{1}{2} u_i^2)}{\partial t} + u_k \frac{\partial (e + \frac{1}{2} u_i^2)}{\partial x_k}\right] = -\frac{\partial q_k}{\partial x_k} + \frac{\partial (u_i \Sigma_{ki})}{\partial x_k} + \rho u_i g_i \tag{2.10}$$

---

[2] We can include the potential energy as part of the intrinsic energy. In that case, however, the work done by the body forces should not be included in the energy balance.

An equation for the kinetic energy per unit volume $\frac{1}{2}u_i u_i$ can be derived simply by taking the dot product of u_i with the momentum equation (2.7). The resulting equation reads

$$\rho\left[\frac{\partial\left(\frac{1}{2}u_i^2\right)}{\partial t} + u_k\frac{\partial\left(\frac{1}{2}u_i^2\right)}{\partial x_k}\right] = u_i\frac{\partial\Sigma_{ki}}{\partial x_k} + \rho u_i g_i \tag{2.11}$$

and states that the rate of increase of kinetic energy following a fluid material element = rate of work done by the net surface forces + rate of work done by body forces. The thermal energy equation is readily obtained by subtracting Eq. (2.11) from Eq. (2.10) to yield

$$\rho\left(\frac{\partial e}{\partial t} + u_k\frac{\partial e}{\partial x_k}\right) = -\frac{\partial q_k}{\partial x_k} + \Sigma_{ki}\frac{\partial u_i}{\partial x_k} \tag{2.12}$$

The second term in the right-hand side is the deformation work done (per unit time) by the surface forces and contains both reversible and irreversible contributions (as dictated by the second law of thermodynamics) to internal energy. Body forces do not contribute directly to internal energy, and Eq. (2.12) states that the rate of increase of internal energy of a fluid material element is proportional to the sum of the net rate of heat inflow and the net rate of work done by surface forces.

2.2 Closing the Equations

Equations (2.3), (2.7), and (2.12) constitute five differential equations for the 17 unknowns ρ, u_i, Σ_{ki}, e, and q_k. Absent any body couples, the stress tensor is symmetric, having only six independent components, which reduces the number of unknowns to 14. Obviously, the continuum flow equations do not form a determinate set. To close the conservation equations, the relation between the stress tensor and deformation rate, the relation between the heat flux vector and the temperature field, and the appropriate equations of state relating the different thermodynamic properties are needed. The stress–rate of strain relation and the heat flux–temperature relation are approximately linear if the flow is not too far from thermodynamic equilibrium. This is a phenomenological result but can be rigorously derived from the Boltzmann equation for a dilute gas if it is assumed that the flow is near equilibrium (see Chapter 13).

For a Newtonian, isotropic, Fourier,[3] ideal gas, for example, these relations read

$$\Sigma_{ki} = -p\,\delta_{ki} + \mu\left(\frac{\partial u_i}{\partial x_k} + \frac{\partial u_k}{\partial x_i}\right) + \lambda\left(\frac{\partial u_j}{\partial x_j}\right)\delta_{ki} \tag{2.13}$$

$$q_i = -\kappa_{\rm f}\frac{\partial T}{\partial x_i} + \text{Heat flux due to radiation} \tag{2.14}$$

$$de = c_v\,dT \quad\text{and}\quad p = \rho\,\mathcal{R}T \tag{2.15}$$

[3] Newtonian implies a linear relation between the stress tensor and the symmetric part of the deformation tensor (rate of strain tensor). The isotropy assumption reduces the 81 constants of proportionality in that linear relation to two constants. Fourier fluid is that for which the conduction part of the heat flux vector is linearly related to the temperature gradient, and again isotropy implies that the constant of proportionality in this relation is a single scalar.

where p is the thermodynamic pressure, μ and λ are the first and second coefficients of viscosity, respectively, δ_{ki} is the unit second-order tensor (Kronecker delta), κ_f is the fluid thermal conductivity, T is the temperature field, c_v is the specific heat at constant volume, and $\mathcal{R}$ is the gas constant given by the Boltzmann constant divided by the mass of an individual molecule ($\mathcal{R} = k/m$). The Stokes' hypothesis relates the first and second coefficients of viscosity, $\lambda + \frac{2}{3}\mu = 0$, although the validity of this assumption has occasionally been questioned (Gad-el-Hak, 1995). Given the preceding constitutive relations, and if radiative heat transfer is neglected,[4] Eqs. (2.3), (2.7), and (2.12), respectively, read

$$\frac{\partial \rho}{\partial t} + \frac{\partial}{\partial x_k}(\rho u_k) = 0 \tag{2.16}$$

$$\rho\left(\frac{\partial u_i}{\partial t} + u_k\frac{\partial u_i}{\partial x_k}\right) = -\frac{\partial p}{\partial x_i} + \rho g_i + \frac{\partial}{\partial x_k}\left[\mu\left(\frac{\partial u_i}{\partial x_k} + \frac{\partial u_k}{\partial x_i}\right) + \delta_{ki}\lambda\frac{\partial u_j}{\partial x_j}\right] \tag{2.17}$$

$$\rho c_v\left(\frac{\partial T}{\partial t} + u_k\frac{\partial T}{\partial x_k}\right) = \frac{\partial}{\partial x_k}\left(\kappa_f\frac{\partial T}{\partial x_k}\right) - p\frac{\partial u_k}{\partial x_k} + \phi \tag{2.18}$$

The three components of the vector equation (2.17) are the Navier–Stokes equations expressing the conservation of momentum for a Newtonian fluid. In the thermal energy equation (2.18), ϕ is the always positive[5] dissipation function expressing the irreversible conversion of mechanical energy to internal energy as a result of the deformation of a fluid element. The second term on the right-hand side of Eq. (2.18) is the reversible work done (per unit time) by the pressure as the volume of a fluid material element changes. For a Newtonian, isotropic fluid, the viscous dissipation rate is given by

$$\phi = \frac{1}{2}\mu\left(\frac{\partial u_i}{\partial x_k} + \frac{\partial u_k}{\partial x_i}\right)^2 + \lambda\left(\frac{\partial u_j}{\partial x_j}\right)^2 \tag{2.19}$$

There are now six unknowns, ρ, u_i, p, and T, and the five coupled equations (2.16), (2.17), and (2.18) plus the equation of state relating pressure, density, and temperature. These six equations together with sufficient numbers of initial and boundary conditions constitute a well-posed, albeit formidable, problem. The system of equations (2.16)–(2.18) is an excellent model for the laminar or turbulent flow of most fluids such as air and water under most circumstances, including high-speed gas flows for which the shock waves are thick relative to the mean free path of the molecules.[6]

Considerable simplification is achieved if the flow is assumed to be incompressible, which is usually a reasonable assumption provided that the characteristic flow speed is less than 0.3 of the speed of sound.[7] The incompressibility assumption is readily satisfied for almost all liquid flows and many gas flows. In such cases, the density is assumed to be either a constant or a given function of temperature (or species

---

[4] A reasonable assumption when dealing with low to moderate temperatures because the radiative heat flux is proportional to T^4.

[5] As required by the second law of thermodynamics.

[6] This condition is met if the shock Mach number is less than 2.

[7] Although, as will be demonstrated in the following section, there are circumstances when even a low-Mach-number flow should be treated as compressible.

concentration).[8] The governing equations for such flow are

$$\frac{\partial u_k}{\partial x_k} = 0 \tag{2.20}$$

$$\rho \left(\frac{\partial u_i}{\partial t} + u_k \frac{\partial u_i}{\partial x_k} \right) = -\frac{\partial p}{\partial x_i} + \frac{\partial}{\partial x_k} \left[\mu \left(\frac{\partial u_i}{\partial x_k} + \frac{\partial u_k}{\partial x_i} \right) \right] + \rho g_i \tag{2.21}$$

$$\rho c_p \left(\frac{\partial T}{\partial t} + u_k \frac{\partial T}{\partial x_k} \right) = \frac{\partial}{\partial x_k} \left(\kappa_f \frac{\partial T}{\partial x_k} \right) + \phi_{incomp} \tag{2.22}$$

where ϕ_{incomp} is the incompressible limit of Eq. (2.19). These are five equations for the five dependent variables u_i, p, and T. Note that the left-hand side of Eq. (2.22) has the specific heat at constant pressure c_p and not c_v. It is the convection of enthalpy—and not internal energy—that is balanced by heat conduction and viscous dissipation. This is the correct incompressible-flow limit of a compressible fluid, as discussed in detail in Section 10.9 of Panton (1996), which is a subtle point, perhaps, but one that is frequently missed in textbooks. The system of equations (2.20)–(2.22) is coupled if either the viscosity or density depends on temperature; otherwise, the energy equation is uncoupled from the continuity and momentum equations and can therefore be solved *after* the velocity and pressure fields are determined.

In nondimensional form, the incompressible flow equations read

$$\frac{\partial u_k}{\partial x_k} = 0 \tag{2.23}$$

$$\left(\frac{\partial u_i}{\partial t} + u_k \frac{\partial u_i}{\partial x_k} \right) = -\frac{\partial p}{\partial x_i} + \frac{Gr}{Re^2} T \delta_{i3} + \frac{\partial}{\partial x_k} \left[\frac{F_v(T)}{Re} \left(\frac{\partial u_i}{\partial x_k} + \frac{\partial u_k}{\partial x_i} \right) \right] \tag{2.24}$$

$$\left(\frac{\partial T}{\partial t} + u_k \frac{\partial T}{\partial x_k} \right) = \frac{\partial}{\partial x_k} \left(\frac{1}{Pe} \frac{\partial T}{\partial x_k} \right) + \frac{Ec}{Re} F_v(T) \phi_{incomp} \tag{2.25}$$

where $F_v(T)$ is a dimensionless function that characterizes the viscosity variation with temperature, and Re, Gr, Pe, and Ec are, respectively, the Reynolds, Grashof, Péclet, and Eckert numbers. These dimensionless parameters determine the relative importance of the different terms in the equations.

For both the compressible and the incompressible equations of motion, the transport terms are neglected away from solid walls in the limit of infinite Reynolds number (i.e., zero Knudsen number). The fluid is then approximated as inviscid and nonconducting, and the corresponding equations read as follows (for the compressible case):

$$\frac{\partial \rho}{\partial t} + \frac{\partial}{\partial x_k} (\rho u_k) = 0 \tag{2.26}$$

$$\rho \left(\frac{\partial u_i}{\partial t} + u_k \frac{\partial u_i}{\partial x_k} \right) = -\frac{\partial p}{\partial x_i} + \rho g_i \tag{2.27}$$

$$\rho c_v \left(\frac{\partial T}{\partial t} + u_k \frac{\partial T}{\partial x_k} \right) = -p \frac{\partial u_k}{\partial x_k} \tag{2.28}$$

---

[8] Within the so-called Boussinesq approximation, density variations have negligible effect on inertia but are retained in the buoyancy terms. The incompressible continuity equation is therefore used.

The Euler equation (2.27) can be integrated along a streamline, and the resulting Bernoulli's equation provides a direct relation between the velocity and pressure.

All the field equations recited thus far are valid at all points in space and time and can be used for either laminar or turbulent flows. Three forms of the equations of motion are particularly useful to flow control: the x, y, and z momentum equations written at the wall, the ensemble average of the instantaneous equations for turbulent flows, and the integral form of the differential equations. Those forms will be given in the three sections immediately following the section on compressibility.

2.3 Compressibility

The issue of whether to consider the continuum flow compressible or incompressible seems to be rather straightforward but is in fact full of potential pitfalls. If the local Mach number is less than 0.3, then the flow of a compressible fluid like air can—according to the conventional wisdom—be treated as incompressible. But the well-known $Ma < 0.3$ criterion is only a necessary but not a sufficient condition to allow a treatment of the flow as approximately incompressible. In other words, there are situations in which the Mach number can be exceedingly small, whereas the flow is compressible. As is well documented in heat transfer textbooks, strong wall heating or cooling may cause the density to change sufficiently and the incompressible approximation to break down, even at low speeds. Less known is the situation encountered in some microdevices in which the pressure may strongly change owing to viscous effects even though the speeds may not be high enough for the Mach number to go above the traditional threshold of 0.3. Corresponding to the pressure changes would be strong density changes that must be taken into account when writing the continuum equations of motion. In this section, we systematically explain all situations in which compressibility effects must be considered.

Let us rewrite the full continuity equation (2.3) as follows

$$\frac{D\rho}{Dt} + \rho \frac{\partial u_k}{\partial x_k} = 0 \tag{2.29}$$

where $\frac{D}{Dt}$ is the substantial derivative $(\frac{\partial}{\partial t} + u_k \frac{\partial}{\partial x_k})$ expressing changes following a fluid element. The proper criterion for the incompressible approximation to hold is that $(\frac{1}{\rho} \frac{D\rho}{Dt})$ is vanishingly small. In other words, if density changes following a fluid particle are small, the flow is approximately incompressible. Density may change arbitrarily from one particle to another without violating the incompressible flow assumption. This is the case, for example, in the stratified atmosphere and ocean where the variable-density/temperature/salinity flow is often treated as incompressible.

From the state principle of thermodynamics, we can express the density changes of a simple system in terms of changes in pressure and temperature,

$$\rho = \rho(p, T) \tag{2.30}$$

From the chain rule of calculus,

$$\frac{1}{\rho} \frac{D\rho}{Dt} = \alpha \frac{Dp}{Dt} - \beta \frac{DT}{Dt} \tag{2.31}$$

where α and β are, respectively, the isothermal compressibility coefficient and the bulk expansion coefficient—two thermodynamic variables that characterize the fluid susceptibility to change of volume, which are defined by the following relations:

$$\alpha(p, T) \equiv \frac{1}{\rho} \frac{\partial \rho}{\partial p}\bigg|_T \qquad (2.32)$$

$$\beta(p, T) \equiv -\frac{1}{\rho} \frac{\partial \rho}{\partial T}\bigg|_p \qquad (2.33)$$

For ideal gases, $\alpha = 1/p$, and $\beta = 1/T$. Note, however, that in the following arguments it will not be necessary to invoke the ideal gas assumption.

The flow must be treated as compressible if pressure or temperature changes, or both are sufficiently strong. Equation (2.31) must of course be properly nondimensionalized before deciding whether a term is large or small. Here in, we follow closely the procedure detailed in Panton (1996).

Consider first the case of adiabatic walls. Density is normalized with a reference value ρ_o, velocities with a reference speed v_o, spatial coordinates and time with, respectively, L and L/v_o, and the isothermal compressibility coefficient and bulk expansion coefficient with reference values α_o and β_o, respectively. The pressure is nondimensionalized with the inertial pressure scale $\rho_o v_o^2$. This scale is twice the dynamic pressure (i.e., the pressure change as an inviscid fluid moving at the reference speed is brought to rest).

Temperature changes for the case of adiabatic walls result from the irreversible conversion of mechanical energy into internal energy via viscous dissipation. Temperature is therefore nondimensionalized as follows:

$$T^* = \frac{T - T_o}{\left(\frac{\mu_o v_o^2}{\kappa_o}\right)} = \frac{T - T_o}{Pr\left(\frac{v_o^2}{c_{p_o}}\right)} \qquad (2.34)$$

where T_o is a reference temperature; μ_o, κ_o, and c_{p_o} are, respectively, reference viscosity, thermal conductivity and specific heat at constant pressure; and Pr is the reference Prandtl number $(\mu_o c_{p_o})/\kappa_o$.

The scaling used above for pressure is based on the Bernoulli's equation and therefore neglects viscous effects. This particular scaling guarantees that the pressure term in the momentum equation will be of the same order as the inertia term. The temperature scaling assumes that the conduction, convection, and dissipation terms in the energy equation have the same order of magnitude. The resulting dimensionless form of Eq. (2.31) reads

$$\frac{1}{\rho^*} \frac{Dp^*}{Dt^*} = \gamma_o\, Ma^2 \left\{\alpha^* \frac{Dp^*}{Dt^*} - \frac{Pr B \beta^*}{A} \frac{DT^*}{Dt^*}\right\} \qquad (2.35)$$

where the superscript * indicates a nondimensional quantity, Ma is the reference Mach number ($Ma \equiv v_o/a_o$, where a_o is the reference speed of sound), and A and B are dimensionless constants defined by $A \equiv \alpha_o \rho_o c_{p_o} T_o$, and $B \equiv \beta_o T_o$. If the scaling is properly chosen, the terms having the * superscript in the right-hand side should be of order one, and the relative importance of such terms in the equations of motion is determined by the magnitude of the dimensionless parameter(s) appearing to their left (e.g., Ma, Pr, etc.). Therefore, as $Ma^2 \to 0$, temperature changes due to viscous dissipation are neglected (unless Pr is very large, as for example in the case of highly viscous polymers and oils). Within the same order of approximation, all thermodynamic properties of the fluid are assumed constant.

Pressure changes are also neglected in the limit of zero Mach number. Hence, for $Ma < 0.3$ (i.e., $Ma^2 < 0.09$), density changes following a fluid particle can be neglected and the flow can then be approximated as incompressible.[9] However, there is a caveat in this argument. Pressure changes due to inertia can indeed be neglected at small Mach numbers, and this is consistent with the way we nondimensionalized the pressure term above. If, on the other hand, pressure changes are mostly due to viscous effects, as is the case, for example, in a long microduct or a microgas bearing, pressure changes may be significant even at low speeds (low Ma). In that case the term $\frac{Dp^*}{Dt^*}$ in Eq. (2.35) is no longer of order one and may be large regardless of the value of Ma. Density then may change significantly, and the flow must be treated as compressible. Had pressure been nondimensionalized using the viscous scale ($\frac{\mu_o v_o}{L}$) instead of the inertial one ($\rho_o v_o^2$), the revised Eq. (2.35) would have had Re^{-1} appearing explicitly in the first term in the right-hand side, accentuating the importance of this term when viscous forces dominate.

A similar result can be gleaned when the Mach number is interpreted as follows:

$$Ma^2 = \frac{v_o^2}{a_o^2} = v_o^2 \frac{\partial \rho}{\partial p}\bigg|_s = \frac{\rho_o v_o^2}{\rho_o} \frac{\partial \rho}{\partial p}\bigg|_s$$

$$\approx \frac{\Delta p}{\rho_o} \frac{\Delta \rho}{\Delta p} = \frac{\Delta \rho}{\rho_o} \tag{2.36}$$

where s is the entropy. Again, the preceding equation assumes that pressure changes are inviscid, and therefore a small Mach number means negligible pressure and density changes. In a flow dominated by viscous effects—such as that inside a microduct—density changes may be significant even in the limit of zero Mach number.

Experiments in gaseous microducts confirm the arguments above. For both low- and high-Mach-number flows, pressure gradients in long microchannels are non-constant, consistent with the compressible flow equations. Such experiments were conducted by, among others, Prud'homme, Chapman, and Bowen (1986); Pfahler et al. (1991); van den Berg, Seldam, and Gulik (1993); Liu et al. (1993, 1995); Pong et al. (1994); Harley, Bau, and Zemel (1995); Piekos and Breuer (1996); Arkilic (1997); and Arkilic, Schmidt, and Breuer (1995, 1997a,b). We will return to this issue in Chapter 13.

Identical arguments can be made in the case of isothermal walls. Here strong temperature changes may be the result of wall heating or cooling even if viscous dissipation is negligible. The proper temperature scale in this case is given in terms of the wall temperature T_w and the reference temperature T_o as follows:

$$\hat{T} = \frac{T - T_o}{T_w - T_o} \tag{2.37}$$

where $\hat{T}$ is the new dimensionless temperature. The nondimensional form of

[9] With an error of about 10% at $Ma = 0.3$, 4% at $Ma = 0.2$, 1% at $Ma = 0.1$, and so on.

Eq. (2.31) now reads

$$\frac{1}{\rho^*}\frac{D\rho^*}{Dt^*} = \gamma_0\, Ma^2\alpha^* \frac{Dp^*}{Dt^*} - \beta^* B\left(\frac{T_w - T_0}{T_0}\right)\frac{D\hat{T}}{Dt^*} \tag{2.38}$$

Here we notice that the temperature term is different from that in Eq. (2.35). The Ma is no longer appears in this term, and strong temperature changes such as large $(T_w - T_0)/T_0$ may cause strong density changes regardless of the value of the Mach number. Additionally, the thermodynamic properties of the fluid are not constant but depend on temperature, and as a result the continuity, momentum, and energy equations all couple. The pressure term in Eq. (2.38), on the other hand, is exactly as it was in the adiabatic case, and the same arguments made before apply: the flow should be considered compressible if $Ma > 0.3$ or if pressure changes due to viscous forces are sufficiently large.

There are three additional scenarios in which significant pressure and density changes may take place without inertial, viscous, or thermal effects. First is the case of quasi-static compression and expansion of a gas in, for example, a piston–cylinder arrangement. The resulting compressibility effects are, however, compressibility of the fluid and not of the flow. Two other situations in which compressibility effects must also be considered are problems with length scales comparable to the scale height of the atmosphere and rapidly varying flows, as in sound propagation (see Lighthill 1963b).

2.4 Equations of Motion at the Wall

In wall-bounded flows, the streamwise (x) and spanwise (z) momentum equations written at the wall give very useful expressions of the wall fluxes of, respectively, the spanwise and streamwise vorticity. For an incompressible fluid over a nonmoving wall of small curvature, and if body forces are neglected, the streamwise, normal, and spanwise momentum equations, written at $y = 0$, follow from Eq. (2.21) and read

$$\rho v_w \frac{\partial u}{\partial y}\bigg|_{y=0} + \frac{\partial p}{\partial x}\bigg|_{y=0} - \frac{\partial \mu}{\partial y}\bigg|_{y=0}\frac{\partial u}{\partial y}\bigg|_{y=0} = \mu \frac{\partial^2 u}{\partial y^2}\bigg|_{y=0} \tag{2.39}$$

$$0 + \frac{\partial p}{\partial y}\bigg|_{y=0} - 0 = \mu \frac{\partial^2 v}{\partial y^2}\bigg|_{y=0} \tag{2.40}$$

$$\rho v_w \frac{\partial w}{\partial y}\bigg|_{y=0} + \frac{\partial p}{\partial z}\bigg|_{y=0} - \frac{\partial \mu}{\partial y}\bigg|_{y=0}\frac{\partial w}{\partial y}\bigg|_{y=0} = \mu \frac{\partial^2 w}{\partial y^2}\bigg|_{y=0} \tag{2.41}$$

These equations are instantaneous and are valid for both laminar and turbulent flows. Here, v_w is the (positive) injection or (negative) suction velocity through the wall. Upward or downward motion of the wall acts analogously to injection or suction. The right-hand side of each of the three equations represents the curvature of the corresponding velocity profile, which in the case of the streamwise–spanwise equation is the same as the flux of spanwise–streamwise vorticity from the wall. Negative curvature implies that the velocity profile is fuller and that the wall is a source (positive flux) of vorticity. Whether the wall is a source or a sink of spanwise–streamwise vorticity depends on whether fluid is sucked or injected from the wall, whether the streamwise–spanwise pressure gradient is favorable or adverse, and whether the wall viscosity is lower or higher than that above the surface. For a canonical turbulent flow, the instantaneous velocity profiles in all three directions change shape continuously as a result of the random changes in the pressure field.

2.5 Turbulent Flows

All the equations thus far are valid for nonturbulent as well as turbulent flows. However, in the latter case the dependent variables are in general random functions of space and time. No straightforward method exists for obtaining stochastic solutions of these nonlinear partial differential equations, and this is the primary reason why turbulence remains as the last great unsolved problem of classical physics. Dimensional analysis can be used to obtain crude results for a few cases, but first-principles analytical solutions are not possible even for the simplest conceivable turbulent flow.

The contemporary attempts to use dynamical systems theory to study turbulent flows have not yet reached fruition, especially at Reynolds numbers far above transition (Aubry et al. 1988), although advances in this theory have helped somewhat with reducing and displaying the massive bulk of data resulting from numerical and experimental simulations (Sen 1989). The recent book by Holmes, Lumley, and Berkooz (1996) provides a useful, readable introduction to the emerging field. It details a strategy via which knowledge of coherent structures, finite-dimensional dynamical systems theory, and the Karhunen–Loève or proper orthogonal decomposition could be combined to create low-dimensional models of turbulence that resolve only the organized motion and describe their dynamical interactions. The utility of the dynamical systems approach as an additional arsenal to tackle the turbulence conundrum has been demonstrated only for turbulence near transition or near a wall, and thus the flow would be relatively simple and a relatively small number of degrees of freedom would be excited. Holmes et al. (1996) summarize the (partial) successes that have been achieved thus far using relatively small sets of ordinary differential equations and suggest a broad strategy for modeling turbulent flows as well as other spatiotemporally complex systems.

The brute-force numerical integration of the instantaneous equations using the supercomputer is prohibitively expensive—if not impossible—at practical Reynolds numbers (Moin and Mahesh 1998). For the present at least, a statistical approach, in which a temporal, spatial, or ensemble average is defined and the equations of motion are written for the various moments of the fluctuations about this mean, is the only route available to get meaningful engineering results. Unfortunately, the nonlinearity of the Navier–Stokes equations guarantees that the process of averaging to obtain moments results in an open system of equations in which the number of unknowns is always greater than the number of equations, and more or less heuristic modeling is used to close the equations. This is known as the closure problem and again makes obtaining first-principle solutions to the (averaged) equations of motion impossible.

To illustrate the closure problem, consider the (instantaneous) continuity and momentum equations for a Newtonian, incompressible, constant density, constant viscosity turbulent flow. In this uncoupled version of the four equations (2.20) and (2.21), for the four random unknowns u_i and p, no general stochastic solution is known to exist. But would it be feasible to obtain solutions for the nonstochastic mean-flow quantities? As was first demonstrated by Osborne Reynolds (1895) over a century ago, all the field variables are decomposed into a mean and a fluctuation. Let $u_i = \overline{U}_i + u_i'$ and $p = \overline{P} + p'$, where $\overline{U}_i$ and $\overline{P}$ are ensemble averages for the velocity and pressure, respectively, and u_i' and p' are the velocity and pressure fluctuations about the respective averages. Note that temporal or spatial averages could be used in place of ensemble average if the flowfield is stationary or homogeneous, respectively. In the former case, the time derivative of any statistical quantity vanishes. In the latter, averaged functions are independent of position. When the decomposed pressure and velocity are substituted into Eqs. (2.20) and (2.21), the equations governing the mean velocity and mean pressure for an incompressible, constant-viscosity turbulent flow become

$$\frac{\partial \overline{U}_k}{\partial x_k} = 0 \tag{2.42}$$

$$\rho \left(\frac{\partial \overline{U}_i}{\partial t} + \overline{U}_k \frac{\partial \overline{U}_i}{\partial x_k} \right) = -\frac{\partial \overline{P}}{\partial x_i} + \frac{\partial}{\partial x_k} \left(\mu \frac{\partial \overline{U}_i}{\partial x_k} - \rho \, \overline{u_i u_k} \right) + \rho \, \overline{g}_i \tag{2.43}$$

where, for clarity, the primes have been dropped from the fluctuating velocity components u_i and u_k.

This is now a system of 4 equations for the 10 unknowns $\overline{U}_i$, $\overline{P}$, and $\overline{u_i u_k}$.[10] The momentum equation (2.43) is written in a form that facilitates the physical interpretation of the turbulence stress tensor (Reynolds stresses), $-\rho \, \overline{u_i u_k}$, as additional stresses on a fluid element to be considered along with the conventional viscous stresses and pressure. An equation for the components of this tensor may be derived, but it will contain third-order moments such as $\overline{u_i u_j u_k}$, and so on. The equations are (heuristically) closed by expressing the second- or third-order quantities in terms of the first- or second-moments, respectively. For comprehensive reviews of these first- and second-order closure schemes, see Lumley (1983, 1987), Speziale (1991), and Wilcox (1993).

2.6 The Kármán Integral Equation

The next useful equation results from integrating in the normal direction (from the wall to the edge of the shear layer) and combining the continuity and streamwise momentum equations, the latter of which is written in terms of a generic shear stress $(\partial \tau_{xy}/\partial y)$. The usual boundary layer approximations are invoked in the momentum equation: (1) neglecting the normal stress terms relative to the shear stress terms, and (2) assuming the pressure to be impressed upon the boundary layer from without. The final result is known as the Kármán integral equation. For a two-dimensional (or axisymmetric) wall-bounded flow of a compressible fluid and with body force

---

[10] The second-order tensor $\overline{u_i u_k}$ is obviously a symmetric one with only six independent components.

neglected, this equation is derived in Kays and Crawford (1993) and reads

$$\frac{C_f}{2} = \frac{1}{U_o^2} \frac{\partial}{\partial t} (U_o \delta^*) + \frac{\partial \delta_\theta}{\partial x} - \frac{\rho_w v_w}{\rho_o U_o}$$

$$+ \delta_\theta \left[\left(2 + \frac{\delta^*}{\delta_\theta} \right) \frac{1}{U_o} \frac{\partial U_o}{\partial x} + \frac{1}{\rho_o} \frac{\partial \rho_o}{\partial x} + \frac{1}{R} \frac{\partial R}{\partial x} \right] \tag{2.44}$$

Similar integral relations can be derived from the kinetic energy and thermal energy differential equations. In the momentum integral equation, the skin-friction coefficient, the displacement thickness, and the momentum thickness are given by, respectively,

$$C_f \equiv \frac{\tau_w}{\frac{1}{2}\rho_o U_o^2} \tag{2.45}$$

$$\delta^* \equiv \int_0^\infty \left(1 - \frac{\rho u}{\rho_o U_o} \right) dy \tag{2.46}$$

$$\delta_\theta \equiv \int_0^\infty \frac{\rho u}{\rho_o U_o} \left(1 - \frac{u}{U_o} \right) dy \tag{2.47}$$

Also, R is the transverse radius of curvature of the body of revolution (the corresponding term in Eq. (2.44) drops for a two-dimensional flow), ρ_o and U_o are the density and velocity outside the boundary layer, respectively, ρ_w and v_w are the density and normal velocity of fluid injected or sucked through the surface (or the upward or downward wall velocity), and τ_w is the shear stress at the wall.

Because the streamwise momentum equation leading to the Kármán integral equation was written in terms of a generic shear stress, the integral momentum equation (2.44) is, in fact, valid for laminar and turbulent flows as well as for Newtonian and non-Newtonian fluids; the only assumption being made is that the flow is two-dimensional (or axisymmetric) in the mean. In case of a turbulent flow, the mean streamwise velocity, $\overline{U}(t, x, y)$, is used in the definition of δ^* and δ_θ. For a two-dimensional flow of a Newtonian fluid,

$$\tau_w = \mu \frac{\partial \overline{U}}{\partial y} \bigg|_{y=0} + \mu \frac{\partial \overline{V}}{\partial x} \bigg|_{y=0} \tag{2.48}$$

The second term on the right-hand side of this equation vanishes if there is no injection or suction (or upward or downward wall motion), or if the transpiration is spatially uniform.

There is a seeming controversy that periodically flares up in the literature: Is the Kármán integral equation accurate for a turbulent flow to within the first-order boundary-layer approximation? In the case of a two-dimensional, incompressible turbulent flow, the streamwise momentum equation contains terms related to the turbulence normal stresses, namely,

$$\rho \frac{\partial \overline{U}}{\partial t} + \rho \overline{U} \frac{\partial \overline{U}}{\partial x} + \rho \overline{V} \frac{\partial \overline{U}}{\partial y} = -\frac{\partial \overline{P}_o}{\partial x} + \rho \frac{\partial}{\partial x} \left(\overline{v^2} - \overline{u^2} \right) + \frac{\partial \tau_{xy}}{\partial y} \tag{2.49}$$

In the last term of this equation, τ_{xy} is the sum of the viscous shear stress ($\mu \frac{\partial \overline{U}}{\partial y}$)

and the Reynolds shear stress $(-\rho\,\overline{uv})$. The second term on the right-hand side has contributions from the fact that the pressure within the boundary layer is not quite equal to that at the edge (the difference being $\rho\,\overline{v^2}$) and from the turbulence normal stress term $(\rho\,\overline{u^2})$. The viscous normal stress terms corresponding to those two Reynolds stress terms are, respectively, $\mu\,\frac{\partial V}{\partial y}$ and $\mu\,\frac{\partial U}{\partial x}$. These clearly have smaller orders of magnitude and are therefore dropped from Eq. (2.49). The question is whether or not to retain the Reynolds normal stress terms. If the turbulence normal stresses are retained, Eq. (2.49) cannot be integrated to yield the Kármán integral equation. Given that the turbulence rms fluctuations within a boundary layer are typically one order of magnitude less than the mean velocity and that the term in question is the difference between two small numbers, careful order-of-magnitude analysis shows that indeed the second term on the right-hand side of Eq. (2.49) can be neglected within the boundary-layer approximation (see also Hinze 1975). The Kármán integral equation (2.44) for a turbulent wall-bounded flow is therefore first-order accurate.

CHAPTER THREE

Unifying Principles

> Mechanics is the paradise of the mathematical sciences because by means of it one comes to the fruits of mathematics.
>> (Leonardo da Vinci, 1452–1519)

> What experience and history teach is this—that people and governments never have learned anything from history, or acted on principles deduced from it.
>> (Georg Wilhelm Friedrich Hegel, 1770–1831)

PROLOGUE

A particular control strategy is chosen based on the kind of flow and the control goal to be achieved. Flow-control goals are strongly, often adversely, interrelated, and there lies the challenge of making the tough compromises. There are several different ways for classifying control strategies to achieve a desired effect. Presence or lack of walls, Reynolds and Mach numbers, and the character of the flow instabilities are all important considerations for the type of control to be applied. All these seemingly disparate issues are what places the field of flow control in a unified framework. They will be discussed in turn in this chapter.

3.1 Control Goals and Their Interrelation

What does the engineer want to achieve when attempting to manipulate a particular flowfield? Typically he or she aims at reducing the drag; at enhancing the lift; at augmenting the mixing of mass, momentum, or energy; at suppressing the flow-induced noise; or at a combination thereof. To achieve any of these useful end results for either free-shear or wall-bounded flows, transition from laminar to turbulent flow may have to be delayed or advanced, flow separation may have to be prevented or provoked, and finally turbulence levels may have to be suppressed or enhanced. All these engineering goals and the corresponding flow changes intended to effect them are schematically depicted in Figure 3.1. None of these are particularly difficult if taken in isolation, but the challenge is to achieve a goal using a simple device that is inexpensive to build as well as to operate, and, most important, has minimum

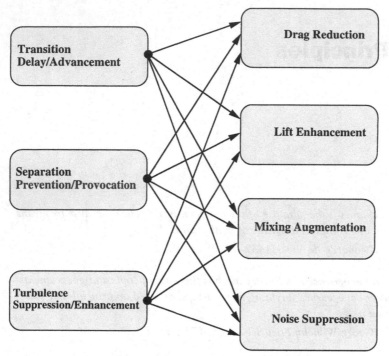

Figure 3.1: Engineering goals and corresponding flow changes.

side effects. For this latter hurdle, the interrelation among control goals must be elaborated, and this is what is attempted below.

Section 3.3 will contrast free-shear and wall-bounded flows. For the purposes of the present argument, we focus on the latter—particularly the technologically very important boundary layers. An external wall-bounded flow, such as that developing on the exterior surfaces of an aircraft or a submarine, can be manipulated to achieve transition delay, separation postponement, lift increase, skin-friction and pressure-drag[1] reduction, turbulence augmentation, heat-transfer enhancement, or noise suppression. These objectives are not necessarily mutually exclusive. The schematic in Figure 3.2 is a partial representation of the interrelation between one control goal and another. To focus the discussion further, think of the flow developing on a lifting surface such as an aircraft wing. If the boundary layer becomes turbulent, its resistance to separation is enhanced, and more lift could be obtained at increased incidence. On the other hand, the skin-friction drag for a laminar boundary layer can be as much as an order of magnitude less than that for a turbulent one. If transition is delayed, lower skin friction as well as lower flow-induced noise is achieved. However, the laminar boundary layer can only support a very small adverse pressure gradient without separation, and subsequent loss of lift and increase in form drag occur. Once the laminar boundary layer separates, a free-shear layer forms and, for moderate Reynolds numbers, transition to turbulence takes place. Increased entrainment of high-speed fluid because of the turbulent mixing may result in reattachment

---

[1] Pressure drag includes contributions from flow separation, displacement effects, induced drag, wave drag, and, for time-dependent motion of a body through a fluid, virtual mass.

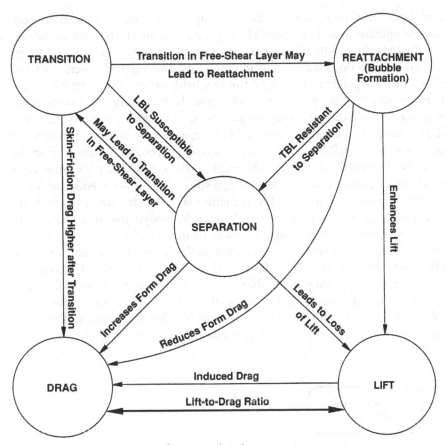

Figure 3.2: Interrelation between flow-control goals.

of the separated region and formation of a laminar separation bubble.[2] At higher incidence, the bubble breaks down, either separating completely or forming a longer bubble. In either case, the form drag increases and the lift-curve's slope decreases. The ultimate goal of all this is to improve the airfoil's performance by increasing the lift-to-drag ratio. However, induced drag is caused by the lift generated on a lifting surface with a finite span. Moreover, more lift is generated at higher incidence, but form drag also increases at these angles.

The preceding discussion points to potential conflicts as one tries to achieve a particular control goal only to adversely affect another goal. An ideal method of control that is simple, inexpensive to build and operate, and does not have any trade-offs does not exist, and the skilled engineer has to make continual compromises to achieve a particular design goal.

3.2 Classification Schemes

There are different classification schemes for flow-control methods. One is to consider whether the technique is applied at the wall or away from it. Surface parameters

---

[2] See Chapter 9 for more details about the formation and control of separation bubbles.

that can influence the flow include roughness, shape, curvature, rigid-wall motion, compliance, temperature, and porosity. Heating and cooling of the surface can influence the flow via the resulting viscosity and density gradients. Mass transfer can take place through a porous wall or a wall with slots. Suction and injection of primary fluid can have significant effects on the flowfield, influencing particularly the shape of the velocity profile near the wall and thus the boundary-layer susceptibility to transition and separation. Different additives, such as polymers, surfactants, microbubbles, droplets, particles, dust, or fibers, can also be injected through the surface in water- or air-wall-bounded flows. Control devices located away from the surface can also be beneficial. Large-eddy breakup devices (also called outer-layer devices or OLDs); acoustic waves bombarding a shear layer from outside; additives introduced in the middle of a shear layer; manipulation of freestream turbulence levels; and spectra, gust, and magneto- and electrohydrodynamic body forces are examples of flow-control strategies applied away from the wall.

A second scheme for classifying flow-control methods considers energy expenditure and the control loop involved. As shown in the schematic in Figure 3.3, a control device can be passive, requiring no auxiliary power and no control loop, or active, requiring energy expenditure. As for the action of passive devices, some prefer to use the term flow management rather than flow control (Fiedler and Fernholz 1990), reserving the latter terminology for dynamic processes. Active control requires a control

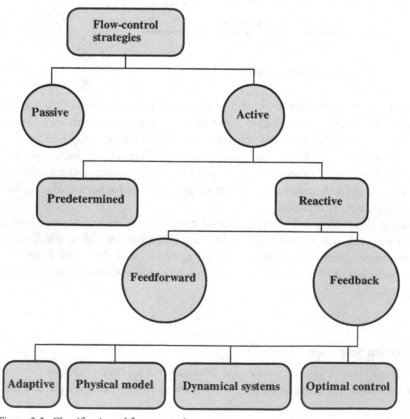

Figure 3.3: Classification of flow-control strategies.

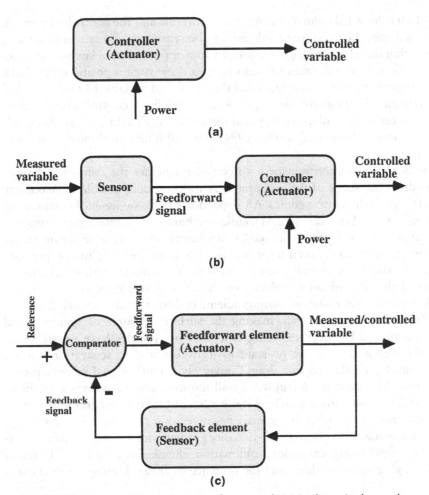

Figure 3.4: Different control loops for active flow control. (a) Predetermined, open-loop control; (b) Reactive, feedforward, open-loop control; (c) Reactive, feedback, closed-loop control.

loop and is further divided into predetermined or reactive categories. Predetermined control includes the application of steady or unsteady energy input without regard to the particular state of the flow. The control loop in this case is open, as shown in Figure 3.4a, and no sensors are required. Because no sensed information is being fed forward, this open control loop is not a feedforward one. This subtle point is often misunderstood in the literature and results in predetermined control being confused with reactive, feedforward control. Reactive[3] control is a special class of active control in which the control input is continuously adjusted based on measurements of some kind. The control loop in this case can either be an open feedforward one (Figure 3.4b) or a closed feedback loop (Figure 3.4c). Classical control theory deals, for the most part, with reactive control.

The distinction between feedforward and feedback is particularly important when dealing with the control of flow structures that convect over stationary sensors and

---

[3] Also termed, by some, interactive flow control. Reactive could be confused with chemically reacting flows but is more proper linguistically to describe this particular control strategy.

actuators. In feedforward control, the measured variable and the controlled variable differ. For example, the pressure or velocity can be sensed at an upstream location, and the resulting signal is used together with an appropriate control law to trigger an actuator, which in turn influences the velocity at a downstream position. Feedback control, on the other hand, necessitates that the controlled variable be measured, fed back, and compared with a reference input. Reactive feedback control is further classified into four categories: adaptive, physical-model-based, dynamical systems-based, and optimal control (Moin and Bewley 1994). We will return to these categories in Chapter 14.

Yet another classification scheme is to consider whether the control technique directly modifies the shape of the instantaneous–mean velocity profile or selectively influences the small dissipative eddies. An inspection of the Navier–Stokes equations written at the surface, Eqs. (2.39)–(2.41), indicates that the spanwise and streamwise[4] vorticity fluxes at the wall can be changed, either instantaneously or in the mean, via wall motion or compliance, suction or injection, streamwise or spanwise pressure gradient (respectively), or normal viscosity gradient. These vorticity fluxes determine the fullness of the corresponding velocity profiles. For example, suction (or downward wall motion), favorable pressure gradient, or lower wall viscosity results in vorticity flux away from the wall, making the surface a source of spanwise and streamwise vorticity. The corresponding fuller velocity profiles have negative curvature at the wall and are more resistant to transition and to separation but are associated with higher skin-friction drag. Conversely, an inflectional velocity profile can be produced by injection (or upward wall motion), adverse pressure gradient, or higher wall viscosity. Such profile is more susceptible to transition and to separation and is associated with lower, even negative, skin friction. Note that many techniques are available to effect a wall-viscosity gradient; for example, surface heating or cooling, film boiling, cavitation, sublimation, chemical reaction, wall injection of lower or higher viscosity fluid, and the presence of shear thinning or thickening additive.

Flow-control devices can alternatively target certain scales of motion rather than globally change the velocity profile. Polymers, riblets, and LEBUs, for example, appear to selectively damp only the small dissipative eddies in turbulent wall-bounded flows. These eddies are responsible for the (instantaneous) inflectional profile and the secondary instability in the buffer zone,[5] and their suppression leads to increased scales, a delay in the reduction of the (mean) velocity-profile slope, and consequent thickening of the wall region. In the buffer zone, the scales of the dissipative and energy-containing eddies are roughly the same and, hence, the energy-containing eddies will also be suppressed, resulting in reduced Reynolds stress production, momentum transport, and skin friction.

3.3 Free-Shear and Wall-Bounded Flows

Free-shear flows, such as jets, wakes, and mixing layers, are characterized by inflectional mean-velocity profiles and are therefore susceptible to inviscid instabilities.

---

[4] Streamwise vorticity exists only if the velocity field is three-dimensional, instantaneously or in the mean.

[5] See Chapter 5 for a description of the different *regions* of a turbulent wall-bounded flow.

Viscosity is only a damping influence in this case, and the prime instability mechanism is vortical induction. Control goals for such flows include transition delay or advancement, mixing enhancement, and noise suppression. External and internal wall-bounded flows, such as boundary layers and channel flows, can also have inflectional velocity profiles, but, in the absence of adverse pressure gradient and similar effects, are characterized by noninflectional profiles and viscous instabilities, which are then to be considered. These kind of viscosity-dominated, wall-bounded flows are intrinsically stable and therefore are generally more difficult to control. Free-shear flows and separated boundary layers, on the other hand, are intrinsically unstable and lend themselves more readily to manipulation.

Free-shear flows originate from some kind of surface upstream, be it a nozzle, a moving body, or a splitter plate, and flow-control devices can therefore be placed on the corresponding walls, albeit far from the fully developed regions. Examples of such control include changing the geometry of a jet exit from circular to elliptical (Gutmark and Ho 1986), using periodic suction or injection in the lee side of a blunt body to affect its wake (Williams and Amato 1989), and vibrating the splitter plate of a mixing layer (Fiedler, Glezer, and Wygnanski 1988). These and other techniques are extensively reviewed by Fiedler and Fernholz (1990), who offer a comprehensive list of appropriate references, and more recently by Gutmark, Schadow, and Yu (1995); Viswanath (1995); Fiedler (1998); and Gutmark and Grenstein (1999).

3.4 Regimes of Reynolds and Mach Numbers

The Reynolds number is the ratio of inertial to viscous forces and, absent centrifugal, gravitational, electromagnetic and other unusual effects, Re determines whether the flow is laminar or turbulent. It is defined as $Re \equiv v_0 L / \nu$, where v_0 and L are, respectively, suitable velocity and length scales, and ν is the kinematic viscosity. For low-Reynolds-number flows, instabilities are suppressed by viscous effects and the flow is laminar, as can be found in systems with large fluid viscosity, small length scale, or small velocity. The large-scale motion of highly viscous volcanic molten rock and air and water flow in capillaries and microdevices are examples of laminar flows. Turbulent flows seem to be the rule rather than the exception and occur in or around most important fluid systems such as airborne and waterborne vessels, gas and oil pipelines, material processing plants, and the human cardiovascular and pulmonary systems.

Because of the nature of their instabilities, free-shear flows undergo transition at extremely low Reynolds numbers as compared with wall-bounded flows. Many techniques are available to delay laminar-to-turbulence transition for both kinds of flows, but these techniques are ineffective in the case of indefinitely high Reynolds numbers. Therefore, for Reynolds numbers beyond a reasonable limit, one should not attempt to prevent transition but rather deal with the ensuing turbulence. Of course early transition to turbulence can be advantageous in some circumstances such as to achieve separation delay, enhanced mixing, or augmented heat transfer. The task of advancing transition is generally simpler than trying to delay it. Chapter 6 specifically addresses the control of laminar-to-turbulence transition. For now, we briefly discuss transition control for various regimes of Reynolds and Mach numbers.

Three Reynolds number regimes can be identified for the purpose of reducing skin friction in wall-bounded flows. First, if the flow is laminar, typically at Reynolds numbers based on distance from leading edge $<10^6$, then methods of reducing the laminar shear stress are sought. These are usually velocity-profile modifiers such as adverse pressure gradient, injection, cooling (in water), and heating (in air) that reduce the fullness of the profile at the increased risk of premature transition and separation. Secondly, in the range of Reynolds numbers from 1×10^6 to 4×10^7, active and passive methods to delay transition as much as possible are sought. These techniques can result in substantial savings and are broadly classified into two categories: stability modifiers and wave cancellation. The skin-friction coefficient in the laminar flat plate can be as much as an order of magnitude less than that in the turbulent case. Note, however, that all the stability modifiers such as favorable pressure gradient, suction, or heating (in liquids) result in an increase in the skin friction over the unmodified Blasius layer. The object is, of course, to keep this penalty below the potential saving, that is, the net drag will be above that of the flat-plate laminar boundary layer but presumably well below the viscous drag in the flat-plate turbulent flow. Thirdly, for $Re > 4 \times 10^7$, transition to turbulence cannot be delayed with any known practical method without incurring a penalty that exceeds the saving.[6] The task is then to reduce the skin-friction coefficient in a turbulent boundary layer. Relaminarization (Narasimha and Sreenivasan 1979) is an option, although achieving a net saving here is problematic at present.

The Mach number is the ratio of a characteristic flow velocity to local speed of sound, $Ma \equiv v_0/a_0$. It determines whether the flow is incompressible ($Ma < 0.3$) or compressible ($Ma > 0.3$). The latter regime is further divided into subsonic ($Ma < 1$), transonic ($0.8 < Ma < 1.2$), supersonic ($Ma > 1$), and hypersonic ($Ma > 5$). Each of these flow regimes lends itself to different optimum methods of control to achieve a given goal. Take laminar-to-turbulence transition control as an illustration (Bushnell 1994). During transition, the field of initial disturbances is internalized via a process termed receptivity, and the disturbances are subsequently amplified by various linear and nonlinear mechanisms. If it is assumed that bypass mechanisms, such as roughness or high levels of freestream turbulence, are identified and circumvented, delaying transition then is reduced to controlling the variety of possible linear modes: Tollmien–Schlichting modes, Mack modes, crossflow instabilities, and Görtler instabilities. Tollmien–Schlichting instabilities dominate the transition, process for two-dimensional boundary layers having $Ma < 4$ and are damped by increasing the Mach number, by wall cooling (in gases), and by the presence of favorable pressure gradient. Contrast this with the Mack modes, which dominate for two-dimensional hypersonic flows. Mack instabilities are also damped by increasing the Mach number and by the presence of favorable pressure gradient but are destabilized by wall cooling. Crossflow and Görtler instabilities are caused by, respectively, the development of inflectional crossflow velocity profile and the presence of concave streamline curvature. Both of these instabilities are potentially harmful across the speed range but are largely unaffected by Mach number and wall cooling. The crossflow modes are enhanced by favorable pressure gradient, whereas the Görtler instabilities are insensitive. Suction suppresses, to different degrees, all the linear modes discussed here.

---

[6] Passive compliant coatings *may* be an exception. See Chapter 7.

3.5 Convective and Absolute Instabilities

In addition to grouping the different kinds of hydrodynamic instabilities as inviscid or viscous, one could also classify them as convective or absolute based on the linear response of the system to an initial localized impulse (Huerre and Monkewitz 1990). A flow is convectively unstable if, at any fixed location, this response eventually decays in time (in other words, if all growing disturbances convect downstream from their source). Convective instabilities occur when there is no mechanism for upstream disturbance propagation, as for example in the case of rigid-wall boundary layers. If the disturbance is removed, then perturbation propagates downstream, and the flow relaxes to an undisturbed state. Suppression of convective instabilities is particularly effective when applied near the point where the perturbations originate.

If any of the growing disturbances has zero group velocity, the flow is absolutely unstable. This means that the local system response to an initial impulse grows in time. Absolute instabilities occur when a mechanism exists for upstream disturbance propagation, as for example in the separated flow over a backward-facing step in which the flow recirculation provides such a mechanism. In this case, some of the growing disturbances can travel back upstream and continually disrupt the flow even after the initial disturbance is neutralized. Therefore, absolute instabilities are generally more dangerous and more difficult to control; nothing short of complete suppression will work. In some flows—for example, two-dimensional blunt-body wakes—certain regions are absolutely unstable whereas others are convectively unstable. The upstream addition of acoustic or electric feedback can change a convectively unstable flow to an absolutely unstable one, and self-excited flow oscillations can thus be generated. In any case, identifying the character of flow instability facilitates its effective control (i.e., suppressing or amplifying the perturbation as needed).

CHAPTER FOUR

Coherent Structures

It was six men of Indostan
To learning much inclined,
Who went to see the Elephant
(Though all of them were blind),
That each by observation
Might satisfy his mind.

The first approached the Elephant,
And happening to fall
Against his broad and sturdy side,
At once began to bawl: "God bless me! but the Elephant
Is very like a wall!"

The second, feeling of the tusk,
Cried: "Ho! what have we here
So very round and smooth and sharp?
To me 'tis mighty clear
This wonder of a Elephant
Is very like a spear!"

The third approached the animal,
And happening to take
The squirming trunk within his hands,
Thus boldly up and spake: "I see," quoth he, "the Elephant
Is very like a snake."

The fourth reached out his eager hand,
And felt about the knee.
"What most this wondrous beast is like
Is mighty plain," quoth he;
" 'Tis clear enough the Elephant
Is very like a tree!"

The fifth, who chanced to touch the ear,
Said: "E'en the blindest man
Can tell what this resembles most: Deny the fact who can,
This marvel of an Elephant
Is very like a fan!"

The sixth *no sooner had begun*
About the beast to grope,
Than, seizing on the swinging tail
That fell within his scope,
"I see," quoth he, "the Elephant
Is very like a rope!"

And so these men of Indostan
Disputed loud and long,
Each in his own opinion
Exceeding stiff and strong,
Though each was partly in the right,
And all were in the wrong!

MORAL

So, oft in theologic wars,
The disputants, I ween,
Rail on in utter ignorance
Of what each other mean,
And prate about an Elephant
Not one of them has seen!

> *(John Godfrey Saxe's, 1816–1887, version of the fable*
> *"The Blind Men and the Elephant," which occurs in the* Udāna,
> *a canonical Hindu scripture)*

PROLOGUE

The relatively recent realization that organized structures play an important role in all turbulent shear flows leads quite naturally to the concept of turbulence control via direct interference with these deterministic events. Active, predetermined, open-loop control can be employed, but perhaps reactive control can be used much more effectively where specific coherent structures are sensed and then targeted for modulation to achieve a useful end result such as drag reduction, lift enhancement, and so forth. Reactive control strategies require sensors, actuators, and appropriate control algorithms, as will be discussed in Chapter 14, but for the present chapter we provide a gentle introduction to the fascinating world of coherent structures in transitional and turbulent flows. First, we briefly describe the different historical perspectives on turbulence. Secondly, we define what is meant by organized motions. This is followed by a summary of what is known about coherent structures in free-shear flows. Lastly, the important topic of coherent motions in wall-bounded flows is covered. This last topic will be further explored from the perspective of Reynolds number effects in the following chapter.

4.1 The Changing Paradigms of Turbulence

Turbulence is the last great unsolved problem of classical physics. Or so it goes for a quote variously attributed to one of the great modern physicists Albert Einstein,

Richard Feynman, Werner Heisenberg, or Arnold Sommerfeld. But in fact the closest sentiments to this quote that could be traced were expressed by the classical physicist Horace Lamb (1895), who actually wrote, starting with the second edition of his celebrated book *Hydrodynamics* under the heading of Turbulent Motion: "It remains to call attention to the chief outstanding difficulty of our subject." A humorous fable, also attributed to several of the great ones, goes as follows. As he lay dying, the modern physicist asked God two questions: Why relativity (or quantum mechanics, depending on who is departing), and why turbulence? "I really think," said the famed physicist, "He may have an answer to the first question." No one knows how to obtain stochastic solutions to the well-posed set of partial differential equations that govern turbulent flows. Averaging those nonlinear equations to obtain statistical quantities always leads to more unknowns than equations, and ad hoc modeling is then necessary to close the problem. So, except for a rare few limiting cases, first-principle analytical solutions to the turbulence conundrum are not possible. In the words of John Lumley (1996a), turbulence is a difficult problem that is unlikely to suddenly succumb to our efforts. We should not await sudden breakthroughs and miraculous solutions but rather keep at it slowly building one small brick at a time.

Our struggle to conquer turbulence has been long and arduous with lots of sweat, few victories, and much frustration. Not surprisingly, the way turbulence is being viewed as a complex physical phenomenon has changed over the years. Indeed, key ideas in the field continue to rise and fall (Liepmann 1979) and perhaps even to rise again! We now know that turbulence is random fluctuations superimposed on mean flow. A turbulent flow contains motions with numerous time and length scales that are random in the sense that there is zero probability of any flow variable having a particular value, and there is zero energy in any one particular frequency or wave number. In other words, the probability density function and spectrum of any flow variable are continuous and finite (Hunt, Carruthers, and Fung 1991). But we also know that the seemingly random mess is at least in part deterministic: a combination of coherent and incoherent motions.

Historically, there are perhaps five doctrines in approaching the five-century-old conundrum: visualization, first principles, the statistical approach, coherent structures, and modern tools. Each of these dogmas is described in turn in the following five subsections.

4.1.1 Visualization

Perhaps more than any other tool available to attack the problem of fluid mechanics in general and turbulence in particular, flow visualization is singly responsible for many of the most exciting discoveries in the field. Because visualization is relatively simple and quick and is capable of giving both global and local behavior, rendering the fluid motion accessible to visual perception can yield invaluable qualitative as well as quantitative information about a complex flow. However, one has to be extremely vigilant to avoid the many possible pitfalls when interpreting visual images of fluid flows, particularly time-dependent flows (Gad-el-Hak 1992).

Leonardo da Vinci pioneered the visualization genre close to 500 years ago. Much of Leonardo's notebooks of engineering and scientific observations were translated into English in a magnificent two-volume book by MacCurdy (1938). Succulent descriptions of the smooth and eddying motions of water alone occupy 121 pages.

Figure 4.1: Leonardo da Vinci's sketch of water exiting from a square hole into a pool; circa 1500.

In them, one can easily discern the Renaissance genius's foreshadowing of some of the turbulence physics to be discovered centuries after his time. Particularly relevant to the present subject, the words eddies and eddying motions percolate throughout Leonardo's treatise on liquid flows.

Figure 4.1 represents perhaps the world's first use of visualization as a scientific tool to study a turbulent flow. Around 1500, Leonardo sketched a free water jet issuing from a square hole into a pool. He wrote, "Observe the motion of the surface of the water, which resembles that of hair, which has two motions, of which one is caused by the weight of the hair, the other by the direction of the curls; thus the water has eddying motions, one part of which is due to the principal current, the other to the random and reverse motion."[1] Reflecting on this passage, Lumley (1992) speculates that Leonardo da Vinci may have prefigured the now famous Reynolds turbulence decomposition nearly 400 years prior to Osborne Reynolds' own flow visualization and analysis!

In describing the swirling water motion behind a bluff body, da Vinci provided the earliest reference to the importance of vortices in fluid motion: "So moving water strives to maintain the course pursuant to the power which occasions it and, if it finds an obstacle in its path, completes the span of the course it has commenced by a circular and revolving movement." Leonardo accurately sketched the pair of quasi-stationary, counterrotating vortices in the midst of the random wake.

Finally, da Vinci's words "...The small eddies are almost numberless, and large things are rotated only by large eddies and not by small ones, and small things are

---

[1] This particular translation from the Italian original was made by Ugo Piomelli, University of Maryland. It differs in several key points from that of MacCurdy ("Watch the movement of the surface of water, how like it is to that of hair, which has two movements, one following the undulation of the surface, the other the lines of the curves: thus water forms whirling eddies, part following the impetus of the chief current, part the rising and falling movement."), but on the basis of careful analysis Piomelli's phrasing appears to be more precise.

turned by both small eddies and large" presage Richardson's cascade,[2] coherent structures, and large-eddy simulations, at least (John L. Lumley, private communication).

4.1.2 First Principles

At the time of Leonardo da Vinci, neither calculus nor the laws of mechanics were available, of course.[3] Little more than a century and half after the incomparable Newton's *Principia Mathematica* was published in 1687, the first principles of viscous fluid flows were firmed in the form of the Navier–Stokes equations with major contributions by Navier in 1823, Cauchy in 1828, Poisson in 1829, Saint-Venant in 1843, and Stokes in 1845. These equations of fluid motion were derived in Chapter 2. With very few exceptions, the Navier–Stokes equations provide an excellent model for both laminar and turbulent flows. But in the latter case no analytical solutions are possible for several reasons: (1) the governing partial differential equations are nonlinear; (2) the dependent variables are random functions of space and time; and (3) the usual simplifications and symmetries do not apply to the instantaneous, three-dimensional flow quantities (although they *may* apply to statistical quantities). Thus, even though the turbulence problem was well defined mathematically, no one could integrate the equations of motion or for that matter do much with them when dealing with stochastic phenomena.

For a century and half only laminar flows and their stability were tractable analytically. Recent advances in computer power made it possible to return to first principles in the case of a turbulent flow and numerically integrate the Navier–Stokes equations for simple geometries and rather low Reynolds numbers. Notwithstanding their rather severe restrictions and limitations, direct numerical simulations (DNS) are therefore the only route available for a direct, brute-force onslaught on the turbulence problem. On the other hand, numerical solutions overwhelm the senses with data while providing little physical understanding.

4.1.3 Statistical View of Turbulence

As is clear from the previous subsection, up until recently not much could be done with first principles for the turbulence problem. But do we really need the kind of detailed information that DNS provides for all the random flow variables? Would a statistical description yielding such quantities as mean or root-mean-square values, correlation functions, spectra, and probability distributions suffice? The answer is an emphatic yes, but there is a price to pay for the lowered expectations.

On the basis of a painstaking flow visualization study, Osborne Reynolds (1883) made the observation that a pipe flow is either direct or sinuous (laminar or turbulent), depending on the value of a dimensionless parameter that we now call the Reynolds number. Eleven years later, on 24 May 1894, Reynolds read to an audience of the British Royal Society a follow-up paper in which he attempted to provide a physical explanation for his earlier observation. That second treatise, published in 1895, is considered by many to have pioneered the modern scientific approach to

---

[2] Richardson's (1920, 1922) poetic description (widely recited but often misquoted) of the turbulence eddies within a cumulus cloud reads: "Big whirls have little whirls that feed on their velocity, and little whirls have lesser whirls and so on to viscosity."
[3] Refer to Figure 1.1.

the turbulence problem and even to have reshaped the direction of fluid mechanics research in general for the next century. Reynolds (1895) asserted that a turbulent flowfield could be decomposed into mean and fluctuating parts. He thus was able to write expressions for the time-averaged momentum, now known as the Reynolds equations, in which convective stress terms appear as new unknowns. This was the dawn of the statistical doctrine of turbulence research. Reynolds (1895) also derived the transport equation for the turbulence kinetic energy and demonstrated that the apparent stresses due to turbulence interacting with the mean velocity gradient lead to a transfer of kinetic energy from the mean motion to the turbulent.

As was illustrated in Section 2.5, in the statistical approach a temporal, spatial, or ensemble average is defined, and the equations of motion are written for the various moments of the fluctuations about this mean. Unfortunately, the nonlinearity of the Navier–Stokes equations guarantees that the process of averaging to obtain moments results in an open system of equations in which the number of unknowns is always greater than the number of equations and more or less heuristic modeling is used to close the equations. This is known as the closure problem, which again makes obtaining rational solutions to the (averaged) equations of motion impossible.

Attempts to close the Reynolds equations are at the heart of the turbulence modeling community. From the simple mixing length ideas of Prandtl, Taylor, and von Kármán to the more involved Reynolds-stress or second-order modeling and beyond, a whole new industry sprang out of the Reynolds decomposition (see, for example, Lumley 1983, 1987; Speziale 1991; Wilcox 1993).

4.1.4 Reemergence of Coherent Structures

The recognition of coherent structures during the last few decades brought us back a full circle to the time of Leonardo. Not only was visualization once again the method of choice for the major discoveries, but it was also the reaffirmation of the importance of eddying motions and the copresence of large, organized motions and small, random ones. In the view of Hussain (1986), the search for coherent events is the embodiment of man's desire to find order in apparent disorder.

The recent history of coherent structures is amply chronicled in the article by Liu (1988). He asserts that the kernel idea germinated as a result of a discussion that took place during the Fifth International Congress of Applied Mechanics held in 1938. There, both Tollmien and Prandtl, responding to a comment by von Kármán regarding the difficulties of reconciling a scalar mixing length with turbulence measurements made in a channel, suggested that the measured turbulence fluctuations include both random and nonrandom elements. Dreyden (1948) pointed out that the boundary layer measurements conducted at a later date at the National Bureau of Standards supported the ideas of Tollmien and Prandtl. Dreyden also lamented that there is no known procedure, experimental or theoretical, that can be used to separate the random processes from the nonrandom ones.

Liepmann (1952), citing measurements in free-shear turbulent flows, in flow between rotating cylinders, and in the far-wake of a cylinder, emphasized the importance of the presence of a secondary, large-scale structure superimposed upon the primary turbulent shear flow. Townsend (1956) thoroughly exploited this concept in the first edition of his famed monograph on the structure of turbulent shear flows. He recognized the quasi-deterministic nature of large eddies and inferred their shapes from the long-time-averaged spatial-correlation tensor measured in a Eulerian frame. Townsend's approach suffers from several shortcomings, including the lack of a unique relationship between the correlation tensor and the unsteady flow that produces it.

In a later article, Liepmann (1962) once again underscored the splitting of seemingly random fluctuations into large-scale, deterministic structures and fine-grained turbulence. Liepmann asserted the importance of large-scale structures in many technological problems in aerodynamic sound, combustion, and so forth. Liepmann (1979) was perhaps the first to suggest that the existence of deterministic eddies in turbulent flows can be exploited for control purposes via direct interference with these large structures.

The modern view of coherent structures resulted from flow visualization studies of low-Reynolds-number boundary layers conducted first at the University of Maryland and then at Stanford University during the late 1950s and 1960s. The new doctrine did not pick up steam, however, until the milestone discovery of Brown and Roshko (1971, 1974) of organized motion in a mixing layer at Reynolds numbers far exceeding transitional ones. The large spanwise vortices, prominent in visual images of the shear layer, were totally missed in the correlation studies of Townsend and others.

4.1.5 The Latest Tools

Finally, dynamical systems and wavelets are the modern (i.e., the 1990s and beyond, at least for now) tools to tackle the last conundrum. A turbulent flow is a complex, nonlinear dynamical system that has an infinite number of degrees of freedom. The issue here is the possibility of representing such a system with a "reasonable" number of degrees of freedom. A Fourier decomposition would not do because it requires a very large number of components.

Mechanical systems with three or more degrees of freedom are capable of chaotic behavior exemplified by a strange attractor in phase space. Such systems are complex, aperiodic, and random and display extreme sensitivity to initial conditions, but they are nevertheless still deterministic. The book by Holmes et al. (1996) provides an excellent introduction to the dynamical system approach to tackle the turbulence problem.

Under rather severe restrictions, chaos theory allows the representation of turbulence as a low-dimensional dynamical system. The flow is modeled as coherent structures plus a parameterized turbulent background. The proper orthogonal decomposition, or Karhunen–Loève decomposition, has been of great use because it is capable of representing the flow with a minimum number of modes. Such a representation is useful on two fronts:

1. It provides an inexpensive[4] surrogate to the turbulent flow and in the process sheds light on its basic physics.
2. Chaos control concepts can be utilized to achieve effective manipulation with minimum energy expenditure once the flow is successfully represented as a dynamical system with a reasonably small number of degrees of freedom. We will return to the latter front in Chapter 14.

---

[4] As compared with DNS.

The dynamical system approach thus far has been successful for flows near transition or near a solid wall, in which cases the associated low Reynolds number implies that a relatively small number of degrees of freedom are excited and that a large fraction of the energy is in the ordered, deterministic component of the flow. More and more degrees of freedom are excited as the flow moves farther from transition or away from a wall. In these cases, the structure of the strange attractor becomes so complex as to negate the dynamical system approach advantages over the classical statistical description.

The second modern tool is wavelets, which was introduced a little more than a decade ago. Wavelet transform is used in many fields, including signal processing, data compression, image coding, and numerical analysis (Wickerhauser 1994). The technique is based on group theory and square integrable representation in terms of basis functions, called wavelets, that are localized in both physical and wave-number spaces. Farge's (1992) review article provides a good introduction to the field and particularly its application to turbulence. The more recent paper by Vasilyev, Yuen, and Paolucci (1997) offers a gentle introduction to the use of wavelets for numerically solving complex, multiscale partial differential equations.

Wavelets allow the unfolding of a flowfield into both space and scale and possibly even direction. Wavelets are localized analyzing functions that are dilated or contracted before convolving with the signal under consideration to achieve scale decomposition. Wavelet analysis can be viewed as a multilevel or multiresolution representation of a function, each level of resolution consisting of basis functions that have the same scale but are located at different positions. In contrast to Fourier transform, wavelet transform is a local one, and the behavior of the signal at infinity plays no role in the analysis. Continuous wavelet transforms offer redundant unfolding in terms of space and scale and are thus suited for tracking coherent motions and their contributions to the energy spectrum. Discrete wavelet transforms, on the other hand, allow orthonormal projection on a minimal number of independent modes and may thus be used to model the flow dynamics.

Basis functions such as Mexican-hat wavelets or Daubechies scaling functions can be used to decompose a velocity field into eddies. Because coherent structures are always of limited spatial extent, wavelet decomposition seems to be a better representation of them than, say, Fourier representation. Fourier modes in the form of (space-filling) trigonometric functions stretch off to infinity and are thus suited for decomposing a velocity field into waves of different, independent wavelengths. On the other hand, a limited-extent eddy is ideally represented by modes that act in groups as the wave number increases. Wavelets offer complete representation at least for homogeneous turbulence.[5] For inhomogeneous flows, a complete representation using wavelets is not as readily achieved, and some difficulties remain to be resolved.[6] Nevertheless, the technique offers some potential advantages from the point of view of controlling turbulent flows. Specifically, wavelet transforms may be used, for example, as an efficient, unbiased strategy for real-time identification of coherent structures from an instantaneous velocity signal. This step is of course at the heart of effective reactive control.

---

[5] As do a variety of conventional techniques with varying degrees of computational efficiency and accuracy.
[6] Those difficulties in fact *are* being resolved as we go to the press (Samuel Paolucci, private communication).

4.2   What Is a Coherent Structure?

The statistical view that turbulence is essentially a stochastic phenomenon having a randomly fluctuating velocity field superimposed on a well-defined mean has been changed in the last few decades by the realization that the transport properties of all turbulent shear flows are dominated by quasi-periodic, large-scale vortex motions (Laufer 1975; Townsend 1976; Cantwell 1981; Fiedler 1988). Despite the extensive research work in this area, no generally accepted definition of what is meant by coherent motion has emerged. In physics, coherence stands for a well-defined phase relationship. We provide here two rather different views, the first is general and the second is more restrictive. According to Robinson (1991), "a coherent motion is defined as a three-dimensional region of the flow over which at least one fundamental flow variable (velocity component, density, temperature, etc.) exhibits significant correlation with itself or with another variable over a range of space and/or time that is significantly larger than the smallest local scales of the flow." The rather restrictive definition is given by Hussain (1986): "A coherent structure is a connected turbulent fluid mass with instantaneously phase-correlated vorticity over its spatial extent." In other words, underlying the random, three-dimensional vorticity that characterizes turbulence, there is a component of large-scale vorticity that is instantaneously coherent over the spatial extent of an organized structure. The apparent randomness of the flowfield is, for the most part, due to the random size and strength of the different type of organized structures comprising that field. Several other definitions are catalogued by Delville et al. (1998). The same authors also provide a cook-book-style approach to coherent structure identification using a variety of classical and modern strategies.

The challenge is to identify a coherent structure well hidden in a sea of random background when such a structure is present either in a visual impression of the flow or in an instantaneous velocity, temperature, or pressure signal. This is of course not a trivial task, and the ancient Hindu fable[7] about the six blind men, each trying from his own limited perspective to identify how an elephant looks, immediately comes to mind. Complicating the issue is that coherent structures change from one type of flow to another and even in the same type of flow as initial and boundary conditions vary. The largest eddies are of the same scale as that of the flow and consequently cannot be universal. Identifying a coherent structure based on certain dynamical properties is more likely to succeed but is quite involved. On the other hand, a kinematic detector based on its creator's perception of the dynamic behavior of the organized motion is simpler to employ but runs the risk of detecting the presence of nonexistent objects (Lumley 1981). Benefiting from hindsight, the few flow-visualization pictures depicted in the next two sections may help in exploring the nature of the whole beast.

4.3   Free-Shear Flows

As indicated above, there is no universal coherent structure. Organized motions in wakes are different from those in boundary layers. In a jet issuing from a nozzle with

---

[7] Recited at the beginning of this chapter with all *six* blind men. This ancient story is often misquoted in the recent literature with three, four, or five of the lads.

a thin, laminar boundary layer on its inner surface, coherent structures are easily observed, whereas a jet issuing from a nozzle with a thick, turbulent boundary layer has organized structures that are nearly undetectable. The proportion of organized turbulence decreases, in general, with the level of disturbances in the incoming flow.

In general again, coherent structures in free-shear flows are easier to detect and characterize than those in wall-bounded flows. According to Liu (1988), for free-shear turbulent flows it is not necessary to conjecture that the local fine-grained turbulence rearranges itself to give bursts of white noise in order to maintain the hydrodynamically unstable waves as in the case of wall-bounded flows. The existence of large-scale, coherent motions in mixing layers, jets, and wakes is instead a manifestation of the dynamic instability associated with the local inflectional mean-velocity profiles. As a result, free-shear flows have pronounced organized motions and wavelike structures (Fiedler 1988, 1998).

In a mixing layer, the dominant structures are rollerlike vortices as large as the shear layer itself (Brown and Roshko 1971, 1974; Roshko 1976). The growth of the mixing layer results from the amalgamation of neighboring large eddies rather than from their individual growth (Winant and Browand 1974). The large transverse eddies in a mixing layer are strung together by a spaghetti-like net of smaller-scale, streamwise, counterrotating vortices. If the mixing layer develops from undisturbed conditions (i.e., thin, laminar boundary layers on the splitter plate), the rollerlike vortices are energetic and only relatively slowly become three-dimensional and less organized. If, on the other hand, the splitter plate has thick, turbulent boundary layers, the proportion of organized motion is considerably less dominant.

The spark-shadowgraph photograph in Figure 4.2 shows the mixing of a fast-moving stream of helium (top) and a slower stream of nitrogen (bottom), both moving from left to right. The two streams have the same mean momentum per unit cross-sectional area, $\rho_1 \overline{U}_1^2 = \rho_2 \overline{U}_2^2$, and originate from a splitter plate with laminar boundary layers. The rollerlike vortices convect at nearly constant speed equal to the average $\frac{1}{2}(\overline{U}_1 + \overline{U}_2)$. Increasing the Reynolds number produces more small-scale structures without significantly altering the large eddies.

The spanwise vortices remain as a permanent dominant feature even at high Reynolds numbers. Brown and Roshko's (1971, 1974) experiment is important because it was the first to show a significant proportion of organized turbulence even at Reynolds numbers far from transitional. Moreover, a much closer tie between

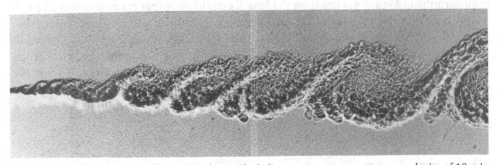

Figure 4.2: High-Reynolds-number mixing layer. The helium stream on top moves at a velocity of 10 m/s, and the nitrogen stream on the bottom moves at a speed of 3.78 m/s. The whole test section is pressurized to $\overline{P} = 8$ atm, giving a Reynolds number based on downstream distance of the order of 10^6 [after Brown and Roshko 1974].

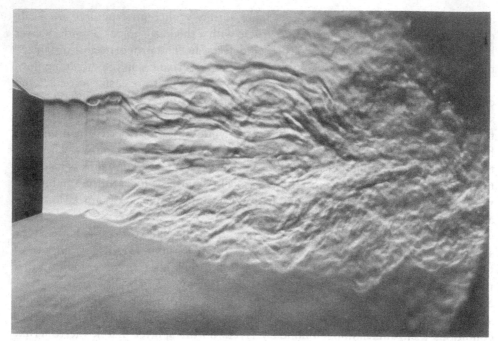

Figure 4.3: Transitional subsonic air jet [after Bradshaw, Ferris, and Johnson 1964].

stability and turbulence has been established: The essentially two-dimensional vortices in the fully turbulent flow are clearly related to the general instability modes of a simple vortex sheet.

The organized structures in jets are not as dominant as those in mixing layers having similar levels of disturbances in the incoming flow. The schlieren photograph in Figure 4.3 depicts a subsonic air jet issuing at a speed of 12 m/s from a 5-cm-diameter nozzle on the left, giving a Reynolds number of $Re \simeq 4 \times 10^4$. The laminar jet undergoes transition shortly after exiting and eventually becomes turbulent. The initial instabilities in the form of vortex rings appear to originate at the interface of the jet and the surrounding potential flow and quickly become three-dimensional as longitudinal striations develop.

In contrast to mixing layers, the spanwise vortices shed in the wake of bluff bodies do not pair. The basic mechanism for wake growth is entrainment. The two photographs in Figure 4.4 contrast the low-Reynolds-number turbulent wake ($Re = 4300$) behind a flat plate at a 45° angle of attack and the wake behind the tanker *Argo Merchant*, which was grounded on the Nantucket shoals in 1976. The laboratory flow is visualized using aluminum flakes suspended in water. Leaking crude oil indicates that the *Argo Merchant* happened to be inclined at about 45° to the current. Although the Reynolds number for the tank ship at $Re \approx 10^8$ is more than four orders of magnitude higher, the two wakes are remarkably similar.

4.4 Wall-Bounded Flows

We now turn our attention to the more enigmatic boundary layers and channel flows. In a wall-bounded flow, a multiplicity of coherent structures have been identified

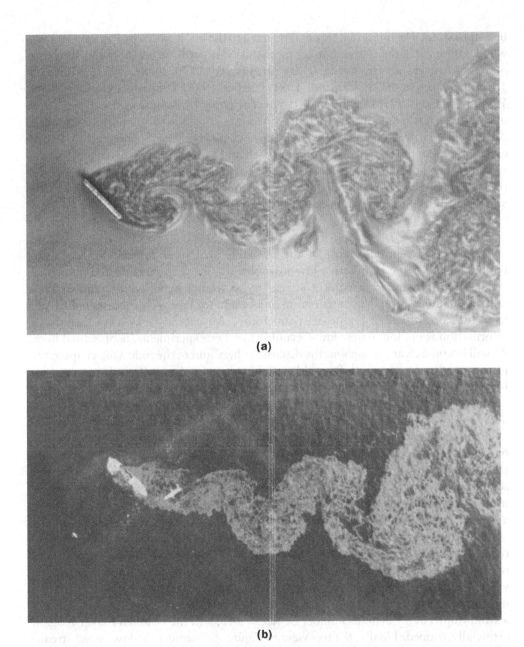

(a)

(b)

Figure 4.4: Turbulent wakes behind bluff bodies. (a) Inclined flat plate in water tunnel; from Cantwell (1981). (b) Grounded tanker [NASA photograph courtesy of O.M. Griffin, Naval Research Laboratory].

mostly through flow visualization experiments, although some important early discoveries have been made using correlation measurements (e.g., Townsend 1961, 1970; Bakewell and Lumley 1967). Although the literature on this topic is vast, no research-community-wide consensus has been reached—particularly on the issues of the origin of, and interaction between, the different structures, regeneration mechanisms, and Reynolds number effects. What follow are somewhat biased remarks addressing these issues. At times diverse viewpoints will be presented, but for the most part particular scenarios, which in my opinion are most likely to be true, will be emphasized. The interested reader is referred to the book edited by Panton (1997a), which emphasizes the self-sustaining mechanisms of wall turbulence, and the large number of review articles available (e.g., Kovasznay 1970; Laufer 1975; Willmarth 1975a,b; Saffman 1978; Cantwell 1981; Fiedler 1986, 1988; Blackwelder 1988, 1998; Robinson 1991; Delville et al. 1998). The paper by Robinson (1991) in particular summarizes many of the different, sometimes contradictory, conceptual models offered thus far by different research groups. Those models are aimed ultimately at explaining how the turbulence maintains itself and range from the speculative to the rigorous, but not one, unfortunately, is self-contained and complete. Furthermore, the structure research dwells largely on the kinematics of organized motion, and little attention is given to the dynamics of the regeneration process.

4.4.1 Overview

With few exceptions, most of the available structural information on wall-bounded flows come from rather low-Reynolds-number experiments and numerical simulations. Organized structures appear to be similar in all wall-bounded flows only in the inner layer. The outer region of a boundary layer is by necessity different from the core region of a pipe or channel flow. An overall view, whose source of information is predominately low-Reynolds-number experiments, is presented here. As will become clear throughout the discussion here and in the following chapter, the picture that emerges at high Reynolds numbers is quite different, and structural information gleaned from low-Reynolds-number physical and numerical experiments may not be very relevant to the more practically important high-Reynolds-number flows.

In boundary layers, the turbulence production process is dominated by three kinds of quasi-periodic eddies: the large outer structures, the intermediate Falco eddies, and the near-wall eddies. Examples of these coherent structures visualized in laboratory-scale boundary layers are depicted in Figures 4.5–4.7. Laser sheet illumination is used in all three photographs. The large eddies forming on a flat plate towed in a water channel are seen in the side view in Figure 4.5. The flow (relative to the plate) is from left to right. The artificially tripped boundary layer has a momentum-thickness Reynolds number at the observation station of $Re_\theta \equiv U_\infty \delta_\theta / \nu = 725$ and is marked with fluorescent dye. The smoke-filled boundary layer shown in the top view in Figure 4.6 depicts the characteristic pockets believed to be induced by the motion of Falco eddies over the wall. In this case, the experiments are conducted in a wind tunnel at a Reynolds number of $Re_\theta = 742$, and the boundary layer is again artificially tripped. Finally, the top view in Figure 4.7 depicts the low-speed streaks in the near-wall region of the same turbulent boundary layer previously shown in side view in Figure 4.5. Flow direction is again from left to right, and the thin sheet of laser used for illumination is parallel to and almost touching the wall.

Figure 4.8, from Gad-el-Hak, Blackwelder, and Riley (1981), shows a top view of an artificially generated turbulent spot evolving in a laminar boundary layer. The displacement thickness Reynolds number at the spot's initiation point is 625, which is well above the critical Re for linear instability. Laser-induced fluorescence is used to visualize different cuts through the growing turbulent structure. In this figure, the laser sheet is parallel to and very near the flat wall. The dynamics within the spot appear to be controlled by many individual eddies similar to those within a turbulent boundary layer. The spot grows in the spanwise direction by an efficient mechanism,

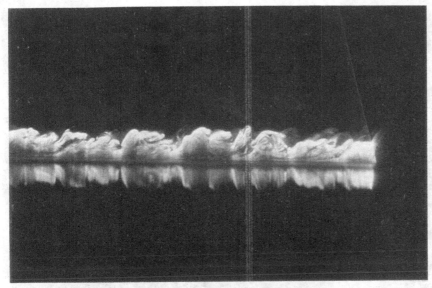

Figure 4.5: Side view of a low-Reynolds-number turbulent boundary layer, $Re_\theta = 725$. Flat plate towed in a water tank. Large eddies are visualized using a sheet of laser and fluorescent dye [after Gad-el-Hak, Blackwelder, and Riley 1984].

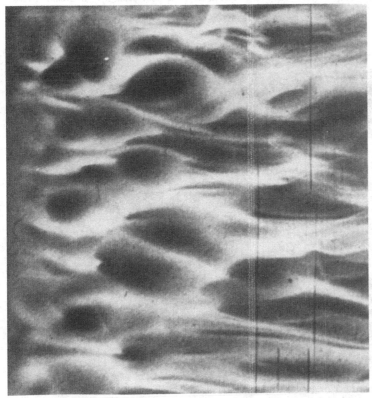

Figure 4.6: Top view of a low-Reynolds-number turbulent boundary layer, $Re_\theta = 742$. Wind tunnel experiment. Pockets, believed to be the fingerprints of typical eddies, are visualized using dense smoke illuminated with a sheet of laser [after Falco 1980].

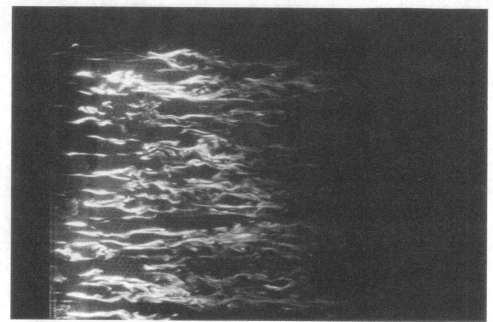

Figure 4.7: Top view of a low-Reynolds-number turbulent boundary layer $Re_\theta = 725$. Flat plate towed in a water tank. Low-speed streaks are visualized using a sheet of laser and fluorescent dye [after Gad-el-Hak, Blackwelder, and Riley 1984].

Figure 4.8: Top view of an artificially generated turbulent spot evolving in a laminar boundary layer. The displacement thickness Reynolds number at the spot's initiation point is $Re_\theta = 625$ [after Gad-el-Hak et al. 1981].

which Gad-el-Hak et al. (1981) have termed the *growth by destabilization* process (see also Riley and Gad-el-Hak 1985). Near the edges of the spot, the dye lines are sharp, which is indicative of the initial breakdown into chaotic motion. Toward the middle, on the other hand, the dye becomes more diffused because the turbulence there is older and more mixing has taken place.

In a turbulent boundary layer, the large, three-dimensional bulges (Figure 4.5) scale with the layer thickness, δ, and extend across the entire boundary layer (Kovasznay et al. 1970; Blackwelder and Kovasznay 1972). These eddies control the dynamics of the boundary layer in the outer region such as entrainment, turbulence production, and so forth. The large eddies are characterized by a sharp interface and a highly contorted surface that exhibits a significant amount of folding (Paiziz and Schwarz 1974) and has a fractal dimension of close to 2.4 (Sreenivasan, Ramshankar, and Meneveau 1989). They appear randomly (quasi-periodically) in space and time and seem to be, at least for moderate Reynolds numbers, the residue of the transitional Emmons spots (Zilberman, Wygnanski, and Kaplan 1977; Gad-el-Hak et al. 1981; Riley and Gad-el-Hak 1985). Note, however, that at higher Reynolds numbers ($Re_\theta > 5000$) the very existence of the large eddy as an isolated coherent structure has been questioned by Head and Bandyopadhyay (1981).

The Falco eddies are also highly coherent and three-dimensional. Falco (1974, 1977) named them typical eddies because they appear in wakes, jets, Emmons spots, grid-generated turbulence, and boundary layers in zero, favorable, and adverse pressure gradients. They have an intermediate scale of about 100 wall units. (Viscous forces dominate over inertia in the wall region. The characteristic scales there are obtained from the magnitude of the mean vorticity in the region and its viscous diffusion away from the wall. Thus, the viscous time-scale, t_v, is given by the inverse of the mean wall vorticity

$$t_v = \left[\frac{\partial \overline{U}}{\partial y} \bigg|_{\mathrm{w}} \right]^{-1} \tag{4.1}$$

and the viscous length scale, ℓ_v, is determined by the characteristic distance by which the (spanwise) vorticity is diffused from the wall, and is thus given by

$$\ell_v = \sqrt{\nu t_v} = \sqrt{\frac{\nu}{\frac{\partial \overline{U}}{\partial y}\big|_{\mathrm{w}}}} \tag{4.2}$$

where ν is the kinematic viscosity. The wall-velocity scale (so-called friction velocity, u_τ) follows directly from the time and length scales

$$u_\tau = \frac{\ell_v}{t_v} = \sqrt{\nu \frac{\partial \overline{U}}{\partial y}\bigg|_{\mathrm{w}}} = \sqrt{\frac{\tau_{\mathrm{w}}}{\rho}} \tag{4.3}$$

where τ_{w} is the shear stress at the wall, and ρ is the fluid density. A wall unit implies scaling with the viscous scales, and the usual $()^+$ notation is used; for example, $y^+ = y/\ell_v = y u_\tau/\nu$.)

The Falco eddies appear to be an important link between the large structures and the near-wall events. In the plan view shown in Figure 4.6, smoke fills the near-wall region of a boundary layer, and the roughly circular regions devoid of marked fluid are called pockets. These undulations are very similar to the so-called folds observed by Perry, Lim, and Teh (1981). Falco (1980) asserted that the pockets are the 'footprints' of some outer structures that induce fluid toward the wall. Robinson, Kline, and Spalart (1989) analyzed the database generated from the direct numerical simulations of Spalart (1986, 1988). They concurred that the pockets are the signature of local wallward motions, evidenced by spanwise divergence of streamlines, above regions of high wall pressure. Low-pressure regions, on the other hand, occur along lines of converging streamlines associated with outward motion. These motions are, respectively, the sweep and ejection events depicted in Figure 4.9.

The third kind of eddies exist in the wall region ($0 \leq y^+ \leq 100$), where the Reynolds stress is produced in an intermittent fashion. Half of the total production of turbulence kinetic energy ($-\overline{uv}\ \partial\overline{U}/\partial y$) takes place near the wall in the first 5% of the boundary layer at typical laboratory Reynolds numbers (smaller fraction of the

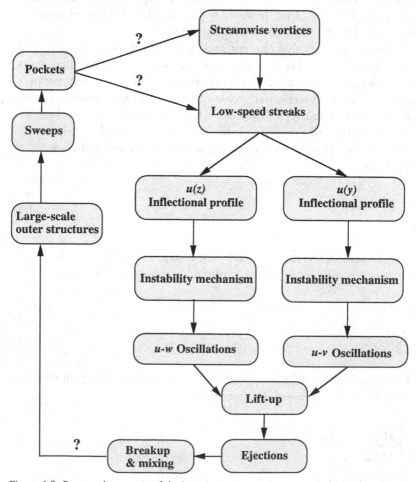

Figure 4.9: Proposed sequence of the bursting process. The arrows indicate the sequential events, and the '?' indicates relationships with less supporting evidence [adapted from Blackwelder 1998].

boundary layer thickness at higher Reynolds numbers), and the dominant sequence of intense organized motions there are collectively termed the bursting phenomenon. This dynamically significant process, identified during the 1960s by researchers at Stanford University (Kline and Runstadler 1959; Runstadler, Kline, and Reynolds 1963; Kline et al. 1967; Kim, Kline, and Reynolds 1971; Offen and Kline 1974, 1975), was reviewed by Willmarth (1975a) and Blackwelder (1978) and most recently by Blackwelder (1998).

To focus the discussion on the bursting process and its possible relationships to other organized motions within the boundary layer, we refer to the schematic in Figure 4.9, adapted from Blackwelder (1998). Qualitatively, the process, according to at least one school of thought, begins with elongated, counterrotating, streamwise vortices having diameters of approximately 40 wall units or $40\nu/u_\tau$. The estimate for the diameter of the vortex is obtained from the conditionally averaged spanwise velocity profiles reported by Blackwelder and Eckelmann (1979). There is a distinction, however, between vorticity distribution and a vortex (Saffman and Baker 1979; Robinson et al. 1989; Robinson 1991), and the visualization results of Smith and Schwartz (1983) may indicate a much smaller diameter. In any case, with reference to Figure 4.10, the counterrotating vortices exist in a strong shear and induce low- and high-speed regions between them. Those low-speed streaks were first visualized by Francis Hama at the University of Maryland (see Corrsin 1957), although Hama's contribution is frequently overlooked in favor of the subsequent and more thorough studies conducted at Stanford University and cited above. The vortices and the accompanying eddy structures occur randomly in space and time. However, their appearance is sufficiently regular that an average spanwise wavelength

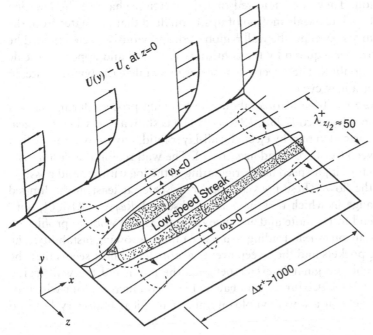

Figure 4.10: Model of near-wall structure in turbulent wall-bounded flows [after Blackwelder 1978].

of approximately 80 to $100\nu/u_\tau$ has been identified by Kline et al. (1967) and others.

It might be instructive at this point to emphasize that the distribution of streak spacing is very broad. The standard of deviation is 30–40 percent of the more commonly quoted mean spacing between low-speed streaks of 100 wall units. Both the mean and standard deviation are roughly independent of Reynolds number in the rather limited range of reported measurements ($Re_\theta = 300$–6500, see Smith and Metzler 1983; Kim, Moin, and Moser 1987). Butler and Farrell (1993) have shown that the mean streak spacing of $100\nu/u_\tau$ is consistent with the notion that this is an optimal configuration for extracting "the most energy over an appropriate eddy turnover time." In their work, the streak spacing remains 100 wall units at Reynolds numbers, based on friction velocity and channel half-width, of $a^+ = 180$–360.

Kim et al. (1971) observed that the low-speed regions (Figure 4.7) grow downstream, lift up, and develop (instantaneous) inflectional $u(y)$ profiles.[8] At approximately the same time, the interface between the low- and high-speed fluid begins to oscillate, apparently signaling the onset of a secondary instability. The low-speed region lifts up away from the wall as the oscillation amplitude increases, and then the flow rapidly breaks up into a completely chaotic motion. The streak oscillations commence at $y^+ \approx 10$, and the abrupt breakup takes place in the buffer layer, although the ejected fluid reaches all the way to the logarithmic region. Because the breakup process occurs on a very short timescale, Kline et al. (1967) called it a burst.

Virtually all of the net production of turbulence kinetic energy in the near-wall region occurs during these bursts. Corino and Brodkey (1969) showed that the low-speed regions are quite narrow (i.e., $20\nu/u_\tau$) and may also have significant shear in the spanwise direction. They also indicated that the ejection phase of the bursting process is followed by a large-scale motion of upstream fluid that emanates from the outer region and cleanses (sweeps) the wall region of the previously ejected fluid. The sweep phase is, of course, required by the continuity equation and appears to scale with the outer-flow variables. The sweep event seems to stabilize the bursting site, in effect preparing it for a new cycle.

Considerably more has been learned about the bursting process during the last two decades. For example, Falco (1980, 1983, 1991) has shown that when a typical eddy, which may be formed in part by ejected wall-layer fluid, moves over the wall, it induces a high uv sweep (positive u and negative v). The wall region is continuously bombarded by pockets of high-speed fluid originating in the logarithmic and possibly the outer layers of the flow. These pockets appear to scale, at least in the limited Reynolds number range in which they have been observed, $Re_\theta = \mathcal{O}[1000]$, with wall variables and tend to promote and enhance the inflectional velocity profiles by increasing the instantaneous shear, leading to a more rapidly growing instability. The relation between the pockets and the sweep events is not clear, but it seems that the former forms the highly irregular interface between the latter and the wall-region fluid. More recently, Klewicki, Murray, and Falco (1994) conducted a four-wire hot-wire probe measurement in a low-Reynolds-number canonical boundary layer to

---

[8] According to Swearingen and Blackwelder (1984), inflectional $u(z)$ profiles are just as likely to be found in the near-wall region and can also be the cause of the subsequent bursting events. See Figure 4.9.

clarify the roles of velocity–spanwise vorticity field interactions regarding the near-wall turbulent stress production and transport.

Other significant experiments were conducted by Tiederman and his students (Donohue, Tiederman, and Reischman 1972; Reischman and Tiederman 1975; Oldaker and Tiederman 1977; Tiederman, Luchik, and Bogard 1985), Smith and his colleagues (Smith and Metzler 1982, 1983; Smith and Schwartz 1983), and the present author and his collaborators. The first group conducted extensive studies of the near-wall region, particularly the viscous sublayer, of channels with Newtonian as well as drag-reducing non-Newtonian fluids. Smith's group, using a unique two-camera, high-speed video system, was the first to indicate a symbiotic relationship between the occurrence of low-speed streaks and the formation of vortex loops in the near-wall region. Gad-el-Hak and Hussain (1986) and Gad-el-Hak and Blackwelder (1987a) have introduced methods by which the bursting events and large-eddy structures are artificially generated in a boundary layer. Their experiments greatly facilitate the study of the uniquely controlled simulated coherent structures via phase-locked measurements.

4.4.2   Open Issues

There are at least four unresolved issues regarding coherent structures in wall-bounded flows, not all of which are necessarily independent: How does a particular structure originate? How do different structures, especially the ones having disparate scales, interact? How does the turbulence continue to regenerate itself? Does the Reynolds number affect the different structures in any profound way? The primary difficulty in trying to answer any of these queries stems from the existence of two scales in the flow that become rather disparate at large Reynolds numbers. The closely related issues of origin, inner–outer interaction, and regeneration will be addressed in the following three subsubsections. Reynolds number effects are deferred to Chapter 5.

Origin of Different Structures

Faced with the myriad of coherent structures existing in the boundary layer, a legitimate question is, Where do they all come from and which one is dynamically significant? Sreenivasan (1988) offers a glimpse of the difficulties associated with trying to answer this question. The structural description of a turbulent boundary layer may not be that complicated, however, and some of the observed structures may simply be a manifestation of the different aspects of a more basic coherent structure. For example, some researchers argue that the observed near-wall streamwise vortices and large eddies are, respectively, the legs and heads of the omnipresent hairpin vortices (Head and Bandyopadhyay 1981). Nevertheless, that still leaves us with a minimum number of building blocks that must be dealt with.

If the large eddies are assumed to be dynamically significant, then how are they recreated? It is easy to argue that the conventional laminar-to-turbulence transition cannot be responsible because the same large eddies appear even in heavily tripped boundary layers in which the usual transition routes are bypassed. Wall events cannot be responsible for creating large eddies because of their extremely small relative scale at high Reynolds number. Furthermore, no hierarchical amalgamation of scales has been observed to justify such proposition.

If, alternatively, wall events are assumed to dominate, then where do the stream-wise vortices or the low-speed streaks come from, and what mechanism sustains the bursting cycle? Mechanisms that assume local instability cannot be valid at large Reynolds numbers for which the wall layer is, say, 0.1% of the boundary layer thickness,[9] and it is difficult to conceive that 99.9% of the boundary layer has no active role in the generation and maintenance of turbulence. On the other hand, assuming the bursting events are triggered by the large eddies brings us back to the original question of where the latter come from.

The preceding difficulties explain the lack of a self-consistent model of the turbulent boundary layer despite the enormous effort expended to establish such a model. None of the existing models are complete in the sense that none accounts for each aspect of the flow in relation to every other aspect. The question marks in Figure 4.9 are just the tip of the iceberg. Developing a complete, self-consistent model is more than an academic exercise, for a proper conceptual model of the flow gives researchers the necessary tools to compute high-Reynolds-number practical flows using the Reynolds-averaged Navier–Stokes equations and to devise novel flow-control strategies as well as to extend known laboratory-scale control devices to field conditions.

Inner–Outer Interaction

There is no doubt that significant interactions between the inner and outer layers take place. On energy grounds alone, it is known that in the outer layer the dissipation is larger than the turbulence kinetic energy production (Townsend 1976). It is therefore necessary for energy to be transported from the inner layer to the outer layer simply to sustain the latter. How that is accomplished and whether coherent structures are the only vehicle to transport energy is not clear, but two distinct schools of thought have emerged. In the first, the large-scale structures dominate and provide the strong buffeting necessary to maintain the low-Reynolds-number turbulence in the viscous region ($y^+ \leq 30$). In the second view, rare, intense wall events are assigned the active role and, through outward turbulence diffusion, provide the necessary energy supply to maintain the outer region. As mentioned in the previous subsubsection, both views have some loose ends.

On the basis of a large number of space–time, two-point correlation measurements of u and v, Kovasznay, Kibens, and Blackwelder (1970) suggested that the outer region of a turbulent boundary layer is dominated by large eddies. The interface between the turbulent flow and the irrotational fluid outside the boundary layer is highly corrugated with a root-mean-square slope in the x–y plane of roughly 0.5. The three-dimensional bulges are elongated in the streamwise direction with an aspect ratio of approximately 2:1 and have a characteristic dimension in the wall-normal direction of between 0.5δ and δ. They appear quasi-periodically and are roughly similar to each other. Kovasznay et al. (1970) allowed that the large eddies are passive in the sense that the wall events and not these eddies are responsible for producing the Reynolds stress. Kovasznay (1970) advanced the hypothesis that wall bursting starts a chain reaction of some sort at all intermediate scales culminating into a sequence of amalgamations that eventually lead to the large structures. As mentioned earlier,

---

[9] This is the near-wall region extending from the surface to $y^+ = 30$ as a percentage of the boundary-layer thickness when the Reynolds number is $Re_\theta \approx 10^5$.

such hierarchical amalgamation of scales has not been directly observed in the laboratory.

Head and Bandyopadhyay (1981), on the other hand, suggested that the very existence of the large eddies at high Reynolds numbers is in doubt. Their combined flow visualization and hot-wire probe experiments are unique in that an unusually large range of Reynolds number was investigated,

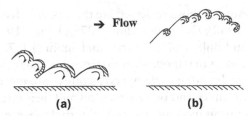

→ **Flow**

(a)   (b)

Figure 4.11: Sketch of large-eddy structures as a collection of smaller-scale hairpin vortices. From Head and Bandyopadhyay (1981). (a) Typical low-Reynolds-number boundary layer. (b) Typical high-Reynolds-number flow.

$Re_\theta = 500$–17,500, allowing them to unambiguously clarify Reynolds number effects on the structure of the boundary layer. Head and Bandyopadhyay maintained that a large structure seen in typical flow-visualization experiments is nothing but the slow overturning of a random collection of smaller-scale hairpin vortices—just a few or even a single isolated vortex loop at low Reynolds numbers (say, $Re_\theta < 1000$) but a large number of them at high Reynolds numbers (say, $Re_\theta > 5000$). This is sketched for typical low- and high-Reynolds-number boundary layers in Figures 4.11a and 4.11b, respectively. A brisker rate of rotation of the isolated (fat) vortex loop is observed at the lowest Reynolds number consistent with prior observations of large eddies in low-speed experiments. The hairpins are inclined at around 45° to the plane of the flow over a major part of the layer thickness. In Head and Bandyopadhyay's (1981) view, the entire turbulent boundary layer consists very largely of vortex loops that become increasingly elongated as the Reynolds number increases. The so-called large eddies, on the other hand, do not appear to exhibit any particular coherent motion beyond a relatively slow overturning or toppling due to shear.

Corroborative evidence for the hairpin angle of inclination of 45° comes from the simultaneous, multiple-point hot-wire measurements of Alving et al. (1990) in a canonical turbulent boundary layer and a boundary layer recovering from the effects of strong convex curvature. Their cross-correlation results are consistent with the observation of large-scale structures spanning the entire shear layer and inclined at angles in the range of 35°–45° near the outer edge of the boundary layer but at continuously decreasing angles as the wall is approached.

The sketch in Figure 4.4.2b for a typical large eddy at high Reynolds number is consistent with the statistical findings of Brown and Thomas (1977), who have shown by using conditional averaging techniques that a typical large structure in a turbulent boundary layer has an upstream rotational–irrotational interface inclined at 18° to the flow direction. Head and Bandyopadhyay (1981) have observed such individual structures only at higher Re_θ (> 5000). It is possible to arrive precisely at this slope by modeling the large structure to be composed of hairpin vortices formed at regular intervals (Bandyopadhyay 1980). Such large structures composed of many hairpin vortices have not been observed in the low-Reynolds-number DNS simulations.

Regeneration Mechanisms

Robinson (1991) summarized many of the conceptual models advanced by different researchers to explain how a wall-bounded turbulent flow maintains itself.

Among those reviewed are the models advocated by Willmarth and Tu (1967); Black (1968); Offen and Kline (1975); Hinze (1975); Praturi and Brodkey (1978); Thomas and Bull (1983); Acarlar and Smith (1987a,b); and Robinson (1990). Some of these conceptual models emphasize a particular aspect of the flow dynamics, as for example the bursting cycle, whereas others are more ambitious and attempt to include both the inner and outer structures and their interaction. Several viewpoints on the regeneration mechanisms of wall turbulence are represented in the books edited by Kline and Afgan (1990) and by Panton (1997a).

Robinson (1991) also lists significant contributions that utilize structural information to predict statistical quantities or invoke a simplified form of the governing equations to model the dynamics of the near-wall turbulence-production process. Among the predictive models discussed are those by Landahl (1967, 1980, 1990); Townsend (1976); Perry and Chong (1982); Perry et al. (1986, 1990); Walker and Herzog (1988); Aubry et al. (1988); Hanratty (1990); and Berkooz, Holmes, and Lumley (1991).

Other potentially useful predictive models not discussed by Robinson (1991) include those based on stability considerations (Malkus 1956, 1979), based on the turbulence energy equation (Bradshaw, Ferriss, and Atwell 1967), based on the u–v velocity-quadrant statistical description of the organized motions (Nagano and Tagawa 1990), and based on a single hairpinlike vortex in a unit domain of turbulence production (Bandyopadhyay and Balasubramanian 1993, 1995, 1996). These models account explicitly for Reynolds number effects and might, therefore, be useful for practical Reynolds numbers.

Inevitably, in almost all the conceptual models the omnipresent hairpin vortex (or horseshoe at low Re) plays a key role. Such a vortex has been proposed earlier by Theodorson (1952) on intuitive grounds as the primary structure responsible for turbulence production and dissipation in the boundary layer. His tornadolike vortices form astride near-wall, low-speed regions of fluid and grow outward with their heads inclined at 45° to the flow direction.

Black (1966, 1968) conducted a more rigorous analytical work to show the fundamental role of hairpin vortices in the dynamics of wall-bounded flows. His basic premise is that the primary role of the random turbulent motion is not to transfer mean momentum directly but rather to excite strong, three-dimensional instability of the sublayer, which is a powerhouse of vorticity. In Black's model, trains of discrete horseshoes are generated by repetitive, localized, nonlinear instabilities within the viscous sublayer. The vortical structures are shed and outwardly migrate from the near-wall region in a characteristic, quasi-frozen spatial array. The horseshoes inviscidly induce an outflow of low-speed fluid from within the vortex loops, creating motions that would be seen by a stationary probe as sharp, intermittent spikes of Reynolds stress. Because of the continuous creation of new vortex loops that replace older elements, the lifetime of the vortical array is much longer than that for its individual members. According to Black (1968), such organized structures are responsible for the efficient mass and momentum transfer within a turbulent boundary layer.

Sreenivasan (1988) offers a similar model to that of Black (1968). The essential structures of the boundary layer, including the hairpin vortices, result from the instability of a caricature flow in which all the mean-flow vorticity has been concentrated in a single fat sheet.

As a parting remark in this subsection, it might be instructive to recall that hairpin vortices play an important role also in the laminar-to-turbulence transition of boundary-layer flows. Essentially, these hairpins are the result of the nonlinear tertiary instability of the three-dimensional peak and valley pattern, which itself is the secondary instability of the primary Tollmien–Schlichting waves (Klebanoff, Tidstrom, and Sargent 1962).

CHAPTER FIVE

Reynolds Number Effects

What is reasonable is real; that which is real is reasonable.
(Georg Wilhelm Friedrich Hegel, 1770–1831)

Science is what you know, philosophy is what you don't know.
(Bertrand Arthur William Russell, 1872–1970)

PROLOGUE

This chapter deals with Reynolds number effects in turbulent shear flows with particular emphasis on the canonical zero-pressure-gradient boundary layer and two-
dimensional channel-flow problems. The Reynolds numbers encountered in many
practical situations are typically several orders of magnitude higher than those studied computationally or even experimentally. High-Reynolds-number research facilities are expensive to build and operate, and the few that exist are heavily scheduled with mostly developmental work. For wind tunnels, additional complications
due to compressibility effects are introduced at high speeds. Likewise, full computational simulation of high-Reynolds-number flows is beyond the reach of current
capabilities. Understanding turbulence and modeling will therefore continue to play
vital roles in the computation of high-Reynolds-number practical flows using the
Reynolds-averaged Navier–Stokes equations. Because the existing knowledge base,
accumulated mostly through physical as well as numerical experiments, is skewed
toward the low Reynolds numbers, the key question in such high-Reynolds-number
modeling as well as in devising novel flow control strategies is, What are the Reynolds
number effects on the mean and statistical turbulence quantities and on the organized
motions? Understanding the Reynolds number effects is important for flow control
on two counts:

1. A passive or active control device developed in a low-Reynolds-number
   facility may perform quite differently at high *Re*.
2. For reactive control, coherent structures are targeted. Those organized motions change scale and even character as the Reynolds number varies.

5.1 The Last Conundrum

Once more, turbulence is the last great unsolved problem of classical physics. Two of the greatest achievements of turbulence research are the Kolmogorov universal equilibrium theory and the universal logarithmic law of the wall. In fact, there is a direct analogy between the two high-Reynolds-number asymptotes, one being concerned with a cascade of energy and an inertial subrange in the frequency domain and the other with a hierarchy of eddies and an inertial sublayer in the physical space. The overall flow dynamics in the energy spectrum subrange and the wall-bounded flow sublayer is independent of viscosity. Dimensional reasoning, similarity, and asymptotic analysis are the tools of choice to derive analytical expressions for certain statistical flow quantities without actually solving the intractable governing equations.

One of the fundamental tenets of boundary-layer research is the idea that any statistical turbulence quantity (mean, rms, Reynolds stress, etc.) measured at different facilities and at different Reynolds numbers will collapse to a single universal profile when nondimensionalized using the proper length and velocity scales (different scales are used near the wall and away from it). This is termed self-similarity or self-preservation and allows convenient extrapolation from the low-Reynolds-number laboratory experiments to the much higher Reynolds number situations encountered in typical field applications. The universal logarithmic profile mentioned above describes the mean streamwise velocity in the overlap region between the inner and outer layers of any wall-bounded flow and is the best-known result of the stated classical idea.

The log-law has been derived independently by Ludwig Prandtl and G.I. Taylor using mixing length arguments, by Theodore von Kármán using dimensional reasoning, and by Clark B. Millikan using asymptotic analysis. Those names belong of course to the revered giants of our field. Questioning the fundamental tenet or its derivatives is, therefore, tantamount to heresy. But the questions and doubts linger as evidenced from the work of Simpson (1970), Malkus (1979), Barenblatt (1979), Long and Chen (1981), Wei and Willmarth (1989), George, Castillo, and Knecht (1992), Bradshaw (1994), Sreenivasan (see the doctoral thesis by Kailasnath 1993), and Smits (see the doctoral thesis by Smith 1994), among others, who at different times challenged various aspects of this law. And those are only the ones who, with varying degrees of difficulty, could get their work published. There is strong suspicion, among the iconoclasts at least, that Reynolds number effects persist indefinitely for both mean velocity and, more pronouncedly, higher-order statistics, and hence that true self-preservation is never achieved in a growing boundary layer. In fairness to the high priests, their logarithmic law was always intended to be a very high-Reynolds-number asymptote. These issues are discussed in greater detail in the survey by Gad-el-Hak and Bandyopadhyay (1994). In this chapter, we give the highlights.

If in fact the log-law is fallible, the implications are far-reaching (Gad-el-Hak 1997). Resolution of the full equations, via direct numerical simulations, at all but the most modest values of Reynolds number is beyond the reach of current or near-future computer capabilities. Modeling will, therefore, continue to play a vital role in the computations of practical flows using the Reynolds-averaged Navier–Stokes equations. Flow modelers, in attempting to provide concrete information for the designers of, say, ships, submarines, and aircraft, heavily rely on similarity principles in order to model the turbulence quantities and circumvent the well-known closure problem. Because practically all turbulence models are calibrated to reproduce the law of the wall in simple flows, failure of this universal relation virtually guarantees that Reynolds-averaged turbulence models will fail too. Finally, developers of flow-control devices to reduce drag, enhance lift, and so forth, often have to extrapolate the widely available low-speed (or more precisely low-Reynolds-number) results to high-speed flows of practical interest for which no data are available. Such extrapolation is not possible if the difficult-to-quantify Reynolds number effects persist indefinitely. For both scientists and engineers the message is essentially back to the drawing board!

In a community of conformists, the heretics never have it easy, of course. The peer review system, although essential for weeding out the charlatans, the misguided, and the fools, is somewhat biased against unorthodox ideas. Nevertheless, the latest two theses to question the infallibility of the log-law are contained in the articles by Barenblatt, Chorin, and Prostokishin (1997) and George and Castillo (1997). The two teams tackle the same problem quite differently and independently. Both papers offer concrete alternatives to the Reynolds-number-independent law of the wall. Barenblatt et al. (1997) use scaling laws that invoke a zero-viscosity asymptote, whereas George and Castillo introduce new tools they term the asymptotic invariance and near asymptotic principles that result in a new law of the wall with explicit Reynolds number dependence. George and Castillo's new "law" is deduced from first principles, fits existing mean-velocity data better, and is extendible to higher-order statistics.

When the log-law and its consequences are challenged, the usual immediate re-action is to doubt the credentials of the blasphemer. Something is wrong with his or her model, experiment, numerical scheme, or with an endless list of potential pitfalls. These are all genuine concerns that turn out to be valid most of the time, but para-noiacs have enemies too! There is also the persistent, albeit misguided, argument that the log-law fits the data well enough for engineering applications. If it isn't broke, don't fix it! This pragmatism is of course simultaneously the curse and the blessing of science conducted by engineers. Moreover, although the errors involved in attempting to fit the log-law to existing mean-velocity data are quite tolerable considering our inability to accurately measure the friction velocity (the velocity scale necessary to collapse the plots), the corresponding errors for higher-order statistics are egregious. The entire enterprise is not unlike the sixteenth-century debate over the Ptolemaic view of the heavens and the Copernican model seeking to replace it. The former the-ory served navigators well for over 1400 years. The Copernican theory made only small corrections but radically changed humankind's view of their universe. And it was a masterpiece in terms of its economy of postulates and assumptions, which is a necessary condition for theoretical elegance.

In judging the alternatives to the log-law, one should of course ask the usual barrage of questions that only a vibrant collection of skeptical neurons could muster. Is the theory simple, elegant, and self-consistent? Does the model provide a better fit to the data? Does it explain previous contradictions? Is the theory based on a minimum number of assumptions? Is it extendable to more complex situations? Are the results asymptotically correct? Is the logic sound? Is the mathematics free of errors? A fitting end to this editorializing is to quote my (elder) academic sibling John Lumley: "... A theory that does all that in an effortless way is often called elegant. Tomorrow, it may be wrong. Even so, it deserves to be regarded as one of the better things of which man is capable" (Lumley 1992).

5.2 The Issues

5.2.1 Field versus Laboratory Flows

It is difficult to overstate the technological importance of the turbulent wall-bounded flow. Vast amounts of energy are spent in overcoming the turbulence skin-friction drag in pipelines and on air, water, and land vehicles. For blunt bodies (e.g., trucks and trains), the pressure drag resulting from boundary layer separation can be several orders of magnitude higher than the skin friction, and even more energy is wasted. Heat transfer and mixing processes crucially depend on the turbulence transport for their efficient attainment. The Reynolds numbers encountered in many practical situations are typically several orders of magnitude higher than those studied computationally or even experimentally (Figure 5.1). Yet, our knowledge of high-Reynolds-number flows is very limited, and a complete understanding is yet to emerge. The existing knowledge base, accumulated mostly through physical as well as numerical experiments, is clearly skewed toward low Reynolds numbers. For many practical applications the key question is then, What are the Reynolds number effects on the mean and statistical turbulence quantities and on the organized motions of turbulence? One always hopes that the flow characteristics become invariant at sufficiently high Reynolds number. That merely shifts the question to What is high enough?

Consider the simplest possible turbulent wall-bounded flow, that over a smooth flat plate at zero incidence to a uniform, incompressible flow or its close cousin, the two-dimensional channel flow. Leaving aside for a moment the fact that such idealized flow does not exist in practice, where three-dimensional, roughness, pressure-gradient, curvature, wall-compliance, heat-transfer, compressibility, stratification and other effects may be present individually or collectively, the canonical problem itself is not well understood. Most disturbing from a practical point of view are the unknown effects of Reynolds number on the mean flow, the higher-order statistical quantities, and the flow structure. The primary objective of this chapter is to review the state of the art of Reynolds number effects in shear-flow turbulence with particular emphasis on the canonical boundary layer and channel-flow problems.

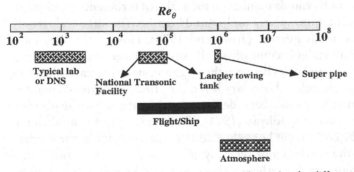

Figure 5.1: Ranges of momentum-thickness Reynolds number for different facilities and for field conditions.

5.2.2 Reynolds Number

Reynolds number effects are intimately related to the concept of dynamic similarity. In a given flow geometry, if L and v_0 are the length and velocity scales, respectively, the nondimensional equation of motion for an effectively incompressible, Newtonian fluid with negligible body forces is given by

$$\frac{\partial \mathbf{u}}{\partial t} + (\mathbf{u} \cdot \boldsymbol{\nabla})\mathbf{u} = -\boldsymbol{\nabla} p + \frac{1}{Re}\nabla^2 \mathbf{u} \qquad (5.1)$$

where $Re = v_0 L/\nu$, p is pressure, and ν is the constant kinematic viscosity. This seemingly superficial nondimensionalization reveals two important properties. The first is the concept of dynamic similarity. No matter how L, v_0 and ν are varied, as long as Re is the same in two geometrically similar flows, they have the same solution. Small-scale model testing of large-scale real-life flows is based on this property. Secondly, for a given geometry and boundary conditions, the effect of changing L, v_0, or ν, or any combination of them, can be described uniquely by the change of Re alone. Although, the importance of Re was recognized earlier by Stokes, it has come to be termed Reynolds number in recognition of Osborne Reynolds' (1883) telling demonstration of its effect on the onset of turbulence. Even today, the laminar-to-turbulence transition is one of the most dramatic Reynolds number effects, and its rational computation continues to be a research challenge. Equation (5.1) shows that Re represents the relative importance of viscous and inviscid forces. Because three forces (inertial, pressure, and viscous) are in equilibrium, the balance can be described by the ratio of any two, although it has become customary to characterize the flow by the ratio of inertial to viscous forces.

In this chapter, only turbulent flows are considered because they are widely prevalent. The report by Bushnell et al. (1993) treats Reynolds number similarity and scaling effects in laminar and transitional flows. The understanding of the effects of Reynolds number relies on our understanding of viscous forces. For a wall-bounded flow, this is true no matter how high the Reynolds number is. Experience shows that, aside from rarefied gas flows, there is no practical Reynolds number for which the no-slip boundary condition, which owes its origin to viscous effects, switches off. Because the net viscous force on an element of incompressible fluid is determined by the local gradients of vorticity, the understanding of the vorticity distribution is the key to determining Reynolds number effects. Vorticity can be produced only at a solid boundary and cannot be created or destroyed in the interior of a homogeneous fluid under normal conditions.

The qualitative effects of Reynolds number on the scales of turbulence are demonstrated by the velocity-fluctuations spectra for low and high Re. The large scale is only weakly dependent on Reynolds number (Townsend 1976). However, as Reynolds number increases, the small scales become physically smaller (larger wave numbers), and the diversity of intermediate scales between the large and small increases. In terms of organized motions in a turbulent boundary layer, the effect of Reynolds number on the omnipresent vortex loops has been demonstrated in the flow-visualization experiments of Head and Bandyopadhyay (1981). With increasing Reynolds number, the aspect ratio of the constituent horseshoe vortices increases while the vortices become skinnier. Thus, the vortex loops typically observed at low Reynolds number are gradually supplanted by horseshoe vortices and then hairpin vortices as Re increases.

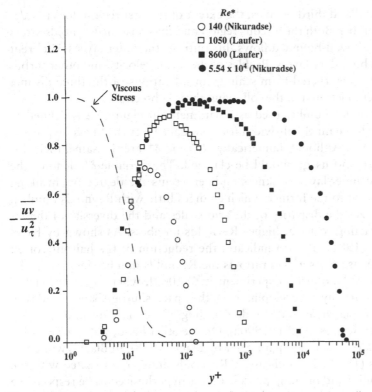

Figure 5.2: Distribution of viscous and turbulence shear stresses in wall-bounded flows [after Sreenivasan 1989].

5.3 Flow Regimes

An inspection of the distribution of viscous and turbulence shear stresses in a typical wall-bounded flow demonstrates the presence of three distinct regions. Figure 5.2, adapted by Sreenivasan (1989) from the smooth-pipe-flow data of Nikuradse (1932) and Laufer (1954), shows such distribution in wall units (friction velocity, $u_\tau = (\tau_w/\rho)^{1/2}$, used as velocity scale, and the ratio of kinematic viscosity to friction velocity used as length scale; see Section 4.4.1). Pipe (or channel) flow data are preferable to flat-plate boundary layer experiments because the Reynolds stress, a rather difficult quantity to measure accurately, can be computed exactly for fully developed channel flows from the relatively simpler measurements of mean-velocity profile and pressure gradient. The semilog plot in the figure enhances the importance of the thin near-wall region relative to the rest of the shear layer.

The broken line in Figure 5.2 is the time-averaged viscous stress distribution computed by differentiating the mean-velocity profile. Note that this laminar flow concept of shear may not be relevant to the time-dependent turbulent flow because turbulence models based on the mean velocity gradients have not been widely successful (e.g., Bradshaw et al. 1967). Nevertheless, it is clear from the figure that the mean viscous stress, $\mu(\partial \overline{U}/\partial y)$, is important only near the wall. This wall layer is followed by a region of approximately constant Reynolds stress. Finally, an outer layer[1] is characterized by a diminishing turbulence shear stress, reaching zero at the centerline of the

---

[1] Called core region in internal flows.

pipe. Unlike the second and third regimes, the extent of the first region does not depend on Reynolds number. Both the viscous region and the constant-Reynolds-stress region are similar in all wall-bounded flows. In contrast, the outer layer is different in internal flows and boundary layers. Profiles of the mean velocity and other turbulence statistics can be constructed from scaling considerations of the three distinct regimes, as will be demonstrated in the following three subsections.

Note that the Reynolds number used as a parameter in Figure 5.2 is defined as $Re^* = u_\tau a/\nu$; that is the channel half-width (or boundary-layer thickness) expressed in wall units: a^+ (or δ^+). Although numerically Re^* and δ^+ are the same, their difference in significance and usage should be clarified. The variable δ^+ denotes the ratio of the outer- to inner-layer thickness and represents the degree of shrinking of the latter with respect to the former, which changes little with Reynolds number. The variable emphasizes the disparity of the two scales and the diversity of the intermediate and interacting scales at higher Reynolds numbers. As shown by Head and Bandyopadhyay (1981), δ^+ also indicates the reduction of the hairpin vortex diameter and the increase in its aspect ratio as the Reynolds number increases. The value of δ^+ in a typical laboratory experiment is $\mathcal{O}[1000]$, whereas it approaches 100,000 in the boundary layer developing over the space shuttle (Bandyopadhyay 1991). This variable is pertinent to the understanding of the mechanism of drag reduction by outer-layer devices (Anders 1990a). On the other hand, Re^* is a Reynolds number that is also called a stability parameter by Black (1968). In Black's work and later in Sreenivasan's (1988), Re^* indicates a Reynolds number associated with the quasi-periodic instability and breakup process that is hypothesized to be responsible for the regeneration of turbulence in a wall-bounded flow. Note that, for a smooth wall, Re^* increases monotonically with Re_θ and never reaches an asymptote.

5.3.1 Viscous Region

Viscosity appears to be important only up to $y^+ = 30$. The viscous region can be subdivided into two subregions: the viscous sublayer and the buffer layer (Figure 5.3). Very close to the wall, $0 \leq y^+ \leq 5$, the turbulence shear stress is nearly zero, which implies that the only relevant quantities there are the kinematic viscosity ν and friction velocity u_τ.[2] In this viscous sublayer, several turbulence statistics can be asymptotically estimated from considerations of the no-slip condition and continuity and dynamical equations. Following Monin and Yaglom (1971) and using experimental data, Sreenivasan (1989) gives the following Taylor's series expressions, in wall units, for the mean streamwise velocity, for the root-mean-square value of the three fluctuating velocity components, and for the Reynolds stress, respectively:

$$\overline{U}^+ = y^+ - 1 \times 10^{-4} y^{+^4} + 1.6 \times 10^{-6} y^{+^5} + \cdots \tag{5.2}$$

$$u'^+ = 0.3 y^+ + c_1 y^{+^2} + \cdots \tag{5.3}$$

$$v'^+ = 0.008 y^{+^2} + c_2 y^{+^3} + \cdots \tag{5.4}$$

$$w'^+ = 0.07 y^+ + c_3 y^{+^2} + \cdots \tag{5.5}$$

$$-\overline{uv}^+ = 4 \times 10^{-4} y^{+^3} - 8 \times 10^{-6} y^{+^4} + \cdots, \tag{5.6}$$

[2] For hydraulically rough walls (i.e., where the average roughness height is greater than the viscous sublayer thickness), the relevant scaling parameters are the characteristic roughness height and friction velocity.

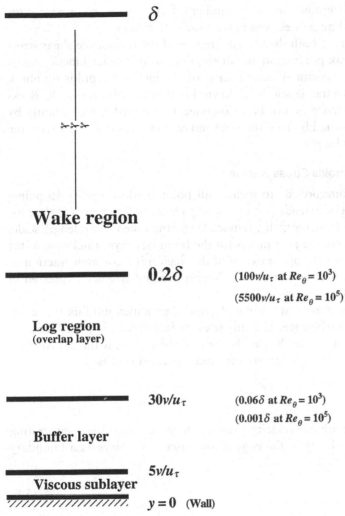

Figure 5.3: Schematic of the different regions within a wall-bounded flow at typical laboratory and field Reynolds numbers.

where the overbar and prime denote, respectively, mean and root-mean-square values.

For $y^+ < 5$, the leading term of each of the preceding expansions suffices. Note, however, that the experimentally determined leading coefficients in Eqs. (5.3)–(5.5) are lower than those computed from direct numerical simulations. Mansour, Kim, and Moin (1988) analyzed the channel-flow database generated by Kim et al. (1987) and reported the following leading-term coefficients for u'^+, v'^+, and w'^+: 0.36, 0.0086, and 0.19, respectively. The last coefficient, in particular, is almost three times that of the corresponding leading term determined experimentally. The reason may be due to the rapid drop in the spanwise velocity fluctuations as the wall is approached, and thus a very small hot-wire probe or LDV focus would be needed to realize the true value.

With three terms, Eq. (5.2) for the mean velocity is valid up to $y^+ = 20$. Note that the constants in the preceding equations are not necessarily universal. As will be discussed later, clearly discernible Reynolds number effects will be demonstrated for all higher-order statistical quantities even in the near-wall region.

The buffer layer is where both the viscous stress and the turbulence shear stress are important and the peak production and dissipation of turbulence kinetic energy occur (at about $y^+ = 12$, seemingly independent of the global Reynolds number). In the buffer layer, the characteristic local Reynolds number of $yu_\tau/\nu = 30$ is exceedingly low, and turbulence cannot be maintained unless buffeted constantly by strong disturbances, presumably from the outer layer. This region merges with the constant-Reynolds-stress layer.

5.3.2 Constant-Reynolds-Stress Region

This region, loosely interpreted[3] to include all points within the -3 dB points of the peak Reynolds stress, extends from $y^+ = 30$ to $y/a = 0.2$, where a is the pipe radius. Here, the distance from the wall y is much larger than the viscous length scale, ν/u_τ, but much smaller than the pipe radius (or the boundary layer thickness, δ, for an external flow). Note that the upper extent of this region is a constant fraction of the boundary layer thickness but varies with Reynolds number when expressed in wall units (see Figure 5.3).

In this region, viscous stresses are negligible, and the momentum flux is accomplished nearly entirely by turbulence. The only relevant length scale is y itself, and the square root of the nearly constant Reynolds stress, $(-\overline{uv}_{\max})^{1/2}$, is the appropriate velocity scale. Thus, the mean-velocity gradient may be expressed as

$$\frac{\partial \overline{U}}{\partial y} \sim \frac{(-\overline{uv}_{\max})^{\frac{1}{2}}}{y}. \tag{5.7}$$

The well-known logarithmic velocity profile follows directly from integrating Eq. (5.7) and using the velocity at the edge of the viscous sublayer as a boundary condition

$$\overline{U}^+ = \frac{1}{\kappa} \ln y^+ + B \tag{5.8}$$

where κ is the von Kármán constant. Both κ and B are presumably universal constants and are determined empirically for flat-plate boundary layers to be approximately 0.41 and 5.0, respectively. Slightly different values are used for the two constants in the case of pipe or channel flows. In that case, Eq. (5.8) holds almost up to the centerline of the channel. As the Reynolds number increases, the extent of the logarithmic region (in wall units) increases, and the maximum Reynolds stress approaches the value of the viscous stress at the wall

$$\frac{(-\overline{uv}_{\max})^{\frac{1}{2}}}{u_\tau^2} \longrightarrow 1 \tag{5.9}$$

Several other methods can be used to derive the logarithmic velocity profile. A mixing length, based on momentum transport, that simply varies linearly with distance from the wall, $\ell = \kappa y$, again yields Eq. (5.8). Millikan's (1939) asymptotic analysis recovers the log-relation by assuming the existence of a region of overlap where both the inner and outer laws are simultaneously valid (see also the rarely cited albeit relevant

---

[3] Strictly speaking, the Reynolds stress is not really constant anywhere in a pipe or channel flow.

article by Izakson 1937). All models invariably rely on the presence of the constant-stress layer experimentally observed at high Reynolds number. Despite copious evidence for the existence of a logarithmic region in the mean-velocity profile, the whole log-law scenario has been periodically questioned (see, for example, Barenblatt 1979, 1993; Malkus 1979; Long and Chen 1981; George et al. 1992, 1994; Barenblatt and Prostokishin 1993). We will return to this point later in Section 5.5.1.

The arguments used by Millikan (1939) to derive the logarithmic relation for the boundary layer are analogous to those employed to establish the universal equilibrium theory of turbulence, which was called the theory of local similarity by its originator Kolmogorov (1941a,b,c; 1962).[4] For the boundary layer, an inertial sublayer exists at sufficiently large Reynolds numbers, and the overall flow dynamics is independent of viscosity, which merely provides a momentum sink of prescribed strength at the wall. Similarly, an inertial subrange exists in the turbulence energy spectrum when the Reynolds number is large enough. There, the wave number is larger than that for the large eddies but smaller than the dissipative wave numbers. The viscosity again provides the dissipative sink for kinetic energy at the small-scale end of the turbulence spectrum. At high Re, the spectral shape in the inertial subrange is nearly completely determined by the energy flux across the wave number domain. The amount of this energy flux, and therefore the eventual rate of dissipation, is determined by the dynamics of the large, energy-containing eddies. Changing the viscosity (i.e., Re) has no influence on the rate of energy dissipation—it simply changes the scale at which the conversion of kinetic energy to heat takes place.

Scaling arguments similar to those leading to Eqs. (5.7) and (5.8) can be used in the constant-Reynolds-stress region to show that[5]

$$\frac{u'}{u_\tau} = \text{constant} = 2.0 \tag{5.10}$$

$$\frac{v'}{u_\tau} = \text{constant} = 1.0 \tag{5.11}$$

$$\frac{w'}{u_\tau} = \text{constant} = 1.4 \tag{5.12}$$

$$\mathcal{P}_{\text{K.E.}} = \mathcal{D}_{\text{K.E.}} = \frac{u_\tau^3}{\kappa y} \tag{5.13}$$

where $\mathcal{P}_{\text{K.E.}}$ and $\mathcal{D}_{\text{K.E.}}$ are the production and dissipation of turbulence kinetic energy, respectively. Additionally, a portion of the power spectrum for each of the three velocity components exhibits a -1 power-law in this same region governed by a constant turbulence shear stress transmitted across its different fluid layers.

The total stress is approximately constant throughout the viscous layer and the constant-Reynolds-stress region. This is the so-called inner layer (see Figure 5.4), and for a smooth wall the mean-velocity profile there is given by the unique similarity law of the wall first formulated by Prandtl (1925a)

$$\overline{U}^+ = f(y^+) \tag{5.14}$$

---

[4] The book by Frisch (1995) is a fascinating reading of the legacy of Andrei Nikolaevich Kolmogorov and his contribution to turbulence theory.

[5] The values of the constants in Eqs. (5.10)–(5.13) are determined empirically from mostly low-Reynolds-number experiments. Again, Section 5.6 will reveal that these constants depend, in fact, on the Reynolds number.

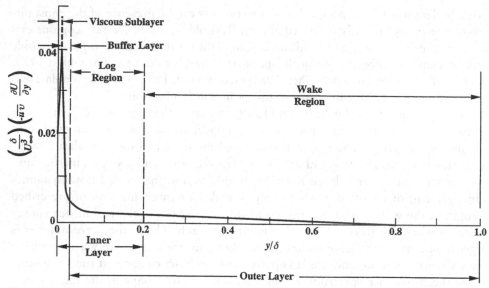

Figure 5.4: Normalized turbulence kinetic energy production rate as a function of normal distance from the wall. Data from a typical laboratory flat-plate boundary layer [after Kline et al. 1967].

where f is a universal function presumably independent of Reynolds number and streamwise location. The inner law is the same for both internal and external flows.

5.3.3 Outer Layer

Beyond the constant-stress region, an outer layer is characterized by a diminishing turbulence shear stress. Note that some researchers include the constant-Reynolds-stress region as part of the outer region. This is perhaps an accurate inclusion because the part of the boundary layer where the logarithmic law is valid is, strictly speaking, the region of overlap between the inner and outer laws (see Figure 5.4).

In internal flows, intermittency of turbulence and interaction with potential freestream are absent. There is, however, an interaction of turbulence from the opposite wall in the case of a two-dimensional channel, and this is even more complex in the case of a circular pipe. Furthermore, fully developed conditions for pipes and channels are defined as all time-averaged flow quantities (except static pressure) being independent of x. Therefore, the core region of a pipe or channel flow differs from the outer layer of a growing boundary layer.

The appropriate length scale in the core region is the pipe radius a (or the boundary-layer thickness δ for an external flow). The mean-velocity profile is characterized by the velocity defect $(U_0 - \overline{U})$, where U_0 is the velocity at the edge of the shear layer (centerline velocity U_c for a fully-developed pipe flow). The velocity-defect (or, more appropriately, momentum-defect) law, formulated by von Kármán (1930), is given by a second universal function

$$\frac{(U_0 - \overline{U})}{u_\tau} = g\left(\frac{y}{\delta}\right) \tag{5.15}$$

This equation is valid even in the logarithmic region and appears to be well confirmed experimentally.

For a turbulent boundary layer, Coles (1956) combined the defect law and the inner law to give the following empirical velocity profile valid throughout the entire wall-bounded shear layer:

$$\overline{U}^+ = f(y^+) + \frac{\Pi}{\kappa} W\left(\frac{y}{\delta}\right) \tag{5.16}$$

where κ is the von Kármán constant and Π is a profile parameter that depends strongly on Re for small Reynolds numbers. Coles' idea is that a typical boundary-layer flow can be viewed as a wakelike structure constrained by a wall. Intermittency and entrainment give rise to the wakelike behavior of the outer part of the flow, which is sensitive to pressure gradient and freestream turbulence. The wall constraint is closely related to the magnitude of the surface shear stress and is sensitive to the wall roughness and other surface conditions.

For equilibrium flows (Clauser 1954), the profile parameter Π is independent of streamwise location. The universal wake function $W(y/\delta)$ is the same for all two-dimensional boundary-layer flows. Its form is similar to that describing a wake flow or, more precisely, the mean-velocity profile at a point of separation or reattachment. For example, the wake function can be adequately represented by

$$W\left(\frac{y}{\delta}\right) = 2\sin^2\left(\frac{\pi}{2} \cdot \frac{y}{\delta}\right) \tag{5.17}$$

Note, however, that this simple expression obtained empirically by Coles (1969) does not yield zero velocity gradient at the edge of the boundary layer, as it should. Lewkowicz (1982) proposed an alternative quartic polynomial that removes this deficiency.

At the same Reynolds number, deviation of the actual mean-velocity distribution from the logarithmic profile in the core region of a pipe or channel flow is smaller than that in the outer region of a boundary layer. In fact, as mentioned in Section 5.3.2, the logarithmic velocity profile, Eq. (5.8) with slightly modified constants κ and B, holds approximately up to the centerline of the pipe.

5.4   Comparison with Other Shear Flows

Before proceeding to investigate the specific effects of Reynolds number on the mean and turbulence quantities of wall-bounded flows, it is instructive to give a coarse comparison between such flows on the one hand and free-shear flows on the other. As will be illustrated, the presence of the wall is of paramount importance to the issue at hand. No matter how large the Reynolds number is, viscosity must be important in a progressively shrinking region close to the wall, and Reynolds number dependence persists indefinitely.

Wakes, jets, and mixing layers are profoundly different from channel and pipe flows and boundary layers. The absence of the wall in free-shear flows implies that at sufficiently high Reynolds numbers, the flow is nearly inviscid and by implication Reynolds-number-independent (Dimotakis 1991, 1993). For wall-bounded flows, on the other hand, there is always a small, progressively shrinking region near the surface where viscosity must be important, no matter how large the Reynolds number is.

In boundary layers and channel flows, the overall behavior and gross structure of turbulence are always affected by viscosity near the wall, whereas the direct effect of viscosity gradually diminishes away from the surface. This implies that the velocity and length scales must be different near the wall and away from it. The disparity of scales for wall-bounded flows increases with Reynolds number, and true self-preservation may never be achieved unless the inner and outer scales are forced to be proportional at all Reynolds numbers. This latter scenario can be realized, for example, in the very special case of flow between two planes converging at a prescribed angle.

In wall-bounded flows, very large levels of turbulence fluctuations are maintained close to the wall despite the strong viscous as well as turbulence diffusion. As indicated in Figure 5.4, at a typical laboratory Reynolds number of say $Re_\theta = \mathcal{O}[10^3]$, more than about a third of the total turbulence kinetic energy production (and dissipation) occurs in the 2% of the boundary-layer thickness adjacent to the wall. The fraction of this thickness decreases as the Reynolds number increases. The near-wall region is directly affected by viscosity,[6] and its importance to the maintenance of turbulence is clearly disproportional to its minute size.

The thinness of the viscous sublayer presents a great challenge to both physical and numerical experiments. Because this region is closest to the wall and is where drag acts, it is extremely important at all Reynolds numbers. Yet in contemporary direct numerical simulations, the viscous sublayer of five wall units is resolved only up to 1.4 wall units. In measurements, probe resolutions are even worse; other than in low-Reynolds-number or oil-channel experiments, a probe length of $\ell^+ < 7$ is indeed rare (Bandyopadhyay 1991).

In free-shear layers, on the other hand, energy production peaks near the inflection points of the mean-velocity profile. Production and dissipation are spread over the entire flow width, as shown in Figure 5.5 depicting the turbulence energy balance for a typical two-dimensional wake. Above a reasonably modest Reynolds number, $\mathcal{O}[10^3]$, all turbulence quantities become invariant to additional changes in Reynolds number.

Despite these differences between boundary layers and free-shear flows, there are also some similarities. The outer region of a boundary layer is characterized by an intermittent rotational–irrotational flow, much the same as that observed in all free-shear flows. Moreover, the outer flow is more or less inviscid at sufficiently high Reynolds numbers, which again is similar to jets, wakes, and mixing layers. The interaction between the outer, or wake, region of a turbulent boundary layer and the potential flow in the freestream is also similar to that in wakes and other free-shear flows. The preceding discussion has addressed observational similarities and differences between the wall-bounded turbulent flows and the free-shear layers. In what follows, they are compared based on dynamic issues, such as the applicability of an inflectional inviscid breakdown mechanism, and it is shown that the subject is still wide open.

Free-Shear Flows

Inviscid stability theory has been successfully used to predict the observed coherent structures in turbulent free-shear flows, but it is not clear that similar arguments can be made when a wall is present. In other words, it is not obvious that the same

---

[6] Through the action of viscous stresses for a smooth wall, or through the action of pressure drag resulting from the separated flow around discrete elements of sufficient size for rough walls.

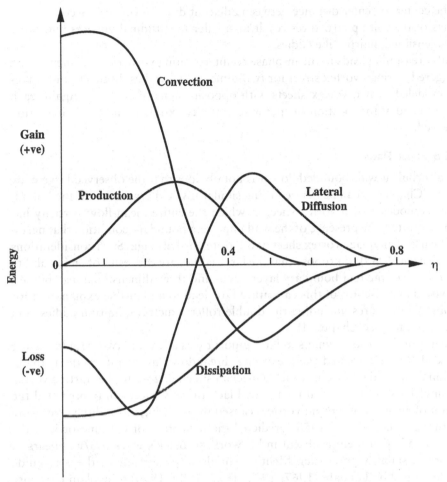

Figure 5.5: Turbulence energy balance in a typical plane wake [after Townsend 1976].

inviscid, mean-flow breakdown mechanism responsible for generating the large eddies in, say, a mixing layer is operable in the case of a boundary layer. Consider the mean streamwise velocity and mean spanwise vorticity distributions for a plane mixing layer, jet, and wake and for a two-dimensional boundary layer.

A two-dimensional mixing layer may be modeled as a single vortex sheet placed at the location of peak vorticity (at the point of inflection of the mean-velocity profile). The local circulation per unit length of the vortex sheet, $\gamma(x)$, is equal to the integral of the mean spanwise vorticity, $\overline{\Omega}_z$, across the shear layer

$$\gamma(x) = \int_{-\infty}^{+\infty} \overline{\Omega}_z \, dy = \int_{-\infty}^{+\infty} \frac{d\overline{U}}{dy} \, dy = U_1 - U_2 \qquad (5.18)$$

where U_1 and U_2 are, respectively, the velocities of the high- and low-speed sides of the mixing layer. The vortex sheet is inviscidly unstable to two-dimensional perturbations, and the resulting Kelvin–Helmholtz instability eventually evolves into the omnipresent two-dimensional vortices observed even in high-Reynolds-number mixing layers (Chapter 4). The resulting vortex blobs correspond to the saturation state of this instability. The total circulation in the blobs is conserved, $\Gamma = \gamma(x)\Delta x$, where

Δx is the center-to-center distance between adjacent discrete vortices. Secondary instabilities of the roll-up structures result in smaller longitudinal vortices and other three-dimensional, hairpinlike eddies.

Similar reasoning leads to the in-phase counterrotating vortices for plane jets and the staggered Kármán vortex street for two-dimensional wakes. Both jets and wakes can be modeled as two vortex sheets with opposite signs of vorticity. Again, each sheet is situated at the location of spanwise vorticity extrema, and total circulation is conserved.

Wall-Bounded Flows

For a turbulent wall-bounded flow it is not obvious that the observed large-eddy structures (Chapter 4) can be derived using similar inviscid arguments. If the boundary layer is modeled by a vortex sheet in which the entire mean-flow vorticity has been concentrated, the presence of the wall imposes a boundary condition that necessitates the use of an image vortex sheet of opposite vorticity sign. Such considerations led Sreenivasan (1988) to propose that the large eddies are the result of an instability of a *caricature* of the real boundary layer. Two- and three-dimensional instabilities, both inviscid and viscous, of this caricature flow lead to a plausible explanation for many observed features, including the double-roller structures, hairpin eddies, and low-speed streaks (see Chapter 4).

It turns out that a somewhat similar argument was advanced two decades earlier by Black (1968). He treated the mean-flow breakdown as an intermittent, three-dimensional, inviscid process in which a mechanism analogous to the starting vortex of an impulsively started airfoil is in play. Black thereby successfully predicted the formation of an array of hairpin vortices caused by a passing instability wave. Note that, although Theodorsen (1955) predicted the formation of hairpin vortices, the aspect of an array of them is absent in his work. In Black's work $u_\tau \delta / \nu$ appears as an important stability parameter. Mention should also be made of the waveguide theory developed by Landahl (1967, 1972, 1977, 1980, 1990) to explain the cause and effect relationships for the variety of coherent structures observed in turbulent boundary layers.

In Sreenivasan's (1988) model, a fat vortex sheet and its image are located on either side of the wall at a distance corresponding to the position of the peak Reynolds stress. Because of the absence of inflection points in the interior of the canonical turbulent flow, Sreenivasan (1988) chose the alternative location of peak $-\overline{uv}$ based upon experience with transitional boundary layers.[7] From all available boundary-layer as well as channel-flow data, this location appears to scale with the geometric mean of the inner and outer scales.[8] Sreenivasan (1988) termed that position the critical layer,[9] although any evidence for the existence of such a two-dimensional layer, where small perturbations are presumed to grow rapidly, is lacking. Using linear stability theory, Sreenivasan successfully showed that the primary instability of the vortex

---

[7] Note that in the past, both Clauser (1956) and Corrsin (1957) have also attempted to treat the turbulence regenerative process as primarily similar to the breakdown mechanism of a critical laminar layer.

[8] This powerful result is contrary to the universal law of the wall and will be analytically demonstrated in Section 5.6.2.

[9] In a transitional wall-bounded flow, the location of peak Reynolds stress coincides with the critical-layer position. Because this location shows similar trends with Reynolds number to that in a turbulent flow, Sreenivasan (1988) chose to place his proposed vortex sheet at this position.

sheet and its image yields two-dimensional roll-up structures, which in turn excite low-speed streaks and bursting. Subsequent instability of the roll-up structures leads to hairpin eddies and double-roller structures. One problem with this picture is that, unlike the case of free-shear flows, the predicted two-dimensional structures have never been observed in an actual turbulent wall-bounded flow. Sreenivasan (1988) himself allows that his simplistic model is unfinished and has a number of weaknesses but offers it as a target for useful criticism.

Experience with turbulence modeling, however, suggests that the turbulence in a wall-bounded flow is not derived directly from the mean flow. In the earliest turbulence models, shear stress is derived from the mean-velocity profile. Such models have not been widely successful. Townsend (1976) and Bradshaw et al. (1967) have argued that instead there is a much closer connection between the shear stress and the turbulence structure. Townsend's work was limited to the near-wall region, whereas Bradshaw et al. have extended the argument to the entire shear layer. Direct measurements of typical eddies have supported their assertion (Falco 1974, Newman 1974). There is a first-order direct connection between the mean flow and turbulence in a a free-shear flow but not in a wall-bounded flow. In a mixing layer, for example, the experimentally observed two-dimensional rollers are the direct result of an inviscid instability of the mean-velocity profile. Their characteristic dimension is equal to the layer thickness, and they contain almost all of the mean-flow vorticity. In a wall-bounded flow, on the other hand, the three-dimensional hairpin vortices are the result of a secondary or a tertiary instability, and their diameters are typically much smaller than the boundary layer thickness. The hairpins contain only a portion of the mean-flow vorticity—that is, they are farther removed from the mean flow.

The arguments above indicate that the existence of an inviscid breakdown mechanism responsible for the self-sustenance of the turbulence has not been firmly established. In other words, it is not clear that the observed coherent structures in a boundary layer or channel flow are the result of an instability of the mean flow or its caricature. Until this issue is resolved, progress in the understanding of wall-bounded flows will continue to lag behind that of free-shear flows. Despite the importance of this dynamical issue, research on the organized nature of turbulent boundary layers has remained largely confined to the kinematics, and high-payoff turbulence control strategies are yet to be developed.

5.5 Mean Flow

Before investigating the issue of Reynolds number effects on coherent structures, available data for the mean velocity and higher-order statistics of wall-bounded flows are reviewed in the present and following sections. The mean-flow velocity in the streamwise direction is a relatively easy quantity to measure, and almost every paper on wall-bounded flows has such measurements (see, for example, Preston and Sweeting 1944; Laufer 1951; Comte-Bellot 1965; Eckelmann 1974; Purtell, Klebanoff, and Buckley 1981; Andreopoulos et al. 1984; Wei and Willmarth 1989). Requirements for probe resolution are modest and, except very near the wall,[10]

---

[10] But here lies the caveat. It is the slope of the velocity profile at the wall that is needed to compute the velocity scale used to collapse the data.

most published data are reliable to better than 1%. This is obviously not the case for measurements of higher-order statistics, and this point will be revisited in the following section.

For a turbulent wall-bounded flow, the region directly affected by viscosity, the viscous sublayer plus the buffer layer, occupies the progressively smaller proportion of the boundary-layer thickness as the Reynolds number increases (see Figure 5.3). The rest of the flow is dominated by inertia, and the effect of viscosity enters only as an inner boundary condition set by the viscous region. It is not surprising, therefore, that the Reynolds number has a considerable effect on the velocity profile. As Re_θ increases, the mean-velocity profile becomes fuller, and the shape factor H (ratio of displacement thickness to momentum thickness) decreases accordingly. For example, at $Re_\theta = 2000$, $H = 1.41$, and at $Re_\theta = 10,000$, $H = 1.33$. The effect is even more pronounced at Reynolds numbers lower than 2000. In a laminar flat-plate flow, in contrast, viscosity is important across the entire layer, and the shape factor is independent of Reynolds number.

5.5.1 The Rise and Fall of the Log-Law

Available data appear to indicate that the wall-layer variables universally describe the streamwise mean velocity in the inner layer of smooth flat-plates, pipes, and channels at all Reynolds numbers. The boundary-layer data of Purtell et al. (1981) and Andreopoulos et al. (1984) and the channel flow data of Wei and Willmarth (1989) are typical and cover a wide range of Reynolds numbers. In Figure 5.6 we depict the former set of low-Reynolds-number measurements as published in *Physics of Fluids*. The graph indicates the presence of a logarithmic region for Re_θ as

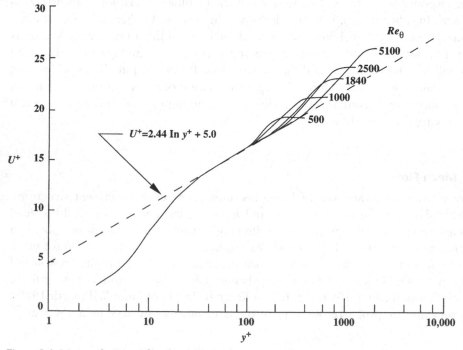

Figure 5.6: Mean-velocity profiles for different Reynolds numbers. Plot from the article by Purtell et al. (1981) as it appears in *Physics of Fluids*.

low as 500. This is rather surprising when one considers that at this Reynolds number a constant-stress region is virtually nonexistent and the maximum Reynolds stress is substantially less than u_τ^2. However, in deriving the log-law, Eq. (5.8), the presumably constant velocity scale has a weak, square-root dependence on the Reynolds stress. The extent of the log-region, expressed in wall units, increases with Reynolds number but is a constant fraction of the boundary layer thickness (see Figure 5.3).

Purtell et al.'s (1981) measurements in low-Reynolds-number boundary layers appear to indicate that the law of the wall does not vary with Reynolds number, thus implying a truly constant value of the von Kármán's constant κ. Close to a decade earlier, Huffman and Bradshaw (1972) analyzed the data from several other low-Reynolds-number experiments and arrived at the same conclusion. At the other

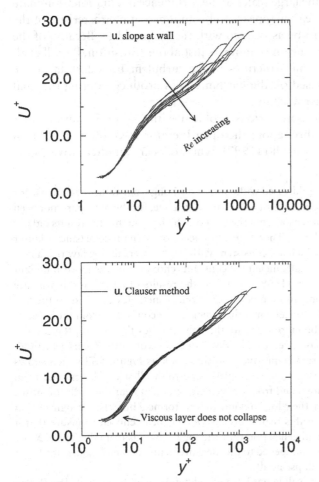

Figure 5.7: Mean-velocity profiles for different Reynolds numbers. Data of Purtell et al. (1981) as replotted by George and Castillo (1997). Note in particular how the curves in the bottom figure do not collapse in the viscous sublayer contrary to the original plots that were published in *Physics of Fluids*. In fairness to Purtell, Klebanoff, and Buckley (1981), this particular practice of plotting the analytical result $\overline{U}^+ = y^+$ in the viscous sublayer region instead of actual data is all too common in the open literature.

extreme, Grigson (1992) used model and field data to show the constancy of κ for Reynolds numbers up to $Re_x = 4 \times 10^9$. Both the low- and high-Reynolds-number results refute an earlier claim by Simpson (1970, 1976) that the von Kármán constant varies with Reynolds number.

The single straight line in Figure 5.6 does not support the assertion by Landweber (1953), Preston (1958), and Granville (1977) that the logarithmic region disappears altogether at low Reynolds numbers. An important question is, What is the minimum Re_θ at which a log-region is first established? Coles' (1962) analysis of the wake component indicates that it is zero at $Re_\theta \leq 600$. The data of Bandyopadhyay and Ahmed (1993) indicate that Clauser's outer-layer shape parameter C reaches zero at $Re_\theta = 425$. These can be regarded as two indications of a minimum value supporting the experimental findings of Purtell et al. (1981).

At low Reynolds numbers, the large scales of the turbulence fluctuations dominate its dynamics. The logarithmic region appears to be an inherent characteristic of the turbulent boundary layer and to be associated with the large eddies. Because of the persistence of the log-region to Reynolds numbers just above transition, Purtell et al. (1981) suggest that the large-scale structures in the turbulent boundary layer are related to, if not simply composed of, the hairpin eddies produced during the final stages of laminar-to-turbulence transition.

There is, however, a much more serious problem with Figure 5.6 and similar mean-velocity plots abundant throughout the open literature. We quote here from the recent work of George and Castillo (1997), who advocate an alternative to the log-law.[11]

> The near wall region can not be addressed without first deciding what values of u_* to use for a given data set. Unfortunately, direct measurements of τ_w have, for the most part, not been very satisfactory for a variety of reasons, not the least of which being that the results did not confirm the Millikan/Clauser theory. Therefore it has been common practice (since Clauser 1954) to choose values for u_* which are consistent with the "universal" log law results for the matched layer, by which is usually meant the constants chosen from the more abundant pipe and channel flow data. Coles (1969) notes that the results obtained in this way are seldom consistent with the momentum integral equation. Nonetheless, in spite of the fact the measurement errors tend to cancel when the integrals are computed from the velocity profiles and therefore should be more accurate than the profiles themselves, the friction results calculated from them have been usually discarded in favor of the Clauser method. Typical of such results is Figure 5 [reproduced in this book as Figure 5.7] which shows the data of Purtell et al. (1981) in inner variables using the wall stress determined from the Clauser method (bottommost) and from the velocity gradient near the wall (topmost). Note the excellent collapse in the "log" region of the former, but also recognize that the data should collapse best in this region because u_* has been chosen to insure that it does—in fact the choice of the tangency point for the log is obvious from the figure. More interesting, however, is what has happened to the measurements for small values of $y^+ < 8$ or so where the data do not collapse at all.
>
> This problem with the inner scaling has been noted before (e.g., Kline et al. 1967), and it seems to have been customary to opt for the method which collapses the "log layer," and rationalize the problems this presents by assuming the measurements near the wall to be in error. While certainly all measurement sets are subject to inaccuracies near the wall, the choice represents more an expression of faith in the log law than a consequence of

---

[11] George and Castillo's nomenclature is slightly different from the one used in this book.

careful error analysis. In fact, curiously, measurements in the linear regions of pipes and channels seem to have been routinely made over the past 50 or so years with little difficulty as long as the ratio of wire diameter to distance from the wall was greater than about 100 (v. Monin and Yaglom, 1971, vol. I, figure 25). Perry and Abell (1975) measure to within 0.4 mm with 4 μm hot-wires in a pipe flow before noticing effects of wall proximity. By the same criterion, the data for Purtell et al. should be expected to be valid to values of y^+ from 2 to 5 for 8 of the 11 data sets. Thus, either hot-wires have behaved very differently in boundary layers than in pipes, or the experimenters have been unwilling to accept the answers the wires were providing.

The values of u_* from the velocity gradient were made using a linear fit for the data in the region $3 < y^+ < 6$, and points closer to the wall were ignored since they were clearly contaminated by near wall effects on the wires. (It will be seen below that there is evidence that the shear stress has been underestimated by about 20% by this method, but with relative errors which are nearly Reynolds number independent.) The data show clearly that the linear region both exists and the data in it collapse nearly perfectly. Moreover the extent of the linear region appears to be greater than for pipes and channels, but decreases with increasing Reynolds number. As expected from the new theory, the data outside the linear layer does not perfectly collapse in inner variables, but shows the Reynolds number dependence which must be present if the inner and outer scale velocities are different.

Similar profiles for the near wall region were published by Blackwelder and Haritonidis (1983) who documented the difference between the shear stress inferred from the velocity gradient near the wall and from a fit to the "log" region. These systematic differences were typically 15% and are tabulated in Table I of their paper. Their plot of the velocity profile (normalized using the shear stress determined from the linear region) is strikingly similar to that of the Purtell data plotted in Figure 5 [Figure 5.7 herein]. Of special interest is the fact that the very near wall data show clearly the onset of the near wall measurement errors well inside the linear region, except for the highest Reynolds numbers.

In the absence of direct force measurements, the fact that the velocity profile near the wall collapses in both these data sets using the wall shear stress inferred from the gradient suggests strongly that both the mean velocity and the shear stress have been reasonably estimated, at least at the lower Reynolds numbers. A similar inference cannot be made using a value of the stress which collapses a "log" region, since (as pointed out above) it is based on the assumptions that such a region exists and that the constants describing it are Reynolds number independent. Clearly the second of these is incorrect if the profiles normalized using the shear stress inferred from the velocity gradient are correct, and the first may be wrong as well.

The first important inference which can be seen from these figures is that the data collapse (if a cross-over can be called "collapse") in the "log" region only if the shear stress is calculated from a method which forces it to by assuming such a layer exists, and then only by compromising the collapse in the linear region. On the other hand, when the shear stress inferred from the mean velocity gradient is used, the profiles collapse well very close to the wall, but not in the "log" layer. The lack of collapse in the linear layer when the "Clauser method" is used is particularly serious since there are no adjustable constants or Reynolds number dependencies here: the measurements must yield $u^+ = y^+$ if they are to be believed.

A second important inference from the figures is that when the data are normalized by a "proper" shear stress, the point of departure from the linear region moves closer to the wall with increasing Reynolds number toward what might be a limiting value. This is clear evidence that if there is an intermediate layer outside the linear region, then it is only asymptotically independent of Reynolds number. The apparent Reynolds number dependence of the extent of the near-linear region is consistent with the new theory presented earlier since only this innermost region of the flow is Reynolds number independent in inner variables. Once the mesolayer is entered, neither u_* nor U_∞ completely collapse the data (except in the infinite Reynolds number limit). Many (Gad-el-Hak and Bandyopadhyay 1994) have noted similar trends for the turbulence moment data near the wall, especially $\langle u^2 \rangle$ and $\langle -uv \rangle$.

5.5.2   Alternatives to the Logarithmic Profile

Despite copious evidence for the existence of the log-law, the whole scenario leading to it has been periodically questioned. From a purely practical point of view, a portion of the streamwise mean-velocity profile could equally well fit either a logarithmic relation or a power-law. If the mean-velocity profile for a pipe flow (or boundary layer) is to be fitted with a power-law of the form

$$\frac{\overline{U}}{U_c} = b\left(\frac{y}{a}\right)^{\frac{1}{n}} \tag{5.19}$$

the value of the exponent $(1/n)$ will decrease as the velocity profile becomes fuller. In other words, n increases as Re increases and H decreases. For example, the smooth-pipe-flow data of Nikuradse (1932) indicate that n changes from 6 to 10 as the Reynolds number varies in the range of $Re_a = 2.53 \times 10^3$ to 1.85×10^6 (Schlichting 1979).

Sreenivasan (1989) argues that although the power-law used by engineers to describe the mean-velocity profile has been discredited by scientists since Millikan (1939) derived the logarithmic law from asymptotic arguments, the basis for the power-law is a priori as sound as that for the log-law—particularly at low Reynolds numbers (see also Barenblatt 1979, 1993; Barenblatt and Prostokishin 1993). The behavior of the exponent $(1/n)$ as $Re_a \to \infty$ is of particular interest. If it tends to zero, the log-law is recovered. If, on the other hand, the limiting value of the exponent is a nonzero constant, the log-law does not strictly hold. This implies that an inertial sublayer is lacking and, therefore, that viscous effects persist even at infinitely large Reynolds number. Such a scenario is consistent with the suggestion by Long and Chen (1981) that the inner flow is the outcome of an interplay between wall effects and outer effects. According to them, it is strange that the matched layer between one characterized by inertia and another characterized by viscosity depends only on inertia. Long and Chen suggest that a "mesolayer" intrudes between the inner and outer regions, preventing the overlap assumed in the derivation of the classical logarithmic velocity profile. Unfortunately, existing experimental or numerical mean-velocity data cannot be readily used to explore this important issue because the difference between a logarithmic relation and a power-law with a large but finite n is imperceptible (see Kailasnath 1993 for a comprehensive review of available data).

George et al. (1992, 1993, 1994) and most recently George and Castillo (1997) have provided the most serious challenge to the validity of the log-law for external wall-bounded flows. As discussed in Section 5.5.1, George and his colleagues assert that boundary layer data taken at different Reynolds numbers collapse in the log-region only if the shear stress is calculated from a method (i.e., the Clauser's method) that forces it to by assuming such a layer exists. Such superficial collapse compromises the collapse in the viscous sublayer where no adjustable constants or Reynolds number dependence should exist. As an alternative, George et al. (1993) used measured shear stress to normalize the data of Purtell et al. (1981) and showed that the profiles collapse well very close to the wall but not in the log-region, where clearly discernible Reynolds number dependence is depicted. To remedy the situation, George et al. (1993) proposed matching a new velocity-defect law with explicit Reynolds number dependence and the traditional law of the wall. The result in the matched region is a power-law velocity profile of the form

$$\frac{\overline{U}}{U_o} = C_o \left(\frac{y}{\delta}\right)^n + B_o \tag{5.20}$$

$$\frac{\overline{U}}{u_\tau} = C_i \left(\frac{u_\tau y}{\nu}\right)^n + B_i \tag{5.21}$$

where the coefficients B_i, B_o, C_i, C_o and the exponent n are Reynolds number dependent, but all are asymptotically constant. In a subsequent article, George et al. (1994) have shown that the additive coefficients B_i and B_o are identically zero. Using the two equations above, the friction coefficient is readily expressed as the Reynolds number to the power $-2n/(1 + n)$.

George et al. (1994) asserted that their approach removes many of the unsatisfying features of the classical Millikan (1939) theory. Furthermore, they argued that a clear distinction should be made between internal and external wall-bounded flows. For fully developed pipe and channel flows, the streamwise homogeneity ensures that the pressure gradient and wall-shear stress are not independent, and thus u_τ is the correct scaling velocity for the entire shear layer, including the core region. This results in a log-law, although the flow has no constant-stress layer. The growing, inhomogeneous boundary layer, in contrast, is governed by a power-law even though it does, at least for zero pressure gradient and high Re, have a constant-stress layer. The matched region of a boundary layer retains a dependence on streamwise distance and hence never becomes Reynolds number independent. The same arguments presented here for the mean flow could be extended to higher-order statistics.

The difference in the inner region between flat-plate boundary layers and fully developed channel flows explored above might have serious implications on the principle and use of the Preston tube, a device widely employed for measuring the local mean friction-coefficient. Such a gauge is commonly calibrated in a pipe flow and relies on the universality of the inner layer to compute the skin friction in a flat-plate boundary layer. Such extrapolation is thus questionable, and direct measurements of wall-shear stress are clearly preferred.

George and Castillo (1993) have also extended the new scaling law described earlier for the flat-plate flow to boundary layers with pressure gradient. Inclusion of roughness or compressibility effects could proceed along similar attempts made in the past for the classical theory (Hama 1954, Coles 1962).

The fresh look at the turbulent boundary layer by George and his colleagues is intriguing and deserves further scrutiny. Independent confirmation of their claims is needed, and carefully controlled boundary layer experiments over a wide range of Reynolds numbers would be most useful. Low-Reynolds-number experiments in which the linear region is resolved and the wall-shear stress is measured directly would be particularly valuable. If validated, their new theory indicates that the boundary layer is governed by a different scaling law than commonly believed. Explicit, albeit weak, Reynolds number dependence is shown for the mean velocity profile all the way down to the edge of the viscous sublayer. The matched region retains a dependence on streamwise distance, and hence Reynolds number effects will always persist for all turbulence quantities.

5.5.3   The Illusory Asymptotic State

Although inner scaling appears[12] to collapse wall-layer mean-flow data onto a single curve regardless of the Reynolds number, the situation is not that simple in the outer layer. As discussed in Section 5.3.3, Coles (1956) proposed to represent the entire mean-velocity profile in any two-dimensional turbulent boundary layer by a linear superposition of two universal functions, the law of the wall and the law of the wake. In fact, Coles suggested a simple extension of his empirical law to represent even yawed and three-dimensional flows. Recall Eq. (5.16)

$$\overline{U}^+ = f(y^+) + \frac{\Pi}{\kappa} W\left(\frac{y}{\delta}\right) \tag{5.22}$$

The first term on the right-hand side is valid for any smooth wall-bounded flow, and available evidence appears to indicate that the function f is independent of Reynolds number, pressure gradient, and freestream turbulence. This term supposedly represents all mixing processes in the wall layer governed primarily by viscosity. The wall constraint is felt mainly in the viscous sublayer, the buffer layer, and the logarithmic portion of the velocity profile. For rough walls, particularly when the roughness is sufficiently pronounced, the viscous length scale is simply replaced by the characteristic roughness height. For both smooth and rough walls, the appropriate velocity scale is derived from the magnitude of the surface shear stress.

The second term, representing turbulence mixing processes dominated by inertia, is the product of the universal wake function $W(y/\delta)$ and the ratio of the profile parameter Π to the von Kármán constant κ. The parameter Π depends on the pressure gradient, the freestream turbulence, and whether the flow is internal or external but is not directly affected by wall conditions such as roughness, and so forth. For a flat-plate boundary layer, the profile parameter increases with Reynolds number but presumably asymptotically approaches a constant value at high enough Re. When Coles (1962) plotted the change of the maximum deviation of the mean velocity from the logarithmic law, ΔU^+, with Reynolds number, it reached a constant value for $Re_\theta \geq 6000$. Coles (1962) termed the flow at this high Reynolds number "equilibrium," which led to the widespread perception that the flow becomes independent of Reynolds number beyond this value. Because the maximum deviation, the so-called strength of the wake component, occurs close to the edge of the boundary layer and $W(y/\delta)$ has been normalized such that $W(1) = 2$, the strength of the wake component is approximately related to the profile parameter by

$$\Delta U^+ \approx 2\frac{\Pi}{\kappa} \tag{5.23}$$

It is clear that the strength of the wake depends upon the somewhat arbitrary way in which the logarithmic portion of the velocity profile is fitted, that is, on the particular values of κ and B chosen.

---

[12] At least until the recent work of George and his colleagues cited in the previous subsection.

Unfortunately, when Coles (1962) extended the plot to larger values of Reynolds numbers, it became clear that the presumed asymptotic state is merely an illusion. The strength of the wake component started decreasing again at about $Re_\theta > 15,000$, although very slowly compared with the rise rate for $Re_\theta < 6000$. After excluding all data containing certain anomalies, Coles (1962) was puzzled by the persistent change in behavior and the variation between data sets for $Re_\theta > 15,000$. The drop cannot be explained from experimental uncertainties such as those caused by probe calibration problems, improper tripping devices, three-dimensional effects, high levels of freestream turbulence, and pressure-gradient effects, and Coles left the issue open.

The rapid rise and the subsequent gradual fall of ΔU^+ with Re_θ appear to be genuine and have been confirmed in several other experiments, as summarized by Mabey (1979). The maximum ΔU^+, reached at about $Re_\theta = 6000$, is 2.7 for subsonic flows (Smith and Walker 1959) but is higher by 30% or more for supersonic flows (Lee, Yanta, and Leonas 1969; Hopkins, Keener, and Polek 1972; Mabey, Meier, and Sawyer 1976).

5.5.4 Is Self-Preservation Ever Achieved?

At approximately $Re_\theta > 15,000$, the gradual departure of ΔU^+ from the apparent low-Re_θ asymptote suggests that some new effects are gradually appearing in the turbulence production process. A new, lower asymptote appears to have been reached when $Re_\theta = 50,000$, but the boundary layer might also continue to change indefinitely, as the inner and outer scales are forever disparting. The paucity of high-Reynolds number reliable data makes it difficult to form a definitive conclusion.

Without definitive experiments at even higher Reynolds numbers, one can never be sure of the universality of the defect law. Very few existing facilities can deliver the required ultra-high-Reynolds-number flows while maintaining relatively low Mach number (in the so-called low-speed, high-Reynolds number tunnels) in order to avoid the added complication of compressibility effects. The world's largest wind tunnel, the 24 m $\times$ 37 m Full-Scale Aerodynamics Facility at NASA Ames, is capable of generating a boundary layer with a momentum-thickness Reynolds number as high as 3.7×10^5 (Saddoughi and Veeravalli 1994).

The largest available water tunnels and towing tanks can deliver momentum thickness Reynolds numbers of approximately 3×10^4 and 9×10^4, respectively (Figure 5.1). Cryogenic tunnels, for example the National Transonic Facility at NASA Langley, typically use nitrogen and run as high as $Re_\theta = 6 \times 10^4$, but their Mach number is near one, and they are rather expensive as well as heavily scheduled (Bushnell and Greene 1991). Tunnels using liquid helium I are an attractive, low-cost alternative to the much larger nitrogen tunnels (Donnelly 1991). Helium facilities can match the Reynolds numbers of the transonic wind tunnels but with essentially zero Mach number and much smaller sizes (e.g., 1 cm $\times$ 1 cm test section). Instrumenting the smaller facilities with high-resolution velocity or pressure probes is at present problematic, although the rapidly developing microfabrication technology has the potential for producing inexpensive megahertz-frequency and micron-size sensors (see, for example, Löfdahl, Stemme, and Johansson 1989, 1991, 1992; Ho and Tai 1996, 1998; Gad-el-Hak 1999; Löfdahl and Gad-el-Hak 1999).

A commonly accessible large-scale, high-Reynolds-number facility is the atmospheric boundary layer. The flow is virtually incompressible, and the momentum-thickness Reynolds number in the atmosphere can be as much as four orders of magnitude higher than that in typical laboratory experiments. Unfortunately, such a natural laboratory has several drawbacks. Firstly, the "wall" in this case is almost always rough, and direct comparison with the canonical boundary layer is difficult. Secondly, the atmospheric experiments are not well controlled, and thus the flow conditions are neither precisely repeatable nor documentable to the needed detail (see however the thesis by Kailasnath 1993, who was able to carry out useful comparison between low-Reynolds-number laboratory data and high-Reynolds-number atmospheric data).

The so-called super-pipe facility is currently operational at Princeton University (Zagarola 1996; Zagarola et al. 1996). The pipe has a diameter of 12.7 cm and a length-to-diameter ratio of 200. This high-pressure-air (200 atm) pipe flow is capable of providing very high Reynolds numbers of up to $Re_a = 2.3 \times 10^7$ at a reasonably large scale and low Mach number and, it is hoped, will help in answering many of the questions raised in this chapter.

For the present at least, it is simply not known if the mean velocity in a wall-bounded flow ever achieves true self-preservation. As will be seen in the next section for higher-order statistics, the evidence for lack of self-preservation is stronger.

5.6 Higher-Order Statistics

Compared with the mean flow, higher-order statistical quantities are more difficult to measure, and the issue of Reynolds number effects is even murkier. For quantities such as root-mean-square, Reynolds stress, skewness, and spectrum, the issues of spatial as well as temporal probe resolutions, three-dimensional effects, and boundary-layer tripping devices become much more critical. In contrast to mean flow, reliable data for higher-order statistics are scarce.

A measurement probe essentially integrates the signal over its active sensing area or volume. This means that velocity or pressure fluctuations having scales smaller than the sensor size are attenuated by the averaging process, and the measured root-mean-square of the fluctuations, for example, is smaller than the true value. Several studies have shown the importance of probe size in the detection of small-scale structures in the near-wall region (for example, Willmarth and Bogar 1977; Schewe 1983; Johansson and Alfredsson 1983; Blackwelder and Haritonidis 1983; Luchik and Tiederman 1986; Karlsson and Johansson 1986; Ligrani and Bradshaw 1987; Wei and Willmarth 1989; Löfdahl et al. 1992). As a rule of thumb, probe length much larger than the viscous sublayer thickness is not acceptable for accurately measuring turbulence levels and spectra anywhere across the boundary layer, and even smaller sensing elements are required to resolve dynamical events within the sublayer itself. In the case of a hot wire, for example, the probe diameter determines its spatial resolution for measuring the time-averaged velocity, whereas the much larger wire length limits the accuracy of measuring the velocity fluctuations.

The issue of sufficient probe resolution is particularly acute when studying Reynolds number effects. A probe that provides accurate measurements at low Re may give erroneous results when the Reynolds number is increased and the scales to be resolved become smaller relative to the probe size. The probe resolution should be expressed in wall units, and as mentioned above ℓ^+ should not be much larger than 5, where ℓ is the probe length.

The classic idea of inner scaling is that any turbulence quantity measured at different Reynolds numbers and in different facilities will collapse, at least in the inner layer, to a single universal profile when nondimensionalized using inner-layer variables. In contrast to mean-velocity profiles, higher-order statistics do not in general scale with wall-layer variables even deep inside the inner layer. In the following two subsections, we review Reynolds number effects on the root-mean-square values of the velocity fluctuations and the Reynolds stress. The cited data plus those for the spectra, skewness and flatness factors, and rms and spectrum of the wall-pressure fluctuations were collected from the original publications and presented in more details in the review article by Gad-el-Hak and Bandyopadhyay (1994). The main conclusion of this survey is that Reynolds number effects on the turbulence quantities persist deep into the viscous region.

5.6.1 Root-Mean-Square Velocity Fluctuations

The intensity of turbulence fluctuations is defined by its root-mean-square value. The streamwise velocity fluctuations are more readily measured using, for example, a single hot-wire probe or a two-beam laser Doppler velocimeter. Measuring the other two velocity components, in contrast, requires two hot wires either in an X-array or a V-array or four intersecting laser beams. Especially very close to the wall, few reliable measurements of the normal velocity components are reported in the literature, and even fewer are available for the spanwise component. A notable exception is the oil-channel data reported by Kreplin and Eckelmann (1979), who measured all three velocity components inside the viscous sublayer. The flat-plate data of Purtell et al. (1981) taken in the range $Re_\theta = 465$–5100 showed clear Reynolds number effects penetrating into the boundary layer much deeper in terms of the turbulence intensity than they do for the mean velocity. Approximate similarity in wall units is maintained only out to $y^+ \approx 15$ compared with mean velocity, which appears to be similar throughout the entire inner layer (but see the discussion in Section 5.5.1). Close inspection of Purtell et al.'s data at the lowest Reynolds number of $Re_\theta = 465$ reveals that even in the viscous region itself some weak dependence on Reynolds number is observed in the rms value of the streamwise velocity fluctuations. Purtell et al. attributed the systematic decrease in u' across most of the boundary layer to the suppression of all but the largest turbulence eddies as the Reynolds number is reduced.

Reynolds number can of course be changed either by varying the freestream velocity at a fixed streamwise location or by holding the tunnel speed constant and conducting the measurements at increasing downstream locations. To check the state of development of the flow, Purtell et al. (1981) measured, for a fixed freestream velocity, the mean-velocity profiles at four downstream distances from the tripping device. At the station closest to the distributed roughness used to trip the boundary layer, they report underdevelopment in the outer region of the mean flow but an undistorted logarithmic region that produces friction velocity values in agreement with directly measured u_τ-values computed from near-wall measurements of $\partial\overline{U}/\partial y$. The same trends were observed at higher Reynolds numbers by Klebanoff and Diehl (1952).

When normalized with the freestream velocity, u' also shows distortion in the outer region for measurements not sufficiently far downstream of the tripping device. However, the rms data plotted in inner variables at a constant freestream velocity and four downstream stations exhibit such a strong Reynolds number dependence in the outer layer that this distortion is obscured. Close inspection of the data again shows a small but systematic Reynolds number effect even below $y^+ = 10$. At a given y^+, u_{rms}^+ increases with x, that is, with Re_θ.

One might argue that the strong Reynolds number effects indicated above will eventually subside at sufficiently high Reynolds number. This is not the case, at least up to Re_θ of 15,406, as shown by Andreopoulos et al.'s (1984) flat-plate data. Their rms values of the longitudinal velocity fluctuations show strong Reynolds number dependence all the way to the edge of the viscous sublayer. In the buffer layer, u'/u_τ decreases as the Reynolds number increases from 3624 to 12,436, thereafter reaching what seems to be a constant value. An opposite trend, which continues for all four Reynolds numbers they tested, is observed in the logarithmic and wake regions. Andreopoulos et al. (1984) indicated that the behavior of their u'-data in the buffer region is in general agreement with the channel-flow results of Laufer (1951), Comte-Bellot (1963, 1965), and Zaric (1972).

Andreopoulos et al. (1984) were able to measure the normal velocity component only at the lowest Reynolds number, $Re_\theta = 3620$. A weak second peak in v'/u_τ appeared at $y^+ = 4$ and seems to be unique to this particular experiment, but the authors do not comment on this. Reliable v' measurements could not be obtained by Andreopoulos et al. for higher Reynolds numbers owing to limitations of the applicable velocity range of their triple-wire sensor.

Relatively more data are available for two-dimensional channel flows. Laufer (1951) conducted his experiments at three Reynolds numbers, based on channel half-width and centerline velocity, of $Re_a = 12,300$, 30,800, and 61,600. Comte-Bellot (1965) covered the higher range of $Re_a = 57,000$ and 230,000. Kreplin and Eckelmann's (1979) experiments were conducted at $Re_a = 3850$. Johansson and Alfredsson (1982) provided data for $Re_a = 6900$, 17,300, and 24,450. Probe resolution problems appear to exist in both Laufer's and Comte-Bellot's high-Reynolds number data. In the former experiment, the hot-wire length increased from 3 wall units at the lowest Reynolds number investigated to 13 wall units at the highest Reynolds number, a fourfold loss in spatial resolution. Laufer (1951) observed (erroneously) a corresponding decrease in the peak value of u'/u_τ with increasing Reynolds number. Comte-Bellot's (1965) probe length increased from 13 to 36 viscous lengths as Re_a increased from 57,000 to 230,000. Correspondingly, she also observed (erroneously) a decrease in the peak value of the nondimensional streamwise turbulence intensity from 2.85 to 2.5. High-quality turbulence data obtained using sufficiently small probes and facilities void of significant trip-memory effects are clearly lacking. In any case, the available data indicate that the turbulence intensity profiles do not collapse even deep into the inner layer.

Similar trends are observed in the rms values of the velocity fluctuations normal to the wall. The peak v' is lower than that for the streamwise fluctuations and occurs farther away from the wall. Validity of inner scaling deep into the viscous region cannot be ascertained from available results because of the scarcity of data for $y^+ < 30$.

In view of the poor quality of most of the data, Wei and Willmarth (1989) systematically investigated the Reynolds number effects using a unique high-resolution, two-component laser-Doppler velocimeter. To reduce the amount of ambient light in the vicinity of the measuring volume and thus to improve the signal-to-noise-ratio of the LDV system, four laser beams were entered and exited into the test section via two narrow slits located at the two side walls of a two-dimensional water channel. Both slots were covered with an extremely thin (17 μm) window of heat-shrinking

Mylar film, which virtually eliminated optical refraction by the window. The laser beams intersected at a single point away from the wall, and the effective probe length ranged from 0.66 to 6.43 wall units as the Reynolds number was varied in the range of $Re_a = 3000$–40,000. Beam refraction in the path between the laser source and the measuring station was minimized using a specially designed optical head.

Wei and Willmarth (1989) ascribed the slight disagreement between the different data sets available in the literature and theirs to a decreased spatial resolution of the hot wires used by Laufer (1951), Comte-Bellot (1965), and Johansson and Alfredsson (1982) at high Reynolds numbers. But, even in the newer data set, the fluctuating turbulence quantities do not scale with wall variables even at as close as 10 viscous lengths from the wall. In fact, inner scaling does not seem to apply to v' across the entire portion of the viscous region where measurements are available.

Additional support for Wei and Willmarth's (1989) basic conclusion that turbulence quantities in the near-wall region do not scale on wall variables comes from the boundary layer experiments of Purtell et al. (1981) and Andreopoulos et al. (1984) referenced earlier in this section as well as from the physical and numerical channel-flow experiments conducted by Antonia et al. (1992) and the flat-plate experiments of Murlis, Tsai, and Bradshaw (1982), Wark and Nagib (1991), Naguib (1992), and Naguib and Wark (1992).[13] Contrary to the prediction of the universal theory of the wall, the peak value of u^+_{rms}, though occurring at the same y^+, increases with Reynolds number. It is, of course, conceivable that wall-layer scaling might apply over the entire inner layer provided that the Reynolds number is high enough, but at the moment at least such ultra-high-Reynolds-number experiments cannot be conducted with sufficient probe resolution.

Coles (1978) summarized the results of fifty different experiments conducted in circular pipes, rectangular channels, and zero-pressure-gradient boundary layers. He did remark that not all experiments are equally reliable. Nevertheless, Figure 3 of Coles' article indicates that, when the ratio of outer to inner length-scales, δ^+, increases from 100 to 10,000, the value of the rms streamwise velocity fluctuations measured at $y^+ = 50$ and normalized with the corresponding peak value measured at $y^+ \approx 15$ systematically increases from about 0.6 to 0.9. This result is consistent with a nonnegligible Reynolds number effect on the turbulence just outside the viscous region.

5.6.2 Reynolds Stress

Reynolds Number Effects

Turbulence shear stress, or Reynolds stress $-\rho\overline{uv}$, is the most important dynamical quantity affecting the mean motion. The major portion of the momentum transported in a two-dimensional turbulent wall-bounded flow is accomplished by $-\overline{uv}$. Therefore, modeling the behavior of Reynolds stress is one of the primary objectives of various prediction schemes. Simultaneous measurements of the streamwise and normal velocity fluctuations are required to compute the Reynolds stress at any particular point in the flowfield. Provided this is done with high fidelity in the low-to-moderate Reynolds-number laboratory experiments, extrapolation to the higher-

---

[13] The channel-flow experiments of Walker and Tiederman (1990) and Harder and Tiederman (1991) should also be mentioned. Their results do not agree with those of Wei and Willmarth (1989), but Gad-el-Hak and Bandyopadhyay (1994) concluded that the latter set of data is more reliable.

Reynolds-number field conditions is only possible if the Reynolds number effects are well understood. In this subsection, we review those effects in boundary layers, channels, and pipes.

In boundary layers, Andreopoulos et al. (1984) measured the normalized cross correlation $-\overline{uv}$ for a single Reynolds number of $Re_\theta = 3664$. Again, they were unable to measure the Reynolds stress reliably at higher Reynolds numbers owing to limitations of the applicable velocity range of their triple-wire probe. The directly measured turbulence shear stress was on the average 10% smaller than the theoretical distribution deduced from the momentum balance and mean-flow data.

More data are available for two-dimensional channel flows. The (kinematic) Reynolds stress is directly computed by averaging the product of the measured u- and v-velocity fluctuations. Eckelmann (1974), using an oil channel, covered the low Reynolds numbers of 2800 and 4100. Alfredsson and Johansson (1984) conducted their experiment at $Re_a = 7500$, whereas Kastrinakis and Eckelmann (1983) conducted theirs at $Re_a = 12,600$. Comte-Bellot (1965) covered the higher range of Reynolds number of 57,000 and 230,000. Wei and Willmarth (1989) covered the Reynolds number range of 2970 to 39,582. Their high-resolution LDV allows measurements very close to the wall where the Reynolds stress is decreasing. When plotted against y/a, the maxima of the nondimensional turbulence shear stress profiles increase in magnitude and are closer to the wall as the Reynolds number increases. But when plotted versus y^+, the maximum value of the normalized turbulence shear stress is not the same for each profile, indicating the lack of inner scaling in the Reynolds number range investigated. When expressed in wall units, the location of peak Reynolds stress moves away from the wall as the Reynolds number increases.

One advantage of investigating fully developed pipe or channel flows is the ability to compare direct measurements with the computed Reynolds stress using the mean-velocity profile and pressure gradient, two quantities that are easier to measure. This has the advantage of enabling the accuracy of the directly measured Reynolds stress to be checked, especially near the wall where probe resolution problems are particularly acute. In a fully developed channel or pipe flow, the average normal and spanwise velocities vanish, there are no mean longitudinal velocity or Reynolds stress variations in the streamwise and spanwise directions, and the pressure gradient is a constant. The longitudinal momentum equation could then be integrated to give an exact relation between the Reynolds stress and mean-velocity distribution

$$-\overline{uv} = -\nu\frac{d\overline{U}}{dy} + u_\tau^2\left(1 - \frac{y}{a}\right) \tag{5.24}$$

where $\overline{U}(y)$ is the streamwise mean-velocity distribution, and a is the channel half-width or pipe radius. The friction velocity, or the slope of the velocity profile at the wall, is related to the constant pressure gradient through

$$u_\tau^2 = \nu\left(\frac{d\overline{U}}{dy}\right)_w = -\left(\frac{a}{\rho}\right)\left(\frac{dp}{dx}\right) \tag{5.25}$$

where p is the static pressure and ρ and ν are the fluid density and kinematic viscosity, respectively. In wall units the momentum balance equation (5.24) reads

$$-\overline{uv}^+ = -\frac{d\overline{U}^+}{dy^+} + \left(1 - \frac{y^+}{a^+}\right) \tag{5.26}$$

Wei and Willmarth (1989) used Eq. (5.26) to compute the Reynolds stress profiles, and again the nondimensional profiles at different Reynolds numbers do not collapse in the outer and logarithmic regions and even well into the viscous region. Except very close to the wall, the agreement between the directly measured Reynolds stress and that computed from the measured mean velocity and pressure gradient is very good and attests to the accuracy of the instantaneous velocity traces reconstructed, filtered, and smoothed from the Doppler burst detector and processor signals. Wei and Willmarth speculated that the divergence between the directly measured and computed Reynolds stresses is due to insufficient spatial and temporal resolution in the direct LDV measurement very close to the wall.

Wei and Willmarth (1989) also computed the turbulence kinetic energy production using both the directly measured Reynolds stress and the momentum balance equation. Except very close to the wall, the two methods of computing $-\overline{uv}(d\overline{U}/dy)$ agree within 10%. Neither method leads to a profile collapse even in the viscous region. The maximum value of kinetic energy production obtained from the momentum balance increases with Reynolds number. Interestingly, although the position of peak Reynolds stress, expressed in wall units, moves away from the wall as the Reynolds number increases, the peak turbulence production seems to be fixed at $y^+ \approx 12$–15. This point will be revisited later in this subsection.

The measurements of mean velocity and pressure drop in the smooth-pipe-flow experiments of Nikuradse (1932) and Laufer (1954) were used to compute the Reynolds stress profiles shown previously in Figure 5.2. The Reynolds number in that figure, $Re^* = u_\tau a/v$, is the ratio of the pipe radius to the viscous length scale and varies over the wide range of 140–55,400. An approximately constant-turbulent-shear-stress region is clear at the highest Reynolds number. As in the channel flows, the peak value of normalized Reynolds stress increases, and its location, relative to the viscous length scale, moves away from the wall as the Reynolds number increases.

Peak Location

Sreenivasan (1989) analyzed several different wall-bounded flow experiments. The distance from the wall, expressed in inner variables, where the streamwise turbulence intensity peaks appears to be independent of Reynolds number (Figure 16 of his article). In contrast, the location where the largest normal fluctuations occur is a strong function of Reynolds number (Figure 17 of his article)

$$[y^+]_{v'_{max}} = [Re^*]^{0.75} \tag{5.27}$$

where Re^* is the pipe radius or boundary-layer thickness in wall units. Available data on the spanwise intensity are scarce, and no conclusion can be reached on the scaling of its peak position. For the total turbulence kinetic energy, however, the position of its peak does scale on wall variables much the same as u'. This is because the near-wall value of the total fluctuation energy is essentially overwhelmed by the streamwise component.

Similar to the normal fluctuations, the peak Reynolds stress occurs at increasingly higher values of y^+ as the Reynolds number increases, as shown in Figures 5.8a and 5.8b, compiled from directly measured and computed turbulence shear-stress data, respectively.[14] The lowermost two data points in Figure 5.8b correspond to

---

[14] These same plots also appeared earlier in the article by Long and Chen (1981).

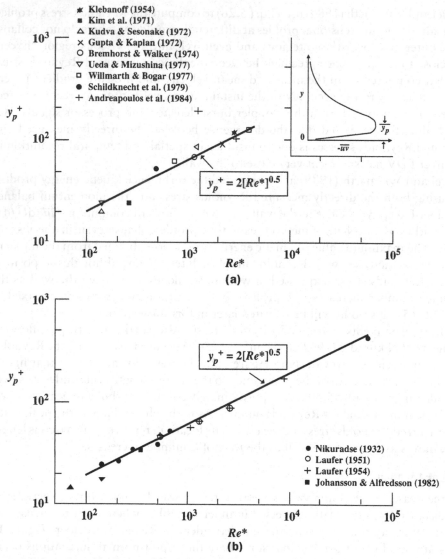

Figure 5.8: Location of peak Reynolds stress as a function of Reynolds number. Data compiled by Sreenivasan (1988) from various wall-bounded flow experiments. Solid lines are least-square fit. (a) Directly measured Reynolds stress in different boundary layers. (b) Computed from measured mean velocity in different channel flows; the lowermost two data points correspond to the critical-layer position in typical transitional flows.

the location of the peak Reynolds stress, or the critical layer position, in typical transitional boundary layer and channel flows. A least-square fit of all the data points in both figures leads to the same equation

$$y_p^+ = 2\,[Re^*]^{0.5} \tag{5.28}$$

where y_p is the location of peak Reynolds stress. Note that although probe resolution has a significant effect on the magnitude of turbulence intensity, Reynolds stress, or other higher-order statistics, a relatively long probe should have less effect on the accuracy of determining the peak location of these quantities. It is therefore not surprising that Sreenivasan (1989) could use a variety of data sources, including some with insufficient probe resolution, to arrive at the correlations in Eqs. (5.27) and (5.28).

Equation (5.28) indicates that the location of the peak turbulence stress scales on the geometric mean of the inner and outer scales. Recall that, because this is the position in Sreenivasan's (1988) model discussed in Section 5.4, where all the mean vorticity of a turbulent boundary layer has been assumed to be concentrated into a single sheet, the correlation in Eq. (5.28) and its applicability to the critical-layer position in transitional flows give some credence to his hypothesis.

Note that if a velocity profile is assumed for the case of fully developed channel flow, exact expressions for the location of the peak Reynolds stress and turbulence kinetic energy production can be derived from Eq. (5.24). For example, if the mean profile is approximately logarithmic in the vicinity of y_p, the peak Reynolds stress occurs at

$$y_p^+ = \left[\frac{Re^*}{\kappa}\right]^{0.5} = 1.56\,[Re^*]^{0.5} \tag{5.29}$$

whereas the peak production occurs at a fixed y^+. Furthermore, the value of the Reynolds stress at its peak can also be computed from the streamwise momentum equation (5.24)

$$\frac{[-\overline{uv}]_{max}}{u_\tau^2} = 1 - 2\,[\kappa\,Re^*]^{-0.5} \tag{5.30}$$

Thus, both the maximum value of the Reynolds stress and the location of its peak depend directly upon the outer lengthscale a (or δ), and the peak location scales with the geometric mean of the inner and outer scales.

At high Reynolds numbers the peak Reynolds stress occurs substantially outside the viscous region. Note however that, owing to the shrinking of the inner layer as the Reynolds number increases, this peak location moves closer to the wall as a fraction of the boundary-layer thickness. Interestingly, although the most significant Reynolds-stress-producing activity does not occur at a universal value of y^+, the production of turbulence kinetic energy, $-\overline{uv}(d\overline{U}/dy)$, does always peak at $y^+ \approx$ 12–15. This implies that, at high Reynolds numbers where the two positions dispart, the scales producing the Reynolds stress are quite different from those responsible for the turbulence kinetic energy production. It is this observation that led Townsend (1961) to hypothesize the existence of an active motion and an inactive motion within the inner layer. The former is due to the vorticity field of the inner layer proper and is responsible for turbulence kinetic energy production. The statistical properties of the active motion are presumably universal functions of the distance from the wall. The inactive, larger-scale motion is partly due to the irrotational field sloshing associated with the pressure fluctuations in the outer layer and partly the large-scale vorticity field of the outer-layer turbulence, which the inner layer sees as an unsteady external stream (see also the substantiative measurements of Bradshaw 1967). The inactive motion does not scale with inner variables and is characterized by intense velocity fluctuations. The effect of increasing the Reynolds number can then be thought of as the increasing significance of the inactive motion (see also Naguib 1992).

The primary conclusion of this and the previous subsubsections is that Reynolds number does have an effect on the turbulence shear stress even in the inner layer. Inner scaling fails to collapse the Reynolds stress profiles. The peak value of $-\overline{uv}$

increases with Reynolds number, and its position moves outward when expressed in wall units.

Asymptotic Theory

The results discussed in this section thus far indicate that inner scaling fails to collapse the profiles for the Reynolds stress and for the root-mean-square values of the velocity fluctuations. Considerable Reynolds number effects are exhibited even for y^+ values less than 100. Panton (1990a) pointed out that a turbulent wall-bounded flow is fundamentally a two-layer structure, a classical single perturbation situation. At finite Reynolds numbers, neither the inner representation nor the outer representation is a uniformly valid approximation to the true answer in the matching region. As Re_θ varies, the overlap layer changes size, and the proportions of inner and outer effects are altered.

A uniformly valid answer for the present singular perturbation problem can be obtained by forming an additive composite expansion from the inner and outer expansions. Matching essentially replaces the two lost boundary conditions at $y = 0$ and $y = \infty$, and the additive composite expansion is simply the sum of the inner and outer ones minus the common part (see, for example, Van Dyke 1964). Systematic changes with Reynolds number considered anomalies when turbulence quantities are expressed as inner expansions can then be regarded as proper first-order trends that are expected when viewed in the proper light. Such treatments were demonstrated for the mean flow (Panton 1990b), for the rms turbulent fluctuations (Panton 1991), and for the Reynolds stress (Panton 1990a). Root-mean-square values of the velocity fluctuations or Reynolds stress expressed as additive composite expansions are equivalent in accuracy to the mean velocity expressed as the law of the wall plus the law of the wake.

For the mean flow, Panton (1990a) suggested the use of an inner velocity scale that is different from the friction velocity (see Eq. (5.7) of this volume), the two being equal only in the limit of infinite Reynolds number. Within the framework of an asymptotic theory (Yajnik 1970; Afzal 1976; Afzal and Bush 1985), the lowest-order equation for the mean flow shows weak Reynolds number dependence, whereas that for the Reynolds stress indicates a much stronger effect. According to Panton (1990a), the logarithmic nature of the inner, outer, and composite expansions for the mean flow dictates minimal Reynolds number effects. On the other hand, the Reynolds stress behaves algebraically, and the inner–outer effects are mixed in different proportions and occur at different locations, resulting in strong Reynolds number dependency even in the first-order theory. Moreover, the additive composite expansion for the Reynolds stress does not evince a constant-stress region; only the inner expansion does so.

5.7   Coherent Structures

The next question to be discussed in this chapter relates to Reynolds number effects on the coherent structures in wall-bounded flows. Reference is made to Section 4.4 of the previous chapter. There are several facets to that issue, and the present section is divided into four subsections. To be addressed below are proper scaling for the period between bursts, the possibility of profound structural changes after the well-known Reynolds number limit of $Re_\theta = 6000$, the ratio of outer to inner scales, and Reynolds number effects on inner structures.

5.7.1 Bursting Period

Because of the problems of threshold setting and probe resolution, bursting frequency and its scaling have become the source of continuing controversy. Cantwell (1981), based on a review of available literature, has concluded that this frequency scales on outer variables, thus establishing a strong link between the inner and outer regions of a wall-bounded flow. On the other hand, Blackwelder and Haritonidis (1983) have shown that the frequency of occurrence of these events scales with the viscous parameters, which is consistent with the usual boundary-layer scaling arguments. Their results, obtained with a hot-wire probe whose length varied in the range of $\ell^+ = 4.5\text{–}20$ as the Reynolds number increased in the range of $Re_\theta = 1000\text{–}10,000$, are shown in Figure 5.9. In Figure 5.9a, outer variables are used to normalize the bursting frequency (or its inverse, the period between bursts). The nondimensional frequency increases with Reynolds number, thus clearly indicating that outer scaling is not applicable. On the other hand, the same data plotted in Figure 5.9b using the viscous time scale to normalize the frequency indicate the validity of inner scaling. Thus, the properly nondimensionalized bursting period is independent of the Reynolds number in agreement with the observations of Kline et al. (1967), Corino and Brodkey (1969), Donohue et al. (1972), Achia and Thompson (1977), and Blackwelder and Eckelmann (1978). Blackwelder and Haritonidis (1983) have suggested that past erroneous results are caused by insufficient spatial resolution of the sensors used to detect the bursts.

On the basis of measurements in the atmospheric boundary layer, where the Reynolds number is several orders of magnitude higher than in typical laboratory experiments, Narasimha and Kailas (1986, 1987, 1990) still maintain that bursting events scale on outer variables. To do otherwise, the insufficient time resolution of the atmospheric data would simply not have allowed the detection of any dynamically significant events. Narasimha and Kailas cite other laboratory experiments to support their position (e.g., Rao, Narasimha, and Badri Narayanan 1971; Ueda and Hinze 1975; Willmarth 1975a; Shah and Antonia 1989; Rajagopalan and Antonia 1984).

Adding to the present confusion, Bandyopadhyay (1982) has shown that the bursting period is not a universal function and that both inner and outer variables are involved in its scaling with Re_θ. He reviewed existing data and concluded that a universal value of the bursting frequency scaled with either inner or outer variables in various boundary layers ranging from relaminarized to separated does not exist. Because a turbulent boundary layer is characterized by three integral variables (C_f, H, and Re_θ), verification of universality with Re_θ alone is clearly inadequate, and the apparent confusion stems in part from the lack of experiments over a sufficiently wide range of shape factors H. Johansson and Alfredsson (1982) have also suggested that the bursting period scale with intermediate scaling proportional to the geometric mean of the inner and outer scales. It should be noted, however, that, within the framework of an asymptotic theory, mixed variables have little or no physical significance.

The arguments by both Blackwelder and Haritonidis (1983) and Narasimha and Kailas (1987) are compelling, and the issue of scaling of the bursting events must, for the moment at least, stay open. The laboratory experiments of the former group

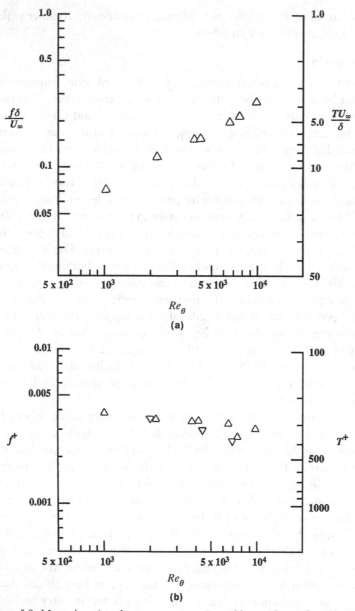

Figure 5.9: Mean bursting frequency versus Reynolds number [after Blackwelder and Haritonidis 1983]. (a) Scaling using outer-flow variables. (b) Scaling using inner-flow variables; the inverted triangles represent three additional data points from an untripped boundary layer.

were well controlled, but the range of Reynolds numbers and range of shape factors investigated are not large enough. The latter group experiments were conducted with sufficient probe resolution, but the atmosphere can neither be controlled nor fully documented. Moreover, the effects of roughness on the scaling are simply not known. Controlled high-Reynolds-number experiments using smooth walls and probes with sufficient resolution should settle the question.

5.7.2 High Reynolds Number

Does the boundary layer structure change when $Re_\theta > 6 \times 10^3$? The Reynolds number variations of ΔU^+ attributed to Coles (1962), discussed in Section 5.5.3, at first suggest that an asymptotic state is reached approximately when $Re_\theta > 6 \times 10^3$. But, the higher-Reynolds-number data beyond that limit show that ΔU^+ drops, although very slowly compared with the rise rate for $Re_\theta < 6 \times 10^3$. The gradual departure of ΔU^+ from the apparent low asymptote suggests that some new effects are appearing in the turbulence production process at approximately $Re_\theta > (6 \text{ to } 15) \times 10^3$. Experiments conducted in several different facilities are briefly described below, and they show that profound changes in the coherent structures of different wall-bounded flows may indeed take place at very high Reynolds numbers.

Relevant to the issue of structural changes when $Re_\theta > 6 \times 10^3$ is the recent assertion by Kailasnath (1993) that the skin friction, the pressure fluctuations, and the mean-velocity profiles all show a distinct change of behavior at about the same Reynolds number. For example, a power-law fit to existing skin-friction data for both boundary layers and pipe flows indicates a break point, at $Re_\theta \approx 5000$, that separates two ranges of Reynolds numbers. This and other evidence prompted Kailasnath (1993) to propose a transitional behavior for wall-bounded flows from the low- to the high-Reynolds-number regimes and to suggest further that the turbulence regeneration mechanism is different in the two regimes.

The flow visualization results of Head and Bandyopadhyay (1981) briefly mentioned in Section 5.2.2 indicate that the hairpin structures exhibit strong dependence on Reynolds number for $Re_\theta < 7000$, and hence the hairpins are atypical. At higher Reynolds numbers, on the other hand, the hairpin vortex is found unambiguously. In Head and Bandyopadhyay's view, Falco's (1977) typical eddies are merely the longitudinal cross sections of the tips of the hairpins. Perry and Chong (1982) and Perry et al. (1986) concur with this view. Their model of the turbulent boundary layer emphasizes a hierarchy of hairpin eddies as the quintessential structure of the outer region. In wave number space, the analogous idea of a hierarchy of interacting scales and energy transfer from large eddies to smaller ones is, of course, not new and has been proposed as early as 1920 by Richardson and formalized by Kolmogorov (1941a).

Turbulent boundary layers ranging from relaminarized to separated cover the entire range of possible shape factor H. The statistical properties of the turbulent–irrotational fluid interface as well as the bursting period in such diverse layers can be described by H (Fiedler and Head 1966, Bandyopadhyay 1982). As can be expected, the location of the maximum deviation of the mean velocity from the logarithmic law also correlates with the mean location of the intermittent layer. Changes in the properties of the intermittent layer can take place when H drops below 1.3, which is approximately when $Re_\theta > 10 \times 10^3$. This is supported by the flow visualization results of Head and Bandyopadhyay (1981) at $Re_\theta = 17.5 \times 10^3$, which shows that the outer part of the boundary layer is noticeably sparser: fewer of the hairpin vortices reach δ, although more of them are produced per unit (dimensional) wall area.

In high-speed, high-Reynolds-number turbulent boundary layers, the mean location of the intermittent layer and its standard deviation change significantly according to the results of Owen et al. (1975) at $Ma_\infty = 7.0$ and $Re_\theta = 85 \times 10^3$. The intermittency profile for the supersonic boundary layer is clearly fuller. Furthermore, at these high Reynolds numbers, the boundary layer structures do not exhibit much over-turning motion, which is typical of lower Reynolds numbers (see the recent book by Smits and Dussauge, 1996, for useful information on supersonic shear layers). In the statistical measurements of conventional boundary-layer properties at high Reynolds numbers, these changes may not always seem dramatic, but their critical importance may lie in the efficiency of outer-layer or other control devices for drag reduction.

Morrison, Bullock, and Kronaner (1971) compared the sublayer spectra of the longitudinal velocity fluctuations for low- and high-Reynolds-number pipe flows. Over the sufficiently wide range of Reynolds numbers $Re_a = 10,000–100,000$, the shape of the two-dimensional spectra expressed in wall-layer variables is not universal. This result contradicts the earlier low-Reynolds-number one- and two-dimensional spectral observations made by Bakewell and Lumley (1967) and Morrison (1969). As the Reynolds number is increased, more energy appears in the low-frequency, low-wave-number region. The additional energy results from disturbances that convect at twice the characteristic velocity of the sublayer of $8u_\tau$. The high Reynolds numbers appear to have the effect of randomizing the phase velocity whereby the disturbances are no longer phase-correlated in the sublayer. This additional evidence also suggests much change in the turbulence production mechanism at very high Reynolds numbers. In fact, Morrison et al. (1971) have strongly suggested that the low-speed streaks are unique to low-Reynolds-number wall-bounded flows. Streaks would no longer appear at very high Reynolds numbers at which a phase-correlated, wavelike turbulence might not exist within the viscous sublayer.

Using a rake of X-wires and conditional averaging techniques, Antonia et al. (1990) have examined the effects of Reynolds number on the topology of the large structures in the range $1360 \leq Re_\theta \leq 9630$. The instantaneous longitudinal sectional streamlines in a moving frame of reference contain many rotational structures $\mathcal{O}[\delta/2]$ at the lowest Reynolds number. Very significant Re_θ effects can be observed in the instantaneous frames (see their Figure 5). As Re_θ is gradually increased to 9630, the large rotational structures become much smaller and no longer dominate the outer layer. When the large structures are selectively sampled and averaged, their foci are found to be more circular at lower Reynolds numbers. As Re_θ is increased from 1360 to 9630, the location of the foci moves closer to the wall from 0.83δ to 0.78δ. This is consistent with the effect of Reynolds number on the mean location of the intermittent layer for similar values of the shape factor H (Fiedler and Head 1966).

5.7.3 Small Structures in Outer Layer

In this subsection, a relation is developed relating the ratio of outer to inner scales to Reynolds number changes. The boundary-layer thickness in wall units, δ^+, is related to the Reynolds number Re_θ via the skin-friction coefficient and the ratio of boundary-layer thickness to momentum thickness

$$\delta^+ \equiv \frac{\delta u_\tau}{\nu} = \frac{u_\tau}{U_\infty} \cdot \frac{\delta}{\delta_\theta} \cdot Re_\theta \tag{5.31}$$

$$\delta^+ \equiv \left(\frac{C_f}{2}\right)^{\frac{1}{2}} \cdot \frac{\delta}{\delta_\theta} \cdot Re_\theta \tag{5.32}$$

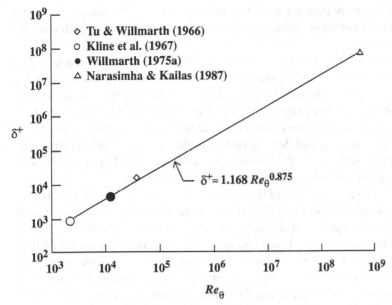

Figure 5.10: Reynolds number dependence of ratio of outer to inner scale. The straight line is computed from the modified pipe-resistance formula and power-law mean-velocity profile. Data compiled by Bandyopadhyay (1991) from different experiments.

For a smooth flat plate, an approximate empirical relation can be obtained by using the modified pipe-resistance formula

$$C_f = 0.0296(Re_x)^{-\frac{1}{5}} \tag{5.33}$$

and the 1/7th-power-law velocity profile. The ratio of the outer scale to inner scale is thus given by

$$\delta^+ = 1.168(Re_\theta)^{0.875} \tag{5.34}$$

Figure 5.10, taken from Bandyopadhyay (1991), shows that Eq. (5.34) describes the data over the entire Reynolds number range for which measurements are available.

5.7.4 Inner Structures

In this subsection, Reynolds number effects on the inner structures are discussed. It will be argued that vortex stretching is enhanced at higher Reynolds numbers and that the low-speed streaks, commonly observed in low-Reynolds number experiments, may become less important at higher speeds.

Wei and Willmarth (1989) have argued that in turbulent channel-flow literature, where an inner-layer scaling has been claimed to hold, errors have crept into measurements that are ascribable to large invasive probes and that sometimes small but systematic variations with Reynolds number have been overlooked because such a scaling was assumed a priori. Their measurements, discussed previously in Section 5.6, show that u_{rms}/u_τ turbulence intensity scales on inner variables only up to $y^+ \approx 10$, which is well inside the inner region. The v_{rms}/u_τ turbulence intensity and the Reynolds shear stress $-\overline{uv}/u_\tau^2$ distributions do not scale on inner variables anywhere in the channel. Interestingly, the maximum normalized Reynolds stress and normal turbulence intensity increase with Reynolds number. This they attributed to the enhancement of the vortex-stretching mechanism in the inner region with increasing Reynolds number.

Because the Reynolds number Re_a or Re_θ is a dimensionless grouping of outer variables, failure of the turbulence quantities in the inner region to scale only on inner variables is an indication that the dynamics of the inner region structure are affected by outer as well as inner variables. Wei and Willmarth (1989) argue that these Reynolds-number dependencies are caused by changes in the coherent structure of the turbulence close to the wall and that the turbulent flow structure at high Reynolds number near solid boundaries (i.e., the hairpin vortex structure and interactions) will differ significantly from lower-Reynolds-number inner structures.

Morrison et al. (1971) have studied the organized motion in the sublayer region of a pipe flow. As discussed in Section 5.7.2, they have measured the two-dimensional frequency–wave-number spectra of the longitudinal velocity fluctuations at $10.6 \times 10^3 \leq Re_a \leq 96.5 \times 10^3$. An appeal of this data lies in its long-time averaged statistical nature and the absence of any troublesome subjective threshold setting as used in VITA or VISA techniques. For $Re_a < 30 \times 10^3$, the streamwise phase velocity c_x^+ in the sublayer is independent of wave number and remains a constant throughout the sublayer. Because this constant velocity coordinates the phase of the periodic motions at different wall-normal locations, Morrison et al. concluded that the sublayer turbulence is wavelike, and, in fact, at low Reynolds numbers it is likely that the sublayer consists of relatively periodic waves.

The critical-layer height is estimated to be $9v/u_\tau$ because at that location the average fluid velocity equals $c_x^+ = 8$. At $Re_a < Re_{\text{crit}} = 30 \times 10^3$, the characteristic spanwise wavelength λ_z^+ of 135 agrees with Kline et al.'s (1967) streak-spacing estimate of 130. However, for $Re_a > Re_{\text{crit}}$, the frequency/wave-number spectra for the laterally spaced points lose their universal shape, and the relative amount of low-frequency, low-wave-number energy increases with Reynolds numbers. This additional energy, which becomes significant at higher Reynolds numbers, results from disturbances that convect at velocities much greater (the average being $16u_\tau$) than the characteristic sublayer velocity of $8u_\tau$. This led Morrison et al. (1971) to conclude that, at higher Reynolds numbers, the character of the sublayer will be substantially altered inasmuch as an increasing amount of low-frequency, low-wave-number energy will be introduced. The disturbances responsible for this additional energy have propagation velocities much larger than that which characterizes the sublayer at low Reynolds numbers. The "streaky" structure that has been assumed to be characteristic of the sublayer will become less important as the Reynolds number is increased, and it is probable that the "streaks" may not be apparent at all at sufficiently large values.

By trial and error, Walsh (1990) has optimized the dimensions of drag-reducing V-groove riblets ($h^+ = s^+ = 12$, where h and s are the riblet height and spanwise spacing, respectively) at low Reynolds numbers ($Re_\theta < 6 \times 10^3$). The value of s^+ does not scale with the spanwise mean-streak-spacing $\lambda_z^+ \approx 100$. The findings of Walsh, Sellers, and McGinley (1989) and Walsh (1990) that the riblet performance does not change at transonic speeds and high Reynolds numbers ($Ma_\infty = 0.7$ and $20 \times 10^3 \leq Re_\theta \leq 50 \times 10^3$) do not, therefore, invalidate Morrison et al.'s (1971) conclusion that the low-speed streaks will gradually become unimportant at high Reynolds numbers.

Grass (1971) investigated the nature of inner and outer interaction in smooth as well as rough wall-bounded flows. He maintains that the essential features of this interaction do not change despite the presence of three-dimensional roughness elements that protrude as much as 80 wall units into the inner layer—well outside the viscous region. Because low-speed streaks are not observed on walls with three-dimensional roughness, Grass' results minimize the importance of the streaks in the maintenance of turbulence.

A related issue is the importance of the intense but rare bursting events at high Reynolds numbers. A partial answer is given by Kailasnath (1993), who used a statistical approach to obtain useful information on the structure of the instantaneous momentum flux, thus sidestepping analysis conditioned on specific episodes and focusing instead on the contribution to the momentum flux associated with various magnitudes of velocity fluctuations. Kailasnath's nonepisodic approach reveals that the contribution to the flux is dominated by medium amplitude velocity fluctuations in the range of $\pm 1.5u'$, which are not rare events. This implies a diminishing importance of the rare, intense events taking place in a progressively shrinking near-wall region as the Reynolds number increases.

5.8 Flow Control

5.8.1 Introductory Remarks

The ability to actively or passively manipulate a flowfield to effect a desired change is of immense technological importance. The term boundary-layer control includes any mechanism or process through which the boundary layer of a fluid flow is made to behave differently than it normally would were the flow developing naturally along a smooth, flat surface. A boundary layer could be manipulated to achieve transition delay, separation postponement, lift enhancement, drag reduction, turbulence augmentation, or noise suppression. These objectives are not necessarily mutually exclusive. For example, by maintaining as much of a boundary layer in the laminar state as possible, the skin-friction drag and the flow-generated noise are reduced. However, a turbulent boundary layer is in general more resistant to separation than a laminar one. By preventing separation, lift is enhanced, and the form drag is reduced. An ideal method of control that is simple, inexpensive to build and operate, and does not have any trade-off does not exist, and the skilled engineer has to make continuous compromises to achieve a particular goal.

Of all the various types of shear-flow control now extant, control of flow separation is probably the oldest and most economically important. The tremendous increases in the capability of computational fluid dynamics, which have occurred as a direct result of increases in computer storage capacity and speed, are transforming flow separation control from an empirical art to a predictive science. Control techniques such as mitigation of imposed pressure gradients, blowing, and suction are all readily parameterized via viscous CFD. Current inaccuracies in turbulence modeling can severely degrade CFD predictions once separation has occurred; however, the essence of flow separation control is the calculation of attached flows and estimation of separation location (and indeed whether or not separation will occur), which are tasks CFD can in fact perform reasonably well within the uncertainties of the transition location estimation.

The knowledge of the Reynolds number effects is useful to flow control. This is because experimental investigations at low Reynolds numbers (i.e., lower speeds, smaller length scales, or both) are less expensive. Most flow-control devices are, therefore, developed and tested at rather low Reynolds numbers, say $Re_\theta = \mathcal{O}[1000]$. Extrapolation to field conditions is not always straightforward though, and it often comes to grief. Although the riblets results seem to extrapolate favorably to field conditions, the verdict on large-eddy breakup devices is disappointing. These points will be discussed in the following three subsections.

5.8.2 Riblets

Properly optimized, longitudinally grooved surfaces, called riblets, can lead to a modest skin-friction drag reduction, in the range of 5–10%, in turbulent boundary layers. The subject dates back to the mid-1960s but has attracted much attention during the 1970s and 1980s. Walsh (1990) and Pollard (1998) provide up-to-date reviews. The exact mechanism through which riblets achieve net drag reduction despite the substantial increase in wetted surface area is still controversial. For the present purpose, however, the issue of Reynolds number effects on riblets performance is more pertinent. Flight tests and transonic tunnel experiments all indicate that the inner variables are the proper scaling for the dimensions of the riblets, as discussed in Section 5.7.4.

Choi's (1989) spectrum measurements show that the energy of skin-friction fluctuations in the riblet groove drops by a decade compared with that of the smooth surface over more than a decade of the flow-frequency range. Typically, over time expressed in wall units of $\ell^+ = 170$, the skin friction in the groove can remain below average and at a quiescent state as if the fluid in the groove is partially relaminarized. The dye flow visualization of Gallagher and Thomas (1984) also shows that the tracer remained quiescent and viscous-pool-like between the ribs, and it leaves the groove only when a burst passes overhead. Like Gallagher and Thomas, Narasimha and Liepmann (1988) have also suggested that the riblets create pools of slow viscous flow in the valleys and thereby modify the interaction of the wall flow with the outer flow. Black (1968) has analytically described the dynamics of the mean-velocity profile of a canonical turbulent boundary layer in terms of a periodic competition between the wall and outer layers whereby the thickness of the sublayer changes with phase. The maximum thickness of the sublayer that the outer layer will allow can be obtained from the extrapolation of the log-law to the sublayer profile. Consider the equality

$$\overline{U}^+ = y^+ = 2.41 \ln(y^+) + 5.4 \tag{5.35}$$

This is satisfied by $y^+ = 11$, which is nearly the same as the optimized riblet height.

In a recent paper, Choi, Moin, and Kim (1993) used direct numerical simulations to study the turbulent flow over a riblet-mounted surface. Quadrant analysis indicates that drag-reducing riblets mitigate the positive Reynolds-shear-stress-producing events. Choi et al. suggest that riblets with sufficiently small spacing reduce viscous drag by restricting the location of the streamwise vortices above the wetted surface such that only a limited area of the riblets is exposed to the downwash of high-speed fluid induced by these vortices (see also the corroborating numerical results of Kravchenko, Choi, and Moin 1993).

Once it is accepted that the riblet performance is unrelated to streaks, it comes as no surprise that sand-grain roughness also has a drag-reducing behavior exactly like riblets. Tani (1987, 1988) has reanalyzed Nikuradse's (1933) experimental data on sand-grain roughness and has shown that the performance of the roughness also changes from drag reducing to drag increasing with increasing h^+, where h is the characteristic roughness height. The skin friction remains lower than that of the smooth wall for $h^+ < 6$. Compared with the optimized riblets, the drag reduction is lower in magnitude but is still of the same order. Tani has also suggested that the mechanism of drag reduction is likely to originate in the nearly quiescent regions of the flow within the interstices of the roughness elements, as observable deep within riblets.

Grass (1971) has shown in a channel flow that the inrush and outrush phases of the production cycle are also present when the wall has a three-dimensional roughness. Note, however, that walls with three-dimensional roughness elements do not have smooth, wavelike, low-speed streaks, and although the outer-layer structure is similar to that in a smooth wall, the near-wall stress flux has a different behavior (Bandyopadhyay and Watson 1988).

5.8.3 Recovery Response

There are at least two modes of interaction between the inner and outer regions of a boundary layer. In the first, the outer structures obtain at least part of their energy by convection and turbulence transport from the inner region of the upstream part of the boundary layer. This view is supported by the near-constancy of the ratio between turbulence stress and twice the turbulence kinetic energy, $-\overline{uv}/q^2$, across a major portion of the boundary layer. This is true even in the wake region (i.e., $y/\delta \geq 0.2$), where the Reynolds stress and the rms velocity fluctuations are rapidly decreasing. According to Townsend (1976), the turbulent fluid in that region has been sheared sufficiently long to attain its equilibrium structure. The second mode of inner–outer interaction involves the pressure effects of the inactive motion. Compared with the convective mode, the pressure mode is much less extended in the streamwise direction. In other words, the first mode points toward long memory, whereas the second is associated with short memory. This may be relevant to the performance of different control devices, as discussed in the following two subsubsections.

Disturbances in Outer Layer

The long memory associated with the outer structure dependence on upstream conditions contrasts with the short memory of the inner region. This was demonstrated by Clauser (1956), who has shown that in a turbulent boundary layer at a given Reynolds number disturbances survive much longer in the outer layer ($y/\delta > 0.2$) than in the inner layer. He demonstrated this by placing a circular rod in the outer and inner regions of a fully developed wall layer. For the rod placed at $y/\delta = 0.16$, the decay of the maximum deviation of the distorted mean-velocity profile from the equilibrium value was reduced to half of its initial value at a downstream distance of 2δ. In contrast, the outer-layer rod at $y/\delta = 0.6$ caused a distortion in the velocity profile that lasted four times longer, 8δ, and that did not completely disappear even at 16δ downstream of the rod (see Figure 13 of Clauser's article). Note that Clauser (1956) compared the response of the inner and outer layers at a low Reynolds number and did not consider any Reynolds number effects. Incidentally, some consider Clauser's demonstration as the predecessor of the modern-day drag-reducing experiments employing modifications to the outer layer (Bushnell and Hefner 1990).

In viscous drag reduction techniques in which a device drag penalty is involved, as with outer-layer devices (OLD), a recovery length $\mathcal{O}[100\delta]$ is desirable to achieve a net gain. To date, drag reduction has been achieved only at low Reynolds numbers, $Re_\theta < 6 \times 10^3$ (Anders 1990a). However, when Anders examined his outer-layer devices at higher Reynolds numbers, to his surprise, the drag reduction, performance was reduced, and the device was no longer a viable candidate for viscous drag reduction. Anders' experiments in the range of Reynolds numbers of $2500 \leq Re_\theta \leq 18,000$ were conducted by towing a slender, axisymmetric body in a water channel. Two outer-layer devices were used; the first consisted of two NACA-0009 airfoil-section rings placed in tandem 1.5 m downstream of the nose of the 3.7-m-long body. The second device used consisted of two Clark Y low-Reynolds-number airfoil-section rings, again placed in tandem. Both devices were optimized to yield lowest skin friction downstream. Although both devices used by Anders (1990a) consistently led to lower skin friction at all Reynolds numbers tested, net drag will, of course, be increased by the device drag penalty. This penalty depends on, among other factors, the thickness and angle of attack of the device; whether the boundary layer on the device itself is laminar, transitional, or turbulent; and the presence and extent of any separation bubble that might form on the outer-layer ribbon or airfoil. Anders (1990a) reported a very modest net drag reduction for his airfoil devices of around 2% at the lowest Reynolds number but a net drag increase of 1–5% at higher Reynolds numbers.

Bandyopadhyay (1986a) used a large-area drag balance to investigate systematically the Reynolds number effects on both single- and tandem-ribbon devices. His Reynolds number range of $1300 \leq Re_\theta \leq 3600$ is lower than that of Anders (1990a), but the loss-of-performance trends are the same. Note that the drag penalty for the thin-ribbon devices used by Bandyopadhyay should be far smaller than that for the airfoil devices used by Anders. We conclude that for both low Re_θ ($<6 \times 10^3$) and high Re_θ ($>6 \times 10^3$), the effectiveness of OLD diminishes with the increase of Reynolds number.

The continued drop in the skin-friction reduction with Reynolds number comes as a surprise because the mean flow analysis of Coles (1962) indicates an asymptotic state of the outer layer to have been reached above $Re_\theta > 6 \times 10^3$. The slow drop in ΔU^+ does not start until $Re_\theta > 15,000$. Anders (1990a) attributed the irreproducibility of the low-Reynolds-number behavior at higher values to a significant change in the turbulence structure at higher Re_θ, as discussed by Head and Bandyopadhyay (1981). The structural changes as the Reynolds number increases provide a simple explanation for the performance deterioration of outer-layer devices. These devices presumably work by selectively suppressing the normal velocity fluctuations and thus decorrelating the streamwise and normal velocities. As discussed earlier, at high Reynolds numbers, fewer hairpin vortices reach the edge of the boundary layer because of increased interactions among these vortices. The overturning motion of the large eddies observed at low Reynolds numbers is less at higher Reynolds numbers, which reduces the u-turbulence suppression role for the OLD.

Disturbances Close to Wall

It is clear from the preceding subsubsection that knowledge of Reynolds number effects on the mean turbulent flow alone does not allow one to address all practical problems. This can be further demonstrated in the posttransition unexpected result described herein. Klebanoff and Diehl (1952) have made measurements on artificially thickened boundary layers at zero pressure gradient. The first 60 cm of their splitter plate was covered with No. 16 floor sanding paper. The measurements were carried out over a length of 320 cm at freestream velocities of 11, 17, and 33 m/s, giving three ranges of Reynolds numbers and producing a maximum Re_θ of 14,850. The Reynolds number at the end of the sand roughness (at $x = 60$ cm) was $Re_\theta = 2640, 4050$, and 7990 at the three above-mentioned speeds, respectively. The recovery response of the turbulent boundary layer to the same wall disturbance (meaning the same sand roughness) at the three different reference Reynolds numbers (that is freestream speeds) is characterized by plotting ΔU^+ versus Re_θ. One normally expects the recovery from wall disturbances to be the quickest (in x) at the highest Reynolds numbers. Therefore, it comes as a surprise that, to the contrary, the return to the apparent "equilibrium" state (given by the behavior of ΔU^+ in the canonical boundary layer) is clearly slowed down as the reference Reynolds number is increased and not decreased! Even at the last station, where $x = 320$ cm and $\Delta x/\delta = 100$, the recovery was not yet complete. The recovery length for an incoming Re_θ of about 8×10^3 for the near-wall disturbance case is similar to the outer-layer disturbance case observed in typical OLD at a similar Reynolds number. This puzzling behavior leads to the question, Why are the near-wall transition-trip disturbances surviving even beyond an x/δ of 100 at such high Reynolds numbers as 15×10^3 in a way similar to the behavior of outer-layer devices at much lower Reynolds numbers? This question is clearly important to model testing in wind tunnels and code validation data, where roughness is used to trip and thicken the boundary layer to simulate high Reynolds numbers or flight conditions (Bushnell et al. 1993).

5.8.4 Control of High-Reynolds-Number Flows

The preceding discussion indicates that posttransition memory is longer at higher Reynolds numbers for certain trips. Wall-layer control may, therefore, have a long-lasting effect, say $\mathcal{O}[100\delta]$, if applied during transition. On the other hand, as per Clauser (1956), if the wall control is applied in the fully developed turbulent region of the flow, the effect does not last long. The relevance of Reynolds number effects to flow control is particularly telling in the case of full numerical simulation because it is currently limited to Reynolds numbers that are not that far from transitional values.

It is instructive to recall typical Reynolds numbers encountered in the laboratory and in the field. Other than a handful of large-scale facilities, boundary layers generated in wind tunnels and water tunnels typically have Reynolds numbers of $Re_\theta = \mathcal{O}[1000]$. A commercial aircraft traveling at a speed of 300 m/s at an altitude of 10 km would have a unit Reynolds number of $10^7/$m. Owing to the much smaller kinematic viscosity of water, a nuclear submarine moving at a modest speed of 10 m/s ($\approx$20 knots) would have the same unit Reynolds number of $10^7/$m. This unit Reynolds number translates to a momentum thickness Reynolds number near the end of either vehicle of roughly $Re_\theta = 300{,}000$. The Reynolds number on the space shuttle is as high as $Re_\theta = 430{,}000$, an aircraft carrier can have a maximum of $Re_\theta = 1.5 \times 10^6$, and in the atmospheric boundary layer the Reynolds number is typically $Re_\theta = 10^6$–10^7. These ranges of Reynolds numbers together with the scopes of operation of typical wind and water tunnels; direct numerical simulations; and three large-scale facilities, the National Transonic Tunnel, NASA–Langley towing tank, and the super-pipe are schematically shown in Figure 5.1.

It is clear from the discussion thus far in this section and from the strong Reynolds number effects on the mean flow, higher-order statistics and coherent structures demonstrated in Sections 5.5–5.7 that control devices developed and tested in the laboratory cannot in general be readily extrapolated to field conditions. Detailed knowledge of high-Reynolds-number consequences is required prior to attempting to control practical wall-bounded flows.

5.9   Summary of Reynolds Number Effects

The classical similarity theory of wall-bounded flows that asserts a universal description for the near-wall flow is found to be increasingly deficient as the questions become more detailed. We summarize below the Reynolds number effects on the mean motion, higher-order statistics and coherent structures. A few items below were not discussed directly in this chapter but can be found in more detail in the survey by Gad-el-Hak and Bandyopadhyay (1994).

- The widely accepted "asymptotic" state of the wake component is present only in the range of $6 \times 10^3 < Re_\theta < 1.5 \times 10^4$. At higher values, it drops, although at a much slower rate than that in the range of $Re_\theta < 6 \times 10^3$.
- The Clauser's shape parameter C is strongly Reynolds-number-dependent at $Re_\theta < 10^3$ and weakly above that.
- Alternatives to the logarithmic mean-velocity profile have been periodically proposed. Such heretical ideas deserve further scrutiny. Independent confirmation via well-controlled experiments that cover a wide range of Reynolds numbers, resolve the linear region, and directly measure the wall-shear stress is needed.
- The freestream turbulence effect is dependent on Reynolds number.
- Turbulence measurements with probe lengths greater than the viscous sublayer thickness (≈ 5 wall units) appear to be unreliable, particularly near the wall.
- Unlike the mean flow, the statistical turbulence quantities do not scale accurately with the wall-layer variables over the entire inner layer. Such scaling applies over only a very small portion of the inner layer adjacent to the wall.
- At low Reynolds numbers, the peak u-turbulence intensity increases slightly with Reynolds number in both channels and flat plates.
- The distance from the wall where the streamwise turbulence intensity peaks appears to scale with inner variables.
- In contrast, the corresponding distances, expressed in wall units, for both the normal fluctuations and the Reynolds stress move away from the wall as the Reynolds number increases. At high Re, the peak normal turbulence

intensity and the peak Reynolds stress occur substantially outside the viscous region.

- The wall-pressure rms increases slightly with Reynolds number.
- Systematic changes in the mean and higher-order statistics as the Reynolds number varies could be considered as proper first-order trends within the framework of an asymptotic theory. At finite Reynolds numbers, the additive composite expansion formed from the inner and outer expansions of any turbulence quantity provides the only uniformly valid approximation in the matched region.
- In flat plates, trip memory can survive the statistical turbulence quantities at even $Re_\theta > 6 \times 10^3$, where the mean flow is said to have reached an asymptotic state.
- The Reynolds number dependence of the posttransition relaxation length of both the mean and turbulence quantities is not well understood.
- In pipe flows, the wave nature of the viscous sublayer, which is observable at low Reynolds numbers, gives way to a poorly understood random process at high Reynolds numbers.
- Although the variously defined (small) length scales differ greatly from each other at low Reynolds numbers, they all asymptotically approach the mixing length at much higher Reynolds numbers ($Re_\theta > 10^4$).
- The outer-layer structure changes continuously with Reynolds numbers, and very little is known about the structure of very high-Reynolds-number turbulent boundary layers.
- The aspect ratioof the hairpin vortices increases with Reynolds number as they also become skinnier. In a large structure, the number of constituent hairpin vortices per unit wall area increases with Reynolds number.
- Changes in the wall-bounded flow physics could be described as due to changing the scale ratio, δ^+ or a^+, and not the Reynolds number per se. In a given boundary layer, δ^+ changes downstream at a rate slightly lower than Re_θ. The influence of the wall changes from nonlocal to local as this scale ratio increases.
- There is a dire need for high-resolution, reliable measurements of mean and statistical turbulence moments at high Reynolds numbers in smooth, flat-plate turbulent boundary layers.
- Reynolds number effects in canonical flows cannot always be extrapolated to noncanonical cases in a simple, straightforward manner.

CHAPTER SIX

Transition Control

I have no special talents. I am only passionately curious.
(Albert Einstein, 1879–1955)

That is the essence of science: ask an impertinent question, and you are on the way to a pertinent answer.
(Jacob Bronowski, 1908–1974)

PROLOGUE

Delaying laminar-to-turbulence transition of a boundary layer has many obvious advantages. Depending on the Reynolds number, the skin-friction drag in the laminar state can be as much as an order of magnitude less than that in the turbulent condition (Figure 6.1). For an aircraft or an underwater body, the reduced drag means longer range, reduced fuel cost and volume, or increased speed. Flow-induced noise results from the pressure fluctuations in the turbulent boundary layer and, hence, is virtually nonexistent in the laminar case. Reducing the boundary layer noise is crucial to the proper operation of an underwater sonar. On the other hand, turbulence is an efficient mixer, and rates of mass, momentum, and heat transfer are much lower in the laminar state; thus, early transition may be sought in some applications as, for example, when enhanced heat transfer rates are desired in heat exchangers or when rapid mixing is needed in combustors. This chapter focuses on transition delay, particularly for wall-bounded flows. Transition advancement will be discussed in Chapter 11.

6.1 Free-Shear versus Wall-Bounded Flows

Free-shear flows, such as jets, wakes, and mixing layers, are characterized by inflectional mean-velocity profiles and are therefore susceptible to inviscid instabilities. Viscosity is only a damping influence in this case, and the prime instability mechanism is vortical induction. Free-shear flows and separated boundary layers are, therefore, intrinsically unstable, and their critical Reynolds numbers are typically at very low values, $Re_{crit} = \mathcal{O}[10]$. Fiedler (1998) provides a comprehensive review of the control of free-shear flows, including transition delay and advancement. However, technologically important free-shear flows will be basically turbulent, and one's efforts should

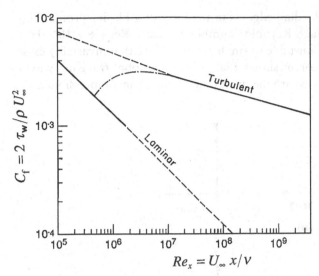

Figure 6.1: Skin-friction coefficient for a smooth flat plate at zero incidence to uniform, incompressible, isothermal flow.

be directed toward controlling the resulting random flow to improve mixing, inhibit shear-layer growth, reduce flow-induced noise, and so forth. We will return to these topics in Chapters 11 and 12.

External and internal wall-bounded flows, such as boundary layers and channel flows, can also have inflectional velocity profiles, but, in the absence of adverse pressure-gradient and similar effects, are characterized by noninflectional profiles, and viscous instabilities are then to be considered. Such viscosity-dominated wall-bounded flows are intrinsically stable and therefore are generally more difficult to control. Their critical Reynolds numbers are typically $Re_\delta|_{crit} = \mathcal{O}[10^3]$, but transition can be delayed by pushing this Re to one or two orders of magnitude higher values using a variety of active and passive control strategies. This is a fruitful approach for devices having moderate Re values such as aircraft wings and torpedoes. But, for the fuselage of a large airplane, the Reynolds number becomes so high downstream that, except for limited regions of the forebody, delaying transition becomes an exercise in futility, and efforts to control the inevitable turbulent flow are then worth pursuing. The present chapter emphasizes transition delay strategies for boundary layers. Methods to advance transition as well as enhance turbulence are discussed in Chapter 11.

6.2 Routes to Transition

The routes to transition are many, but some are more understood than others. Readers who work in the field are surely familiar with the instructive, albeit a bit messy, *Map of the Roads to Wall Turbulence* drawn and redrawn during numerous presentations by Mark Morkovin of the Illinois Institute of Technology. (See Figure 22.11 of the book by Panton 1996 for a tamed reproduction of at least one version of this map.)

On a semi-infinite, smooth flat plate placed in a clean, uniform, incompressible, isothermal flow with as little external disturbances as possible, the laminar flow (somewhat downstream of the leading edge) is in the form of a Blasius profile (Figure 6.2b) and exists at low enough Reynolds number, typically $Re_x < 6 \times 10^4$. This laminar shear layer is, however, unstable to small perturbations that invariably exist in any flow. Squire's (1933) theorem shows that two-dimensional traveling waves (Tollmien–Schlichting [T–S] waves) are the most dangerous for incompressible-flow

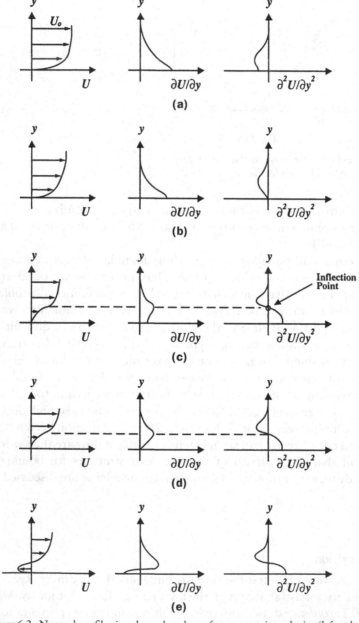

Figure 6.2: Normal profiles in a boundary layer for streamwise velocity (left column), its derivative with respect to y (central column), and its second derivative (right column). The profiles are less full (more inflectional) from top to bottom. (a) Full profile; (b) Blasius profile; (c) Inflectional profile; (d) Profile at the point of steady, two-dimensional separation; (e) Profile with flow reversal.

instability and become unstable when the Reynolds number exceeds a critical value. However, as soon as the T–S waves are amplified, gain a certain amplitude, and nonlinear effects take hold, three-dimensional disturbances can no longer be excluded (Itoh 1987). Following the linear step, the originally two-dimensional waves inherently acquire a nearly periodic spanwise modulation and three-dimensional structures evolve as a result of a secondary instability. Hairpin vortices develop, perhaps due to a tertiary instability, and finally breakdown to turbulence occurs (Klebanoff et al. 1962). Not surprisingly, these nonlinear processes are far less understood than the linear step. The linear amplification step is, however, the slowest of the successive multiple steps in the transition process and, hence, factors that affect the linear amplification determine the magnitude of the transition Reynolds number.

Reshotko (1976, 1985, 1987) asserted that transition is a consequence of the nonlinear response of the laminar shear layer (a very complicated oscillator) to random forcing disturbances that result from freestream turbulence, radiated sound, surface roughness, surface vibrations, or combination of these environmental factors. If the initial disturbances are small, transition Reynolds number depends upon the nature and spectrum of these disturbances, their signature and excitation of the normal modes in the boundary layer (receptivity; see, for example, Morkovin 1969; Goldstein and Hultgren 1989), and the linear amplification of the growing normal modes. Once wave interaction and nonlinear processes set in, transition is quickly completed. If the initial disturbance levels are large enough, the relatively slow linear amplification step mentioned above is bypassed (Morkovin 1984, 1988), and transition can occur at much lower Reynolds numbers. In fact, a sufficiently violent disturbance, $u_{\mathrm{rms}}/U_{\infty} \approx 10\%$, can cause transition of a laminar boundary layer to advance to the position upstream at which perturbations of all wave numbers decay (Klebanoff et al. 1955).

Other routes to transition include the Taylor–Görtler vortices forming on a concave surface as a set of counterrotating, streamwise standing eddies (Stuart 1963; DiPrima and Swinney 1985), the crossflow instabilities occurring on a swept wing or a rotating disk (Reed and Saric 1987, 1989), and any other situation in which an instability will occur as a result of disturbing the equilibrium of body forces, inertia, and surface forces (Drazin and Reid 1981). Once again, these instabilities will generally occur at Reynolds numbers lower than the critical Reynolds number for the growth of Tollmien–Schlichting waves.

To delay transition to as far a downstream position as possible, the following steps may be taken. First, because factors that affect the linear amplification of Tollmien–Schlichting waves determine the magnitude of the transition Reynolds number, these waves may be either inhibited or canceled. In the former method of control, the growth of the linear disturbance is minimized by using any, or a combination of, the so-called stability modifiers that alter the shape of the velocity profile. These include increased length of favorable pressure gradient, wall transpiration, wall motion, surface heating and cooling, wall curvature, and body forces. Wave cancellation of the growing perturbation is accomplished through exploiting, but not altering, the stability characteristics of the flow. Secondly, the forcing disturbances in the environment in which the laminar shear layer develops may be reduced. This is accomplished by using smooth surfaces, reducing the freestream turbulence and the radiated sound, minimizing body vibration, and ensuring a particulate-free incoming flow or, in case of a contaminated environment such as the ocean, using a particle-defense mechanism.

Practically achieved surface smoothness and levels of radiated noise place an upper limit for the unit Reynolds number required for a successful laminar flow-control system. For aircraft, this typically translates into a requirement for high-altitude operation (above 10 km). Thirdly, one may provide a flow where other kinds of instabilities such as Taylor–Görtler vortices or crossflow instabilities, will not occur or at least will not grow at a rapid rate. This is done by avoiding as much as possible concave surfaces or concave streamlines, minimizing the sweep on lifting surfaces, and so forth.

In the next section, the elements of linear stability theory are briefly recalled, again for the benefits of newcomers to the field. This is followed by a discussion of the first of the three steps outlined above, stability modifiers and wave cancellation methods. Most of the arguments presented here are for incompressible flows. A brief comparison between compressible and incompressible shear flows is given in Section 3.4.

6.3 Linear Stability Theory

The linear stability of a laminar flow is governed by the Orr–Sommerfeld equation and appropriate boundary conditions. For an isothermal, incompressible flow the Orr–Sommerfeld equation reads

$$(U - c)(\phi'' - \alpha^2 \phi) - U''\phi = \left(\frac{-i}{\alpha\,Re}\right)(\phi'''' - 2\alpha^2\phi'' + \alpha^4\phi) \tag{6.1}$$

This fourth-order, linear, ordinary differential equation is derived from the Navier–Stokes equation (2.21) of Chapter 2 by assuming a two-dimensional small disturbance superimposed upon a steady, unidirectional mean flow, $U(y)$. The stream function for the perturbation is given by

$$\psi(x, y, t) = \phi(y)e^{i\alpha(x-ct)} \tag{6.2}$$

In Eqs. (6.1) and (6.2), c is the wave speed, ϕ is the amplitude function of the fluctuations, the superscript $'$ denotes derivative with respect to y, α is the wave number, i is the imaginary number ($i = \sqrt{-1}$), and Re is the Reynolds number based upon the freestream velocity and the boundary layer thickness. The Orr–Sommerfeld equation has been nondimensionalized using the boundary layer thickness as a length scale and the freestream velocity as a velocity scale. Though linear, this equation is notoriously difficult to solve analytically.

The order of the Orr–Sommerfeld equation increases as forces additional to inertia and viscous ones are incorporated into the momentum balance. For example, in case of a rotating-disk flow, Coriolis and streamline-curvature terms are needed, leading to a sixth-order stability equation.

In Eq. (6.1), either the temporal or the spatial growth of instability waves is considered as an eigenvalue problem. In the former case, a disturbance oscillates in space but either grows or decays exponentially with time. A complex eigenvalue $c = c_r + ic_i$ is determined for each pair of values of the real parameter α and the Reynolds number. The real part of c is the phase velocity of the prescribed disturbance, and the sign of the imaginary part determines whether the wave is temporary amplified ($c_i > 0$), temporary damped ($c_i < 0$), or neutrally stable ($c_i = 0$).

The more realistic spatial stability problem involves disturbances that oscillate in time but either grow or decay exponentially with downstream distance. In this case, a complex eigenvalue $\alpha = \alpha_r + i\alpha_i$ is determined for each pair of values of the real radian frequency $\omega = 2\pi f = \alpha c$ and the Reynolds number. The real part of α is the wave number, and the sign of α_i determines whether the wave is spatially amplified ($\alpha_i < 0$) or spatially damped ($\alpha_i > 0$).

In either the temporal or spatial instability studies, the major difficulty in numerically integrating the Orr–Sommerfeld equation lies in its being highly stiff and unstable (Lin 1945), which makes the use of conventional numerical schemes virtually impossible. Explicit numerical methods with a step size commensurate with the global behavior of the solution cannot be used to integrate Eq. (6.1) because of the numerical instabilities that characterize this ordinary differential equation. The papers by Orszag (1971) and Scott and Watts (1977) give examples of special codes to handle stiff eigenvalue problems.

In deriving Eq. (6.1), the fluid viscosity in Eq. (2.21) is assumed constant. If viscosity varies in space as a result of, for example, surface heating or cooling, additional terms containing the first and second derivatives of viscosity with respect to y result, and one obtains the so-called modified Orr–Sommerfeld equation (Wazzan, Okamura, and Smith 1968)

$$(U - c)(\phi'' - \alpha^2\phi) - U''\phi = \left(\frac{-i}{\alpha\,Re}\right)[\mu(\phi'''' - 2\alpha^2\phi'' + \alpha^4\phi)$$
$$+ 2\mu'(\phi''' - \alpha^2\phi') + \mu''(\phi'' + \alpha^2\phi)] \qquad (6.3)$$

where $\mu(y)$ is the fluid viscosity nondimensionalized using the freestream value, and Re is defined using the freestream viscosity. This is the governing equation for studying the stability of, for example, wall-bounded flows with surface heating or cooling.

6.3.1 Critical *Re* and Transition

The Reynolds number below which (linear) perturbations of all wave numbers decay is termed the critical Reynolds number, Re_{crit}, or the limit of stability. For a given velocity profile $U(y)$, the critical Reynolds number and the rate of growth of perturbations depend strongly on the shape of the velocity profile. A profile with an inflectional point ($\partial^2 U/\partial y^2 = 0$) above the wall provides a necessary and sufficient condition for inviscid instability (Rayleigh 1880; Tollmien 1935). Such profiles must have a positive curvature at $y = 0$ because $\partial^2 U/\partial y^2$ is negative at a large distance from the wall (Figure 6.2). Even when viscous effects are included (right-hand side of the Orr–Sommerfeld equation $\neq 0$), a velocity profile becomes more stable as its second derivative near the wall becomes negative, $[\partial^2 U/\partial y^2]_o < 0$. The profile is then said to be more full (Figure 6.2a), having a smaller ratio of displacement thickness to momentum thickness than, for example, an inflectional velocity profile (Figure 6.2c–e). In the former case, the critical Reynolds number is increased, the range of amplified frequencies is diminished, and the amplification rate of unstable waves is reduced.

In a spatially growing boundary layer, transition to turbulence involves the integrated effect of the growth of unstable waves as they are convected downstream by the mean flow. If $|A|$ represents the amplitude of a wave, then its local spatial growth is given by

$$\frac{d|A|}{dx} = -\alpha_i |A| \tag{6.4}$$

The net growth of a wave with a given real frequency, from the location x_0, where it first becomes unstable, to the location x_{oo}, where transition to turbulence occurs, is given by

$$\ln\left(\frac{|A|}{|A_0|}\right) = -\int_{x_0}^{x_{oo}} \alpha_i dx = n \tag{6.5}$$

Empirically, a modal amplification n of between 8 and 11 is found at transition to turbulence (e.g., Smith and Gamberoni 1956; Van Ingen 1956; Jaffe, Okamura, and Smith 1970).

Despite the apparent success of the amplitude-ratio method to predict transition, Reshotko (1976) maintained that the method is defective in principle and perhaps also in practice. Transition location depends strongly on the freestream turbulence levels and other environmental factors. Moreover, nonlinear effects that are physically significant in the transition process cannot be accounted for by procedures based on linear stability theory. In general, the e^n-method works fairly well for situations in which the freestream turbulence level is below 1% (e.g., cruising aircraft), but the strategy fails for noisy environments (e.g., turbine blades). Transition processes dominated by T–S instabilities can accurately be predicted using the e^n-method, whereas transition caused by crossflow instabilities and the like cannot be forecast. Notwithstanding these drawbacks, the e^n-rule provides a practical scheme for estimating the effectiveness of a transition-delay method of control (Bushnell and McGinley 1989; Bushnell 1989).

6.4 Stability Modifiers

Stability modifiers are those methods of laminar flow control that alter the shape of the velocity profile to minimize the linear growth of unstable waves. For a boundary layer flow along a nonmoving wall[1] of small curvature, the three momentum equations at the surface were listed in Chapter 2. Herein, we allow both streamwise and normal motions of the wall, and the instantaneous streamwise momentum equation at $y = 0$ reads

$$\rho_w \frac{\partial u}{\partial t}\bigg|_{y=0} + \rho_w u_w \frac{\partial u}{\partial x}\bigg|_{y=0} + \rho_w v_w \frac{\partial u}{\partial y}\bigg|_{y=0} + \frac{\partial p}{\partial x}\bigg|_{y=0}$$

$$- \frac{\partial \mu}{\partial y}\bigg|_{y=0} \frac{\partial u}{\partial y}\bigg|_{y=0} = \mu \frac{\partial^2 u}{\partial y^2}\bigg|_{y=0} \tag{6.6}$$

Equation (6.6) is valid for a fluid with variable properties ρ and μ. For a two-dimensional laminar boundary layer, vorticity is only in the spanwise direction and is given by $\omega_z = -\partial u/\partial y$. The right-hand side of Eq. (6.6), therefore, represents the

---

[1] Wall was not allowed to move in the streamwise or spanwise directions. Normal wall motions are analogous to wall transpiration, v_w, and were, therefore, included in Eqs. (2.39)–(2.41).

flux of vorticity at the surface, as demonstrated by Lighthill (1963a). Any of the terms on the left-hand side of Eq. (6.6) can affect the sign of the second derivative of the velocity profile (or the direction of the vorticity flux) at the wall and, hence, both the flow stability and its resistance to separation. Stability modifiers do just that and include wall motion such that the first, second, or third term on the left-hand side of Eq. (6.6) is negative, wall suction ($v_w < 0$), favorable pressure gradient ($\partial p/\partial x < 0$), or lower wall viscosity ($\partial \mu/\partial y > 0$). Any one or a combination of these methods will cause the curvature of the velocity profile at the wall to become more negative and, hence, increase the lower critical Reynolds number and reduce the spatial or the temporal amplification rates of unstable waves. Curvature effects and body forces were not included in Eq. (6.6) but can also be very effective in advancing or delaying transition. A discussion of Lorentz forces[2] is deferred to Chapter 14.

For historical reasons, boundary layers, which are stabilized by extending the region of favorable pressure gradient, are known as natural laminar flow (NLF), whereas other methods to modify the stability of a shear flow are termed laminar flow control (LFC). It is clear from Eq. (6.6) that the effects of all those strategies are additive. The term hybrid laminar flow control normally refers to the combination of NLF and one of the LFC techniques. The recent review article by Joslin (1998) discusses NLF and LFC specifically as applied to aircraft.

The most common method to lower the viscosity near the surface is to heat the wall for liquids ($d\mu/dT < 0$) or to cool it for gases ($d\mu/dT > 0$). Other strategies to effect a near-wall viscosity gradient include surface-film boiling, cavitation, sublimation, chemical reaction, wall injection of lower or higher viscosity fluid, and the presence of shear thinning or thickening additive. The following four subsections describe in turn the four stability modifiers to delay the transition of a boundary layer: wall motion, suction, shaping, and wall heating or cooling.

6.4.1 Wall Motion

Wall motion can be generated by either actively driving the surface or by using a flexible coating whose modulus of rigidity is low enough so that surface waves are generated under the influence of the stress field in the fluid. In the former case, the wall motion is precisely controlled and can be made to affect the shape of the velocity profile in a desired manner. For example, a transverse or longitudinal wave can be forced to propagate on the surface of the solid and thus modulate the instantaneous values of the first three terms on the left-hand side of Eq. (6.6). This method of control is, however, quite impractical and is used mainly to provide controlled experiments to determine what type of wave motion is required to achieve a given result. The more practical passive flexible walls can be broadly classified into two categories: truly compliant coatings that have extremely small damping and modulus of elasticity and can therefore respond with little phase lag to the boundary layer flow, and the more readily available resonant walls in which vibration modes in the solid, acting as an active vibration damper, are excited by the flow-disturbance forcing function (Bushnell, Hefner, and Ash 1977). The flow stabilization in this case may be a result of altering the phase relation between the instantaneous streamwise and normal velocity components (i.e., the Reynolds stress $-\overline{uv}$) in the viscous region rather than changing the curvature of the mean velocity profile at the wall.

---

[2] Body forces that result when a conducting fluid moves in the presence of a magnetic field.

Theoretical studies of boundary-layer stability in the presence of a flexible wall started in the early 1960s, stimulated in part by the pioneering experimental work of Kramer (1960a). In both the classical work (e.g., Benjamin 1960a, Landahl 1962; Kaplan 1964) and the more recent research (e.g., Carpenter and Garrad 1985; Willis 1986; Yeo and Dowling 1987), steady-state stability theory has been used with an assumed velocity profile that is not allowed to change as the wall moves. It is clear, however, that the wall displacement will induce a traveling pressure signal that, in turn, will modulate the mean velocity profile. Either a quasi-steady stability analysis of the modulated flow or true time-dependent calculations would be preferable to conventional linear-stability theory, but neither has been attempted yet owing to the obvious complexities of the problem. Notwithstanding the shortcoming of present stability calculations, Willis (1986) obtained a very impressive agreement between linear theory and a carefully conducted experiment. Eigenvalue calculations were performed to predict the amplification factors for a range of modal frequencies. The flexible coating was a silicon-rubber and silicon-oil mix covered by a thin latex rubber skin stretched across the surface. The experiments were conducted in a water towing tank using a flat plate, and controlled, harmonic, two-dimensional disturbances were introduced upstream of the compliant surface. A reduction of the wave growth by an order of magnitude is feasible when this rather simple coating is used, almost eliminating transition due to Tollmien–Schlichting type of instability. The flexible wall itself is, however, susceptible to other kinds of instability, and care must be taken to ensure that these surface waves will not grow to an amplitude that will promote transition through a roughness-like effect (see the review article by Riley, Gad-el-Hak, and Metcalfe 1988).

Although the original Kramer's (1960a) experiments were discredited up until a few years ago, more recent theoretical and experimental evidence confirms Kramer's results (Carpenter and Garrad 1985). Passive flexible coatings with density the order of the fluid density appear to be capable of considerable transition postponement. All of Chapter 7 is devoted to this simple control method that does not require energy expenditure, slots, ducts, or internal equipment of any kind.

6.4.2 Suction

A second method for postponing transition is the application of wall suction. As seen from Eq. (6.6), small amounts of fluid withdrawn from the near-wall region of the boundary layer change the curvature of the velocity profile at the wall and can dramatically alter the stability characteristics of the boundary layer. Additionally, suction inhibits the growth of the boundary layer, and thus the critical Reynolds number based on thickness may never be reached.

Although laminar flow can be maintained to extremely high Reynolds numbers provided that enough fluid is sucked away, the goal is to accomplish transition delay with the minimum suction flow rate. Not only will this reduce the power necessary to drive the suction pump but also the momentum loss due to suction; hence, the skin friction, is minimized. This latter point can easily be seen from the momentum integral equation. If one rewrites the Kármán integral equation (2.44) for a steady, incompressible flow ($\rho = \rho_0 = $ constant) over a flat plate $(dU_0/dx = 0; dR/dx = 0)$ with uniform suction through the wall (v_w negative), the equation reads

$$\frac{C_f}{2} = \frac{d\delta_\theta}{dx} + \frac{|v_w|}{U_0} \tag{6.7}$$

The second term on the right-hand side is the suction coefficient, C_q, and although withdrawing the fluid through the wall leads to a decrease in the rate of growth of the momentum thickness, C_f increases directly with C_q. Fluid withdrawn through the wall has to come from outside the boundary layer where the streamwise momentum per unit mass is at the relatively high level of U_∞. The second term is proportional to the rate of momentum loss due to withdrawing a mass per unit time and area of $\rho |v_w|$. Note that this term does not exist for pipe flows because of the mass-flow constraint. Hence, this momentum penalty is not paid for channel flows with wall transpiration, which is an important distinction between internal and external flows.

Although Prandtl (1904) used suction to prevent flow separation from the surface of a cylinder near the beginning of this century, the first experimental demonstration that boundary-layer transition can be delayed by withdrawing near-wall fluid did not take place until about four decades later. Holstein (1940); Ackeret, Ras, and Pfenninger (1941); Ras and Ackeret (1941); and Pfenninger (1946) used carefully shaped, single- and multiple-suction slits to demonstrate the decrease in drag associated with delaying transition. Braslow et al. (1951) used continuous suction through a porous wall to maintain laminar flow on an airfoil to a chord Reynolds number of 2.0×10^7. Raspet (1952) conducted unique, noise-free experiments that confirmed the large decrease in drag when suction is applied through the wings of a sailplane. In the early 1960s, test flights of two X–21 aircraft (modified U.S. Air Force B–66) indicated the feasibility of maintaining a laminar flow on a swept wing to chord Reynolds numbers as high as 4.7×10^7 (Whites, Sudderth, and Wheldon 1966). The wing surfaces contained many thin and closely spaced spanwise suction slots, and the total airplane drag was reduced by 20% as compared with the no suction case.

Although discrete suction slots were used first (because of the unavailability of suitable porous surfaces in the early 1940s), the theoretical treatment of the problem is considerably simplified by assuming continuous suction through a porous wall where the characteristic pore size is much smaller than a boundary layer thickness. In fact, the case of a uniform suction from a flat plate at zero incidence is an exact solution of the incompressible Navier–Stokes equation (2.21). If one assumes suction weak enough that the potential flow outside the boundary layer is unaffected by the loss of mass at the wall (sink effects), the asymptotic velocity profile in the viscous region is exponential and has a negative curvature at the wall

$$U(y) = U_\infty \left[1 - \exp\left(-\frac{|v_w| y}{\nu} \right) \right] \tag{6.8}$$

The displacement thickness has the constant value $\delta^* = \nu/|v_w|$, where ν is the kinematic viscosity and $|v_w|$ is the absolute value of the normal velocity at the wall. In this case, Eq. (6.7) reads

$$C_f = 2C_q \tag{6.9}$$

Bussmann and Münz (1942) computed the critical Reynolds number for the preceding asymptotic velocity profile to be $Re_{\delta^*}|_{\text{crit}} \equiv U_\infty \delta^*/\nu = 70,000$. From the value of δ^* given above, the flow is stable to all small disturbances if $C_q \equiv |v_w|/U_\infty > 1.4 \times 10^{-5}$. The amplification rate of unstable disturbances for the asymptotic profile is an order of magnitude less than that for the Blasius boundary layer (Pretsch 1942). This treatment ignores the development distance from the leading edge needed to reach the asymptotic state. When this is included in the computation, a higher $C_q(1.18 \times 10^{-4})$ is required to ensure stability (Iglisch 1944; Ulrich 1944). Wuest (1961) presented a summary of transpiration boundary layer computations up to the early 1960s.

The more complicated analysis for the stability of a boundary layer with suction through discrete spanwise strips was only carried out satisfactorily relatively recently. Reed and Nayfeh (1986) conducted a numerical-perturbation analysis of a linearized, triple-deck, closed-form basic state of a flat plate boundary layer with suction through a finite number of spanwise porous strips. Their results were compared with interacting boundary-layer calculations (Ragab and Nayfeh 1980) as well as with the carefully conducted experiments of Reynolds and Saric (1986). Suction applied through discrete strips can be as effective as suction applied continuously over a much longer streamwise length. Reed and Nayfeh (1986) suggested a scheme for optimizing the strip configuration. Their results showed that suction should be concentrated nearer the leading edge (branch I of the neutral stability curve) when disturbances are still small in amplitude.

Suction may be applied through porous surfaces, perforated plates, or carefully machined slots. It is of course structurally impossible to make the whole surface of an aircraft wing or the like out of porous material, and often strips of sintered bronze or steel are used. A relatively inexpensive woven stainless steel, Dynapore, is now available and provides some structural support (Reynolds and Saric 1986). Superior surface smoothness and rigidity are obtained by drilling microholes in titanium using the recently developed electron-beam technology. The lower requirement for a pressure drop in the case of a perforated plate translates directly into pumping-power saving. However, outflow problems may result from regions of the wing having strong pressure gradients (Saric and Reed 1986). Outflow in the aft region of a suction strip can cause large destabilizing effects and local three-dimensionality.

Although structurally a surface with multiple slits is more rigid than a porous surface, slots are more expensive to fabricate accurately. Moreover, the higher mass-flow rates associated with them may result in high-Reynolds-number instabilities such as separation and backflow, which adversely affect the stability of the basic flow. The rule of thumb is that the Reynolds number, based on slot width (or hole diameter in the case of a perforated plate) and the local suction velocity, should be kept below 10 to avoid adverse effects on the boundary-layer stability. Nevertheless, Saric and Reed (1986) claimed a hole Reynolds number an order of magnitude higher than that without destabilization of the basic flow.

Delaying transition using suction is a mature technology, in which most of the remaining problems are in the maintainability and reliability of suction surfaces and the optimization of suction rate and distribution. To protect the delicate suction surfaces on the wing of an aircraft from insect impacts and ice formation at low altitudes, special leading edge systems are used (Wagner and Fischer 1984; Wagner et al. 1984, 1988, 1990). Suction is less suited for underwater vehicles because of the abundance of suspended ocean particulate that can clog the suction surface as well as destabilize the boundary layer.

6.4.3 Shaping

The third method of control to delay laminar-to-turbulence transition is perhaps the simplest and involves the use of suitably shaped bodies to manipulate the pressure distribution. In Eq. (6.6), the pressure gradient term can affect the sign of the curvature of the velocity profile at the wall and, hence, change the stability characteristics of the boundary layer. According to the calculations of Schlichting and Ulrich (1940), the critical Reynolds number based on displacement thickness and freestream velocity changes from about 100 to 10,000 as a suitably nondimensionalized pressure gradient varies from $\Lambda = -6$ (adverse) to $\Lambda = +6$ (favorable). The Pohlhausen parameter, Λ, is the ratio of pressure forces to viscous forces and is given by

$$\Lambda \equiv -\frac{dp}{dx} \cdot \frac{\delta^2}{\mu U_o} = \frac{\delta^2}{v} \cdot \frac{dU_o}{dx} \tag{6.10}$$

where δ is the boundary layer thickness, U_o is the velocity outside the boundary layer, v is the kinematic viscosity, μ is the dynamic viscosity, and dp/dx is the pressure gradient.

For the case of a favorable pressure gradient, no unstable waves exist at infinite Reynolds number. In contrast, the upper branch of the neutral stability curve in the case of an adverse pressure distribution tends to a nonzero asymptote so that a finite region of wavelengths at which disturbances are always amplified remains even as $Re \rightarrow \infty$.

Streamlining a body to prevent separation and reduce form drag is quite an old art going back to prehistoric times, as indicated in Chapter 1, but the stabilization of a boundary layer by pushing the longitudinal location of the pressure minimum as far back as possible dates to the 1930s and led to the successful development of the NACA 6-Series NLF airfoils. Newer, low-Reynolds-number lifting surfaces used in sailplanes, low-speed drones, and executive business jets have their maximum thickness point far aft of the leading edge. The recent success of the *Voyager's* 9-day, unrefueled flight around the world was due in part to a wing design employing natural laminar flow to approximately 50% chord. Application of NLF technology to underwater vehicles is feasible but somewhat more limited (Granville 1979).

For a lifting surface, the favorable pressure gradient extends to the longitudinal location of the pressure minimum. Beyond this point, the adverse pressure gradient becomes steeper and steeper as the peak suction is moved farther aft. For an airfoil, the desired shift in the point of minimum pressure can only be attained in a certain narrow range of angles of incidence. Depending on the shape, angle of attack, Reynolds number, surface roughness, and other factors, the boundary layer either becomes turbulent shortly after the point of minimum pressure or separates first and then undergoes transition. One of the design goals of NLF is to maintain attached flow in the adverse pressure gradient region, and some method of separation control (Chapter 8) may have to be used there.

Factors that limit the utility of NLF include crossflow instabilities and leading edge contamination on swept wings, insect and other particulate debris, high unit Reynolds numbers at lower cruise altitudes, and performance degradation at higher angles of attack due to the necessarily small leading edge radius of NLF airfoils. Reductions of surface waviness and smoothness of modern production wings, special leading edge systems to prevent insect impacts and ice formation, higher cruise altitudes of newer airplanes, and higher Mach numbers all favor the application of NLF (Runyan and Steers 1980). To paraphrase an older statement by Holmes (1988), an NLF airfoil is no longer as finicky as Morris the cat. It is true that a boundary layer that is kept laminar to extremely high Reynolds numbers is very sensitive to environmental factors such as roughness, freestream turbulence, radiated sound, and so forth. However, the flow is durable and reliable within certain conservative design corridors that must be maintained by the skillful designer and eventual operator of the vehicle. Current research concentrates on understanding the achievability and maintainability of natural laminar flow, expanding the practical applications of NLF technology, and extending the design methodology to supersonic aviation (Bushnell and Malik 1988; Bushnell 1989; Joslin 1998; Becker, Durst, and Lienhart 1999).

6.4.4   Wall Heating and Cooling

The last of the stability modifiers to be considered in this chapter is the addition or removal of heat from a surface, which causes the viscosity to vary with distance from the wall. In general, viscosity increases with temperature for gases, whereas the opposite is true for liquids. Thus, if heat is removed from the surface of a body moving in air, the fifth term on the left-hand side of Eq. (6.6) is negative. In that case, the velocity gradient near the wall increases and the velocity profile becomes fuller and more stable. The term containing the viscosity derivative will also be negative if the surface of a body moving in water is heated. With heating in water or cooling in air, the critical Reynolds number is increased, the range of amplified frequencies is diminished, and the amplification rate of unstable waves is reduced. Substantial delay of transition is feasible with a surface that is only a few degrees hotter (in water) or colder (in air) than the freestream. Note that for gases in particular, surface cooling affects both viscosity and density. The increased near-wall density has additional beneficial effects via the stable buoyancy (for certain orientations) and enhanced near-wall momentum it induces.

The first indirect evidence of this phenomenon was the observation that the drag of a flat plate placed in a wind tunnel increases by a large amount when the plate is heated (Linke 1942). Both Frick and McCullough (1942) and Liepmann and Fila (1947) showed that the transition location of a flat-plate boundary layer in air at low subsonic speeds is moved forward as a result of surface heating. The stability calculations of Lees (1947) confirmed these experiments and, moreover, showed that cooling has the expected opposite effects. The critical Reynolds number based on distance from the leading edge increases from 10^5 to 10^7 when the wall of a flat plate placed in an air stream is cooled to 70% of the absolute ambient temperature. Even a modest cooling of the wall to $0.95T_\infty$ results in doubling of the critical Reynolds number (Kachanov, Koslov, and Levchenko 1974).

With cooling, the range of amplified frequencies is diminished and the growth rate of T–S waves is reduced, resulting in a substantial increase in transition Reynolds number. These same trends were dramatically confirmed in subsonic and supersonic flights[3] by Dougherty and Fisher (1980), who studied the transition on an airborne cone over the Mach number range of 0.55–2.0. They reported a transition Reynolds number that varied approximately as T_w^{-7}, where T_w is the wall temperature. For aircraft, this method of transition delay is feasible only for a vehicle that uses a cryofuel such as liquid hydrogen or liquid methane. In that case, a sizable heat sink is readily available, the idea being that the fuel is used to cool the major aerodynamic surfaces of the aircraft as it flows from the fuel tanks to the engines. Reshotko

---

[3] In hypersonic flows, a different mode of stability, Mack's second mode, dominates the transition process and cooling and in fact promotes earlier transition to turbulence.

(1979) examined the prospects for the method and concluded that, particularly for a hydrogen-fueled aircraft, substantial drag reductions are feasible. His engineering calculations indicated that the weight of the fuel saved is well in excess of the weight of the required cooling system.

The preceding effects are more pronounced in water flows owing to the larger Prandtl number (good thermal coupling) and the stronger dependence of viscosity on temperature.[4] In a typical low-speed situation, a surface heating of 1°C in water has approximately the same effect on the curvature of the velocity profile at the wall as a surface cooling of 20°C in air or a suction coefficient (in air or water) of 0.0003 (Liepmann, Brown, and Nosenchuck 1982). Wazzan et al. (1968, 1970) used the modified fourth-order Orr–Sommerfeld equation (6.3) combined with the e^9-method of Smith and Gamberoni (1956) and confirmed that wall heating can produce large increases in the transition Reynolds number of water-boundary layers. They predicted a transition Reynolds number, based on freestream velocity and distance from the leading edge of a flat plate, as high as 2×10^8 for wall temperatures that are only 40°C above the ambient water temperature.

Lowell and Reshotko (1974) refined Wazzan et al.'s (1968, 1970) calculations by introducing a coupled sixth-order system of vorticity and energy disturbance equations. The predicted critical Reynolds numbers for wall overheats of up to 2.8°C were confirmed by the experiments of Strazisar, Reshotko, and Prahl (1977), who measured the growth rates of small disturbances generated by a vibrating ribbon in a heated flat plate in water. These experiments did not yield data on transition or on stability at higher overheats.

The transition predictions of Wazzan et al. (1968, 1970) at higher overheats were partially confirmed by the very carefully conducted experiments of Barker and Gile (1981), who used the entrance region of an electrically heated pipe. The displacement thickness was much smaller than the pipe radius and, thus, the boundary-layer development was approximately the same as that of a zero-pressure gradient flat plate. Barker and Gile reported a transition Reynolds number of 4.7×10^7 for a wall overheat of 8°C. No further increase in $(Re|_{\text{transition}})$ was observed as the wall was heated further, which is in contradiction to the computations of Wazzan et al. (1968, 1970). Barker and Gile (1981) investigated possible causes of this discrepancy, including buoyancy effects, wall roughness, effects of geometry, flow asymmetries, and suspended particulate matter. Their analysis and numerous related work (e.g., Kosecoff, Ko, and Merkle 1976; Chen, Goland, and Reshotko 1979; Hendricks and Ladd 1983; Lauchle and Gurney 1984) indicate that increased concentration and size of suspended particulate diminish the stabilizing effect of surface heating until at some point surface heating no longer stabilizes the boundary layer but is in fact a destabilizing influence.

On a heated body of revolution in a high-speed water tunnel, Lauchle and Gurney (1984) observed an increase in transition Reynolds number from 4.5×10^6 to 3.6×10^7 for an average overheat of 25°C. Clearly, surface heating in water can be an extremely effective method of transition delay, which accounts for drag reduction for small, high-speed underwater vehicles in which the rejected heat from their propulsion

---

[4] For water at room temperature, $Pr \approx 7$, and absolute viscosity is decreased by approximately 2% for each 1°C rise in temperature. For room temperature air, $Pr \approx 0.7$, and absolute viscosity is decreased by approximately 0.2% for each 1°C drop in temperature.

system can be used to increase the surface temperature along the body length. The detrimental effects of freestream particulate alluded to earlier are, however, a major obstacle at present for a practical implementation of this method of control. Suspended particulate having wide-band concentration spectra are abundant in the oceans, and "particle-defense" mechanisms must be sought before using any of the transition-delay methods in a contaminated environment.

In addition to surface heating, several other techniques are available to lower the near-wall viscosity ($\partial\mu/\partial y > 0$) in a liquid boundary layer and thus favorably affect the stability of the flow. These include, as mentioned earlier, film boiling, cavitation, sublimation, chemical reaction, or wall injection of a gas or lower-viscosity liquid. Finally, a shear-thinning additive can be introduced into the boundary layer. Because the shear increases as the wall is approached, the effective viscosity of the non-Newtonian fluid decreases there, and ($\partial\mu/\partial y$) becomes positive.

6.5   Wave Cancellation

An alternative approach to the four stability modifiers used for increasing the transition Reynolds number of a laminar boundary layer is wave cancellation. If the frequency, orientation, and phase angle of the dominant element of the spectrum of growing linear disturbances in the boundary layer are detected, a control system and appropriately located disturbance generators may then be used to effect a desired cancellation or suppression of the detected disturbances. In this case, the stability characteristics of the boundary layer are exploited but not altered (Reshotko 1985). Wave cancellation is feasible only when the disturbances are still relatively small, their growth is governed by a linear equation, and the principle of superposition is still valid.

The first reported use of wave cancellation is that due to Schilz (1965, 1966). He used a vibrating ribbon to excite a T–S wave on a test plate that had a flexible surface. A unique wall-motion device flush mounted into the plate moved the flexible wall in a transverse, wavelike manner with a variety of frequencies and phase speeds. A significant amount of cancellation resulted when the flexible wall motion had the opposite phase but the same frequency and phase speed as the T–S wave. Both Milling (1981) and Thomas (1983) used two vibrating wires, one downstream of the other, to generate and later cancel a single frequency T–S wave. Thomas (1983) observed that interaction between the primary disturbance and background excitations prevented complete cancellation of the primary wave. To further study the consequences of wave interactions, Thomas applied the same method of control to eliminate two interacting waves of different frequency. Although the primary waves were behaving linearly, a nonlinear interaction gave rise to a low-amplitude difference frequency that could only be partially reduced and ultimately led to transition. Thomas (1983) concluded that it is not possible to return the flow completely to its undisturbed base state because of wave interactions and that it is perhaps more appropriate to describe this control method as wave superposition rather than wave cancellation.

The same principle of wave superposition can be applied using wall heating and cooling (Liepmann et al. 1982; Liepmann and Nosenchuck 1982; Ladd and Hendricks 1988), plate vibration (Gedney 1983), compliant wall (McMurray, Metcalfe, and Riley 1983), or periodic suction and blowing (Biringen 1984).

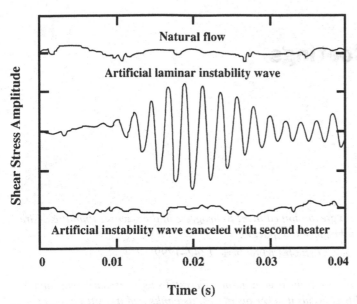

Figure 6.3: Cancellation of artificial instability waves using adaptive heat [after Ladd and Hendricks 1988].

Liepmann and Nosenchuck (1982) used flush-mounted hot-film probes to sense natural T–S waves in a flat-plate boundary layer in a water tunnel. A feed-forward control loop was then used to synthesize and introduce disturbances of equal amplitude but of opposite phase via flush-mounted wall heaters. Ladd and Hendricks (1988) performed their experiment in a water tunnel on a 9:1 fineness-ratio ellipsoid. Strip heaters were again used to create and actively attenuate T–S waves. Ladd and Hendricks applied digital filtering techniques to synthesize the attenuation signal. The filter was able to adapt the attenuation signal actively to changes in amplitude and frequency of the artificially introduced instability wave with no loss in attenuation downstream. A sample of Ladd and Hendrick's results is depicted in Figure 6.3. Time records of a flush-mounted hot-film probe are shown for the natural flow, for the artificially introduced T–S wave using a heating ring, and for the successfully canceled wave using a second downstream heating ring.

The transition delay achieved by active wave cancellation is modest (typically a factor of two or less increase in the transition Reynolds number based on distance from the leading edge). Reshotko (1985) maintained that to achieve significant delay in transition using this technique would require an extensive array of disturbance detectors and generators as well as prohibitively complicated control system that could cancel both the primary and residual disturbance spectra. Significant delay in transition is more readily achieved via the stability modifiers summarized in Section 6.4.

CHAPTER SEVEN

Compliant Coatings

There is no greater impediment to progress in the sciences than the desire to see it take place too quickly.

　　(George Christoph Lichtenberg, 1742–1799)

There is a river in the ocean: in the severest droughts it never fails, and in the mightiest floods it never overflows; its banks and its bottom are of cold water, while its current is of warm; the Gulf of Mexico is its fountain, and its mouth is the Arctic Seas. It is the Gulf stream. There is in the world no other such majestic flow of waters.

　　(Matthew Fontaine Maury, 1806–1873)

PROLOGUE

Boundary layer manipulation via reactive control strategies is now in vogue. The payoffs are handsome, but the difficulties involved are daunting. This topic is deferred to the last chapter of the book. There are, however, much simpler alternatives to such sophisticated flow alteration devices, and the present chapter discusses one such alternative: passive compliant walls. We particularly review the important developments in the field of compliant coatings that took place during the past decade or so. During this period, progress in theoretical and computational methods somewhat outpaced that in experimental efforts. There is no doubt that compliant coatings can be rationally designed to delay transition and to suppress noise on marine vehicles as well as other practical hydrodynamic devices. Transition Reynolds numbers that exceed by an order of magnitude those on rigid-surface boundary layers can readily be achieved. Recent theoretical work indicates that transition to turbulence can be delayed indefinitely, at least in principle, provided that optimized multiple-panel compliant walls are used. There is renewed experimental evidence of favorable interactions of compliant coatings even for air flows and even for turbulent boundary layers, but more research is needed to confirm these latest results.

7.1　Introduction

A compliant wall, as opposed to a rigid one, offers the potential for favorable interference with a wall-bounded flow. Laminar-to-turbulence transition may be delayed or advanced, boundary layer separation may be prevented or triggered, flow-induced noise may be modulated, and skin-friction drag in both laminar and turbulent flows may be altered. The challenge is of course to find a coating with the right physical properties to achieve a desired goal.

Passive compliant coatings were around long before reactive flow control was even contemplated. For better or worse, hydrodynamically speaking, the epidermis of most nekton is pliable at their typical swimming speeds. For close to half a century the science and technology of compliant coatings has fascinated, frustrated, and occasionally gratified Homo sapiens searching for methods to delay laminar-to-turbulence transition, to reduce skin-friction drag in turbulent wall-bounded flows, to quell vibrations, and to suppress flow-induced noise. Compliant coatings offer a rather simple method to delay laminar-to-turbulence transition as well as to interact favorably with a turbulent wall-bounded flow. In its simplest form, the technique is passive, relatively easy to apply to an existing vehicle or device, and perhaps not too expensive. Unlike other drag-reducing techniques such as suction, injection, polymer or particle additives, passive compliant coatings do not require slots, ducts, or internal equipment of any kind. Aside from reducing drag, other reasons for the perennial interest in studying compliant coatings are their many other useful applications, for example, as sound-absorbent materials in noisy flow-carrying ducts in aero-engines and as flexible surfaces to coat naval vessels for the purposes of shielding their sonar arrays from the sound generated by the boundary-layer pressure fluctuations and of reducing the efficiency of their vibrating metal hulls as sound radiators.

The original interest in the field was spurred by the experiments of Kramer (1957), who demonstrated a compliant coating design based on dolphin's epidermis and claimed substantial transition delay and drag reduction in hydrodynamic flows. Those experiments were conducted in the seemingly less-than-ideal environment of Long Beach Harbor, California. Subsequent laboratory attempts to substantiate Kramer's results failed, and the initial interest in the idea fizzled. A similar bout of excitement and frustration that dealt mostly with the reduction of skin-friction drag in turbulent flows for aeronautical applications followed. Those results were summarized in the comprehensive review by Bushnell et al. (1977). During the early 1980s, interest in the subject was rejuvenated mostly as the result of modest investment in resources by the Office of Naval Research in the United States and the Procurement Executive of the Ministry of Defence in Great Britain.[1] Significant advances were made during this period in numerical and analytical methods to solve the coupled fluid-structure problem. New experimental tools were developed to measure the minute yet important surface deformation caused by the unsteady fluid forces. Coherent structures in turbulent wall-bounded flows were routinely identified, and their modulation by the surface compliance could readily be quantified.

Careful analyses by Carpenter and Garrad (1985) and Willis (1986) as well as the well-controlled experiments reported by Daniel, Gaster, and Willis (1987)

---

[1] Through all the ups and downs in the West, compliant coating research continued at more or less steady pace in the former Soviet Union, but open-literature publications resulting from this work, either in English or in Russian, are rather scarce. For some valuable references, see, for example, the books by Aleyev (1977), and Choi (1991), and the article by Carpenter and Garrad (1985).

and Gaster (1988)[2] have, for the first time, provided direct confirmation of the transition-delaying potential of compliant coatings, convincingly made a case for the validity of Kramer's original claims, and offered a plausible explanation for the failure of the subsequent laboratory experiments. There is little doubt now that compliant coatings can be rationally designed to delay transition and to suppress noise on marine vehicles and other practical hydrodynamic devices. Transition Reynolds numbers that exceed by an order of magnitude those on rigid-surface boundary layers can readily be achieved. Although the number of active researchers in the field continues to dwindle, new and promising results are being produced. Recent theoretical work by Davies and Carpenter (1997a) and Carpenter (1998) indicates that transition to turbulence can be delayed indefinitely, at least in principle, provided that optimized multiple-panel compliant walls are used and that the freestream is a low-disturbance environment. There is also recent evidence of favorable interactions of compliant coatings even for air flows (Lee, Fisher, and Schwarz 1995) and even for turbulent boundary layers (Lee, Fisher, and Schwarz 1993a; Choi et al. 1997).

This chapter emphasizes the significant compliant coating research that took place during the last 10–15 years and suggests avenues for future research. The reader is referred to prior reviews for more classical work on the subject such as those by Bushnell et al. (1977), Gad-el-Hak (1986a, 1987a), Riley et al. (1988), Carpenter (1990), and Metcalfe (1994). Following these introductory remarks, a somewhat sketchy history of the subject, particularly prior to 1985, is recalled. This will help place more recent developments in proper perspective.

7.2 Compliant Coatings prior to 1985

Before embarking on describing the recent accomplishments in the field of compliant coatings, we first elaborate on its history before 1985. This seemingly arbitrary date is chosen because it demarks the time after which tools for rationally designing a compliant coating to delay transition became more readily available. The victories and defeats of the subject matter will become clear through the discussion that follows. The idea of using compliant coatings for drag reduction motivated much of the earlier work in this area and was first introduced by Kramer (1957) based on his earlier observation, while crossing the Atlantic ocean in 1946, of dolphins swimming in water. He advanced the concept that the stability and transition characteristics of a boundary layer may be influenced by coupling it hydroelastically to a compliant coating. In his pioneering paper and several subsequent publications, Kramer (1960a, b,c; 1961; 1962; 1965; 1969) reported substantial drag reduction for towed underwater bodies covered with compliant coating modeled after the dolphin skin. He hypothesized that by tuning the elastic wall damping to a frequency near that of the most unstable Tollmien–Schlichting wave, it would be possible to dissipate partially the instability waves, thus delaying the transition to turbulence. Kramer's tests were performed by towing a test model behind a motor boat in Long Beach Harbor. Unfortunately, many attempts by other investigators to repeat Kramer's experiments under more controlled conditions failed to yield similar conclusions (e.g., Puryear's 1962

---

[2] Originally reported in the thesis by Willis (1986) and later contrasted to theory in the paper by Lucey and Carpenter (1995).

experiment in a towing tank). This so-called Kramer controversy will be revisited in Section 7.5.

Theoretical work by Benjamin (1960a,b), Betchov (1960), Landahl (1962), and Kaplan (1964) indicated that drag reduction by delaying transition is possible. However, the theoretically predicted successful coatings had specific characteristics that would be extremely difficult to match in practice. It is important to stress that almost all this early work addressed the delay of transition and ignored the potential for reducing turbulence skin friction with compliant coatings.

During the mid-1960s, Benjamin (1966) explored the possibility that a compliant coating may affect the skin-friction drag in a fully developed turbulent boundary layer without necessarily delaying transition. Dinkelacker (1966) conducted careful tests of a compliant surface in a water-pipe flow. He systematically attempted to determine the repeatability of rigid-tube data, the influence of small steps in the tube wall, and the possible occurrence of organ-pipe acoustic modes. Dinkelacker's results seemed to indicate a modest reduction in drag by using a compliant wall.

Blick and his coworkers at the University of Oklahoma experimentally demonstrated significant reductions in turbulence skin friction for compliant surfaces in air (Fisher and Blick 1966; Looney and Blick 1966; Smith and Blick 1966; Blick and Walters 1968; Chu and Blick 1969). Subsequent tests by Lissaman and Harris (1969), who attempted to substantiate Blick's conclusions, yielded only extremely modest gains. In another study, McMichael, Klebanoff, and Meese (1980) demonstrated that the apparent reduction in turbulence skin friction in the University of Oklahoma's experiments could be a consequence of experimental deficiencies coupled with the improper interpretation of data. McMichael et al. (1980) concluded that drag reduction via compliant coating in gaseous flows would not be as successful as in liquids.

During the 1970s various compliant materials were tested in water at the Naval Ocean Systems Center, the Naval Research Laboratory, the Naval Undersea Systems Center, and the Advanced Technology Center of the LTV Corporation, all in the United States. In no case was a statistically significant reduction in drag measured. Fischer and Ash (1974) presented a general review of concepts for reducing skin friction, including the use of compliant coatings. Bushnell et al. (1977), in summarizing the work conducted at the NASA Langley Research Center and the general status of compliant surface drag reduction, stated that, although it was possible to increase the transition Reynolds number by perhaps a factor of two, there was no definitive reduction of drag for *turbulent* flows in air. They also stated that drag reduction in turbulent flows in water is potentially feasible and can be accomplished using surfaces that can be practically built. It is of particular interest to note that much of the research on compliant coatings has been based on materials that attempt to replicate dolphin skin. Yet, in the Russian book *Nekton* (Aleyev 1977) it is indicated that the "wrinkling" (Figure 7.1) of the dolphin skin has no hydrodynamic-drag advantage. Other characteristics of the dolphin's skin may, however, be beneficial. This subject will be revisited in Section 7.6.4.

Bushnell et al. (1977) put forward the possibility of a feedback mechanism in turbulent wall-bounded flows through which the quasi-periodic, coherent structures termed bursts regenerate (Chapter 4). Older bursts grow, migrate away from the wall, and interact to produce a pressure field that contains pulses of sufficient duration and amplitude to induce new bursts in the near-wall region. This model is supported by the measurements of Burton (1974), who reported a strong correlation between the

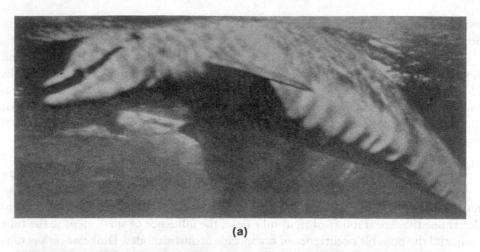

(a)

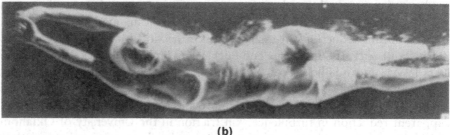

(b)

Figure 7.1: Wavelike folds in compliant skin appearing during rapid swimming of a *delphinidae* (top photograph) and a *Homo sapiens* (bottom photograph) [after Aleyev 1977].

occurrence of a burst and the imposition on the wall flow of a large, moving adverse pressure gradient followed by a favorable pressure gradient. Bushnell et al. (1977) hypothesized that a successful compliant coating would modulate the preburst flow in the turbulent boundary layer by providing a pressure field that would tend to block the feedback mechanism and, thus, inhibit burst formation. This would result in a reduction in the number of bursts occurring per unit time and also in the skin-friction drag. Orszag (1979) assumed this conceptual model and performed numerical calculations of wall boundary layer instability to explore the effects of complaint surfaces. His results, although preliminary, indicated that turbulence drag reduction may be possible for certain classes of materials. He concluded that compliant walls that support only short wavelengths may have an appreciable effect in inhibiting further bursts in a turbulent boundary layer.

During the U.S. Navy-sponsored research program conducted over the period 1980–1985, the subject of boundary layer interaction with compliant coatings has been reexamined to answer the question: Can compliant coatings delay transition, or significantly reduce turbulence skin friction on bodies at high Reynolds numbers, or both? Several significant developments have been achieved by the many investigators participating in this research program. Although unrefutable experimental evidence of compliant coating drag reduction was still lacking by 1985, our understanding of boundary-layer flow over a compliant surface has increased dramatically over this period. That understanding proved crucial to the subsequent successes in the field, which is a subject that will be emphasized throughout this chapter.

7.3 Free-Surface Waves

Before addressing the complex issue of stability of the coupled fluid-structure system, it is instructive to study the solid as a wave-bearing medium when no fluid (another wave-bearing medium) is present. Some understanding of how a compliant surface will respond to the flow above it can be obtained by examining the free-surface waves of the coating. As pointed out by Rayleigh (1887), the surface waves can be modeled as a linear combination of waves having displacements perpendicular and parallel to the propagation direction. These are called transverse and longitudinal displacement waves, respectively. If it is assumed that the coating is a single-layer, elastic solid of thickness d attached to a rigid half-space at its lower boundary and bounded by vacuum on its upper surface, both wave systems satisfy the wave equation (see, for example, Landau and Liftshitz 1987)

$$\frac{\partial^2}{\partial t^2} \eta - c^2 \left(\frac{\partial^2}{\partial x^2} + \frac{\partial^2}{\partial y^2} \right) \eta = 0 \tag{7.1}$$

where η is a component of the displacement vector, and x and y are coordinates parallel and normal to the undisturbed surface, respectively. The propagation velocities are $c = c_t = \sqrt{G/\rho_s}$ for the transverse waves and $c = c_\ell = \sqrt{(\Theta + 2G)/\rho_s}$ for the longitudinal waves, where G and Θ are elastic constants, and ρ_s is the density of the solid.

The free-surface wave dispersion relationship is obtained by assuming that the wave solutions have exponential dependence on the distance measured normal to the surface y and have harmonic dependence on the distance measured along the surface x, and on time t

$$\xi = (A \sinh \alpha y + B \cosh \alpha y) e^{i(k_\ell x - \omega t)} \tag{7.2}$$

$$\eta = (C \sinh \alpha y + D \cosh \alpha y) e^{i(k_\ell x - \omega t)} \tag{7.3}$$

where ξ and η are, respectively, the displacement in the x- and y-direction, and k_ℓ is the longitudinal wave number. For real α the waves decay exponentially with depth, whereas for imaginary α they oscillate. Substituting these solutions into the wave equation (7.1), and applying the boundary conditions, no normal or tangential stress at the surface $y = 0$, and no displacement at the bottom $y = -d$, respectively,

$$(c_\ell^2 - 2c_t^2) \frac{\partial \xi}{\partial x} + c_\ell^2 \frac{\partial \eta}{\partial y} = 0 \quad \text{at} \quad y = 0 \tag{7.4}$$

$$\frac{\partial \xi}{\partial y} + \frac{\partial \eta}{\partial x} = 0 \quad \text{at} \quad y = 0 \tag{7.5}$$

$$\xi = \eta = 0 \quad \text{at} \quad y = -d \tag{7.6}$$

Gad-el-Hak, Blackwelder, and Riley (1984) obtained the following dispersion relationship:

$$M(\zeta) \equiv 4 \frac{\alpha_t \alpha_\ell}{k_\ell^2} (2 - \zeta^2) - \frac{\alpha_t \alpha_\ell}{k_\ell^2} [4 + (2 - \zeta^2)^2] \cosh \alpha_t d \cosh \alpha_\ell d$$

$$+ \left[4 \frac{\alpha_t^2 \alpha_\ell^2}{k_\ell^4} + (2 - \zeta^2)^2 \right] \sinh \alpha_t d \sinh \alpha_\ell d = 0 \tag{7.7}$$

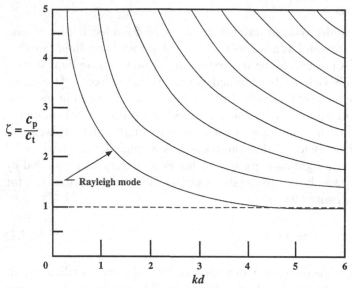

Figure 7.2: Solutions of the dispersion equation for free-surface waves on a single-layer, homogenous coating [after Gad-el-Hak, Blackwelder, and Riley 1984].

where $\alpha_t = k_\ell \sqrt{1 - \zeta^2}, \alpha_\ell = k_\ell \sqrt{1 - \zeta^2 R^2}$, R is the ratio of transverse to longitudinal wave speed, and $\zeta = c_p/c_t$ is the ratio of the surface wave speed to transverse wave speed.

Plots of the dispersion curves resulting from Eq. (7.7) are given in Figure 7.2 for the case $R = 0$, which is close to typical experimental conditions. Only one solution has a surface wave speed below c_t ($\zeta < 1$). This solution approaches its asymptote $\zeta = 0.956$ for large $k_\ell d$, which is the value of Rayleigh's infinite half-layer solution. Also note that as ζ increases for a given $k_\ell d$, the possibility exists for a richer range of interactions between the coating and a flow. Typical values for $k_\ell d$ in experiments, with the wavelength $2\pi/k_\ell$ taken as the boundary layer thickness, are in the range of 1–5. Figure 7.2 indicates that interaction between a fluid flow and a compliant surface may not be expected for flow speeds much below c_t and that the best opportunity for interactions will be for flow speeds well above c_t. Unfortunately, hydroelastic instability waves appear for freestream speeds somewhat above c_t, thus causing large surface deformations and limiting the opportunities for favorable interactions. The waveform can be obtained by determining the constants A, B, C, and D in Eqs. (7.2) and (7.3) using the four boundary conditions (7.4)–(7.6).

Duncan and Hsu (1984) calculated the dispersion relations for a two-layer coating by finding the zeroes of the determinant of the boundary condition coefficients. Their results for a thin, stiff coating placed on a much thicker, soft coating are shown in Figure 7.3. The upper coating has about 100 times the modulus of rigidity of the lower surface. In this figure, the wave number is normalized by the total coating thickness, whereas the phase speed of the free-surface waves is normalized by the transverse wave speed of the lower layer. This normalization allows an easy comparison with the single-layer case when the upper portion of the coating is replaced by a layer of stiffer material. The gross shapes of the curves in Figure 7.3 are similar to those in Figure 7.2, although a wrinkle is clearly seen in the two-layer case. The dashed line in

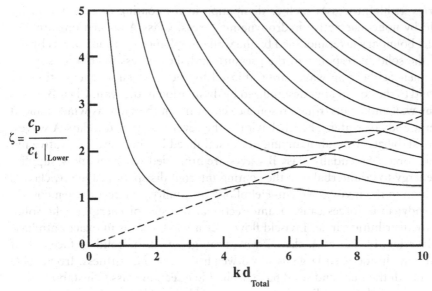

Figure 7.3: Solutions of the dispersion equation for free-surface waves on a two-layer coating [after Duncan and Hsu 1984].

Figure 7.3 connects the maxima of the wrinkle in each modal curve and is a solution of the equation for long bending waves in a free plate. Duncan and Hsu (1984) have hypothesized that the upper coating acts like a plate that is influenced by the lower layer. They have attributed the cutoff of the response of the bending waves in the upper layer at small wave numbers to the lack of slow-moving waves in the lower layer.

7.4 System Instabilities

From a fundamental viewpoint, a rich variety of fluid–structure interactions exists when a fluid flows over a surface that can comply with the flow. Not surprisingly, instability modes proliferate when two wave-bearing media are coupled. Some waves are flow-based, some are wall-based, and some are a result of the coalescence of both kind of waves. What is most appealing about compliant coatings is their potential to inhibit, or to foster, the dynamic instabilities that characterize both transitional and turbulent boundary layer flows and in turn to modify the mass, heat, and momentum fluxes and change the drag and the acoustic properties. Although it is relatively easy to suppress a particular instability mode, the challenge is of course to prevent other modes from growing if the aim is, say, to delay laminar-to-turbulence transition. From a practical point of view, it is obvious that an in-depth understanding of the coupled system instabilities is a prerequisite to the design of a coating that meets a given objective.

There are at least three classification schemes for the fluid–structure waves, each with its own advantages and disadvantages. The original scheme was introduced by Benjamin (1963) and divides the waves into three classes according to their response to irreversible energy transfer to and from the compliant wall. Both class **A** and class **B** disturbances are essentially oscillations involving conservative energy exchanges between the fluid and solid, but their stability is determined by the net effect of irreversible processes such as dissipation in the coating or energy transfer to the solid by nonconservative hydrodynamic forces. Class **A** oscillations are T–S waves in the boundary layer modified by the wall compliance (in other words by the motion of the solid in response to the pressure and shear-stress fluctuations in the flow). Tollmien–Schlichting waves are stabilized by the irreversible energy transfer from the fluid to the coating but destabilized by dissipation in the wall. Class **B** waves reside in the wall and result from a resonance effect much the same as wind-induced waves over a body of water. Their behavior is the reverse of that for class **A** waves, for they are stabilized by wall damping but destabilized by the nonconservative hydrodynamic forces. Essentially class **B** waves are amplified when the flow supplies sufficient energy to counterbalance the coating internal dissipation. Finally, class **C** waves are akin to the inviscid Kelvin–Helmholtz instability and occur when conservative hydrodynamic forces cause a unidirectional transfer of energy to the solid. The pressure distribution in an inviscid flow over a wavy wall is in exact antiphase with the elevation. In that case, class **C** waves can grow on the solid surface only if the pressure amplitude is so large as to outweigh the coating stiffness. Irreversible processes in both the fluid and solid have negligible effect on class **C** instabilities.

If one considers the total disturbance energy of the coupled fluid–solid system, a decrease in that energy leads to an increase in the amplitude of class **A** instabilities, class **B** is associated with an energy increase, and virtually no change in total energy accompanies class **C** waves. In other words, any nonconservative flow of activation energy from or to the system must be accompanied by disturbance growth of class **A** or **B** waves, whereas the irreversible energy transfer for class **C** instability is nearly zero.

The second classification scheme was introduced by Carpenter and Garrad (1985, 1986). It simply divides the waves into fluid-based (Tollmien–Schlichting instabilities [TSI]) and solid-based (flow-induced surface instabilities [FISI]). The latter are closely analogous to the instabilities studied in hydro- and aeroelasticity and include the traveling-wave flutter that moves at speeds close to the solid free-wave-speed (class **B**) and the essentially static, and more dangerous, divergence waves (class **C**).[3] The main drawback of this classification scheme is that under certain circumstances the fluid-based T–S waves and the solid-based flutter can coalesce to form a powerful new instability termed transitional mode by Sen and Arora (1988). According to the energy criterion advanced by Landahl (1962), this latest instability is a second kind of class **C** waves. In a physical experiment, however, it is rather difficult to distinguish between the static-divergence waves and the transitional ones.

The third scheme to classify the instability waves considers whether they are convective or absolute (Huerre and Monkewitz 1990). As discussed in Section 3.5, an instability mode is considered to be absolute if its group velocity is zero. On the other hand, the unstable development of a disturbance is said to be convective when none of its constituent modes possesses zero group velocity. Both classes **A** and **B** are convective, whereas class **C** divergence and transitional modes are absolute. As Carpenter (1990) points out, the occurrence of absolute instabilities would lead

---

[3] Static-divergence waves were erroneously interpreted in the past by, for example, Gad-el-Hak, Blackwelder, and Riley (1984), Duncan et al. (1985), and Yeo (1992) as being class **A**. The confusion occurs when divergence is treated as a convective instability when in fact it is an absolute one (Carpenter 1990).

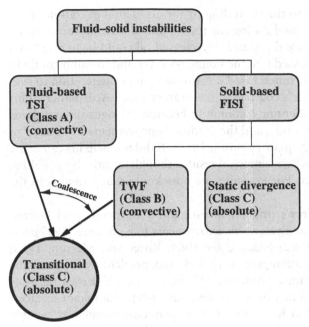

Figure 7.4: Summary of all three classification schemes.

to profound changes in the laminar-to-turbulence transition process. It is therefore pointless to consider reducing their growth rate or postponing their appearance to a higher Reynolds number; nothing short of complete suppression would work. Figure 7.4 combines and summarizes all three classification schemes.

7.5 The Kramer Controversy

It may be worth recalling in more detail the pioneering work of Max O. Kramer and the controversy surrounding it. The entire field of compliant coatings became the Rodney Dangerfield of fluid mechanics research, getting no respect from a skeptical community largely because Kramer's original experiments lost credibility. However, as will be seen below, the most recent evidence resurrects the good name of this ingenious German–American and with it renewed confidence in this waning and waxing field.

As already mentioned, Kramer (1957; 1960a,b,c; 1961; 1962; 1965; 1969) conducted his original experiments by towing a model behind an outboard motor boat in Long Beach Harbor, California. His early tests showed a drag reduction of more than 50% when a dolphinlike skin was used. A typical successful coating used by Kramer consisted of a flexible inner skin, an outer diaphragm, and stubs, all made of soft, natural rubber. The cavity between the outer diaphragm and the inner skin was usually filled with a highly viscous damping fluid, such as silicone oil, which in Kramer's view damped out the Tollmien–Schlichting waves.

Subsequent experiments to confirm Kramer's findings were conducted in a towing tank, a lake, or a water tunnel (Puryear 1962; Nisewanger 1964; Ritter and Messum 1964; Ritter and Porteous 1964). No significant drag reduction was observed in any of these investigations. Since then, many researchers have assumed that Kramer's results were in error and that his observed drag reduction could have come about as a result of favorable changes to the form drag or the accidental excretion of the silicone oil used as the damping fluid during the tests. Surface discontinuities could have favorably altered the pressure drag, and the released oil could have acted as a drag-reducing polymer when released into the boundary layer and the ambient fluid.

Carpenter and Garrad (1985) stated that "it is probably no exaggeration to suggest that the credibility of Kramer's coatings is now rather low." Acceptance of his results was not granted by the scientific community because the rigorous standards of scientific investigation were not met, and the gradual improvements by Kramer to meet these standards were not adequate (Johnson 1980). It did not help his cause any that Kramer's explanations of his own empirical results, though intuitively appealing, were proven physically incorrect. For example, we now know that damping in the solid *destabilizes* TSI.

Almost 30 years after Kramer's original investigation, Carpenter and Garrad (1985) presented a very careful analysis of his experiments (e.g., Kramer 1957) and the subsequent tests (Puryear 1962; Nisewanger 1964; Ritter and Messum 1964; Ritter and Porteous 1964) that attempted to provide independent evidence of the drag-reducing capabilities of Kramer's coatings. On the basis of their own rigorous analysis of the hydrodynamic stability of flows over Kramer-type compliant surfaces, Carpenter and Garrad argued that Kramer's coatings were only marginally capable of delaying transition. Any unfavorable factor, such as an adverse pressure gradient, a step in which the compliant surface is joined to a rigid surface or an unusually high freestream turbulence level, could badly affect the performance of the coating. Also, a particular coating was designed for a restricted range of Reynolds number and was therefore unlikely to delay transition outside that range.

Carpenter and Garrad (1985) contend that one or more of the above adverse factors may have existed in the experiments conducted by Puryear (1962), Nisewanger (1964), Ritter and Messum (1964), and Ritter and Porteous (1964). Puryear's (1962) experiments were conducted using a prolate spheroid in a towing tank. He did not use Kramer's coating with the best performance, and serious problems were encountered in making a smooth joint between the rigid and compliant surfaces. Nisewanger's (1964) tests were conducted by releasing a lighter-than-water body of revolution from the bottom of a lake. His Kramer-like coating contained a fluid with a viscosity below that of the optimum fluid as determined from Kramer's results. Ritter and Messum (1964), and Ritter and Porteous (1964) conducted their experiments in a water tunnel using either a flat plate or a cylindrical model with an elliptic nose. The conventional flume had a relatively high freestream turbulence level, which may have rendered the facility unsuitable for transition experiments.

In addition to these adverse effects, some evidence existed in the tests conducted to confirm Kramer's results for the onset of a hydroelastic instability in the coating. Such large-amplitude waves would certainly lead to drag increase, and their presence may indicate that the boundary layer was already turbulent. On the basis of these experiments, Carpenter and Garrad (1985) concluded that the results presented in these tests should not be taken as conclusive evidence that the Kramer coatings are incapable of delaying transition and that "the case against Kramer's coating may not be so strong as popularly supposed."

Carpenter's (1988) optimization procedure described in Section 7.6.2 results in a compliant coating capable of delaying transition by a factor of 4–6 in Reynolds number. It is therefore quite conceivable to design a Kramer-type coating that may lead to a drag reduction of the order reported by the original inventor himself. The preceding analysis of Kramer's tests illustrates the importance of carefully selecting the flow facility to conduct compliant coating experiments. The background turbulence in the facility should be particularly monitored if transition delay is sought. This is precisely what was done in the successful experiments conducted by Gaster (1988) to confirm the theoretical prediction of Carpenter and Garrad (1985), both of which are described in more detail in the following section.

7.6 Transitional Flows

7.6.1 Linear Stability Theory

The hydrodynamic and the hydroelastic stability theories have reached an impressive level of maturity during the last two decades. The linear theories can be handled, for the most part, analytically, whereas the nonlinear stability theories are more computer intensive. Perhaps no one has contributed more to the recent application of the stability theory to compliant coatings than Peter W. Carpenter, originally with the University of Exeter and presently with the University of Warwick. His list of relevant publications includes 65 papers and is growing; obviously only a selected few will be cited in the present short chapter.

Within the framework of the linear stability theory, two-dimensional small disturbances are assumed to be superimposed upon a steady, unidirectional mean flow. As discussed in Chapter 6, the nonlinear, partial Navier–Stokes equations are then reduced to the well-known Orr–Sommerfeld equation, which is a fourth-order, linear, ordinary differential equation. The order of this equation increases when additional complexities are included in the problem. For rotating-disk flows, for example, Coriolis and streamline-curvature terms are incorporated, leading to a sixth-order stability equation. The major difficulty in integrating the Orr–Sommerfeld equation is that it is highly stiff and unstable, which makes it virtually impossible to apply conventional numerical schemes. Explicit codes with step size that is commensurate with the global behavior of the solution lead to numerical instabilities, and alternative routines have been developed to handle this stiff eigenvalue problem.

An added difficulty when the walls are compliant is the interfacial conditions that require continuity of velocity and stress. Those boundary conditions can also be linearized, but special care should still be exercised in handling them. Appropriate equations must be used for the compliant walls to be able to fully couple the fluid and solid dynamics. Many types of compliant surfaces exist, and thus there are numerous models for the solid. Those models can be either surface-based or volume-based (Figure 7.5). The former model reduces the spatial dimensions by one and is therefore less computationally demanding. An example is the thin plate–spring model used by Garrad and Carpenter (1982), Carpenter and Garrad (1985, 1986), Domaradzki and Metcalfe (1987), Metcalfe et al. (1991), and Davies and Carpenter (1997a,b), among others, to simulate Kramer-type coatings. This model is relatively simple yet contains characteristics representative of a broad range of surfaces. If a coordinate system is chosen with the x-axis lying along the undisturbed free surface and the y-axis normal to this surface, then the equation for the y-component[4] of the

---

[4] The x-component of momentum is usually neglected in such a model.

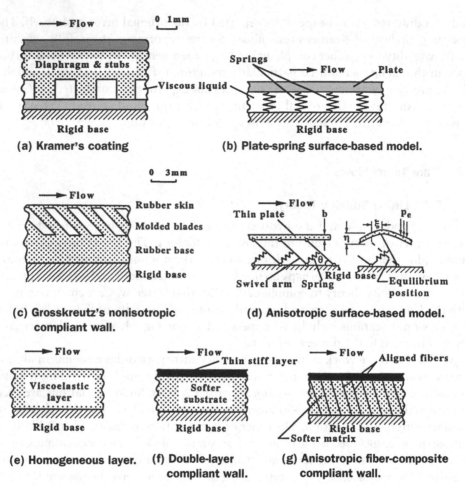

Figure 7.5: Volume-based and surface-based models of compliant coatings [adapted from Carpenter 1990].

momentum of the compliant coating reads

$$\frac{\partial^2 \eta}{\partial t^2} = \frac{T_\ell}{m}\frac{\partial^2 \eta}{\partial x^2} - D\frac{\partial \eta}{\partial t} - \frac{\mathcal{F}}{m}\frac{\partial^4 \eta}{\partial x^4} - \frac{k}{m}\eta + F \qquad (7.8)$$

where $\eta(x, t)$ is the y-displacement of the surface from its equilibrium state at time t and position x, T_ℓ is the longitudinal tension and $\mathcal{F}$ the flexural rigidity of the thin plate, m is the mass per unit area, D is the damping coefficient, k is the spring constant, and F is an external forcing term.

The volume-based models are based on the Navier equation and include single and multilayer coatings (Duncan, Waxman, and Tulin 1985; Fraser and Carpenter 1985; Buckingham et al. 1985; Yeo 1988) as well as isotropic and anisotropic materials (Yeo 1990, 1992; Duncan 1988). The equations describing the stability of the coupled system form a numerical eigenvalue problem for the complex wave number of the disturbance. Duncan (1987) offers a useful comparison between the results obtained from a surface-based model and a corresponding volume-based one.

Compliant walls do suppress the Tollmien–Schlichting waves due to the irreversible energy transfer from the fluid to the solid, but solid-based instabilities proliferate if the coating becomes too soft. For the class **A** T–S waves, the wall compliance reduces the rate of production (via Reynolds stress) of the disturbance kinetic energy. Simultaneously, the viscous dissipation is increased, and thus the balance between the energy production and removal mechanisms is altered in favor of wave suppression.

Experimental validation of the stability calculations is rather difficult and requires well-controlled tests in a quiet water or wind tunnel. Several careful experiments to test the flow stability to two-dimensional as well as three-dimensional controlled disturbances have been reported in the past few years (Daniel et al. 1987; Gaster 1988; Lee et al. 1995, 1997). Figures 7.6 and 7.7 show the remarkable agreement between theory and towing tank experiments for both rigid-wall and compliant-wall cases. The top part of each figure is the predicted amplification factors as function of flow speed for a range of modal frequencies, and the bottom part is the measured growth-decay cycle of artificially induced T–S waves. A simple compliant model predicts a dramatic decrease in the instability of the flow, and this prediction agrees well with the experimental observations when a thick, soft coating is covered with a thin, stiff layer.

The articles by Lee et al. (1995, 1997) report the results of wind tunnel experiments and actually demonstrate the stabilizing potential of compliant coatings in aerodynamic flows—a remarkable achievement that had been deemed impractical in the past (Bushnell et al. 1977; Carpenter 1990). Excellent agreements are reported between the results of the stability theory and the hydrodynamic experiments (Willis 1986; Daniel et al. 1987; Gaster 1988; Riley et al. 1988; Carpenter 1990; Lucey and Carpenter 1995). The article by Lucy and Carpenter, in particular, applies the linear stability theory to predict the experimentally observed evolution of both Tollmien–Schlichting waves and traveling-wave flutter in water flows. For the wind tunnel experiments, Carpenter (1998) has conducted the corresponding calculations, but his preliminary results thus far are negative: the density of an effective coating must be comparable to the fluid density; otherwise, no transition delaying benefits are observed. This theoretical result leaves open the question of explaining the positive experimental findings of Lee et al. (1995). Here, we show one example of the suppression of T–S waves in an air boundary layer developing on top of a silicone elastomer–silicone oil compliant surface. Figure 7.8 presents the wind tunnel results of Lee et al. (1995). The coating in Figure 7.8a was made by mixing 91% by weight of 100 mm^2/s silicone oil with 9% of silicone elastomer. In Figures 7.8b,c, and d, the corresponding mix was 90% and 10%, yielding about 35% higher modulus of rigidity. As compared with the rigid wall, the single-layer, isotropic, viscoelastic compliant coating significantly suppresses the rms amplitude of the artificially generated Tollmien–Schlichting waves across the entire boundary layer for a range of displacement-thickness Reynolds numbers. Reductions in the maximum rms amplitude of as much as 40% are observed for the softer coating (Figure 7.8a), which may lead to delayed transition.

7.6.2 Coating Optimization

If a compliant coating is to be designed for use on an actual vehicle, a relevant question may be, What are the optimum wall properties to give the greatest transition delay? The large number of available parameters makes it imperative that a rational (i.e., one derived from first principles) selection process be conducted. For obvious

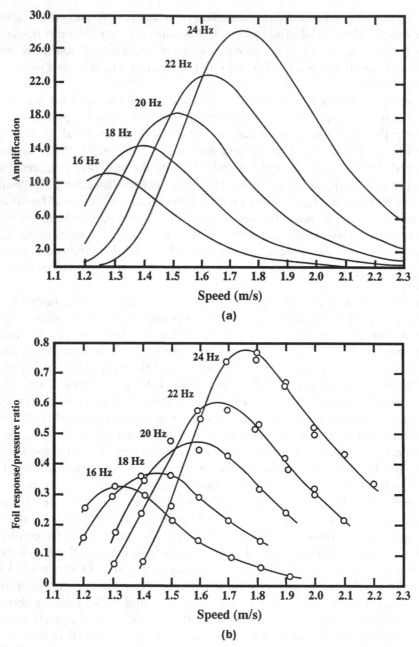

Figure 7.6: Growth curves for the rigid-wall case. (a) Theoretical prediction; (b) Experimental results [after Willis 1986].

reasons, the trial-and-error empirical approach used in the past (if it is soft, let us try it!) should not even be contemplated. This should be particularly true now that rational optimization procedures are becoming readily available, as described below. A wall that is too compliant (i.e., too soft) can substantially delay transition via TSI by shrinking its unstable region in the frequency-Reynolds number plane, but rapid breakdown can occur through the amplification of wall-based instabilities (Lucey and Carpenter 1995). Both kinds of FISI are potentially harmful. The divergence

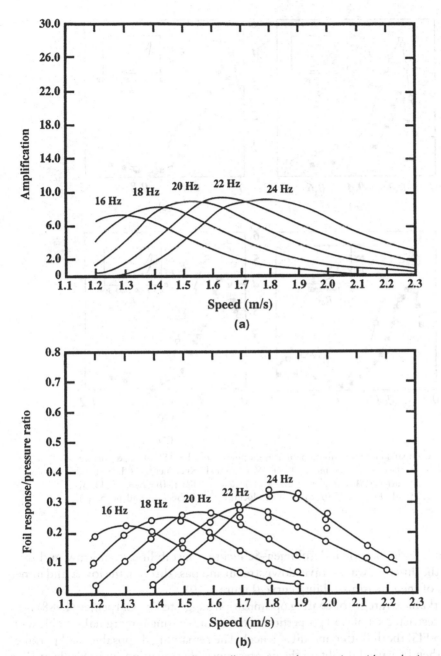

Figure 7.7: Growth curves for the compliant-wall case. A two-layer coating with an elastic modulus of the lower layer of $E = 5000 \, \text{N} \, \text{m}^{-2}$. (a) Theoretical prediction; (b) Experimental results [after Willis 1986].

instabilities are absolute and nearly static and yield to wholesale deformations of the surface, which are likely to trigger premature transition due to a roughnesslike effect (Figure 7.9). Flutter instabilities, though convective, are also dangerous. As shown in the stability diagram in Figure 7.10, their narrow band of unstable frequencies extends indefinitely as Reynolds number increases downstream. Thus, once these instabilities are encountered at some downstream location, sustained growth follows.

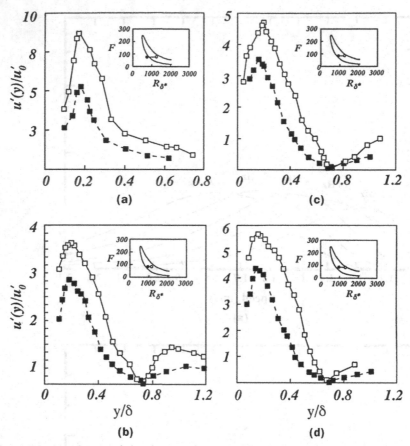

Figure 7.8: Comparison of distribution of rms-amplitude of the TSI of rigid surface and compliant surface across boundary layer. Wind tunnel experiments of Lee et al. (1995). □, rigid surface; ■, compliant surface. (a) $Re_{\delta*} = 1274$; (b) $Re_{\delta*} = 1105$; (c) $Re_{\delta*} = 1225$; (d) $Re_{\delta*} = 1350$. Inset shows locations of ribbon (•) and probe (○) relative to neutral-stability curve.

This is unlike the broadband Tollmien–Schlichting instabilities that grow and then decay as the different waves travel downstream and pass through the lower and upper branches of their neutral-stability curve (Figure 7.10).

A workable strategy for coating optimization suggested by Carpenter (1988) is to choose a restricted set of wall properties such that the coating is marginally stable with respect to FISI (both flutter and divergence). The remaining disposable wall parameters can then be varied to obtain the greatest possible transition (via TSI) delay. For the plate-spring, surface-based model, for example, there are two disposable parameters: the wall damping and the critical wave number for divergence. The down-stream location of the transition region is estimated from an e^n-criterion, where n is typically chosen in the range of 7–10. The lower exponent represents the approximate limit of validity of the linear stability theory for a low-disturbance environment and provides a rather conservative calculation.

Although wall dissipation destabilizes Tollmien–Schlichting waves, a viscoelastic coating with moderate level of damping leads to greater delay in transition as compared with purely elastic surfaces. Apparently the stabilizing effects of wall damping

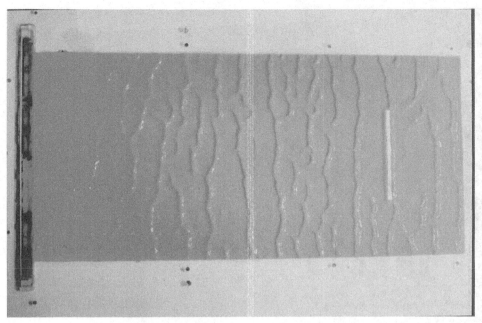

Figure 7.9: Static-divergence waves on a single-layer viscoelastic coating subjected to the pressure fluctuations of a turbulent boundary layer. Freestream velocity is 6.3 times the transverse-wave speed in the solid [after Gad-el-Hak, Blackwelder, and Riley 1984].

on traveling-wave flutter allow a softer wall to be used and thus more than offset the adverse effects of coating dissipation on TSI.

Coating optimization with respect to TSI growth rate is performed at a rather narrow range of Reynolds numbers. On a growing boundary layer, the Reynolds number increases monotonically, and a compliant coating will not be optimum over the whole

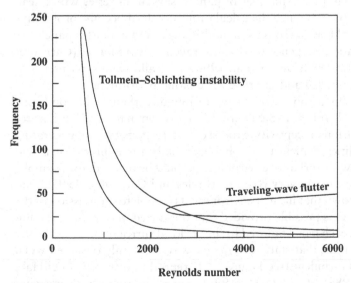

Figure 7.10: Typical neutral-stability curves for Tollmien–Schlichting waves and for traveling-wave flutter. The solid is a double-layer, Gaster-type compliant coating [after Lucey and Carpenter 1995].

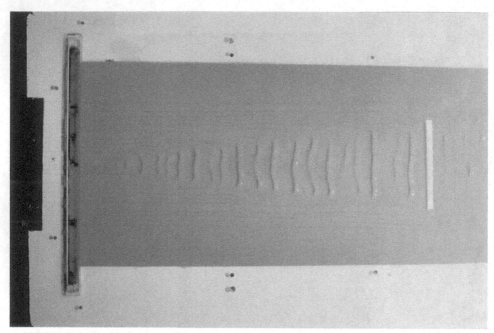

Figure 7.11: Static-divergence waves are clearly seen under the turbulent wedge generated by a single roughness element embedded into an otherwise laminar boundary layer [after Gad-el-Hak, Blackwelder, and Riley 1984].

length of a vehicle. Carpenter (1993) suggested that a multiple-panel wall placed in series, with each panel optimized for a particular range of Reynolds numbers, is likely to produce larger transition delays than a single-panel wall. His calculations for a two-panel, plate-spring-type compliant wall indicate an additional performance improvement of more than 30% over an optimized single-panel wall.

It seems reasonable that a large number of panels, say 10, in series would lead to superior performance, but of course the calculations involved become prohibitive very quickly. An additional benefit from using multipanels is that shorter panels are more resistant to both static-divergence waves and traveling-wave flutter (Carpenter 1993; Lucey and Carpenter 1993; Dixon et al. 1994), thus allowing softer panels to be used that further suppress TSI and improve the coating performance.

In the flat plate and similar boundary layers in low-disturbance environments, the quasi-two-dimensional Tollmien–Schlichting waves dominate the laminar-to-turbulence transition. Various receptivity processes are responsible for generating two-dimensional instabilities, which are probably initially three-dimensional and randomly distributed. Low-disturbance environments could be realized, for example, in free flight and marine vehicles. Very recently, Davies and Carpenter (1997a) and Carpenter (1998) have shown that in such environments complete suppression of the Tollmien–Schlichting waves is possible provided that optimized multiple-panel compliant walls are used with each panel tailored to suit its local surrounding. Assured by the experimental observations that static-divergence waves are only observed when the wall-bounded flow is turbulent (see Figure 7.11, reproduced from Gad-el-Hak, Blackwelder, and Riley 1984),[5] Carpenter's (1998) new assumptions are somewhat

---

[5] In other words, one less instability mode to worry about when the flow is laminar and the objective is to keep it so.

less conservative than those used in earlier theories (Carpenter 1993; Dixon et al. 1994). The new finding raises the possibility of maintaining laminar flow in situations where the T–S instabilities are the primary cause of transition,[6] to indefinitely high Reynolds numbers—a very profound prospect indeed.

Work on nonlinear stability theory has recently been in the forefront and confirms that transition-delaying coatings, optimized using the linear theory, maintain their beneficial effects into the latter stages of transition to turbulence (Metcalfe et al. 1991; Joslin and Morris 1992; Thomas 1992a,b). Lee et al. (1997) experimentally studied the effects of a compliant surface on the growth rates of both the subharmonic and three-dimensional fluid-based instabilities of a laminar boundary layer in air. Their results suggest that a delay of the excitement of the secondary instability can be achieved by suppressing the growth of the primary waves using surface compliance.

7.6.3 Practical Examples

Most of the theoretical as well as experimental compliant coating research has been concerned with canonical boundary layers. Nevertheless, an attempt is made herein to estimate the potential benefit of applying the technique for field applications in which strong three-dimensional and pressure-gradient effects and, for aeronautical applications, compressibility effects may be present. The typical Reynolds numbers, based on vehicle speed and overall length, for a hydrofoil, a torpedo, and a nuclear submarine are, respectively, of the order of 10 million, 50 million, and 1 billion. Applying e^n-type calculations (with the exponent chosen conservatively to be $n = 7$) to an optimum two-panel, plate-spring-type compliant wall, Carpenter (1993) computed a transition Reynolds number of 13.62×10^6 as compared with 2.25×10^6 for a rigid wall.[7] This means that the laminar region that would normally extend over 23%, 5%, and 0.2% of the respective vehicle lengths would, with the use of an optimum coating, extend over a larger length of 100%, 27%, and 1%. When one computes the corresponding overall drag coefficients using standard methods for a mixed laminar-turbulent boundary layer over a flat plate, the potential reduction in skin-friction drag using the optimum two-panel compliant wall can be as much as 83%, 19%, and 0% for the three respective vehicles. Obviously the large submarine does not benefit, as far as drag reduction is concerned, from the use of transition-delaying compliant coating, but the smaller vehicles do. However, extending the laminar region on a submarine even by 1 m can be significant for sonar applications requiring longer quiet regions of the boundary layer.

For aeronautical applications, a cruising commercial jet aircraft has a fuselage Reynolds number of the order of 0.5 billion and a wing Reynolds number of the order of 50 million. Again, increasing the transition Reynolds number by a factor of 5 or so is significant for the wing but not for the fuselage. Skin-friction reduction of the order of 20% is achievable for the wings (whose skin-friction drag accounts

---

[6] Excludes, for example, flows where the walls, streamlines, or both are concave (Görtler instabilities), where the boundary layer is three-dimensional and the crossflow velocity profile is inflectional (crossflow instabilities), or where the wall is rough or the freestream turbulence levels are unusually high (bypass transition).

[7] Contrast this sixfold increase to the 30% higher transition Reynolds number reported in the experiments of Gaster (1988), who did not attempt to optimize his single-panel, two-layer, silicone rubber–latex rubber coating. Theoretical calculations by Dixon et al. (1994) indicate that an optimum Gaster-type coating would provide a 500% higher transition Reynolds number.

for about 50% of the skin friction of the entire aircraft and 25% of the total drag).[8] Finding a compliant coating that would reduce the turbulent skin-friction drag would of course be very beneficial for both the typical fuselage and long submarine.

The estimates above were made for a simple plate-spring model. Using more than two panels can provide further transition delay. More complex compliant surfaces, particularly anisotropic ones designed specifically to suppress the Reynolds stress fluctuations, can conceivably offer more spectacular savings. Such custom-designed coatings can also favorably interact with fully turbulent flows. Even for laminar flows, the calculations involved when complex, wall-based models are used, though straightforward in principle, are quite demanding in practice.

7.6.4 The Dolphin's Secret

The ability to swim or to fly with minimum skin friction and pressure drag is of extreme importance to the Darwinian survival of certain nektonic and avian species. Homo sapiens interested in building the fastest submarine or the most fuel-efficient aircraft have much to learn about alternative drag-reducing approaches from humble earthlings. As Kramer remarked close to half a century ago, a school of porpoises, including the young and the old, the weak and the strong, showing off their seemingly effortless glide along a fast ocean liner is a sight to behold.

Cetaceans appear to possess unusually low overall drag coefficients. This is the basic factor for the so-called Gray's (1936) paradox in which a steady-state energy balance based on the anticipated muscle power of various nektons, including the dolphin, failed to explain their unusually fast swimming speeds. Gray clocked bottlenose dolphins, *Tursiops trucatus*, swimming at speeds exceeding 10 m/s for a period of 7 s. If one assumes that the power output of cetaceans is equal to that of other mammals ($\approx$35 W/kg of body weight), then such speeds are reached under turbulent flow conditions only if dolphins can expend several times more power than their muscles can generate. Lang (1963) concluded on the basis of energy considerations that dolphins can not exceed a speed of 6 m/s for periods greater than 2 hours.

Transition delay is of course an obvious, albeit arduous, technique for achieving about an order of magnitude lower skin-friction drag, but does the dolphin possess an exotic means by which such a difficult flow-control goal can be accomplished? Obviously the dolphin is not sharing its secrets with other fellow mammals. Kramer's (1957, 1961) invention of a compliant coating tried to mimic the dolphin's epidermis and claimed drag reduction of as much as 60%. His explanation for the dolphin's secret is that its skin, like his successful compliant coating, is capable of substantially delaying laminar-to-turbulence transition. Kramer's work was discredited for a while but now seems to be back in vogue, as remarked in Section 7.5. The calculations presented in Section 7.6.3 indicate that it is quite possible to design a Kramer-type coating that delays transition by a factor of 4–6 in Reynolds number and that drag reduction of the order reported by Kramer is also quite possible. Does the dolphin or other similar fast swimmers possess such a coating?

In a recent article, Bushnell and Moore (1991) quoted the relevant energetic and controlled swimming studies but concluded by supporting the explanation offered

---

[8] Note that the dominant mechanism leading to transition on a swept wing is crossflow instability and not T–S waves. However, recent evidence (Cooper and Carpenter 1995, 1997a,b,c) indicates beneficial effects of compliant surfaces even for the former kind of instability.

by Au and Weihs (1980) that dolphins, which must periodically breath air, achieve high-speed swimming by simply *porpoising*, that is, momentarily leaping out of the water, thereby reducing their drag force by a factor of 800 (density ratio of air and water). This more than pays for the additional interfacial or wave drag and accounts for the abnormally low apparent drag coefficients inferred from the assumption of fully submerged travel.

I, however, do not concur with the preceding *final solution* to the Gray's paradox. Dolphins have been clocked at sustained and burst speeds of close to 10 and 20 m/s, respectively. *Delphinus delphis* has a typical length of 2 m. This leads to sustained and burst Reynolds numbers based on the overall length of the order of 20 million and 40 million, respectively. Carpenter (1993) reported the results of optimizing a rather simple plate-spring coating. By using a single panel, as compared with a rigid surface, a 4.6-fold increase in transition Reynolds number is estimated, which leads to a drag reduction of 36% at the typical dolphin's sustained speed and 20% at burst speed. Using a mere two-panel coating, the transition Reynolds number becomes 6.1 times the value for a rigid surface, and the potential drag reductions for the sustained and burst speeds are now 52% and 30%, respectively.

These lower levels of skin friction are compatible with the available muscle power for a dolphin of the size used above. Admittedly, the preceding estimates were made for a flat-plate boundary layer and may not hold when pressure-gradient and other shape effects are taken into account. Additionally, the dolphin also has pressure drag on top of the (much larger) skin friction. On the other hand, cetaceans have had millions of years of evolutionary adaptations to hone their coatings for maximum speed and efficiency, and it is quite conceivable that their epidermis is much more complex, and hydrodynamically beneficial, than the simple ones computed in the examples above. Moreover, each portion of the skin could have been optimized for the appropriate range of local Reynolds numbers. Therefore, the dolphin's apparent success is not incompatible with having optimum compliant coatings to delay laminar-to-turbulence transition substantially and therefore to attain inordinately low coefficients of drag.

Other fascinating questions related to the amazing swimming abilities of the dolphin include the possibility that its excreted mucin is a drag-reducing additive. Is there a hydrodynamic advantage to the warm-blooded cetaceans because their epidermal temperature is higher than the ambient one (in which case the near-wall water viscosity is lowered and the turbulent boundary layer may be relaminarized)? Does the dolphin's particular body shape during coasting (with no attendant overall body deformation) or actual swimming (accompanied by appropriate body oscillations) offer additional drag-reducing advantages? Also, what are the potential benefits to the porpoise when it uses ship-generated bow waves for body surfing? These subjects, though related to the preceding discussion, are outside the scope of the present discussion and are therefore left for another circumstance.

7.7 Turbulent Wall-Bounded Flows

Unlike the laminar and transitional flows investigated in Section 7.6, compliant coating effects on turbulent boundary layers are rather difficult to study theoretically. In fact *any* turbulent flow is largely unapproachable analytically. For a turbulent flow, the dependent variables are random functions of space and time, and no straight-forward method exists for analytically obtaining stochastic solutions to the governing nonlinear, partial differential equations. The statistical approach to solving the Navier–Stokes equations always leads to more unknowns than equations (the closure problem), and solutions based on first principles are again not possible. Direct numerical simulations (DNS) of the canonical turbulent boundary layer have thus far been carried out up to a very modest momentum-thickness Reynolds number of 1410 (Spalart 1988).

How would one go about rationally choosing a coating to achieve a particular control goal for a turbulent boundary layer? Analytical optimization procedures such as those used to delay transition (Section 7.6.2) would not work for fully turbulent flows. To analyze the full problem, direct numerical simulations of the turbulent boundary layer should be coupled to a finite-element model of the compliant coating, a task that is extremely time-consuming and expensive and taxes the fastest supercomputer around. Modeling the turbulence by an eddy-viscosity or even a more sophisticated closure scheme is less computationally demanding, but there is no guarantee that turbulence models developed primarily for rigid surfaces would work for a compliant surface. In fact, it is not difficult to argue that closure models based on mean quantities completely miss the all-important spectral contents of a fluid–solid interaction and will therefore never work.

A turbulent boundary layer is characterized by a hierarchy of coherent structures. Near the wall, the dynamics are dominated by the quasi-periodic bursting events (Robinson 1991). A crude, albeit resourceful, attempt to model a turbulent boundary-layer interaction with a single-layer, isotropic, viscoelastic coating has been advanced by Duncan (1986). He approximates the turbulent flow over the coating by a potential flow with a superimposed pressure pulse, convecting downstream, that mimics the pressure footprint of a single bursting event. To relate the problem to a real turbulent flow, the pressure pulse characteristics are taken from actual boundary layer measurements and the potential flow is modified to incorporate the reduced magnitudes and phase shifts found experimentally in boundary layer flows over moving wavy walls. At low flow speeds (relative to the transverse-wave speed in the solid), the coating response to the pressure pulse is stable and primarily localized under it. At intermediate speeds, the response is still stable but includes a discernible wave pattern tagging along behind the pressure pulse. At the highest speed studied, large-amplitude, unstable waves develop on the compliant surface, much the same as the FISI observed experimentally. Duncan and Sirkis (1992) have recently extended this model to anisotropic compliant coatings. They reported that certain anisotropic surfaces provide more effective control over the amplitude and angular extent of the generated stable response pattern. Larger amplitudes are generated as compared with isotropic surfaces, thus providing for greater potential for modifying the turbulence.

Whenever the flow speed in a turbulent boundary layer becomes sufficiently large compared with the transverse free-wave speed in the solid, flow-induced surface instabilities proliferate. The pressure fluctuations within the flow are an order of magnitude larger than the normal and tangential viscous stresses and drive the coating response. In laminar wall-bounded flows it is difficult to observe the hydroelastic waves in their unstable state. As soon as flutter or divergence waves grow, rapid breakdown to turbulence takes place in the boundary layer, and the flow is no longer laminar.

Most of the experimental studies concerning compliant coating effects on turbulent boundary layers focused on documenting the unstable flow-induced surface instabilities. When divergence waves or flutter are unstable, the effects, though adverse, are pronounced and are somewhat easier to document. Only recently few hardy souls have attempted to investigate the wall-bounded flows when these FISI are stable or neutrally stable. Obviously, the latter kind of studies have to await the development of refined techniques to measure the minuscule surface deformation and the associated coherent structure modulation when the FISI are neutrally stable.

Both Gad-el-Hak (1986a,b) and Hess, Peattie, and Schwarz (1993) introduced nonintrusive methods for the point measurement of the instantaneous vertical surface displacement of a compliant coating, whereas Lee, Fisher, and Schwarz (1993b) offered an optical holographic interferometer, in connection with an interactive fringe-processing system, to capture whole-field random topographic features. The latter technique is more expensive to set up but offers higher spatial resolution, of the order of 1 μm, and yields simultaneous surface displacement information on a large section of the compliant coating. Both the local and global methods were initially employed to document the unstable surface response to the pressure fluctuations in turbulent boundary layers. The holographic interferometer was recently used to record the surface topography in the presence of *stable* flow-induced deformations (Lee et al. 1993a).

The onset speed and wave characteristics of the solid-based class **B** and **C** instabilities were systematically documented in a series of towing-tank experiments (Gad-el-Hak 1986b). Divergence waves were observed on a single-layer viscoelastic coating made from a PVC plastisol. The flutter appeared on an elastic coating made from common household gelatin, but, in the absence of damping, its threshold speed was consistently lower than that for divergence (Figure 7.12). The damping in the PVC coating stabilized the traveling-wave flutter, and hence only divergence was observed there.[9] For the elastic coating, flutter appeared first and dominated the observed surface deformation. For both kind of waves, the threshold speed decreases with coating thickness; in other words, thin surfaces (relative to the displacement thickness of the boundary layer) are less susceptible to hydroelastic instabilities than thick ones.

Typical profiles of unstable class **B** and **C** waves were also recorded in the same hydrodynamic experiments using a laser displacement gauge (Figure 7.13. The vertical displacement at a point associated with the slow-moving, asymmetric, large-amplitude divergence waves contrasts with the faster, more-or-less symmetric, smaller-amplitude flutter. Both waves cause roughness-like effect, but the static divergence is the more dangerous instability. The phase speed of the static-divergence waves is of the order of 1% of the freestream speed, and their wavelength is about 5–10 times the coating thickness. The corresponding quantities for the flutter are 40% and 1.5–3, respectively.

Hess (1990) and Lee et al. (1993a) also investigated compliant coating effects on turbulent boundary layers. Both experiments were conducted in the same water tunnel, but the second article focused on the stable interaction between the fluid and a

---

[9] Parenthetically, this and similar earlier observations led Gad-el-Hak (1986a) and others to the wrong conclusion, as stated in footnote 3 in Section 7.4, regarding the classification of the divergence waves. To reiterate, the class **B** flutter is stabilized by damping, whereas the class **C** divergence is largely unaffected.

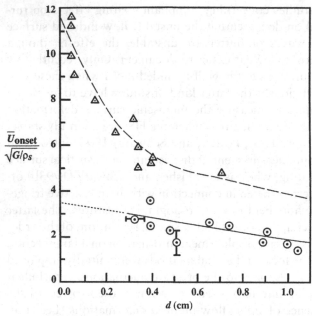

Figure 7.12: Onset speed dependence on thickness. ○, traveling-wave flutter on an elastic coating; △, static-divergence waves on a coating with damping [after Gad-el-Hak 1986b].

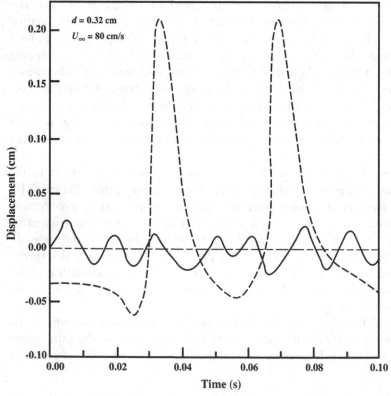

Figure 7.13: Typical unstable displacements of an elastic coating (———) and viscoelastic coating (– – – –) [after Gad-el-Hak 1986b].

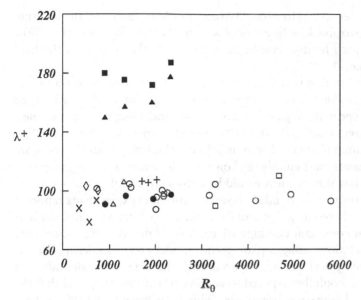

Figure 7.14: Variations of mean dimensionless spanwise spacing of the low-speed streaks in a turbulent boundary layer. Filled ○, rigid surface; filled △, compliant surface, friction velocity u_τ obtained from rigid-surface mean-velocity measurements; ■, compliant surface, u_τ obtained from compliant-surface mean-velocity measurements. These data are from the water tunnel experiments of Lee et al. (1993a). Other symbols in the figure are from classical rigid-surface measurements.

single-layer, homogeneous, viscoelastic coating made of a mixture of silicone rubber and silicone oil. Lee et al.'s coating was chosen based on the criterion established by Duncan (1986). In the presence of a stable wave pattern on the compliant surface, the flow visualization experiments indicated low-speed streaks with increased spanwise spacing (by as much as 80%; see Figure 7.14) and elongated spatial coherence compared with those obtained on a rigid surface. More significantly, for the particular compliant coating investigated, an intermittent relaminarization-like phenomenon was observed at low Reynolds numbers. Lee et al. (1993a) also reported a slight thickening of the buffer region and viscous sublayer and an upward vertical shift in the compliant law-of-the-wall. The streamwise turbulence intensity, the local skin-friction coefficient and the Reynolds stress across the boundary layer were all reduced, indicating a possible interruption of the feedback loop, which allows the turbulence to be self-sustaining. Thus, potentially favorable interaction between a compliant coating and a turbulent boundary layer has been demonstrated for the first time. The more recent hydrodynamic experiments by Choi et al. (1997) provide additional evidence for favorable interaction and indicate a total drag reduction for a long, slender body of revolution of the order of 7%.

7.8 The Future

The diminishing pool of researchers remaining active in the field of compliant coatings includes teams from the University of Warwick, University of Nottingham, Johns

Hopkins University, University of Houston, University of Maryland, and the Institute of Thermophysics in Novosibirsk. A larger pool was involved during the early 1980s, but the realities of research funding combined with the checkered past of the field led to the present decline.

Few suggestions for future research are given in here. The optimization procedures discussed in Section 7.6.2 have not been validated experimentally. Gaster-type experiments should be repeated using optimized coatings, including multi-panel ones. The recent claims by Davies and Carpenter (1997a) and Carpenter (1998) regarding the possibility of maintaining laminar flow to indefinitely high Reynolds numbers are very profound. Experiments, particularly field ones in low-disturbance environments specifically designed to test those claims would be extremely useful.

The results of the transitional-boundary-layer, wind-tunnel experiments reported by Lee et al. (1995, 1997) are intriguing and fly in the face of the conventional wisdom. They indicate that compliant coatings are capable of delaying transition even for air flows. Past calculations using a plate-spring model and considering the extremely large density of typical walls compared with the density of air indicated that very flimsy coatings would be required to achieve transition delay and that the situation gets worse as the air speed increases. This led Carpenter (1990) among others to conclude that the use of wall compliance is impractical for aeronautical applications. But the plate-spring results do not apply in any straightforward way to the homogeneous, single-layer walls studied by Lee et al. (1995). Validating the recent favorable results using both independent experiments and numerical simulations would open the door for aerodynamic applications, which is something that was seriously considered but later abandoned by NASA and the aerospace industry. The optimization procedures developed by Carpenter (1988) for transitional hydrodynamic flows should be extended to air flows. Experiments should be conducted using the resulting optimized coatings.

More complex coatings could potentially yield superior performance as compared with the relatively simple walls studied thus far. Multipanels, multilayers, anisotropic coatings, and combinations thereof should be investigated. In any such research program, experiments have to be guided with theoretical results. As already mentioned, trying to pick a compliant coating by trial and error is a very inefficient use of limited resources and will perhaps never work.

Favorably modulating a fully turbulent flow, in contrast to merely delaying transition, is also of great practical importance. The experimental results reported by Lee et al. (1993a) are very encouraging, but the coating used was chosen based on a rather simplistic model of the turbulence pressure fluctuations. To custom design compliant coatings to achieve particular control goals for turbulent wall-bounded flows, direct numerical simulations of the coupled fluid-structure system have to be performed. Turbulence modeling via classical closure schemes, although sufficient for some simple flows over rigid surfaces, will perhaps not yield reliable results for compliant walls. Direct numerical simulations, on the other hand, require extensive computer resources and are quite expensive to carry out. The bottom line is that relatively large investment in resources is required for this task, but the enormous potential payoffs could easily justify the expenditure.

Most of the research thus far has considered incompressible, zero-pressure-gradient, flat-plate boundary layers. Effects of compressibility, pressure gradient, and three-dimensionality on the performance of compliant coatings are largely unknown.

Such studies will yield invaluable information for field application of the control technique for both air and water flows. Most practical aerodynamic flows are in the moderate-to-high Mach number regime, and compressibility effects must therefore be investigated before compliant coatings are used on actual aircraft. Related to the pressure-gradient effects is the question of separated flows: Does compliant coating affect separation favorably or adversely? Other stability modifiers, such as favorable pressure gradient, suction, or heating or cooling, do delay transition as well as prevent separation. It is not known whether compliant coatings also have this dual benefit, and it may be beneficial to research the possibility. Finally, real flows are three-dimensional and involve complex geometries. A model problem for three-dimensional flows is the rotating disk. Few experiments were conducted using a rotating disk with a compliant face (Hansen and Hunston 1974). More recently, Cooper and Carpenter (1995; 1997a,b,c) analyzed the crossflow (type I; inviscid) as well as the viscous (type II) fluid-based instabilities that develop in the same three-dimensional flow. The preliminary results are encouraging and indicate that compliant coatings can suppress the more dangerous type I instabilities.

Active compliant coatings, or more properly internally driven flexible coatings, though bringing us back to the complexity of reactive control systems, is an emerging area deserving of further research. Energy expenditure is required to drive the wall, but the potential for significant net drag reduction is higher than that for passive coatings. The feasibility of the concept for stabilizing laminar boundary layers has been shown through numerical experiments (Metcalfe et al. 1986). Internally driven flexible coatings could also be used to suppress the Reynolds stress and reduce the skin-friction drag in turbulent wall-bounded flows, but any realistic field application of the technique has to await further development of reasonably priced and rugged microfabricated sensors and actuators (Gad-el-Hak 1994, 1996b). This topic will be revisited in Chapter 14.

Using the subject of compliant coatings to make a point, I would like to end this section with a personal commentary. There is a growing impatience among our fellow citizens with the glacial pace of transferring knowledge from the laboratories to the factories and thus getting back the invested research dollars of yesterday in the form of stronger industrial competitiveness tomorrow. In this era of instant gratification, what matters is not what basic research has done for you over the long haul, but what research has done lately. In his closing remarks on the occasion of the presentation of the 1990 American Physical Society Fluid Dynamics Prize, Lumley (1992) lamented that the United States is a curiously unsympathetic environment for a theoretician, or any scientist interested in fundamental work. At a later pulpit, Lumley (1996a) observed that

> public support of science in the United States has always been characterized by a certain inconstancy, a fickleness, a tendency to unsuitable liaisons and the transfer of affection from one pretty face to another at the drop of a hat. This is nowhere more evident than in the support of research in turbulence; the great need for a solution, coupled with the evident slow rate of progress, means that new ideas sometimes attract more attention than they are subsequently found to deserve. The turbulence community itself is somewhat prone to this kind of behavior.

Through all its ups and downs, compliant coating research provides a good case study of what my academic sibling John Lumley is talking about. Although in the general scheme of things this basic research is but a drop in the ocean, its triumphs and debacles are not untypical. The usual 5-year cycle for academic research is just enough to get off the ground. One should always remember the words of George Lichtenberg recited at the beginning of this chapter.

When the compliant coating research program was reignited in the early 1980s, few veterans from the 1960s and 1970s were around to share their valuable experiences, and the newcomers have had to climb the learning curve from its bottom. Nevertheless, we now know how to carry out stability calculations, solve fully coupled fluid–solid problems, numerically simulate turbulent flows, conduct well-controlled experiments for both transitional and turbulent flows, reliably measure surface deformation, identify as well as quantify coherent structures, optimize a coating for a particular task using first principles, and so forth. In other words, the compliant coating research community has now most of the tools it needs for significant further progress. Unfortunately, this community has recently been forced into an early retirement.

As amply illustrated in this chapter, compliant coating research, despite its checkered history, offers the potential for substantial transition delay and favorable interactions with turbulent boundary layers. It requires modest commitment of resources, but the payoff is extraordinary. It might be worth recalling that a mere 10% reduction in the total drag of an aircraft translates into a saving of $1 billion in annual fuel cost (at 1999 prices) for the commercial fleet of aircraft in the United States alone. Contrast this benefit to the annual expenditure of less than $2 million for the 5-year compliant coating research program that was sponsored by the U.S. Office of Naval Research in 1980. Private capital will not and cannot step in place of the government to support leading-edge research with long-term promise but without short-term return. Do we have the will, desire, patience, and resources to continue the journey toward real-life applications? In so many similar circumstances in the past, the answer was no. Scarce resources were spent for 5 years only to be hastily diverted to newer areas—long before the fruits of our labor have even had a chance of being reaped. But even in these difficult times of trying to reduce the federal budget deficit, one hopes for a different road to prosperity—and I do not mean for the researchers involved—this time around.

And while we are at it, cutting funds for basic research in general may be popular but is certainly unwise. Today's research generates the knowledge from which the future is built, and a responsible government must strike a balance between near-term goals and long-term economic growth and prosperity. Reducing spending on current outlays is one thing, but investing less in the country's future is an entirely different matter. Fundamental knowledge provides the foundation for a nation's productivity and economic growth, sustains its high standard of living, improves its quality of health and environment, and ensures its security. The overall detrimental impact of reducing government funding for basic research probably will not be felt for a generation, but it will be felt.

7.9   Parting Remarks

Passive compliant coatings present a much simpler alternative to reactive flow-control strategies aimed at favorably interfering with wall-bounded flows. The last 10–15 years have witnessed renewed interest in compliant coatings as a means to achieve beneficial flow-control goals. Significant advances were made in numerical and analytical techniques to solve the coupled fluid structure problem. Novel experimental tools were developed to measure the stable as well as the unstable surface deformations caused by the pressure fluctuations in the boundary layer. In turbulent wall-bounded flows, coherent structures were routinely identified, and their modulation by wall compliance could be quantified.

Most significant results in the field thus far were obtained when a strong cooperation existed between theory and experiment. Recent theoretical work indicates that complete suppression of the Tollmien–Schlichting waves may be possible, provided that optimized multiple-panel compliant walls are used. The new finding raises the possibility of maintaining laminar flow to indefinitely high Reynolds numbers, which is a very profound prospect indeed. Recent experiments indicate favorable compliant coating interactions even for aerodynamic flows and even for turbulent boundary layers. More research is needed, however, to confirm these latest results.

The coupled system instabilities are now well understood, and compliant coatings can therefore be rationally designed to achieve substantial, perhaps even indefinite, transition delay in hydrodynamic flows. That recent shift from random to rational search for the right kind of coating is not unlike the great paradigm change in synthetic chemistry that took place near the beginning of the twentieth century. Increased understanding of the molecular geometry of organic compounds changed the scene from a hapless alchemist muddling around hoping to chance the right combination of ingredients, heat, pressure, and catalysts to produce something useful, to a professional chemist figuring out what he or she wants and working backward from the shape of a desired molecule for, say, a synthetic hormone. In fact, the present analogy is apt: the fluid dynamist working with the Navier–Stokes equations can tell the chemist the exact properties of the compliant coating to be synthesized to achieve a given goal. If the needed molecular structure is too complicated, futuristic nanoscale machines can assemble the required molecules directly, element by element.

CHAPTER EIGHT

Separation Control

> Logical consequences are the scarecrows of fools and the beacons of wise
> men.
> (Thomas Henry Huxley, 1825–1895)

> Aristotle discovered all the half-truths which were necessary to the
> creation of science.
> (Alfred North Whitehead, 1861–1947)

PROLOGUE

Under certain conditions, wall-bounded flows separate. To improve the performance
of natural or man-made flow systems, it may be beneficial to delay or advance this
detachment process. This chapter reviews the status and outlook of separation control
for steady and unsteady flows. Passive and active techniques to prevent or to provoke
flow detachment are considered, and suggestions are made for further research.

8.1 Introduction

8.1.1 The Phenomenon of Separation

Fluid particles in a boundary layer are slowed down by wall friction. If the flow
is sufficiently retarded, for example, owing to the presence of an adverse pressure
gradient, the momentum of those particles will be reduced by both the wall shear
and the pressure gradient. In terms of energy principles, the kinetic energy gained at
the expense of potential energy in the favorable-pressure-gradient region is depleted
by viscous effects within the boundary layer. In the adverse-pressure-gradient region,
the remaining kinetic energy is converted to potential energy but is too small to
surmount the pressure hill, and the motion of near-wall fluid particles is eventually
arrested. At some point (or line), the viscous layer departs or breaks away from the
bounding surface. The surface streamline nearest to the wall leaves the body at this
point, and the boundary layer is said to separate (Maskell 1955). At separation,
the rotational flow region next to the wall abruptly thickens, the normal velocity
component increases, and the boundary-layer approximations are no longer valid.

Because of the large energy losses associated with boundary-layer separation, the performance of many practical devices is often limited by the separation location. For example, if separation is postponed, the pressure drag of a bluff body is decreased, the circulation and hence the lift of an airfoil at high angle of attack is enhanced, and the pressure recovery of a diffuser is improved. On the other hand, the high-lift capabilities of delta wings are achieved by provoking separation and forming leading-edge vortices.

Great strides have been made in the past few decades in establishing a firm analytical foundation for steady, two-dimensional separation. On the other hand, theoretical or numerical analysis of three-dimensional or unsteady separation is considerably less developed, and progress to date in these areas has depended crucially on experimental work. From a practical point of view, the local separation over, say, a lifting surface has a very strong effect on the global aerodynamic properties, and a thorough, fundamental understanding of the phenomenon is obviously needed. Separation on such a surface occurs just prior to or at maximum loading, thus greatly influencing the device's optimum performance.

The breakthrough in unsteady separation research was achieved by Moore, Rott, and Sears during the 1950s. Before their work, it was believed that steady and unsteady separations have the same characteristics; namely, the point of vanishing wall shear, the termination of the boundary layer, and the beginning of the wake or bubble of separated fluid. Rott (1956), in analyzing the unsteady flow in the vicinity of a stagnation point, noted that the point of vanishing wall shear does not coincide with the point of boundary-layer detachment. In the same year, Sears made the assumption that the unsteady separation point is characterized by the simultaneous vanishing of the shear and the velocity at a point *within* the boundary layer as seen by an observer in a coordinate convected with the separation velocity. Two years later, Moore (1958), while investigating a steady flow over a moving wall, arrived at the analogous model to that for unsteady separation. He stated that "for a slowly moving wall separation occurs when, at some point in the boundary layer, the profile velocity and shear simultaneously vanish."

8.1.2 Separation Control

Of all the various types of shear flow control now extant, control of flow separation, historically referred to as boundary-layer control or BLC, is probably the oldest and most economically important. Separation control is of immense importance to the performance of air, land, and sea vehicles; turbomachines; diffusers; and a variety of other technologically important systems involving fluid flow. Generally it is desired to postpone separation so that form drag is reduced, stall is delayed, lift is enhanced, and pressure recovery is improved. However, in some instances it may be beneficial to provoke separation. For example, to improve the subsonic high-lift performance of an airfoil optimized for supersonic flight, a flap may be used to initiate leading-edge separation followed by reattachment.

Flow separation causes significant deviations from inviscid pressure distributions. Such deviations can be either favorable, as in the vortex lift associated with lee surface separation on swept leading edges and the use of spoilers to obviate ground-effect-lift during landing or, more commonly, detrimental, resulting, for example, in high form drag and reduced diffuser efficiency. Typical applications and benefits of flow separation control include effective low-Reynolds-number airfoils for remotely piloted vehicles (RPVs), propellers, windmills, helicopters, and so forth; efficient inlets and diffusers; improved axial flow compressors; increased $C_{L_{max}}$ for greater payload, reduced engine power and noise at takeoff, shorter runways and reduced approach speed; supermaneuverability or birdlike flight; efficient and effective stall or spin control; reduced drag on missiles, automobiles, ships, and helicopters; and a myriad of applications in industrial aero- and hydrodynamics. As examples of estimated benefits, a 5% improvement in landing $C_{L_{max}}$ would allow a 25% payload increase (Butter 1984); mitigation of military stall and spin accidents, which from 1977 to 1986 involved over 150 aircraft; application of flow separation control to tractor-trailer truck drag reduction, which could save in excess of 50 million barrels of oil per year (Muirhead and Saltzman 1979); and diffuser suction in supersonic closed-circuit wind tunnels, which would result in an estimated 30–50% power reduction on a device that can draw in excess of 20 MW (Bushnell and Trimpi 1986).

8.1.3 Method of Control

Flow separation control is currently employed, for example, via vortex generators on the wings of most Boeing aircraft; via blown flaps on older generation supersonic fighters or leading-edge extensions and strakes on newer generations; and via passive bleed in the inlets of supersonic engines on, for example, the SR–71 and Concorde. Future possibilities for aeronautical applications of flow separation control include providing structurally efficient alternatives to flaps or slats; cruise application on conventional takeoff and landing aircraft (CTOL), including BLC on thick spanloader wings (Smith and Thelander 1973); and cruise application on high-speed civil transports (HSCT) for favorable interference-wave drag reduction, increased leading-edge thrust, and enhanced fuselage and upper surface lift (Bushnell 1990). In fact, much of the remaining gains to be made in aerodynamics appear to involve various types of flow control, including separated flow control (e.g., Hazen 1984; Chambers 1986).

Typical, in some cases serious, problems associated with flow-separation control include parasitic device drag or energy consumption; system weight, volume, complexity, reliability or cost; performance sensitivity to body attitude or orientation; and, especially in the case of the automobile, styling (Poisson-Quinton 1950; Silhanek 1969; Decken 1971; Sovran, Morel, and Mason 1978). The status of flow-separation control is still typified by two comments made by Darby (1954): "The Germans were experimenting with BLC in the wind tunnel in the 1920s—NACA began wind-tunnel tests in the late 1920s. It was not until 1949 that American agencies became interested—the skepticism seems to be breaking down at long last, but any branch of technology which has gone so long without bearing fruit certainly requires a close scrutiny." "There are an almost bewildering number of systems, at least for separation control, which work."

In general, the field of flow-separation control is far richer than the conventional view, which usually considers only suction, injection, and vortex generators. Many decades of research have proven that separation control, in most of its guises, will work. The task of researchers and designers now is to improve reliability, where necessary, and increase net gains through innovation. This richness of existing approaches for flow-separation control is mirrored in the extensive literature partially available for low-speed flows in the publications by Lachmann (1961), Colin and Williams (1971), Chang (1976), Adkins (1977), Gad-el-Hak (1989, 1990), Lin, Howard, Bushnell, and Selby (1990), and Gad-el-Hak and Bushnell (1991a,b) and references therein as well as herein. For high-speed flows, excellent reviews for separation control in shock-boundary-layer interactions are provided by Delery (1985) and Viswanath (1988).

Given an imposed pressure field, the kernel problem in separation postponement is to add momentum to the very near-wall region of the flow by either transferring momentum from flow regions farther from the wall, which are still momentum rich, or by direct addition of power drawn from the propulsive system. Probably the most popular flow-separation control technique has been to add momentum to the near-wall region either actively (e.g., tangential blowing or wall jets) or passively (e.g., boundary-layer tripping, turbulence enhancement, or vortex generators of various scales).

8.1.4 Outline

The objective of this chapter is to summarize the status and outlook for flow separation control with an emphasis on more recent developments such as optimized vortex generators; various methods of turbulence control, including dynamic inputs; and wall-heat transfer. Direct comparisons between the various approaches on common test problems will be made. The recent advances in computational fluid dynamics are allowing the transformation of much of BLC from an empirical art to a predictive science, and this subject will be briefly discussed. Owing to space limitations, the presentation does not include large-scale-provoked separation or control *of* vortices (e.g., Williams and Amato 1989; Bushnell and Donaldson 1990; Panton 1990c; Williams and Papazian 1991) nor the related problem of asymmetric nose–body vortex control on missiles and similar vehicles at high angles of attack (Vakili 1990). However, control *by* vortices is included in this chapter. Separation control is addressed for both nominally two-dimensional and three-dimensional flows, although the knowledge base concerning the latter is noticeably deficient. Typically, three-dimensional separation occurs sooner but is less catastrophic than the quasi-two-dimensional case (Driver 1989).

In this chapter, we review methods to control separation for steady and unsteady boundary layers. Both passive and active techniques to prevent or to provoke flow detachment are considered. For historical reasons, most of the separation control methods discussed here were developed for aircraft wings. The results, however, pertain to fundamental properties of fluid flow and could readily be extended to a variety of systems such as diffusers, steam turbine blades, wind turbine rotors, pump impellers, and off-shore structure components. To set the appropriate mechanistic framework for discussing the variety of control tools available, differences between steady and unsteady separations are first considered.

8.2 Steady and Unsteady Separation

8.2.1 Steady Separation

Separation of a steady, two-dimensional boundary layer was explained first by Prandtl (1904) in his milestone presentation "Über Flüssigkeitsbewegung bei sehr kleiner Reibung" in which he introduced the boundary-layer theory. Fluid particles near the surface are retarded by the friction of the wall and by any adverse pressure gradient present in the freestream. If the near-wall fluid has insufficient momentum for it to continue its motion, it will be brought to rest at the separation point (line).

Further downstream, the adverse pressure forces will cause reverse flow. Because the velocity at the wall is always zero, the gradient $[\partial u/\partial y]_{y=0}$ must be positive upstream of separation, zero at the point of separation, and negative in the reverse-flow region. For an axially symmetric flow, the line of separation becomes a circle, and the point of vanishing shear still coincides with the point of separation.

In Prandtl's view, the separation point is entirely determined by external conditions. Boundary-layer separation is accompanied by a thickening of the rotational flow region and ejection of vorticity. Downstream of the separation point the shear layer either passes over the region of recirculating fluid and reattaches to the body surface or forms a wake and never reattaches to the body. The characteristic dimension of the recirculating region is quite large in the latter case and is of the order of the body height.

Analytically, the solution of the steady, two-dimensional, laminar boundary-layer equations with a prescribed external-pressure (or external-velocity) distribution breaks down at the point of separation, and this is commonly known as the Goldstein's singularity, in honor of Sydney Goldstein, who, in 1948, first noted the singular behavior of the solution near a point of zero skin friction. (For a review of other singularities occurring in the equations for three-dimensional or unsteady boundary-layer flows, or both, see the article by Williams 1985.) The singularity of the boundary-layer equations at separation is obviously not a physical property of the flow and can be overcome by prescribing either the displacement thickness or the wall-shear distribution instead of the external pressure. This kind of analysis is termed inverse calculation.

8.2.2   Unsteady Separation

For two-dimensional flow over moving walls, two-dimensional unsteady flows, and three-dimensional steady and unsteady flows the point (line) of vanishing wall shear does not necessarily coincide with separation, and this greatly complicates the problem.   This was first observed by Rott (1956) while analyzing the unsteady flow in the vicinity of a stagnation point. He observed that, although the wall shear vanished with an accompanying reverse flow, there was no singularity or breakdown of the boundary-layer assumptions. In seeking a generalized model for separation, Sears (1956) postulated that the unsteady separation point is characterized by the simultaneous vanishing of the shear and the velocity at a point within the boundary later as seen by an observer moving with the velocity of the separation point.

Moore (1958), while investigating a steady flow over a moving wall, arrived at the same model for unsteady separation. On the basis of an intuitive relationship between steady flow over a moving wall and unsteady flow over a fixed wall, Moore was able to sketch the expected velocity profiles for both cases, as seen in Figures 8.1 and 8.2. He considered the possibility that a Goldstein-type singularity occurs at the location where the velocity profile simultaneously has zero velocity and shear at a point above the moving wall. Equivalently, for unsteady separation on a fixed wall, the separation point is the location at which both the shear and velocity vanish in a singular fashion in a frame of reference moving with the separation point. The main drawback of this Moore–Rott–Sears (MRS) model in the fixed-wall case is that the speed of the separation point is not known a priori, making it difficult to locate this point and forcing researchers to rely on more qualitative measures for unsteady separation. Sears and Telionis (1972a, 1975) were the first to prove that the boundary-layer

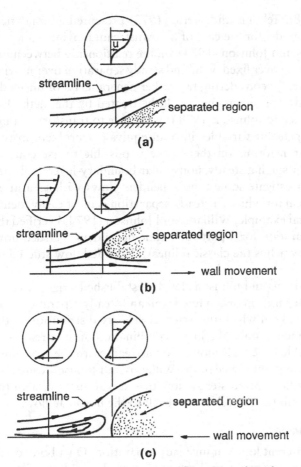

Figure 8.1: Streamlines and velocity profiles when a boundary layer over a fixed or moving wall separates. (a) Fixed wall; (b) Wall moving downstream; (c) Wall moving upstream.

separation singularity accompanies detachment of unsteady flows. They suggested that the existence of such singularity could serve as a criterion for unsteady separation when numerically integrating the boundary-layer equations.

Because of the preceding difficulty, early attempts to verify this important MRS model of unsteady separation considered the more tractable problem of steady separation over moving walls. Both Vidal (1959) and Ludwig (1964) experimentally investigated a shrouded rotating cylinder in steady flow. As expected, separation was delayed (moved downstream) when the wall moved in the freestream direction and was advanced when the wall moved opposite the main flow. The measured velocity profiles corresponded to those

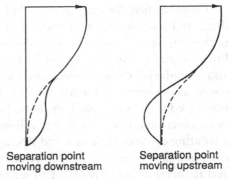

Separation point moving downstream

Separation point moving upstream

Figure 8.2: Moore's (1958) postulated velocity profiles for unsteady separation on a fixed wall. (———) Moving separation point; (–––––) fixed separation point.

hypothesized by Moore (1958). Telionis and Werle (1973) presented an analytical verification of the separation model for the case of a downstream moving wall.

Until the work of Williams and Johnson (1974a,b), the relationship between un-steady boundary-layer separation over fixed walls and steady separation over moving walls has been an intuitive one. By considering the rather general class of unsteady, two-dimensional boundary-layer problems that could be treated by the method of semisimilar solutions, Williams and Johnson (1974a) were able to transform an un-steady problem in three independent variables into an equivalent problem in two independent variables. This transformation then made it possible to use conven-tional numerical techniques for solving steady, nonsimilar boundary-layer problems and allowed the authors to investigate some time-dependent flows with separation occurring in a coordinate system for which unsteady separation is most easily identi-fied and analyzed. As a practical example, Williams and Johnson (1974a) verified the MRS model for unsteady separation for a particular time-dependent retarded flow, the steady flow equivalent of which is the classical linearly retarded flow studied by Howarth (1938).

In a subsequent article, Williams and Johnson (1974b) established a rigorous ana-lytical link between unsteady separation over a fixed wall and steady separation over a moving wall for the special case in which the external velocity distribution in the fixed coordinate system is a function only of a linear combination of the streamwise coordinate and time and in which the wall moves downstream with the (constant) speed of the unsteady separation point. Once more, Williams and Johnson were able to transform a given unsteady flow into a steady flow over a wall moving with the speed of the separation point and then to relate this back to the unsteady flow.

Archetypes of Unsteady Separation

There are two distinctly different kinds of unsteady separation. On a body oscil-lating in pitch, the pressure gradient may vary both in magnitude and form. On the other hand, if the freestream velocity changes periodically with time, the impressed (mean) pressure gradient over a fixed surface varies in magnitude but not in form during the oscillation. Detailed velocity measurements in the first kind are scarce because of the obvious difficulties involved in instrumenting a moving body. Accord-ingly, the exact nature of unsteady separation for this case remains unclear despite the abundance of qualitative flow visualization data (Gad-el-Hak 1987b).

Consider first the case of an airfoil oscillating in pitch. Sudden changes in lift, drag, and pitching moment occur near the onset of separation. The dynamic stall is characterized by two distinctly different flow phenomena (Ericsson 1971, 1988; Ericsson and Reding 1971, 1987). The first is quasi-steady and is due to time-lag and boundary-layer improvement effects. The second phenomenon is transient and concerns the effects of the forward movement of the separation point. As the airfoil leading edge moves upward, the boundary layer between the stagnation and separa-tion points experiences a moving wall–wall jet effect very similar to that observed on a rotating cylinder (Ericsson and Reding 1984, 1986, 1987; Ericsson 1988). Thus, the boundary layer has a fuller velocity profile as compared with the steady case and is therefore more resistant to separation. On the downstroke the effect is the opposite, promoting separation. At stall, a vortex is shed from the leading edge and is convected downstream over the chord. For high reduced frequency, this spilled vortex remains over the airfoil a significant portion of the cycle, and extra lift is produced.

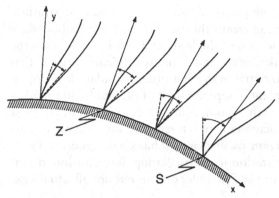

Figure 8.3: Envelopes of velocity profiles for periodic oscillations. The dashed lines and the arrows indicate the extreme positions of the tangents to the profiles at the wall. Schematic representation of Despard and Miller's (1971) data.

The relatively simpler second kind of unsteady separation was investigated by Despard and Miller (1971), who performed a series of experiments to measure the instantaneous velocity profiles in oscillating boundary layers subjected to adverse pressure gradients. An array of 10 hot-wire probes was used to measure the normal profiles of streamwise velocity at several points along the rear portion of an airfoil-like body placed in a low-speed wind tunnel. The freestream velocity was varied periodically by a set of rotating shutter valves located downstream of the test section. The resulting envelopes of the velocity profiles throughout the entire cycle of oscillation are shown schematically in Figure 8.3 for four different stations along the body. Within half a boundary layer thickness of the beginning of the region of wake formation, the point S in the figure exhibits reverse flow or zero velocity gradient at the wall for the entire cycle of profiles. This led Despard and Miller to propose a practical definition of unsteady separation as the farthest upstream point at which there is zero velocity or reverse flow at some point in the velocity profile throughout the entire oscillation cycle. This approach avoids the difficulty of determining the speed of the separation point as required by the MRS model. Point Z in Figure 8.3 is the farthest upstream point at which the wall shear stress passes through zero during a complete cycle of oscillation. Therefore, the point of vanishing shear oscillates between S and Z, but the outer flow remains attached until S. Thus, Despard and Miller's results provided the first experimental evidence that a thin layer of reversed flow can be embedded at the bottom of an otherwise attached boundary layer, as postulated by Moore (1958). Despard and Miller further concluded that the presence of an oscillating velocity component and a time-dependent adverse pressure gradient cause the separation point to be displaced upstream of its steady-flow position. The distance between the two points diminishes as the frequency of oscillation increases but is rather insensitive to variations in the amplitude of the outer flow oscillations.

Shortly after this important experiment, Sears and Telionis (1975) provided a review of the numerical evidence for the validity of the MRS model. Numerical integrations of unsteady laminar and turbulent boundary layers showed no evidence of the Goldstein singularity, even in a region of partially reversed flow and through the point of zero skin friction. The singularity did appear, however, at the point of unsteady separation, as defined by the MRS criterion.

Some controversy still exists over the precise definition of unsteady separation. Telionis (1979) reiterated that separation means the location on the solid boundary where the flow stops creeping over the skin of the body and breaks away from the wall, thus generating a turbulent wake. Sears and Telionis (1972a,b) argued that abrupt changes of boundary-layer properties in the first-order boundary-layer equations may signal the approach of the point of separation or of the Goldstein singularity. Despard and Miller (1971), on the other hand, defined separation as the farthest upstream station at which the shear fluctuates between zero and some negative value throughout an entire cycle of freestream oscillation. Tsahalis and Telionis (1974) numerically studied the unsteady separation in an oscillating flow and found that Despard and Miller's criterion could be verified under certain but not all situations.

Turbulent Boundary Layers

The carefully executed, well-documented experiments of Simpson and his colleagues (Shiloh, Shivaprasad, and Simpson 1981; Simpson, Chew, and Shivaprasad 1981a,b, 1983; Simpson and Shivaprasad 1983) are very useful in understanding some of the basics of steady and unsteady separation in turbulent flows. Simpson et al. investigated the structure of separating, nominally two-dimensional, turbulent boundary layers in a steady freestream (Shiloh et al. 1981; Simpson et al. 1981a,b) as well as in a sinusoidally oscillating freestream (Simpson et al. 1983; Simpson and Shivaprasad 1983). Their experiments illustrate both the similarities and differences between steady and unsteady separations. Upstream of where intermittent backflow begins, the flow behaves in a quasi-steady manner. Downstream, there are non-quasi-steady effects on the ensemble-averaged flow structure. Also, the hysteresis effects in the ensemble-averaged velocity profiles are more pronounced for higher-reduced-frequency freestream oscillations. Simpson (1989) provides a comprehensive review of the topic of turbulent flow separation.

8.2.3 Vortex-Induced Separation

Separation can be triggered experimentally with concentrated vortices moving close to a wall. Harvey and Perry (1971) placed a half-span rectangular wing in a wind tunnel and observed the evolution of the resulting single tip vortex as it passes over a downstream moving floor (Figure 8.4). Total-head surveys in planes across the flow were conducted to observe separation and the formation of a secondary vortex of opposite sense to the main one. The secondary vortex causes a rebounding of the trailing vortex, as sketched in Figure 8.5. Similar effects are present when a ring vortex approaches a wall normal to its propagation direction (Magarvey and McLatchy 1964; Schneider 1980). In the latter case, a downstream-moving, unsteady separation is produced.

Ho (1986) attributed the sparseness of a positive experimental proof (or disproof) of the MRS criterion for unsteady separation to the difficulty of measuring the separated velocity near the wall. The temporal variation of the resulting three-dimensional velocity field necessitates the use of a large number of small, fast-response probes and a colossal data acquisition and storage system. This problem is particularly acute for an upstream moving separation in which flow reversal occurs, leading to additional measuring difficulties. No flow reversal occurs in a downstream moving separation, and hot-wire probes can then be used to survey the flowfield in the vicinity of the wall.

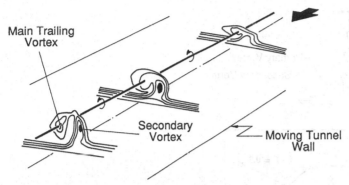

Figure 8.4: Surveys of total head when a tip vortex passes over a downstream moving wall [after Harvey and Perry 1971].

Detailed measurements of the downstream-moving separation caused by periodic ring vortices impinging normally onto a flat plate were conducted by Didden and Ho (1985). They forced an axisymmetric jet to produce a train of primary ring vortices in the air-jet shear layer sketched at the top of Figure 8.6. The phase-locked flow visualization pictures obtained by Didden and Ho using smoke clearly show that the wall-jet boundary layer evolves into a secondary vortex that is counterrotating with respect to the primary vortex and periodically separates. The unsteady separation is induced by the primary vortex and moves downstream in the radial mean-flow direction, thus allowing detailed measurements with hot-wire probes. The phase-averaged measurements of normal and radial velocity components provide data for locating the onset of separation in space and time.

Like all other cases discussed thus far, the unsteady separation phenomenon observed in the impinging jet flow involves a strong viscous–inviscid interaction. The detailed data obtained by Didden and Ho (1985) reveal the following sequence of events. The primary ring vortex in the inviscid region produces a fast-moving stream while it approaches the wall. The resulting accelerating flow causes a negative wall pressure in the low-pressure region downstream from the jet axis. Further downstream, pressure recovery results in an unsteady adverse pressure gradient, which retards the flow in the viscous layer. Hence, a strong shear layer is generated at the viscous–inviscid interface. The instability of this shear layer leads to the rollup of the vortex sheet and the formation of a secondary vortex. The ejection of this secondary vortex, believed to be associated with the onset of unsteady separation, leads to an abrupt increase in the momentum thickness and the formation of a bulge that convects downstream at the same speed as the zero-shear-stress point of the radial velocity profile. In other words, the unsteady separation originates from a local

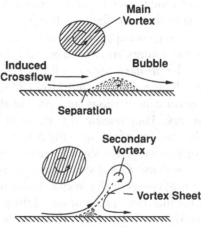

Figure 8.5: Suggested interpretation of the total-head surveys. (a) A section downstream of the initial separation; (b) Subsequent development of the secondary vortex [after Harvey and Perry 1971].

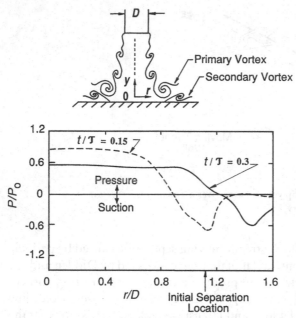

Figure 8.6: Forced axisymmetric jet impinging normally onto a flat plate. Phase-averaged pressure traces at the instant of separation ($t/T = 0.15$) and after a short time interval ($t/T = 0.3$) [after Didden and Ho 1985].

shear layer, which is initiated in the viscous region by the unsteady adverse pressure gradient produced by the primary vortex.

At the instant of apparent separation, a sharp, high-amplitude suction peak occurs in the surface pressure signature of the secondary vortex, as shown with the broken line in Figure 8.6 representing the (normalized) phase-averaged surface pressure trace at $t/T = 0.15$, where t is time and T is the period of the forcing signal. After the passing of both the primary and secondary vortices, the boundary layer reattaches, and a high-level pressure is found in the wake of separation, as shown with the solid line in the same figure a short time after separation ($t/T = 0.3$).

Given the appearance of a sharp, negative peak pressure at about the same time as the secondary vortex and because the relatively more gentle negative pressure induced by the primary vortex as it passed very nearly parallel to the wall was not expected to change, Didden and Ho (1985) conjectured that the observed high-amplitude pressure fluctuations are caused by the unsteady separation rather than by the primary vortex. They reiterated that, although viscosity does not play an important role in the stability of the spatially developing local shear layer, viscous effects are essential for producing the shear layer before the onset of the unsteady separation.

Because a local shear layer has been observed in many other separated flows, Ho (1986) contends that such a shear layer is a generic flow module to all separated flows. This interesting idea of associating unsteady separation with a spatially developing local shear layer allows researchers in the subject area to borrow from the extensive knowledge available in free-shear-layer studies (Ho and Huang 1982; Ho and Huerre 1984). Many examples of this association are provided in the review article by Gad-el-Hak (1987b).

The qualitative criteria of separation, namely, the thickening of the boundary layer, the ejection of vorticity, and the vanishing of the wall shear, do not appear at the same time or convect at the same speed in the unsteady case. The MRS criterion for unsteady separation is a quantitative one. However, as mentioned earlier, the convection speed of the separation point is not known a priori, and this greatly increases the difficulty of the problem. By choosing an appropriate frame of reference based on a well-documented flowfield, Didden and Ho (1985) were able to provide positive proof of the MRS criterion for a downstream-moving separation. Moreover, at the separation position and in a frame moving at the local velocity of the point of zero shear stress (or the convection speed of the boundary layer's bulge), the phase-averaged streamlines were found to form a closed recirculating region in qualitative agreement with the calculations reported by Walker (1978) for a line vortex near a wall.

8.3 Equations of Motion at the Wall

To examine the local and instantaneous effects of many of the separation control tools to be described in this chapter, we need to bring to the forefront three equations from Chapter 2. Recall the Navier–Stokes equations written at the wall from Section 2.4. In wall-bounded flows, the streamwise (x) and spanwise (z) momentum equations written at the wall give very useful expressions of the wall fluxes of, respectively, the spanwise and streamwise vorticity. For an incompressible fluid over a nonmoving wall of small curvature and with body forces neglected, the streamwise, normal, and spanwise momentum equations, written at $y = 0$, read

$$\rho v_{\mathrm{w}} \frac{\partial u}{\partial y}\bigg|_{y=0} + \frac{\partial p}{\partial x}\bigg|_{y=0} - \frac{\partial \mu}{\partial y}\bigg|_{y=0} \frac{\partial u}{\partial y}\bigg|_{y=0} = \mu \frac{\partial^2 u}{\partial y^2}\bigg|_{y=0} \tag{8.1}$$

$$0 + \frac{\partial p}{\partial y}\bigg|_{y=0} - 0 = \mu \frac{\partial^2 v}{\partial y^2}\bigg|_{y=0} \tag{8.2}$$

$$\rho v_{\mathrm{w}} \frac{\partial w}{\partial y}\bigg|_{y=0} + \frac{\partial p}{\partial z}\bigg|_{y=0} - \frac{\partial \mu}{\partial y}\bigg|_{y=0} \frac{\partial w}{\partial y}\bigg|_{y=0} = \mu \frac{\partial^2 w}{\partial y^2}\bigg|_{y=0} \tag{8.3}$$

The preceding equations are instantaneous and are valid for both laminar and turbulent flows. Here, v_{w} is the (positive) injection or (negative) suction velocity through the wall. Upward or downward motion of the wall acts analogously to injection or suction. The right-hand side of each of the three equations represents the curvature of the corresponding velocity profile, which in the case of the streamwise–spanwise equation is the same as the flux of spanwise–streamwise vorticity from the wall. Negative curvature implies that the velocity profile is fuller and that the wall is a source (positive flux) of vorticity. Whether the wall is a source or a sink of spanwise–streamwise vorticity depends on whether fluid is sucked or injected from the wall, whether the streamwise–spanwise pressure gradient is favorable or adverse, and whether the wall viscosity is lower or higher than that above the surface.

8.3.1 Equation for Turbulent Flows

For a turbulent flow, Eqs. (8.1)–(8.3) are still valid instantaneously. For a canonical turbulent flow, the instantaneous velocity profiles in all three directions change shape continuously as a result of the random changes in the pressure field. But what about the mean quantities? Recall from Section 2.5 the equation for the mean momentum for a Newtonian, incompressible turbulent flow. Let $u_i = \overline{U}_i + u_i'$ and $p = \overline{P} + p'$, where $\overline{U}_i$ and $\overline{P}$ are ensemble averages for the velocity and pressure, respectively, and u'_i and p' are the velocity and pressure fluctuations about the respective averages. The continuity and momentum equations governing the mean velocity and mean pressure for an incompressible turbulent flow become

$$\frac{\partial \overline{U}_k}{\partial x_k} = 0 \tag{8.4}$$

$$\rho\left(\frac{\partial \overline{U}_i}{\partial t} + \overline{U}_k \frac{\partial \overline{U}_i}{\partial x_k}\right) = -\frac{\partial \overline{P}}{\partial x_i} + \frac{\partial}{\partial x_k}\left(\mu \frac{\partial \overline{U}_i}{\partial x_k} - \rho\,\overline{u_i u_k}\right) + \rho\,\overline{g}_i \tag{8.5}$$

where, for clarity, the primes have been dropped from the fluctuating velocity components u_i and u_k. At the wall, the equation for the streamwise component of momentum reads

$$\rho v_{\mathrm{w}}\left.\frac{\partial \overline{U}}{\partial y}\right|_{y=0} + \left.\frac{\partial \overline{P}}{\partial x}\right|_{y=0} - \frac{d\mu}{d\overline{T}}\left.\frac{\partial \overline{T}}{\partial y}\right|_{y=0}\left.\frac{\partial \overline{U}}{\partial y}\right|_{y=0} + \rho\left.\frac{\partial \overline{uv}}{\partial y}\right|_{y=0} = \mu\left.\frac{\partial^2 \overline{U}}{\partial y^2}\right|_{y=0} \tag{8.6}$$

where $\overline{T}$ is the mean temperature field, and viscosity is assumed to change as a result of surface heat transfer. The right-hand side of Eq. (8.6) is the flux of mean spanwise vorticity, $\overline{\Omega}_z = -\partial \overline{U}/\partial y$, at the surface. In the absence of wall transpiration, mean pressure gradient, and surface heating or cooling, the first three terms on the left-hand side of Equation (8.6) vanish. The fourth term on the left-hand side is the slope of the normal profile of $\overline{uv}$ at $y = 0$. This term could be asymptotically estimated as the wall is approached. Consider a Taylor's series expansion in powers of y in the neighborhood of the point $y = 0$. As a result of the no-slip condition, the streamwise velocity fluctuations u varies linearly with y. To conserve mass, the normal velocity fluctuations must vary as y^2. It follows then that very near the wall (within the viscous sublayer), the tangential Reynolds stress $\overline{uv}$ varies at least as y^3 and that $\partial \overline{uv}/\partial y$ varies as y^2. At the wall itself, $y = 0$ and $[\partial \overline{uv}/\partial y]_{y=0} = 0$, although close to the wall the slope of the tangential Reynolds stress profile is quite large.

Therefore, it can be deduced from Eq. (8.6) that the mean streamwise velocity profile for the canonical turbulent boundary layer (two-dimensional, incompressible, isothermal, zero-pressure-gradient, over an impervious, rigid surface) will have a zero curvature at the wall. Notwithstanding this common characteristic with the Blasius boundary layer (Figure 6.2b of Chapter 6), the turbulent boundary layer is quite different from the laminar one. As pointed out by Lighthill (1963a), turbulent mixing concentrates most of the mean vorticity much closer to the wall as compared with the laminar case. The mean vorticity at the wall, $[\partial \overline{U}/\partial y]_{y=0}$, is typically an order of magnitude larger than that in the laminar case. This explains the higher skin-friction drag associated with a turbulent flow. Turbulent mixing also causes the mean vorticity to migrate away from the wall, and about 5% of the total is found much farther from the surface. The flux of mean spanwise vorticity is zero at the wall itself but very large close to it, reaching a maximum at about the same location where the root-mean-square vorticity fluctuations peak (near the edge of the viscous sublayer). This trait is responsible for the turbulent boundary layer's resistance to separation.

In the next six sections, available and contemplated flow-control methods to delay or to provoke separation will be discussed. The equations developed in this section for laminar and turbulent boundary layers will help in presenting a unified view of the different control techniques.

8.4   Velocity Profile Modifiers

As mentioned in Section 8.2.1, Prandtl (1904) was the first to explain the mechanics of separation. He provided a precise criterion for its onset for the case of a steady, two-dimensional boundary layer developing over a fixed wall. If such a flow is retarded, the near-wall fluid may have insufficient momentum to continue its motion and will be brought to rest at the point of separation. Fluid particles behind this point move in a direction opposite to the external stream, and the original boundary-layer fluid passes over a region of recirculating flow. Because the velocity at the wall is always zero, the gradient $[\partial u/\partial y]_{y=0}$ will be positive upstream of separation, zero at the point of separation, and negative in the reverse-flow region. The velocity profile at separation must then have a positive curvature at the wall. However, $[\partial^2 u/\partial y^2]_{y=0}$ is negative at a large distance from the wall, which means the velocity profile at separation must have a point of inflection somewhere above the wall, as shown in Figure 6.2d. Because $[\partial^2 u/\partial y^2]_{y=0} > 0$ is a necessary condition for a steady, two-dimensional boundary layer to separate, the opposite (i.e., a negative curvature of the velocity profile at the wall) must be a sufficient condition for the boundary-layer flow to remain attached.

The preceding arguments naturally lead to several possible methods of control to delay (or advance) steady, two-dimensional separation that rely on modifying the shape of the velocity profile near the wall. Namely, the object is to keep $[\partial^2 u/\partial y^2]_{y=0}$ as negative as possible, or in other words to make the velocity profile as full as possible (Figure 6.2a). In this case, the magnitude of the spanwise vorticity decreases monotonically away from the wall, and the surface vorticity flux is in the positive y direction. Not surprisingly, then, methods of control to postpone separation that rely on changing the velocity profile are similar to those used to delay laminar-to-turbulence transition (Chapter 6).

Inspection of the streamwise momentum equation (8.1) indicates that separation control methods may include the use of wall suction ($v_w < 0$), favorable pressure gradient ($\partial p/\partial x < 0$), or lower wall viscosity ($\partial \mu/\partial y > 0$). Obviously, any one or a combination of these methods may be used in a particular situation. For example, beyond the point of minimum pressure on a streamlined body the pressure gradient is adverse and the boundary layer will separate if the pressure rise is sufficiently steep; however, enough suction may be applied there to overcome the retarding effects of the adverse pressure gradient and to prevent separation. Each of these velocity profile modifiers is covered in more detail in the following three subsections.

8.4.1   Shaping

For any two-dimensional, subsonic flow over a closed-surface body, adverse pressure gradient always occurs somewhere in the aft region. Streamlining can greatly reduce the steepness of the pressure rise, leading to the prevention or postponement of separation. Numerous biological species are endowed with body shapes that avoid separation and allow for minimum fluid resistance to their motion in air or water. Archaic Homo sapiens discovered, through scores of trials and errors, the value of streamlining spears, sickle-shaped boomerangs, and fin-stabilized arrows (Williams 1987). In supersonic flows, pressure always rises across a shock wave, and a boundary layer may separate as a result of the wave interaction with the viscous flow (Young 1953; Lange 1954).

Laminar boundary layers can only support very small adverse pressure gradients without separation. In fact, if the ambient incompressible fluid decelerates in the streamwise direction faster than $U_o \sim x^{-0.09}$, the flow separates (Schlichting 1979). On the other hand, a turbulent boundary layer, being an excellent momentum *conductor*, is capable of overcoming much larger adverse pressure gradients without separation. In this case, separation is avoided for external flow deceleration up to $U_o \sim x^{-0.23}$ (Schlichting 1979). The efficient momentum transport that characterizes turbulent flows provides the mechanism for mixing the slower fluid near the wall with the faster fluid particles farther out. The forward movement of the boundary-layer fluid against pressure and viscous forces is facilitated, and separation is thus postponed. According to the experimental results of Schubauer and Spangenberg (1960), a larger total pressure increase without separation is possible in the case of a turbulent flow by having larger adverse pressure gradient in the beginning and continuing at a progressively reduced rate of increase.

To expand the attached flow operational envelope to off-design conditions using the concept of pressure gradient mitigation generally requires some form of variable geometry such as vanes, slats, or flaps. These can be combined with other separation control techniques such as active or passive blowing, a rotating cylinder at the flap knee, or use of an injection-stabilized trapped vortex. For low-speed flows, bodies can now be designed for incipiently separated flow over large surface areas using the so-called Stradford closure to be discussed in the following subsubsection (Smith 1977; Smith, Stokes, and Lee 1981). Although it minimizes skin-friction drag, this approach exacerbates the attached-flow, viscous-induced form or pressure drag, and the minimum drag body is actually a less extreme design. Also, such bodies tend to generate large separated flow regions off-design, and therefore some form of standby flow separation control would probably be required to ensure reasonable off-design performance.

Skin Friction

The skin friction downstream of the separation line is negative. However, the increase in pressure drag that results from flow detachment is far greater than the saving in skin-friction drag. The articles by Stradford (1959a,b) provide useful discussion on the prediction of turbulent-boundary-layer separation and the concept of flow with continuously zero skin-friction throughout its region of pressure rise. By specifying that the turbulent boundary layer be just at the condition of separation, without actually separating, at all positions in the pressure rise region, Stradford (1959b) experimentally verified that such flows achieve a specified pressure rise in the shortest possible distance and with the least possible dissipation of energy. A lifting surface that could utilize the Stradford distribution immediately after transition from laminar to turbulent flow would be expected to have very low skin-friction as well as pressure drag. Liebeck (1978) successfully followed this strategy using a highly polished wing to achieve the best lift-to-drag ratio (over 200) of any airfoil tested in the low-Reynolds-number range of 5×10^5–2×10^6. He argued that the entire pressure-recovery region of an airfoil's upper surface would be operating at its maximum capacity if the adverse pressure distribution were uniformly critically close to separation. By assuming an incipient-separation, turbulent-flow profile, Liebeck calculated the pressure field required and then used an inverse calculation procedure to derive the airfoil shape from the given critical-velocity distribution.

When one attempts to reduce skin-friction drag by driving the boundary layer toward separation, a major concern is the flow behavior at off-design conditions, as discussed earlier. A slight increase in angle of attack, for example, can lead to separation and consequent large drag increase as well as loss of lift. High performance airfoils with lift-to-drag ratio of over 100 utilize carefully controlled adverse pressure gradient to retard the near-wall fluid, but their performance deteriorates rapidly outside a narrow envelope (Carmichael 1974).

Separation Bubbles

Laminar-to-turbulence transition on the upper surface of a lifting surface typically occurs at the first onset of adverse pressure gradient if the Reynolds number exceeds 10^6. The separation-resistant turbulent boundary layer that evolves in the pressure recovery region results in higher maximum lift and a relatively larger angle of stall. At lower Reynolds numbers, and depending on the severity of the initial adverse gradient (hence on the airfoil shape), laminar separation may take place before transition. For sufficiently low Re, the separated flow will not reattach to the surface. However, in the intermediate Reynolds number range of typically 10^4–10^6, transition to turbulence takes place in the free-shear layer owing to its increased susceptibility (Gad-el-Hak 1990). Subsequent turbulent entertainment of high-speed fluid causes the flow to return to the surface, thus forming what is known as a laminar separation bubble, a topic to be discussed in more detail in the next chapter. Regardless of whether the flow subsequently reattaches, the laminar separation leads to higher form drag and lower maximum lift. Delicate contouring of the airfoil near the minimum pressure point to lessen the severity of the adverse pressure gradient may be used to accomplish separation-free transition (Pfenninger and Vemuru 1990).

8.4.2 Transpiration

The second method to avert separation by changing the curvature of the velocity profile at the wall involves withdrawing the near-wall fluid through slots or porous surfaces. Prandtl (1904) applied suction through a spanwise slit on one side of a circular cylinder. His flow visualization photographs convincingly showed that the boundary layer adhered to the suction side of the cylinder over a considerably larger portion of its surface. By removing the decelerated fluid particles in the near-wall region, the velocity gradient at the wall is increased, the curvature of the velocity profile near the surface becomes more negative, and separation is avoided. In the following, an approximate method to compute the amount of suction needed to prevent laminar separation is briefly recalled (Prandtl 1935).

For a laminar boundary layer, the ratio of pressure forces to viscous forces is proportional to the Pohlhausen parameter, Λ, as given by

$$\Lambda \equiv -\frac{\mathrm{d}p_0}{\mathrm{d}x} \cdot \frac{\delta^2}{\mu U_0} = \frac{\delta^2}{\nu} \cdot \frac{\mathrm{d}U_0}{\mathrm{d}x} \tag{8.7}$$

where δ is the boundary layer thickness, U_0 is the velocity outside the boundary layer, ν is the kinematic viscosity, μ is the dynamic viscosity, and $\mathrm{d}p_0/\mathrm{d}x$ is the streamwise pressure gradient. At separation, $[\partial u/\partial y]_{y=0} = 0$, and Equation (8.1) reads

$$\frac{\mathrm{d}p_0}{\mathrm{d}x} = \mu \left[\frac{\partial^2 u}{\partial y^2}\right]_{y=0} \tag{8.8}$$

The Pohlhausen parameter at the point of separation of a laminar boundary layer is $\Lambda = -12$, and from Eqs. (8.7) and (8.8) the expressions for the curvature of the velocity profile at the wall and the boundary-layer thickness become, respectively,

$$\left[\frac{\partial^2 u}{\partial y^2}\right]_{y=0} = \frac{12 \, U_o}{\delta^2} \tag{8.9}$$

$$\delta = \sqrt{\frac{-12 \, \nu}{\left(\frac{dU_o}{dx}\right)}} \tag{8.10}$$

The velocity distribution dU_o/dx is determined from the potential flow solution. As an example, suppose we wish to compute the suction coefficient $C_q \equiv |v_w|/U_o$, which is just sufficient to prevent laminar separation from the surface of a cylinder. By assuming that the velocity profiles in the vicinity of separation are identical with that at the point of separation and computing (dU_o/dx) at the downstream stagnation point, Prandtl (1935) used the momentum integral equation and the preceding results to make a simple estimate of the required suction

$$C_q = 4.36 \, Re^{-0.5} \tag{8.11}$$

where Re is the Reynolds number based on the cylinder diameter and the freestream velocity.

Several researchers have used similar approximate methods to calculate the laminar boundary layer on a body of arbitrary shape with arbitrary suction distribution (see, e.g., Schlichting and Pechau 1959; Chang 1970). A particularly simple calculation was made by Truckenbrodt (1956). He reduced the problem to solving a first-order ordinary differential equation. As an example, for a symmetrical Zhukovskii airfoil with uniform suction, Truckenbrodt predicted a suction coefficient just sufficient to prevent separation of

$$C_q = 1.12 \, Re^{-0.5} \tag{8.12}$$

where Re is the Reynolds number based on the airfoil chord and the freestream velocity.

For turbulent boundary-layers, semi-empirical methods of calculation are inevitably used owing to the well-known closure problem. Suction coefficients in the range of $C_q = 0.002$–0.004 are sufficient to prevent separation on a typical airfoil (Schlichting 1959; Schlichting and Pechau 1959). Optimally, the suction should be concentrated on the low-pressure side of the airfoil just a short distance behind the nose where, at large angles of attack, the largest local adverse pressure gradient occurs.

Passive Suction

For the high-speed, shock-boundary-layer interaction case, a passive porous surface can be used to mitigate the local pressure gradients and obviate separation as well as reduce wave drag and shock losses (Bahi, Ross, and Nagamatsu 1983; Savu and Trifu 1984; Nagamatsu et al. 1985, 1987; Raghunathan 1985; Stanewsky and Krogmann 1985; Barnwell et al. 1985; Koval'nogov, Fomin, and Shapovalov 1987) in both transonic and supersonic flows (Bauer and Hernandez 1988). The basic device, sketched in Figure 8.7, is an empty subsurface plenum covered by a porous surface and located underneath the shock-boundary-layer interaction region. Such a

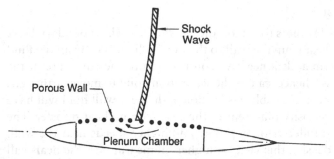

Figure 8.7: Schematic of porous airfoil for passive suction.

passive porous surface allows mass to self-bleed from downstream of the shock to upstream, resulting in a more gradual viscous–inviscid interaction, a series of weaker shock waves, and reduced pressure gradients. Flow separation is then delayed and wave drag is minimized. Suction implemented via passive bleed is employed as an integral part of the technologically important case of high-speed inlet design (Delery 1985; Viswanath 1988).

Suction Optimization

Flow separation control by suction is the other conventional technique which, along with blowing, is within the capacity of contemporary CFD for design and optimization via tailored and perhaps distributed flow profiles and is well reviewed in readily available literature. Suction can be instituted via active or passive systems with the boundary-layer diverter constituting the reductio ad absurdum. The parameter space for separation control by suction (or injection) includes distribution of mass transfer location vis-à-vis adverse pressure gradient regions, spatial distribution (discrete, continuous), exit orientation, tailoring of orifice velocity profiles, active or passive source or sink, and distribution of suction and injection both chordwise and spanwise.

In summary, studies of suction control at high speeds indicate that suction holes should probably be inclined in the upstream direction (Purohit 1987), that upstream control is less effective than direct control of the separated flow region, and that optimum distributed suction requires several independent plenum chambers to avoid local backflows and consequent loss of suction control. An additional possibility for high-speed flows is to utilize the embedded near-wall spanwise vortex structure associated with swept shocks to remove the inner portion of the incident boundary layer (S.M. Bogdonoff, private communication). This may require provision of an additional control shock upstream of the interaction. The increased interest in, and favorable flight experience with, hybrid laminar flow control (where suction is employed in the leading-edge region ahead of wing box) has revived interest in suction control for the subsonic case. Preliminary studies indicate that the LFC suction system for cruise application is not incompatible with leading-edge region suction separation control or high-lift requirements for both CTOL and SST (supersonic transport) applications. Flight experiments of leading-edge suction systems for high lift (e.g., Hunter and Johnson 1954) and LFC (Hefner and Sabo 1987) do not indicate any stoppers. A combined approach using the same suction surface and system for both LFC at cruise and leading-edge-region high lift for takeoff and landing has considerable promise. For LFC, there is the added benefit of eliminating the joints and other factors associated with conventional leading-edge, variable-geometry devices.

8.4.3 Wall Heat Transfer

The third term in Eq. (8.1) points to yet another method to delay boundary-layer separation. By transferring heat from the wall to the fluid in liquids or from the fluid to the wall in gases, this term adds a negative contribution to the curvature of the velocity profile at the wall and, hence, causes the separation point to move farther aft. If the surface of a body in a compressible gas is cooled, the near-wall fluid will have larger density and smaller viscosity than that in the case with no heat transfer. The smaller viscosity results in a fuller velocity profile and higher speeds near the wall. Combined with the larger density, this yields a higher momentum for the near-wall fluid particles and, hence, the boundary layer becomes more resistant to separation. Although this method of control has been successfully applied to delay transition in both water and air flows as discussed in Chapter 6, its use to prevent separation has been demonstrated only for high-speed gaseous flows. These effects are confirmed via the analytical results of Libby (1954), Illingworth (1954), and Morduchow and Grape (1955). Experimental verification is provided by the work of Gadd, Cope, and Attridge (1958); Bernard and Siestrunck (1959); and Lankford (1960, 1961). Excellent summaries of the problem of heat transfer effects on the separation of a compressible boundary layer are available in Gadd (1960) and Chang (1970, 1976).

Active separation control by wall cooling in air is straightforward. The technique might be particularly appropriate for high-altitude, long-endurance vehicles (HALE) having thick, low-Reynolds-number wings and cryogenic fuel to provide the requisite heat sink (e.g., Baullinger and Page 1989). Unfortunately, this method is mainly restricted to cryogenically-fueled aircraft. This makes it particularly appropriate for hypersonic applications such as shock-boundary-layer interactions on hydrogen-fueled vehicles. Cooling in air works according to both experiment and theory for low (Macha, Norton, and Young 1972; Lin and Ash 1986) as well as high (Spaid 1972; Ogorodnikov, Grin, and Zakharov 1972) speeds.

In liquids, surface heating lowers the near-wall viscosity but the density remains essentially unchanged. Using simple asymptotic analysis of the coupled energy and momentum equations, Aroesty and Berger (1975) compared the effectiveness of wall heating with suction as a means of delaying separation for a prescribed adverse pressure gradient in a water boundary layer. They concluded that surface heating can be used in water to delay separation somewhat. However, it seems that this analytical result has not been confirmed experimentally.

In addition to surface heating and cooling, several other methods are available to establish a viscosity gradient in a wall-bounded flow and thus to affect the location of separation. These include film boiling, cavitation, sublimation, chemical reaction, wall injection of a secondary fluid having lower or higher viscosity, and the introduction into the boundary layer of shear-thinning or shear-thickening additive.

8.5 Moving-Wall and Time-Dependent Separations

8.5.1 Moving Walls

The moving-wall effects can be exploited to postpone separation. For a surface moving downstream, the relative motion between the wall and the freestream is minimized, and thus the growth of the boundary layer is inhibited. Furthermore, the surface motion injects additional momentum into the near-wall flow. Prandtl (1925b) demonstrated the effects of rotating a cylinder placed in a uniform stream at right angles to its axis. Separation is completely eliminated on the side of the cylinder where the wall and the freestream move in the same direction. On the other side of the cylinder, separation is developed only incompletely. In fact, for high enough values of circulation, the entire flowfield can be approximated by the potential flow theory. The asymmetry causes a force on the cylinder at a right angle to the mean flow direction. This important phenomenon, known as the Magnus effect (Magnus 1852; Swanson 1961), is exploited in several sports balls (Mehta 1985a) and even in an experimental device used for propelling ships known as the Flettner's (1924) rotor.

From a practical point of view, wall motion for body shapes other than circular cylinders or spheres is prohibitively complicated, although it is feasible to replace a small portion of the surface of, say, an airfoil by a rotating cylinder, thus energizing the boundary layer and avoiding separation (Alvarez-Calderon 1964). Rotating cylinders have been successfully employed to delay separation at the leading and trailing edges of airfoils and control surfaces (Johnson et al. 1975; Mokhtarian and Modi 1988; Mokhtarian et al. 1988a,b; Modi et al. 1989), at flap junctures (Alvarez-Calderon 1964; Lee 1974; Tennant et al. 1975; Modi et al. 1980), and in diffusers (Tennant 1973). Critical parameters include the rotational speed and the gap between the cylinder and fixed surface.

Flight tests were conducted on a YOV-10A STOL-type aircraft having flaps with rotating cylinders at their leading edges (Cichy, Harris, and Mackay 1972; Weiberg et al. 1973; Cook, Mickey, and Quigley 1974). With the flaps in lowered position, the cylinders were rotated at high speed, and lift coefficients as high as 4.3 were recorded at a modest flying speed of 30 m/s along approaches up to $-8°$. Modi and his colleagues (Modi et al. 1980, 1981, 1989; Mokhtarian and Modi 1988; Mokhtarian, Modi, and Yokomizo 1988a,b) carried out a comprehensive wind tunnel test program involving a family of airfoils each having one or more rotating cylinders located at the leading edge, the trailing edge, or the upper surface, as sketched in Figure 8.8. Under optimum conditions, the lift coefficient increased by as much as 200%, and the stall angle was delayed to 48°.

More recently, Modi, Fernando, and Yokomizo (1990) extended the concept of moving surface boundary-layer control to reduce the drag of land vehicles. On a scale model of a typical tractor-trailer truck configuration having a splined rotating

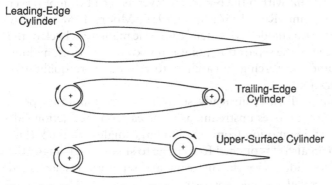

Figure 8.8: Various rotating cylinder configurations used to increase lift and delay stall of an airfoil.

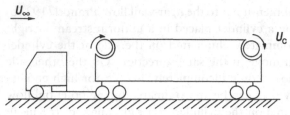

Figure 8.9: Tractor-trailer truck with rotating cylinders.

cylinder at the top leading edge of the trailer (Figure 8.9), drag was lowered by as much as 27% when the cylinder surface velocity was 3.7 times the freestream speed. Modi et al. (1990) maintained that this separation control concept is essentially semi-passive, requiring a negligible amount of power for its implementation.

Creating a wall-slip layer biases the mean flow such that a larger pressure gradient can be tolerated before separation occurs. There are essentially two techniques for establishing a slip layer on the wall to mitigate separation. The first of these is actually to translate the wall itself (e.g., moving belts or embedded rotating cylinders, as discussed above). The other approach is the establishment of stabilized cavity vortex flows. Either small-scale (Migay 1960a,b; 1961; 1962a,b,c; Stull and Velkoff 1975; Howard and Goodman 1985) or large-scale (Ringleb 1961; Adkins 1975; Adkins, Mathaus, and Yost 1980; Burd 1981; Chow, Chen, and Huang 1985; Krall and Haight 1972; Haight et al. 1974) vortex stabilization is necessary; otherwise, the trapped vortex will generally periodically shed downstream, causing disrupted operation and higher drag and losses. Stabilization techniques include injection (Krall and Haight 1972; Haight, Reed, and Morland 1974), suction (Adkins 1975; Adkins et al. 1980; Burd 1981; Chow et al. 1985), and viscous forces via low-cavity Reynolds number (e.g., Howard and Goodman 1985). The vortex flap is the latest version of such a device. These slip-layer separation control strategies work for moderately separated flows, but in extreme cases reverse flow can still occur away from the surface (Zhuk and Ryzhov 1980).

8.5.2 Time-Dependent Separation

For time-dependent flows, the separation point is no longer stationary but rather moves along the surface of the body. The Moore–Rott–Sears (MRS) criterion states that unsteady separation occurs when the shear and velocity vanish simultaneously and in a singular fashion at a point within the boundary layer as seen by an observer moving with the separation point (Rott 1956; Sears 1956; Moore 1958). Steady separation is clearly included in this model as a special case. The main drawback of the MRS criterion is that the speed of the separation point is not known a priori, making it difficult to locate this point and forcing researchers to rely on more qualitative measures for unsteady separation.

In analogy to the moving-wall case, unsteady separation is (temporally) postponed when the separation point moves upstream, as is the case on the suction side of an airfoil undergoing a pitching motion from small to large angles of attack (Figure 8.10). Conversely, when an airfoil is pitched from large to small attack angle, the separation point on the suction side moves downstream, and separation is advanced much the same as the case of a wall moving upstream.

An airfoil oscillating sinusoidally through high angles of attack can produce very high lift coefficients and maintain flow attachment well beyond static stall attack

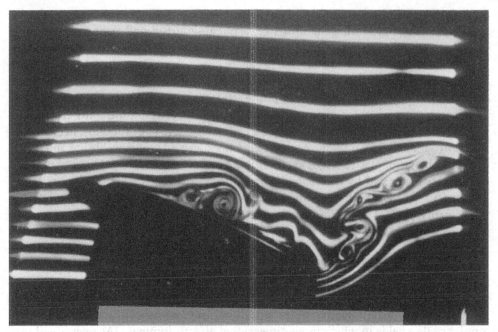

Figure 8.10: A NACA 0012 rectangular wing during a pitching-up motion between 0° and 30°. Aspect ratio = 4; $Re_c = 1.25 \times 10^4$; reduced frequency $\equiv f c/U_\infty = 0.2$ [after Gad-el-Hak 1992].

angles (McCroskey 1977, 1982). During the upstroke, the separation point moves upstream, and reverse flow exists in an attached and mathematically well-behaved boundary layer. The global aerodynamic properties of a pitching airfoil are strongly influenced by the local unsteady separation. Sudden changes of lift, drag, and pitching moment occur near the onset of separation and the spillage of a leading edge vortex. These effects are particularly significant at high frequencies and large amplitudes. Moment stall is observed when the reverse-flow region extends over most of the airfoil and a large-scale vortex is formed near the leading edge. A discontinuous increase in circulation is associated with the spilled vortex. During this phase of the cycle, lift continues to increase. Lift stall follows moment stall and occurs when the separation vortex reaches the latter half of the airfoil and a double-peaked pressure distribution results on the suction side. In other words, the suction on the upper surface of the airfoil continues to increase at the initial stages of separation, and a sudden decrease in suction does not occur until the leading edge is in the wake of separation.

Similar phenomena are observed on three-dimensional lifting surfaces undergoing pitching motion (Gad-el-Hak 1986c, 1988a,b; Gad-el-Hak and Ho 1986a). For highly swept wings, both steady as well as unsteady flows are vortex dominated. The latter flow is characterized by the existence of unsteady large- and small-scale vortices that go through a growth-decay cycle with hysteresis during each period (Gad-el-Hak and Ho 1985, 1986b; Kandil and Chuang 1990a,b; Atta and Rockwell 1990; Huyer, Robinson, and Luttges 1990).

According to Ericsson (1967, 1988), the forces on an airfoil oscillating in pitch will deviate from the static forces realized at the instantaneous angle of attack due to the superposition of two effects. The first is the frequency-induced normal velocity distribution over the airfoil. This so-called q effect can be visualized as a frequency-induced camber. The second is the effect of attack angle rate of change, the so-called

$\dot{\alpha}$ effect. This can be visualized as a frequency-induced change of the mean velocity vector or plunging. During the upstroke, a pitching airfoil will appear as having a positive camber and as plunging. According to the unsteady Bernoulli's equation, the local pressure gradient is less adverse in the dynamic case. Thus, the boundary layer at a particular α during the upstroke has a more favorable upstream time history as compared with the static case. The opposite effects take place during the downstroke.

Insects, most of which mate and eat while airborne, exploit unsteady separation effects to achieve remarkable aerodynamic characteristics. The dragonfly, in existence for approximately 250 million years, presumably survived innumerable life and death aerodynamic struggles (Luttges et al. 1984; Luttges 1989). The enviably large lift coefficients generated by the chalcid wasp during hovering suggest the existence of an efficient, unsteady lift-generation mechanism (Weis-Fogh 1973; Lighthill 1973; Maxworthy 1979, 1981; Ellington et al. 1996; Brookes 1997).

8.6 Three-Dimensional Separation

Three-dimensional boundary layers are more common in practical flow situations than two-dimensional ones. Bodies of revolution at nonzero angle of attack, flow near wing tips, turbine blades, pump impellers, and low-aspect-ratio wings are examples of flowfields in which three-dimensional effects dominate. As mentioned earlier, the point of boundary-layer separation from a three-dimensional body does not necessarily coincide with the point of vanishing wall shear. Instead, the shear stress at the wall is equal to zero only at a limited number of points along the separation line. The number and type of these critical or singular points must satisfy certain topological laws (Lighthill 1963a; Tobak and Peake 1982).

The projection of the limiting streamlines as the distance from the wall goes to zero coincides with the skin-friction lines on the surface of the body. Oil-streak techniques and the like are usually used to obtain separation and attachment patterns for steady, three-dimensional flows (Maltby 1962). A necessary condition for the occurrence of flow separation is the convergence of skin-friction lines onto a particular line. Because of the three-dimensionality of the flow, the near-wall fluid may move in a direction in which the pressure gradient is more favorable and not against the adverse pressure in the direction of the main flow, as is the case for two-dimensional flows. Consequently, three-dimensional boundary layers are in general more capable of overcoming an adverse pressure gradient without separation.

Three-dimensional relief of the streamwise adverse pressure gradient may be exploited to delay separation. Properly designed corrugated trailing edges can provide sufficient easement to postpone the separation at higher angles of attack. In nature, three-dimensional serrated geometry is to be found in the trailing edges of the fins and wings of many aquatic animals and birds (Norman and Fraser 1937; Lighthill 1975). For man-made lifting surfaces, the same concept was tested in the low-Reynolds-number regime by Vijgen et al. (1989). They reported a modest 5% increase in the maximum lift-to-drag ratio when triangular serrations were added to the trailing edge of a natural-laminar-flow airfoil (Gad-el-Hak 1990).

8.7 Turbulators

A turbulent boundary layer is more resistant to separation than a laminar one, and, mostly for that reason, transition advancement may be desired in some situations.

In low-Reynolds-number terminology, the transition-promoting devices are called turbulators. For a zero-pressure-gradient boundary layer, transition typically occurs at a Reynolds number based on distance from leading edge of the order of 10^6. The critical Reynolds number Re_{crit} below which perturbations of all wave numbers decay is about 6×10^4. To advance the transition Reynolds number, one may attempt to lower the critical Re, increase the growth rate of Tollmien–Schlichting waves, or introduce large disturbances that can cause *bypass* transition. The first two routes involve altering the shape of the velocity profile—making it less full—by using wall motion, injection, adverse pressure gradient, surface heating in gases or cooling in liquids, or other strategies to increase the near-wall viscosity. The third route, exposing the boundary layer to large disturbances, is much simpler to implement though more difficult to analyze (Smith and Kaups 1968; Cebeci and Chang 1978; Nayfeh, Rageb, and Al-Maaitah 1986; Cebeci and Egan 1989).

Morkovin (1984) broadly classified the large disturbances that can cause bypass transition into steady or unsteady ones originating in the freestream or at the body surface. The most common example is single, multiple, or distributed roughness elements placed on the wall. The mechanical roughness elements, in the form of serrations, strips, bumps or ridges, are typically placed near the airfoil's leading edge. If the roughness characteristic length is large enough, the disturbance introduced is nonlinear and bypass transition takes place. For a discrete three-dimensional roughness element with a height-to-width ratio of one, Tani (1969) reports a transition Reynolds number of $Re_{\delta*} \simeq 300$ for a roughness Reynolds number of $Re_{\Upsilon} \simeq 10^3$. Here, $Re_{\delta*}$ is based on the velocity outside the boundary layer and the displacement thickness ($Re_{\delta*} \equiv U_o \delta^*/\nu$), and Re_{Υ} is based on the height of the roughness element Υ and the velocity in the undisturbed boundary layer at the height of the element ($Re_{\Upsilon} \equiv \overline{U}(\Upsilon)\,\Upsilon/\nu$). Note that the transition Reynolds number, $Re_{\delta*}$, indicated above is below the critical $Re_{\delta*}|_{\text{crit}} = 420$ predicted from the linear stability theory. For a roughness Reynolds number of about 600, transition occurs at $Re_{\delta*} \simeq 10^3$. For a smooth surface, transition typically takes place at $Re_{\delta*} \simeq 2.6 \times 10^3$. An important consideration when designing a turbulator is to produce turbulence and suppress laminar separation without causing the boundary layer to become unnecessarily thick. A thick turbulent wall-bounded flow suffers more drag and is more susceptible to separation than a thin one. Consistent with this observation, available data indicate that a rough airfoil has higher lift-to-drag ratio than a smooth one for $Re_c < 10^5$ but that this trait is reversed at higher Reynolds numbers.

For low-Reynolds-number airfoils, performance may be improved by reducing the size of the laminar separation bubble through the use of transition ramps (Eppler and Somers 1985), boundary layer trips (Davidson 1985; Van Ingen and Boermans 1986), or even pneumatic turbulators (Pfenninger and Vemuru 1990). Donovan and Selig (1989) provide extensive data using both methods for 40 airfoils in the Reynolds number range of $Re_c = 6 \times 10^4$–3×10^5. A long region of roughly constant adverse pressure gradient on the upper surface of a lifting surface (termed a bubble ramp) achieves a lower drag than the more conventional laminar-type velocity distribution in which initially the pressure remains approximately constant and then quickly recovers. Trips were also used in Donovan and Selig's experiments to shorten the separation bubble. A simple two-dimensional trip performed as well or better than zig-zag tape, hemisphere bumps, and normal blowing. Donovan and Selig concluded that an airfoil that performs poorly at low Reynolds numbers can be improved through the use of turbulators. Trips were less effective, however, at improving airfoils, which normally had low drag. We will return to low-Reynolds-number situations in the following chapter.

Other large disturbances that can lead to early transition include high turbulence levels in the freestream, external acoustic excitations, particulate contamination, and surface vibration. These are often termed environmental tripping. Transition can also be effected by detecting naturally occurring T–S waves and artificially introducing in-phase waves. Transition can be considerably advanced, on demand, using this wave superposition principle.

Early transition can also be achieved by exploiting other routes to turbulence such as Taylor–Görtler or crossflow vortices (Taylor 1923; Görtler 1955; Gregory, Stuart, and Walker 1955; Reed and Saric 1987, 1989). For example, a very mild negative curvature of $0.003/\delta^*$ results in the generation of strong streamwise vortices. In this case, transition Reynolds number is lowered from $Re_{\delta^*} \simeq 2600$ for the flat-plate case to $Re_{\delta^*} \simeq 700$ for the concave surface (Tani 1969). For high-Mach-number flows, the general decay in spatial amplification rate of T–S waves makes conventional tripping more difficult as the Mach number increases (Reshotko 1976). For these flows, trips that generate oblique vorticity waves of wavelength appropriate to cause resonant growth of three-dimensional modes may be most effective to advance the transition location. For example, on a sharp cone at $Ma = 3.5$, Corke and Cavalieri (1997) and Corke, Cavalieri, and Matlis (1999) used an array of plasma (corona-discharge) electrodes to excite oblique vorticity-mode pairs and successfully advanced the transition of the supersonic boundary layer. A similar strategy to generate pairs of oblique waves was employed by Corke and Mangano (1989) in an incompressible Blasius boundary layer. There, Corke and Mangano used a spanwise array of heaters placed at the position of the critical layer to generate time-periodic, spanwise-phase-varying velocity perturbations.

The last issue to be considered in this section is augmentation of the turbulence for a shear flow that has already undergone transition. Notwithstanding that the turbulence levels and the Reynolds stresses are highest immediately following transition (Harvey, Bushnell, and Beckwith 1969), the newly developed turbulent flow is in general less capable of resisting separation than a corresponding flow at higher speeds (Lissaman 1983). Turbulence augmentation in the low-Reynolds-number case is then a useful control goal to energize the flow and to enhance its ability to resist separation at higher angles of attack. Roughness will enhance the turbulence, but its associated drag must be carefully considered. Other devices to enhance the turbulence mixing include vane-type vortex generators, which draw energy from the external flow, or Wheeler-type or Kuethe-type generators, which are fully submerged within the boundary layer and presumably have less associated drag penalty (Rao and Kariya 1988). These and other devices will be detailed in the next section.

8.8 Momentum Addition to Near-Wall Flow

8.8.1 Introductory Remarks

Near-wall momentum addition is the usual approach of choice for control of residual flow separation remaining after mitigation of the causative pressure field or for off-design conditions. Common to all these control methods is the supply of additional energy to the near-wall fluid particles that are being retarded in the boundary layer. The additional longitudinal momentum is provided either from an external source or through local redirection into the wall region. Passive techniques do not require auxiliary power but do have an associated drag penalty. These include intentional tripping of transition from laminar to turbulent flow upstream of what would be a laminar separation point (Mangalam et al.

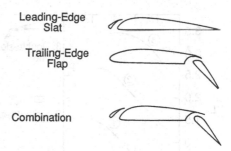

Figure 8.11: Passive blowing through leading-edge slats and trailing-edge flaps.

1986; Harvey 1986); using boundary-layer fences to prevent separation at the tips of swept-back wings; placing an array of vortex generators on the body to raise the turbulence level and enhance the momentum and energy in the neighborhood of the wall (Mehta 1985b; Rao and Kariya 1988); employing a rippled trailing edge (Werle, Paterson, and Presz 1987), streamwise corrugations (Mabey 1988), or stepped afterbodies to form a system of captive vortices in the base of a blunt body (Kentfield 1985a,b; Kidd, Wikoff, and Cottrell 1990); and using a screen to divert the flow and increase the velocity gradient at the wall.

Active methods to postpone separation require energy expenditure. Obviously, the energy gained by the effective control of separation must exceed that required by the device. In addition to suction or heat transfer reviewed in Sections 8.4.2 and 8.4.3, fluid may be injected parallel to the wall to augment the shear-layer momentum or normal to the wall to enhance the mixing rate (Horstmann and Quast 1981). Either a blower is used or the pressure difference that exists on the aerodynamic body itself is utilized to discharge the fluid into the retarded region of the boundary layer. The latter method is found in nature in the thumb pinion of a pheasant, the split-tail of a falcon, or the layered wing feathers of some birds.

In man-made devices, passive blowing through leading-edge slots and trailing-edge flaps is commonly used on aircraft wings (Smith 1975; Lin et al. 1992), as sketched in Figure 8.11. Although in this case direct energy expenditure is not required, the blowing intensity is limited by the pressure differentials obtainable on the body itself. Nevertheless, the effect of passive blowing on lift and drag can be dramatic. This is shown convincingly in Figure 8.12 for the NACA 23012 airfoil section with no flap, with a single trailing-edge flap, and with a double-slotted flap. Compared with the clean (no flap) case, when a single trailing-edge flap is used the maximum lift is increased by about 175%, whereas the section drag at $C_{L_{max}}$ is increased by more than 180%. The corresponding numbers when a double-slotted flap is used are, respectively, 230% and 500%. The induced drag, which must be considered in addition to the section drag, is proportional to $C_{L_{max}}^2$, and the total drag, therefore, rises sharply at low aircraft speeds. Very recently, Thomas et al. (1997, 1999) emphasized the importance of unsteady effects caused by the confluent boundary layer[1] on the performance of multielement airfoils.

Many of the control methods discussed in this section can be used for external as well as internal flows. For example, Viets (1980) used an asymmetrical rotating

---

[1] This term denotes the (typically) turbulent flow generated by the interaction of the wake of a leading-edge slat with the boundary layer on the primary lifting surface.

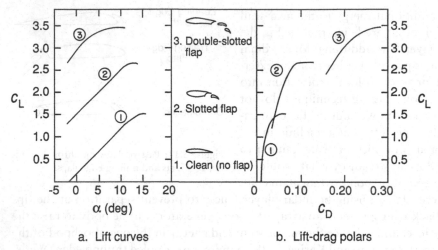

a. Lift curves   b. Lift-drag polars

Figure 8.12: Effects of trailing-edge flaps on lift and section drag. NACA 23012 airfoil section [adapted from data by Abbott and von Doenhoff 1959].

cam embedded in the wall to produce large eddies in a turbulent boundary layer with zero- and adverse-pressure gradients. By using this device in a wide-angle diffuser, Viets, Ball, and Bougine (1981a) were able to postpone the natural separation and dramatically improve the diffuser's performance.

8.8.2 Passive Vortex Generators

Passive momentum addition is most commonly carried out via one of two general approaches: either macro overturning of the mean flow using embedded streamwise vortices generated by fixed lifting surfaces or Reynolds stress amplification, which leads to increased cross-stream momentum transfer. Conventional passive vortex generators (VGs) date from the 1940's (Taylor 1948a) and are simple and effective and therefore generally the first tried as a fix to an existing flow separation problem. Passive VGs have been applied, for example, to compressor blades (Staniforth 1958), diffusers (Henry, Wood, and Wilbur 1956; Feir 1965; Brown, Nawrocki, and Paley 1968), airfoils (Pearcey 1961; Nickerson 1986; Bragg and Gregorek 1987), and the afterbody of aircraft fuselages (Calarese, Crisler, and Gustafson 1985; Wortmann 1987). What these embedded vortices do is cause overturning of the near-wall flow via macro motions. Fluid particles with high streamwise momentum are swept along helical paths toward the surface to mix with and, to a certain extent, to replace the retarded near-wall flow. The vortex influence upon the turbulence can actually be debilitating owing to streamline-curvature-induced stabilization.

Passive vortex generators are essentially small-aspect-ratio airfoils mounted normal to the surface. Their individual parameterization includes planform shape, section profile and camber, yaw angle, aspect ratio, and height with respect to the boundary-layer thickness. The spatial relationship of the devices is also critical (e.g., corotating versus contrarotating biplane and wing, use of downstream reenforcers, and spacing). Counterrotating vortices force large regions of vorticity to rise above the surface and hence are often not as efficient as corotating devices. However, corotating vortices with too close a spacing undergo mutual vorticity cancellation; therefore, spacing is critical in this case.

Nominal guidelines for conventional VGs are given in the articles by Taylor (1948b), Henry et al. (1956), and Pearcey (1961): aspect ratio of 0.5–1; rectangular or triangular planform with either simple flat plate or low-Reynolds-number airfoil cross section, yaw angle less than 15°, and individual height of the order of the boundary-layer thickness. For corotating devices, the spanwise spacing should be greater than three times the device height to avoid interactive cancellation or annihilation of the vortical structures. The recommended nominal spanwise spacing for contrarotating vortex production devices is on the order of 4–5 device heights. These conventional vortex-generating devices produce sizable parasitic drag. When employed to handle off-design problems, the VGs cause a reduction in cruise performance unless the devices are retracted when not needed. The articles by Schubauer and Spangenberg (1960); Gadetskii, Serebriiskii, and Fomin (1972); Liandrat, Aupoix, and Cousteix (1986); Cutler and Bradshaw (1986, 1989); Inger and Siebersma (1988); Mehta (1988); and Briedenthal and Russell (1988) provide generic studies of vortex generator physics and operation. Application to supersonic flows is summarized by Gartling (1970).

Several approaches are now available to optimize the performance of passive vortex generators. The first of these is the use of downstream reenforcers, which are vortex generators of the same sense located in the path of the upstream-generated vortex to maintain the strength of the overturning motion (Kuethe 1973; Wheeler 1984; Lin and Howard 1989; Rao and Kariya 1988). A second optimization approach is surprisingly recent and consists of simply reducing the device height from the order of δ to $\mathcal{O}[\delta/5]$ or less (Rao and Kariya 1988; Lin and Howard 1989; Lin, Howard, Bushnell, and Selby 1990a). This size reduction significantly diminishes the parasitic drag and is enabled by the extreme fullness of the mean velocity profile in a turbulent boundary layer. In other words, with sizable longitudinal momentum levels readily available quite close to the surface, it is not necessary to have $\mathcal{O}[\delta]$ devices. Such sub-δ devices must be placed closer to the nominal separation location and therefore may be less suitable than larger devices for situations in which the separation region is not relatively localized.

Alternative approaches for separation control by streamwise vortices are the V-shaped cutouts of the NACA's flush inlet, leading-edge serrations (Harris and Bartlett 1972; Soderman 1972; Barker 1986), and use of large-scale (flowfield versus boundary-layer scale) vortical motions to control separation on highly swept wings. Such motions can be generated by either auxiliary lifting devices (e.g., canards) or by simple abrupt planform variations such as wing leading-edge extensions (LEXs). Such large-scale vortical motions alter both the near-wall momentum and the basic pressure field, generating increased lift. Problems with such an approach include vortex bursting, which is usually caused primarily by adverse pressure gradients. Control of bursting is actually control *of* vortices and is a subsidiary problem of separation control *by* vortices (Vakili 1990).

8.8.3 Passive Turbulence Amplification

The fundamental flow-separation-control approach associated with turbulence amplification is augmentation of the cross-stream momentum via Reynolds stresses inasmuch as the overall pressure rise for incipient separation is directly proportional to the square root of the skin-friction coefficient of the undisturbed boundary layer (Hayakawa and Squire 1982). The zeroth-order consideration for most separation control is to ensure a turbulent rather than laminar boundary-layer state. For Reynolds numbers less than $\mathcal{O}[10^7]$, this may necessitate use of various boundary-layer tripping devices such as roughness and waviness, passive or active mass transfer, acoustic fields, body vibration, or even elevated freestream disturbance levels.

Once a turbulent boundary-layer flow is established, large-scale dynamic transverse vortical entities can be generated to augment the innate turbulent Reynolds stress field beyond the usual amplification concomitant with the adverse pressure gradients usually associated with separation. Applicable devices include transverse cylinders or flow control rails mounted just above, or in the outer part of, the boundary layer (Sajben, Chen, and Kroutil 1976). Airfoils with chord of $\mathcal{O}[\delta]$ mounted parallel to the body in the outer part of the boundary layer and set at a nonzero angle of attack also generate large transverse dynamic vortical motions (Corke, Guezennec, and Nagib 1980), as can embedded cavity (Helmholtz) resonator surfaces. Dynamic three-dimensional horse-shoe or hairpin eddies can be generated in the tip flow of three-dimensional stubs mounted normal to the surface. Except for the flow-control rail, which works quite well, these passive devices have received only limited attention and development for flow-separation control.

Howard and Goodman (1985, 1987) recently investigated the effectiveness of two passive techniques to reduce flow separation: transverse rectangular grooves and longitudinal V-grooves placed in the aft shoulder region of a bluff body. Both types of grooves were beneficial in reducing the form drag on a body at zero and moderate angles of yaw. Basically, the grooves work by redirecting the outerflow momentum to the near-wall region through the three-dimensionalization of a two-dimensional flow. This is an example of locally mitigating the adverse pressure gradient through partial boat-tailing. Lin, Howard, and Selby (1990b) extended the longitudinal groove approach to the low-speed, two-dimensional, rearward-facing ramp. Their closely packed grooves resulted in a reduction of the reattachment distance by up to 66%. Selby and Miandoab (1990) reported that the base pressure of a blunt trailing-edge airfoil with surface grooves increased with increasing groove depth and angle. They speculated that minimally attached flow in the grooves is the mechanism by which fluid of higher momentum is redirected to the base flow region to effect an increase in the pressure.

Additional turbulence amplification ploys include destabilizing longitudinal surface curvature to increase wall shear: concave for boundary layers and convex for wall jets (Coanda effect), and for high speeds, use of upstream shock-wave-interaction, which can amplify turbulence significantly (Anyiwo and Bushnell, 1982; Zang, Hussaini, and Bushnell 1984) while also delaying separation (Schofield 1985; Gol'd Fel'd and Zatoloka 1979). The use, generation, or both of freestream turbulence and disturbance fields with intermediate scale motions also amplify turbulence in the boundary layer and delay separation (Hoffmann 1981; Sasaki and Kiya 1985; Hoffmann, Kassir, and Larwood 1988; Isomoto and Honami 1989).

8.8.4 Active Wall-Region Momentum Addition

Active momentum addition techniques include streamwise vortex generation via discrete blowing or injection, turbulence or Reynolds stress amplification through use of dynamically activated or driven devices, and direct tangential injection of high-velocity fluid. The two passive techniques discussed in Sections 8.8.2 and 8.8.3, vortex generation and turbulence amplification, can also be employed via active systems. Of particular interest is the use of discrete jet injection for streamwise vortex generation.

This concept arose in the 1950s (Wallis and Stuart 1958; Pearcey 1961; Kukainis 1969; Wimpenny 1970) and is the subject of current research (Zhang and Sheng 1987; Reynolds et al. 1988; Johnston and Nishi 1989, 1990; Compton and Johnston 1991; McManus et al. 1994, 1996; McManus and Magill 1996, 1997; Johari and McManus 1997). What is particularly appealing is the ability to deploy or retract the vortex generators as required, which eliminates parasitic drag in the retracted nonblowing condition. The associated fluid supply lines operate at high pressure and are thus relatively small. They might even be utilizable as structural reenforcement elements. The status of research in this area is such that, although it is clear that discrete jets will generate vortices and delay separation, the approach has not yet been optimized. Research is required to determine optimal injection orientation, spacing, individual hole geometry and size, velocity, pressure, and finally location vis-à-vis the adverse pressure gradient regions. Papell (1984) addresses vortex generation within the injection jet itself. Another use of injection for vortex generation is spanwise injection along the leading edges of swept wings for upper surface separation control (e.g., Bradley and Wray 1974). Employing spanwise arrays of small, skewed, pitched jets from holes in the surface, Johnston and Nishi (1990) have shown that the jets can produce longitudinal vortices strong enough to reduce substantially and nearly eliminate a large stalled region of a turbulent separated flow.

A relatively recent development in flow separation control is active turbulence amplification in the bounding shear layer for flows already separated. The fundamental concept is excitation and, through phasing, enhanced interaction of large transverse eddy structures near and downstream of the separation point, thereby amplifying the mixing in the shear layer bounding the separated flow region. This increases entrainment and generally reduces the extent of separation. For initially laminar flows, the zeroth-order influence of dynamic forcing is to trip transition and enhance the eddy dynamics of the low-Reynolds-number shear-layer (e.g., Collins 1979, 1981; Mullin, Greated, and Grant 1980; Sigurdson and Roshko 1985; DeMeis 1986; Durbin and McKinzie 1987; Huang, Maestrello, and Bryant 1987; Neuburger and Wygnanski 1988; Zaman and McKinzie 1989; Bar-Sever 1989).

Various dynamic devices have been tried for the initially turbulent flow case, including acoustic drivers (Bhattacharjee, Scheelke, and Troutt 1985), oscillating embedded plates, spoilers, and flaps (Reisenthel, Nageb, and Koga 1985; Roos and Kegelman 1986; Miau, Chen, and Chow 1988; Chen and Shi Ying 1989; Katz, Nishri, and Wygnanski 1989a,b); both chordwise and spanwise dynamic blowing (Oyler and Palmer 1972; Ely and Berrier 1975; Vakili, Wu, and Bhat 1988; Vakili 1990); rotating cams (Viets et al. 1979; 1981a,b,c; 1984); and even dynamic motions of the entire body. One of the few large-scale dynamic experiments thus far is also one of the earliest (Oyler and Palmer 1972). The same incremental increase in lift was achieved with only 50% of the blowing mass flow required for the steady-state case. What is obvious from the initially turbulent, dynamic-input separation control research thus far is that, once again, the method works. What is not so obvious is how well it would work in engineering applications and the nature and operating range of the optimal dynamic devices.

Wall Jets

Direct tangential injection—wall jet—was and still is the preferred and straight-forward flow separation control technique that has been applied to military fighters and STOL transports. High-pressure air can be used, enabling relatively small interior lines as opposed to suction control which, although generally more energy efficient than blowing at both low and high speeds, usually requires larger interior ducting (Gratzer 1971). In some applications, a lighter gas is introduced to reduce the rate at which heat is exchanged between the wall and the external stream and, thus, to provide thermal protection at high supersonic speeds.

High-pressure bleed air was readily available from the early jet engines but less so for modern high-bypass-ratio turbofans. Computational fluid dynamics (CFD) can now be used to design the system and optimize the injection velocity profile for a given mass flow for optimal separation delay (see, for example, Saripalli and Simpson 1980). Performance can be further enhanced by convex longitudinal curvature (Coanda effect). Tangential blowing is also used to stabilize a trapped vortex, particularly in the knee region of wing flaps. The literature for the steady blowing case is both extensive and readily available, and most information dates from the 1950s. Separation control by blowing at high speeds is covered in the reviews by Delery (1985) and Viswanath (1988). Of possible interest for separation control via direct tangential injection is the application of turbulence control techniques to reduce the mixing between the injected and incident flows and thereby preserve the high near-wall momentum for a larger extent downstream (McInville, Hassan, and Goodman 1985).

Tangential jet blowing over the upper surface of a rounded trailing-edge airfoil sets an effective Kutta condition by fixing the location of separation. This circulation control concept was initially described by Cheeseman and Seed (1967), and a substantial database has been gathered since then for the purpose of performance evaluation (Kind 1967; Wood and Nielsen 1985; Novak, Cornelius, and Roads 1987). More recently, McLachlan (1989) conducted an experimental study of the flow past a two-dimensional circulation control airfoil under steady leading- and trailing-edge blowing. In the range of chord Reynolds numbers of 1.2×10^5–3.9×10^5, McLachlan observed a dramatic increase in the lift coefficient when trailing-edge blowing was used to control the location of the rear separation points. He reported a gain in the lift coefficient of the order of 80 times the injected momentum coefficient. When leading-edge blowing was employed simultaneously, a slight decrease in lift was observed.

Quite recently, McManus et al. (1994, 1996), McManus and Magill (1996, 1997), and Johari and McManus (1997) conducted a series of laboratory experiments in which pulsed jets were employed to improve the leading-edge flap effectiveness on high-performance fighters. Their separation control technique exploits the remarkable vorticity-generation capabilities of impulsively started jets to enhance boundary-layer momentum transport and thus suppress stall for both compressible and incompressible flows. In addition to the customary streamwise vortices, large-scale turbulent vortex rings are also generated in the flow as a result of pulsing the wall jets. The net effect is to achieve fairly good stall suppression with substantially lower jet-mass-flow rate.

Although not directly related to separation control, jets in the form of thin sheets exiting in the spanwise direction from the tips of a straight wing can be used to enhance the lift through an effective enlargement of the wing span (Wu et al. 1983, 1984; Tavella et al. 1988; Lee et al. 1989; Vakili 1990). This application was first reported in 1956 by Ayers and Wild and suggests the possibility of using such an arrangement in place of conventional ailerons or flaps to alter the aerodynamic forces acting on an aircraft (Tavella, Wood, Lee, and Roberts 1986). Either a single long slot (Lee et al. 1986; Tavella, Lee, and Wood 1986) or several short ones (Wu et al.

1983, 1984) are used along the entire chord at the wing tips. The enlargement of the aspect ratio associated with the lateral displacement of the tip vortices via blowing leads also to a reduction of the induced drag as well as to a beneficial effect on stall.

Additional Active Control Methods

Other active methods for controlling boundary-layer separation and reattachment include acoustic excitations (Collins and Zelenevitz 1975; Ahuja, Whipkey, and Jones 1983; Ahuja and Burrin 1984; Zaman, Bar-Sever, and Mangalam 1987; Huang et al. 1987), periodic forcing of the velocity field via an oscillating flap or wire (Koga, Reisenthel, and Nagib 1984; Sigurdson and Roshko 1985; Reisenthel et al. 1985; Roos and Kegelman 1986; Katz et al. 1989a,b; Bar-Sever 1989), and oscillatory surface heating (Maestrello, Badavi, and Noonan 1988).

As early as 1948, Schubaur and Skramstad observed that sound at particular frequencies and intensities could enhance the momentum exchange within a boundary layer and could, therefore, advance the transition location. Collins and Zelenevitz (1975) and Collins (1979, 1981) introduced the external acoustic excitation technique to enhance the lift of an airfoil. In this case, sound is radiated onto the wall from a source outside the boundary layer. Using the same technique at chord Reynolds numbers up to 1×10^6, Ahuja et al. (1983) and Ahuja and Burrin (1984) successfully demonstrated that sound at a preferential frequency and sufficient amplitude can postpone the separation of a turbulent boundary layer developing on an airfoil in both pre- and poststall regimes. The optimum frequency was found to be $4\,U_\infty/c$ (i.e., Strouhal number $St = 4$), where U_∞ is the freestream velocity and c is the airfoil chord. Goldstein (1984) speculated that the delay in separation in Ahuja et al.'s (1983) experiment resulted from enhanced entrainment promoted by instability waves that were triggered on the separated shear layer by the acoustic excitation.

Zaman et al. (1987) conducted further study of the beneficial interaction between external acoustic excitation and the separated flow around an airfoil at high angles of attack. They found that the most effective separation control is achieved at frequencies at which the acoustic standing waves in their wind tunnel induce transverse velocity fluctuations in the vicinity of the lifting surface. The loudspeakers used by Zaman et al. (1987) as well as by Ahuja et al. (1983) essentially excited the resonant modes in their respective wind tunnels, and one of these modes forced the shear layers to separate from their respective airfoils. Extremely high level of excitation was required, however, to maintain these wind-tunnel resonance modes, making the external acoustic excitation technique impractical for field applications. Zaman et al. (1987) speculated that a more effective separation control can be obtained by direct introduction of velocity disturbances.

Huang et al. (1987) introduced the internal acoustic excitation technique in which sound is emanated from a hole or a slot on the surface of a lifting surface. The loudspeaker in this case is essentially used as a piston to produce localized vorticity perturbations at the leading edge of the airfoil. More recently, Hsiao et al. (1990) reported improved aerodynamic performance of a two-dimensional airfoil using sound emitted from three narrow wall slots located near the leading edge. The sound pressure levels used in their experiment was substantially lower than that used externally by other researchers. Additionally, a correct Strouhal number scaling was achieved. Hsiao, Liu, and Shyu (1990) concluded that the enhancement of momentum transport resulting from the sound excitation produces a suction peak at the leading edge, an increase of lift, and a narrower wake as long as the excitation frequency is locked in to the most unstable frequency of the separated shear layer.

To disturb the velocity field directly, Koga et al. (1984) used a computer-controlled spoilerlike flap in a flat-plate turbulent boundary layer with and without modeled upstream separation. They were able to manipulate the separated flow region and its reattachment length characteristics by varying the frequency, amplitude, and waveform of the oscillating flap. Reynolds and Carr (1985) offered a plausible explanation, from the viewpoint of a vorticity framework, for the experimental observations of Koga et al. It seems that the large-scale vortical structures produced by forcing play a major role in enhancing mixing and entrainment, thus leading to reattachment. The active flap controls the size of the separated region by providing an additional mechanism for removing vorticity from this zone; namely, large-scale vortex convection. More recent experiments by Nelson, Kogan, and Eaton (1987, 1990) seem to confirm that the dominant mechanism of vorticity transport behind an oscillating spoiler is convective.

Periodic forcing of the velocity field has been shown to reduce reattachment length in both laminar and turbulent flows on a number of other basic geometrical configurations (Sigurdson and Roshko 1985; Roos and Kegelman 1986; Katz et al. 1989a,b; Bar-Sever 1989). At a chord Reynolds number in the range of 1×10^5–3×10^5, Bar-Sever (1989) used an oscillating wire to introduce transverse velocity fluctuations into a separated shear layer on an airfoil at high incidence. The effectiveness of this separation control technique is depicted in Figure 8.13, showing the variation of unforced and forced lift coefficients as a function of angle of attack. For each angle, the forced case represents the best lift achieved at any combination of forcing frequency and amplitude. At $\alpha = 20°$, the controlled forcing moved the separation from the leading edge to about $0.8c$. A wide band of forcing reduced frequencies (0.7–2.7) was found to be effective, although diminished influence occurred at lower frequencies.

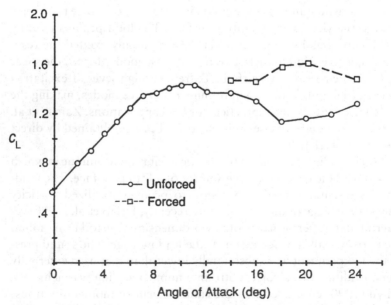

Figure 8.13: Lift coefficient variation with angle of attack for unforced and best forced case; $Re_c = 1.5 \times 10^5$ [after Bar-Sever 1989].

8.9   Separation Provocation

Although most of the control methods reviewed thus far are designed to prevent separation, under certain circumstances the designer may wish to provoke separation. Controlled separation is associated with the formation of jets, throttling action in household faucets, and noisy as well as musical acoustic effects generated by flow. During takeoff and landing of a supersonic aircraft, a freestream flap may be used to provoke leading-edge separation followed by reattachment at the leading edge of the flap, thus forming a thicker pseudobody with the desired aerodynamic shape for subsonic speeds, as shown

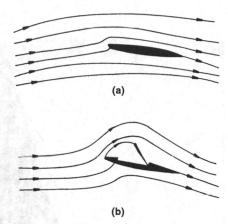

Figure 8.14: Freestream flap. (a) During supersonic flight; (b) During takeoff and landing.

in Figure 8.14 adapted from Hurley (1961). The airfoil has then a relatively high lift coefficient, although its lift-to-drag ratio is quite low.

The detached bow shock forming upstream of a blunt body in supersonic flight may be changed into a weaker, attached oblique shock by placing a spike in front of the body. The pressure rise and the presence of a solid surface on which a boundary layer forms causes the flow to separate downstream of the spike tip. A properly designed spike may result in lower drag, higher lift, and corresponding change in pitching moment (Wood 1961).

Periodic separation may also be provoked by changes in the wall geometry. Francis et al. (1979) initiated separation on an airfoil by periodically inserting and removing a spoiler at the wall. Viets et al. (1984) studied separation inducement and control by use of a cam-shaped rotor mounted on an airfoil. The cam, either driven or free-wheeling, periodically extended out into the flow, causing the boundary layer to separate. Large-scale coherent spanwise structures were periodically generated and were responsible for the flow detachment, as explained by Reynolds and Carr (1985).

8.9.1   Delta Wings

On a sharp-leading-edge delta wing, the separation position is fixed, and a strong shear layer is formed along the entire edge (Lee and Ho 1989, 1990). The shear layer is wrapped up in a spiral fashion, which results in a large-bound vortex on each side of the wing. The two vortices appear on the suction surface of the wing in the form of an expanding helix when viewed from the apex. The low pressure associated with the vortices produces additional lift on the wing, often called nonlinear or vortex lift, which is particularly important at large angles of attack. The experiments of Gad-el-Hak and Blackwelder (1985) have indicated that small discrete vortices are shed parallel to the leading edge at a repeatable frequency determined by the angle of attack and Reynolds number (Figure 8.15). Repeated vortex pairings result in the formation of progressively larger vortices. This process can be modulated by weak, periodic suction or injection through a leading edge slot. In particular, when the perturbation frequency is a subharmonic of the natural shedding frequency, the evolution of the bounded shear layer is dramatically altered (Gad-el-Hak and Blackwelder 1987a; Blackwelder, Gad-el-Hak, and Srnsky 1987).

(a)

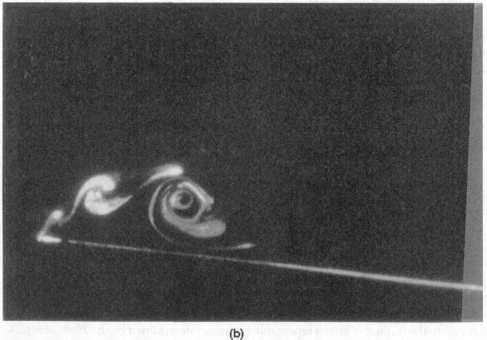

(b)

Figure 8.15: Sharp-leading-edge, $60°$-sweep delta wing in steady flight. Reynolds number based on root chord $Re_c = 1.3 \times 10^4$; $\alpha = 10°$. (a) Top view; (b) End view [after Gad-el-Hak and Blackwelder 1985].

Wood and Roberts (1986, 1988) examined the feasibility of vortex control by tangential mass injection at the leading edge of a 60°-sweep delta wing. Their initial experimental results indicated that modest continuous blowing is capable of extending the regime of stable, controlled vortical flow over the upper surface of the wing by approximately 30° angle of attack. Increases in maximum normal force of 30% were achieved, and significant rolling moments were produced at attack angles of 35–60°. On a delta wing with rounded leading edges, the blowing seems to control the location of the crossflow separation points and hence the trajectories of the ensuing vortices. Wood and Roberts (1988) proposed that this blowing scheme may be a practical solution for changing the normal force without changing attitude for the production of both steady and transient control moments at extremely high angles of attack and for increasing the lift-to-drag ratio of very slender bodies at modest attack angles. In a later experiment, Wood, Roberts, and Celik (1990) have shown that the effects of asymmetric leading-edge blowing are uncoupled at prestall angles of attack. In this case, the overall forces and moments for symmetric blowing can simply be deduced by superposition of asymmetric blowing situations. On the other hand, the response of the vortical flowfield is strongly coupled for asymmetric blowing at poststall conditions. Wood et al.'s results imply that tangential leading-edge blowing may result in substantial rolling moments at conditions under which other control devices cease to be effective.

Controllable leading-edge flaps provide an active method of influencing the large vortices on a delta wing (Rao 1979; Marchman, Manor, and Plentovich 1980). These devices appear to be capable of improving L/D ratio of a given wing primarily through drag reduction. Unlike the control by blowing discussed above, the use of flaps results in only a modest improvement in the angle-of-attack envelope.

8.10 Parting Remarks

8.10.1 Recapitulation

The performance of many practical devices is often controlled by the separation location. For a steady, two-dimensional boundary layer, the streamwise velocity and its normal gradient vanish at the point (or line) of separation. For three-dimensional or unsteady flows, the point of vanishing shear does not necessarily coincide with separation. Under some circumstances, a thin layer of reverse flow can be embedded at the bottom of an otherwise attached and mathematically well-behaved boundary layer.

Advancing or delaying boundary layer detachment is of immense importance to the performance of biological and man-made systems involving fluid flow. Postponing separation may lead to pressure-drag reduction, lift enhancement, stall delay, and pressure-recovery improvement. Provoking flow detachment, on the other hand, may improve the subsonic performance during takeoff and landing of a thin airfoil optimized for supersonic cruising.

Methods of separation control that rely on modifying the shape of the velocity profile near the wall include shaping, transpiration, and establishing a normal viscosity gradient through surface heating or cooling, film boiling, cavitation, sublimation, chemical reaction, wall injection of a secondary fluid having lower or higher viscosity, and the introduction into the boundary layer of shear-thinning or shear-thickening additive. To delay separation, any one or a combination of the preceding techniques is configured to make the velocity profile as full as possible. The wall in this case is a source of spanwise vorticity. To advance separation, the reverse technique is employed to make the wall a sink of vorticity.

For a moving wall, separation is delayed when the motion of the surface is in the same direction as that of the freestream. This phenomenon is exploited in sports balls, Flettner's rotor, and some experimental STOL-type aircraft. In analogy with the moving wall case, unsteady separation is postponed when the separation point moves upstream and is advanced when the point of detachment moves downstream. Airfoils oscillating through high angles of attack can produce very high lift and maintain flow attachment well beyond their static stall angles. Insects exploit these unsteady separation effects to achieve remarkable aerodynamic characteristics.

In three-dimensional flows, the near-wall fluid may move in a direction in which the pressure gradient is more favorable and not against the adverse pressure in the direction of the main flow, as is the case for two-dimensional flows. Consequently, three-dimensional boundary layers are in general more capable of overcoming an adverse pressure gradient without extensive or catastrophic separation. A turbulent boundary layer is more resistant to separation than a laminar one, and mostly for that reason transition advancement may be desired in some situations. In low-Reynolds-number terminology, the transition-promoting devices are called turbulators and are in the form of single, multiple, or distributed roughness elements placed on the wall. The mechanical roughness elements, in the form of serrations, strips, bumps or ridges, are typically placed near the airfoil's leading edge.

Many other passive and active methods to postpone separation for low- and high-speed flows are available. Common to all these control methods is an attempt to supply additional energy to the near-wall fluid particles that are being retarded in the boundary layer. Passive techniques include intentional tripping of transition from laminar-to-turbulent flow upstream of what would be a laminar separation point, using boundary-layer fences to prevent separation at the tips of swept-back wings, placing an array of vortex generators on the body to enhance the momentum and energy in the neighborhood of the wall, utilizing a rippled trailing edge, employing streamwise corrugations, and using a screen to divert the flow and increase the velocity gradient at the wall. Active methods to postpone separation include fluid injection parallel to the wall to augment the shear-layer momentum or normal to the wall to enhance the mixing rate, the use of acoustic excitations, periodic forcing of the velocity field via an oscillating flap or wire, and oscillatory surface heating.

8.10.2   Computational Fluid Dynamics

The tremendous increases in CFD capability that have occurred as a direct result of increases in computer storage capacity and speed are transforming flow separation control from an empirical art to a predictive science. Control techniques such as mitigation of imposed pressure gradients, blowing, and suction are all readily parameterized via viscous CFD. Current inaccuracies in turbulence modeling can severely degrade CFD predictions once separation has occurred; however, the essence of separation control is the calculation of attached flows, estimation of separation location, and indeed whether or not separation will occur. These tasks can in fact be performed reasonably well via CFD within the uncertainties of the transition location estimation (e.g., Smith 1975). This latter uncertainty has been significantly reduced for low-disturbance freestreams and smooth surfaces using CFD along with the e^n-method discussed in Chapter 6 (Bushnell and McGinley 1989). Therefore it is now possible, at least to first order, to design bodies over which the flow will not separate (Cooke and Brebner 1961; Pinebrook and Dalton 1983; Yang et al. 1984; Dillner, May, and McMasters 1984; Waggoner and Allison 1987; Szodruch and Shneider 1989), and this is generally done, for aircraft, for the cruise condition. This is not currently done, for example, for helicopters and automobiles.

The cab-top fairings on tractor-trailer trucks (Kirsch 1974) and fairings at after-body junctures (Howard et al. 1981) are useful experimental attempts at reducing the causative adverse pressure fields. An exploratory study documented by White-head and Bertram (1971) indicated that continuous-curvature surface geometry can, presumably via minimization of pressure gradients, profoundly reduce three-dimensional separation. Current CFD capability allows the design of intersection regions that are properly filleted to obviate formation of the usual organized horse-shoe vortices found in three-dimensional separated flows (e.g., Lakshmanan et al. 1988).

8.10.3 Applications and Comparisons

It should be evident from the discussion in this chapter thus far that there are a large number of strategies that can control or mitigate flow separation. The choice of which technique to employ is a function of the particular flow situation and purpose for the control. Considerations include system aspects such as volume, weight, complexity, cost, reliability, overall energy and drag budgets, and any requirements for dynamic response and styling. Aside from design for separation minimization, significant applications thus far have been relatively limited though important. Examples of successful separation control systems include active injection on several 1950's and 1960's era fighters, STOL transports, the ubiquitous vortex generators, passive bleed for supersonic inlets, leading edge extensions for the 1970's and 1980's fighters, cab top fairing for tractor-trailer trucks, and various types of blown flaps on transport airplanes.

Particularly intriguing in the future are possible applications at cruise, especially at high speeds, as well as standby techniques for off-design situations, thereby allowing extremely tight designs. In addition, increased knowledge of, and control innovations for, vortical flows should allow solutions to the problems of high-angle-of-attack ma-neuvering and stall and spin prevention or recovery. Alternate techniques should also be sought for either supplementation or replacement of conventional high-lift flap systems. Candidate approaches include use of the LFC leading-edge suction system along with air jet vortex generators and perhaps either rotating wings or airport ski jumps. Also, tremendous energy saving potential exists in further separation control or form drag reduction for vehicles that have notoriously high drag: automobiles, helicopters and tractor-trailer trucks.

Various comparisons have been made between candidate separation control tech-niques on common test beds. Vakili et al. (1985) reported that vortex generators perform better than a flow-control rail for a diffusing S-duct. For a swept wing, experiments indicated that slot blowing is more effective than either vane or air-jet vortex generators (Kukainis 1969). For CTOL wings, improvements in the design of mechanical high-lift systems have tended to keep pace with the trend toward higher wing loading, and thus benefits from BLC techniques have not appeared sufficiently attractive (Gratzer 1971). Butter (1984) and Dillner et al. (1984) provided excellent analyses and discussions of current CTOL high-lift design practice, problems, and performance. As a suggestion for further work, a technique that has not yet been researched sufficiently is the three-dimensionalization of a nominally two-dimensional problem which, from the work of Ball (1971), McLean and Herring (1974), and Lin and Howard (1989) tends to reduce the extent of separation.

8.10.4 Recommendations

Flow-separation control will be of increasing importance in the future as declining petroleum reserves, concern over the greenhouse effect, economic development of the Third World, continued population growth, and increasing economic competition force stringent energy conservation and efficiency improvements. Much of the remaining gains in aerodynamics involve some form of viscous flow control, including flow-separation control. Particularly intriguing is the possibility of replacing or at least augmenting conventional high-lift devices (i.e., flaps and slats). In the limit, this may require wing rotation because some increase in wing angle of attack would probably be required.

With the exception of vortex generators, the rapid development of CFD has transformed much of conventional flow separation control from art to science, including the capability of mitigating the causative pressure fields and optimizing several control approaches. However, the effects of combinations of separation control approaches and the potential for synergistic benefits require considerable further research.

The use of flow-separation control at cruise for high-speed, supersonic civil transports may allow significant increases in lift-to-drag ratio via increased leading-edge thrust, upper surface and fuselage lift contribution, and effective favorable interference wave drag reduction from shock-boundary-layer interaction separation control. The art and science of flow separation control is far richer than the conventional view of blowing or suction and vortex generators. The recent research regarding miniaturized and jet-injection vortex generators allows reduction of parasitic cruise drag for such devices and is particularly intriguing for several applications, including replacement or supplementation of conventional high-lift systems. Microelectromechanical systems (MEMS) offer the potential for extremely small actuators and thus even more effective and less expensive means for flow control (Ho and Tai 1996, 1998). We will return to this topic in Chapter 14.

Suggested future research directions include further work on microdevices designed specifically for separation control; optimization of air-jet vortex generators, including pulsed injection; three-dimensionalization of nominally two-dimensional surfaces; use of weak upstream shocks for turbulence amplification and swept shocks for spanwise removal of the near-wall low-momentum region; downstream vortex reenforcers and miniaturized near-wall vortex generators; and techniques to force momentum toward the wall (e.g., downwash from embedded lifting surfaces).

CHAPTER NINE

Low-Reynolds-Number Aerodynamics

Oh, how much is today hidden by science! Oh, how much it is expected to hide!

(Friedrich Wilhelm Nietzsche, 1844–1900)

My philosophy of life is work. Bringing out the secrets of nature and applying them for the happiness of man. I know of no better service to render during the short time we are in this world.

(Thomas Alva Edison, 1847–1931)

PROLOGUE

Among the goals of external flow control are separation postponement, lift enhancement, transition delay or advancement, and drag reduction. These objectives are not necessarily mutually exclusive. For low-Reynolds-number lifting surfaces, where the formation of a laminar separation bubble may have a dominant effect on the flowfield, the interrelation between the preceding goals is particularly salient, presenting an additional degree of complexity when flow control is attempted to achieve, say, maximum lift-to-drag ratio. This chapter discusses the aerodynamics of low-Reynolds-number lifting surfaces—particularly the formation and control of separation bubbles.

9.1 Introduction

Insects, birds and bats have perfected the art of flight through millions of years of evolution. Man's dream of flying dates back to the early Greek myth of Daedalus and his son Icarus, but the first successful heavier-than-air flight took place less than a century ago. Today, the Reynolds numbers for natural and man-made fliers span the amazing range from 10^2 to 10^9, insects being at the low end of this spectrum and huge airships occupying the high end (Carmichael 1981).

The function of the airfoil section on those fliers is to produce lift. Inevitably, viscous effects, compressibility effects, and the finite span of the lifting surface all ensure that drag is also produced. A thrust must be generated by some sort of a power plant to overcome this streamwise resistance to the motion. The lift-to-drag

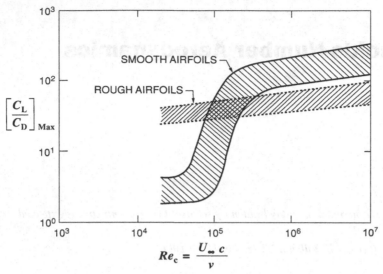

Figure 9.1: Airfoil performance as function of chord Reynolds number [after McMasters and Henderson 1980].

ratio is a measure of the effectiveness of the airfoil. In general, this ratio is very low at low Reynolds numbers and improves with increases in this parameter. As shown in Figure 9.1, reproduced from McMasters and Henderson (1980), the maximum $[C_L/C_D]$ improves dramatically in the range of chord Reynolds numbers of $Re_c = 10^4$–10^6. Below 10^4, typical of insects and small model airplanes, the boundary layer around the lifting surface is laminar. Stalling in this case is caused by an abrupt separation of the laminar flow near the leading edge as the angle of attack is increased to modest values.[1] The maximum lift is limited, and the drag increases significantly when the lifting surface stalls.

For $Re_c > 10^6$, typical of large aircraft, boundary-layer transition to turbulence usually takes place ahead of the theoretical laminar separation point. A turbulent boundary layer can negotiate quite severe adverse pressure gradients without separation, and this kind of lifting surface often experiences a trailing-edge stall at relatively high angles of attack. The stall is preceded by a movement of the separation point forward from the trailing edge with increasing incidence (McCullough and Gault 1951).

In the range of Reynolds numbers of 10^4–10^6, termed low Reynolds number for the purpose of this chapter (Lissaman 1983), many complicated phenomena take place within the boundary layer. Separation, transition and reattachment can all occur within a short distance and dramatically affect the performance of the lifting surface. The laminar separation bubble that commonly forms in this range of Reynolds numbers plays an important role in determining the boundary layer behavior and the stalling characteristics of the airfoil (Tani 1964). As indicated in Figure 9.1, the maximum lift-to-drag ratio for a smooth airfoil increases by two orders of magnitude in this Reynolds number regime. Remotely piloted aircraft and turbine blades are examples of lifting surfaces having this range of Reynolds numbers.

---

[1] The arguments and the figure herein are for steady-state flows. Insects and other very low-Reynolds-number fliers exploit unsteady effects to achieve remarkable performances, as was discussed in Chapter 8.

The skilled designer has available a variety of passive and active techniques to effect a beneficial change in the complex flowfield that characterizes this intermediate range of Reynolds numbers. Roughness and shaping are among the simplest passive methods to ensure flow attachment beyond a critical angle of attack and, thus, an improved performance. Wall transpiration and heat transfer are examples of active control methods to improve the lift-to-drag ratio. Although these broad flow-control strategies are similar to those discussed elsewhere in this book, the emphasis here is on the low-Reynolds-number regime dominated by the formation of laminar separation bubbles.

This chapter discusses the aerodynamics of low-Reynolds-number lifting surfaces, particularly the formation and control of separation bubbles. Flow-control goals in those cases are strongly interrelated leading to certain difficulties in choosing the best control strategy for a particular end result. These potential conflicts will be elaborated.

9.2 Low-Reynolds-Number Airfoils

In the range of Reynolds numbers of $Re_c = 10^4$–10^6, a substantial improvement in the lift-to-drag ratio of an airfoil takes place. According to Carmichael (1981), this is the Reynolds number regime in which we find man and nature together in flight: large soaring birds; large radio-controlled model aircraft; foot-launched ultralight, man-carrying hang-gliders; human-powered aircraft; and the more recently developed remotely-piloted-vehicles (RPVs) used for military and scientific sampling, monitoring, and surveillance. Three review articles on low-Reynolds-number aerodynamics by Tani (1964), Lissaman (1983), and Mueller (1985a) are particularly recommended. There is also a wealth of information available in the proceedings of the following conferences dedicated to the subject matter: (1) Conference on Low Reynolds Number Airfoil Aerodynamics, University of Notre Dame, Notre Dame, Indiana, 17–19 June 1985 (Mueller 1985b); (2) International Conference on Aerodynamics at Low Reynolds Numbers $10^4 < Re < 10^6$, Royal Aeronautical Society, London, Great Britain, 15–18 October 1986; (3) Conference on Low Reynolds Number Aerodynamics, University of Notre Dame, Notre Dame, Indiana, 5–7 June 1989 (Mueller 1989); and (4) Eighth International Conference on Remotely Piloted Vehicles, Bristol, Great Britain, 2–4 April 1990.

In this range of Reynolds numbers, very complex flow phenomena take place within a short distance on the upper surface of an airfoil at incidence. Unless artificially tripped, the boundary layer remains laminar at the onset of pressure recovery, and the airfoil's performance is then entirely dictated by the laminar flow's poor resistance to separation. The separated flow forms a free-shear layer, which is highly unstable, and transition to turbulence is readily realized. Subsequent reattachment of the separated region may take place because of the increased entrainment associated with the turbulent flow. Provided that the high-speed fluid entrained into the wall region supplies sufficient energy to maintain the circulating motion against dissipation, a separation bubble forms.

The precise conditions for the occurrence of separation, transition, and reattachment—in other words for the formation of a laminar separation bubble—depend on the Reynolds number, the pressure distribution, the surface curvature, the surface

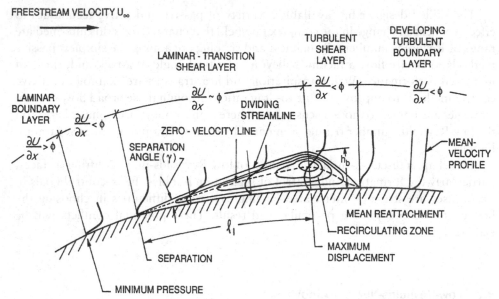

Figure 9.2: Sketch of a laminar separation bubble.

roughness, and the freestream turbulence as well as other environmental factors. If the Reynolds number is sufficiently high, transition takes place near the minimum pressure point ahead of the location at which separation would have occurred if the boundary layer had remained laminar. For moderate Reynolds numbers, separation takes place before transition. The laminar boundary layer can only support very small adverse pressure gradient without separation. As stated in Chapter 8, if the ambient incompressible fluid decelerates in the streamwise direction faster than $U_o \sim x^{-0.09}$, the flow separates. The separated flow will not reattach to the surface, and no bubble will be formed if the Reynolds number is sufficiently low. However, for the intermediate Reynolds number range (typically 10^4–10^6), the separated flow proceeds along the direction of the tangent to the surface at the separation point (von Doenhoff 1938a) and transition to turbulence takes place in the free-shear layer owing to its increased transition susceptibility. Subsequent turbulent entrainment of high-speed fluid causes the flow to return to the surface, thus forming what is known as a laminar separation bubble, as sketched in Figure 9.2. Downstream of the point of reattachment, the newly formed turbulent boundary layer is capable of negotiating quite severe adverse pressure gradients without separation. The ability of a turbulent boundary layer to resist separation improves as the Reynolds number increases (Lissaman 1983).

9.3 Conditions for Bubble Formation

It is clear from the preceding arguments that bubble formation is confined to a certain range of Reynolds numbers and that this range changes from one airfoil to another as well as from one environment to another. A rough rule according to Carmichael (1981) is that the Reynolds number based on freestream velocity and the distance from separation to reattachment is approximately 5×10^4. In general, then, an airfoil with chord Reynolds number less than 5×10^4 will experience laminar separation

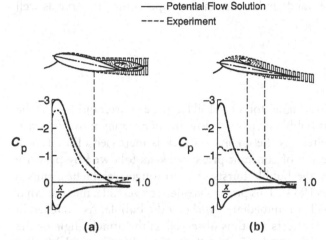

Figure 9.3: Qualitative pressure distributions for two airfoils at incidence. (———) Potential flow solution; (————) experiment. (a) Short bubble on upper surface and subsequent rear separation of turbulent boundary layer; (b) Long bubble.

with no subsequent reattachment. For chord Reynolds numbers slightly higher than 5×10^4, a long bubble is expected. Shorter bubbles are formed at higher Reynolds numbers. Tani (1964) asserted that a Reynolds number typical of local conditions in the boundary layer is more appropriate to characterize a separation bubble than the chord Reynolds number. Typically, the Reynolds number, based on the boundary-layer's displacement thickness[2] and the velocity just outside the rotational flow region at the point of separation, is more than 500 for a short bubble[3] and less than 500 for a long one. The corresponding bubble's streamwise extent, normalized with the displacement thickness at the point of separation, is 10^2 and 10^4, respectively (Tani 1964).

The short separation bubble generally has a length of the order of a few percent of the chord. It merely represents a transition-forcing (tripping) mechanism to allow reattachment of an otherwise separated shear layer. Such a bubble does not greatly affect the peak suction, as determined from the potential flow solution around the airfoil. Except for the appearance of a minute bump in the lift curve (C_L versus α curve, where C_L is the lift coefficient and α is the angle of attack), the presence of a short bubble has no significant effect on the pressure distribution around the lifting surface, as depicted in Figure 9.3a. On the other hand, a long bubble may be as much as $0.2c$–$0.3c$, where c is the airfoil chord, and significantly changes the pressure distribution by effectively altering the shape over which the outer potential flow is developed. In this case, the sharp suction-peak near the leading edge is generally not realized, and a suction plateau of a reduced level extends over the region occupied by the bubble (Figure 9.3b). A long bubble tends to increase in length as incidence

---

[2] The displacement thickness δ^* for a laminar boundary layer near the leading edge of an airfoil is extremely small and cannot accurately be measured. Instead the momentum thickness δ_θ is computed from the pressure distribution using Thwaites' (1949) formula, and the ratio (δ^*/δ_θ) at the separation point is assumed to be 3.7.

[3] The bubble gets shorter at higher Re_{δ^*}; and at $Re_{\delta^*} \approx 6000$, bubble formation is precluded (bubble's streamwise extent approaches zero) by transition to turbulence in the boundary layer.

is increased, leading to a corresponding decrease in the slope of the lift curve as well as an increase in the pressure drag.

9.4   Bubble's Breakdown

In general, the lift-to-drag ratio is higher for an airfoil having a shorter bubble. As the result of many factors, a short bubble forming at low incidence may move forward and contract in streamwise extent as the angle of attack is increased (Tani 1964). Within the bubble, a small region of constant pressure exists followed by pressure recovery. At higher incidence, the bubble bursts and no longer reattaches, thus a leading-edge stall ensues (Jones 1934). This process is often irreversible, meaning that reducing the angle of attack will not immediately *unburst* the bubble. As sketched in Figure 9.4, strong lift hysteresis effects are thus observed as the attack angle or the Reynolds number is recycled (Schmitz 1967; Mueller 1985c; Bastedo and Mueller 1985; Brendel and Mueller 1988). Similar hysteresis effects are also seen in drag.

For thin airfoils of small nose-radius, pressure recovery commences very near the leading edge, and the adverse pressure gradients are severe at high angles of attack. Separation bubble may occur on these airfoils even at chord Reynolds numbers exceeding 10^6. At large incidence, the short bubble breaks down into a long one. With

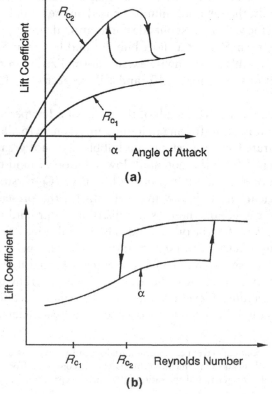

(a)

(b)

Figure 9.4: Illustration of lift hysteresis as the angle of attack (top plot) or the Reynolds number (bottom plot) is recycled.

increasing angles of attack, the reattachment point moves progressively backward until it reaches the trailing edge, at which stage its maximum thickness is typically 3% of the chord. A further increase in incidence leads to completely detached flow and the so-called thinairfoil stall. A comprehensive review of the different kinds of stall on thin airfoils is given by Crabtree (1957).

Aerodynamics data for both thin and thick airfoils in the low-Reynolds-number regime are accessible in several recent articles and conference proceedings (e.g., Burns 1981; Mueller and Burns 1982; Render 1984; Render, Stollery, and Williams 1985; Leibeck and Camacho 1985; Bastedo and Mueller 1986; Stollery and Dyer 1989; Mueller 1985b, 1989). Available experimental data on bubble's formation and bursting indicates that transition to turbulence in the separated shear-layer and subsequent reattachment will occur if the Reynolds number based on displacement thickness at the point of laminar separation exceeds a critical value that is not necessarily universal (Tani–Owen–Klanfer criterion). A lower limit for this Reynolds number seems to be $Re_{\delta*} \approx 350$. Bursting occurs if the pressure recovered in the reattachment process in terms of the dynamic pressure at separation (pressure recovery coefficient) exceeds a certain critical value (Crabtree criterion). Again, this critical value changes from one airfoil to another, but an upper limit of 0.35 appears to be valid for many shapes. Crabtree (1954) assumes that bubble's breakdown occurs because a maximum possible value of pressure exists that can be recovered in the turbulent entrainment process that causes the flow reattachment. This implies the existence of a maximum possible value of the shear stress setup in the turbulent entrainment region so as to counteract the pressure gradient. At breakdown, caused by either an increase in incidence or a decrease in Reynolds number, the Tani–Owen–Klanfer criterion is satisfied, but the Crabtree criterion is about to be violated.

The question of primary concern to us in this chapter is how to control the flow around a low-Reynolds-number airfoil to achieve an improved performance. The interrelation between the different control goals is particularly salient when a separation bubble exists, and this issue will be tackled in the next section.

9.5 Control Goals and Their Interrelation

According to Tani (1964), all three kinds of stall, trailing-edge stall, leading-edge stall, and thin-airfoil stall, may occur for a given airfoil at different Reynolds numbers or for different airfoils at a given Reynolds number. A particular lifting surface produces higher lift at higher incidence limited by the angle at which the airfoil stalls. At that point, drag increases dramatically, and the lifting surface performance deteriorates rapidly. Flow control is aimed at improving this performance. Among the practical considerations that must be taken into account for both active and passive control devices are their cost of construction and operation, complexity, and potential trade-offs or penalties associated with their use. It is this latter point in particular that presents an additional degree of complexity for controlling low-Reynolds-number lifting surfaces. Achieving a beneficial effect for one control goal may very well adversely affect another goal, and design compromises must often be made.

Among the desired goals of external flow modification are separation or reattachment control, transition delay or advancement, and lift enhancement and drag reduction. These objectives are not necessarily mutually exclusive, and for

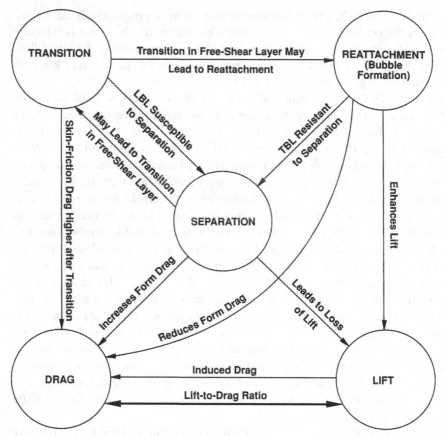

Figure 9.5: Interrelation between flow-control goals.

low-Reynolds-number lifting surfaces the interrelations between these goals are particularly conspicuous, presenting an additional degree of complexity. As mentioned before, in the range of Reynolds numbers of 10^4–10^6, a laminar separation bubble may form and may have a dominant effect on the flowfield and the airfoil's performance.

Recall the discussion in Section 3.1. The schematic in Figure 9.5 is a partial representation of the interrelation between one control goal and another for a lifting surface. If the boundary layer becomes turbulent, its resistance to separation is enhanced, and more lift could be obtained at increased incidence. On the other hand, the skin-friction drag for a laminar boundary layer can be as much as an order of magnitude less than that for a turbulent one. If transition is delayed, lower skin friction as well as lower flow-induced noise is achieved. However, the laminar boundary layer can only support very small adverse pressure gradient without separation, and subsequent loss of lift and increase in form drag occur. Once the laminar boundary layer separates, a free-shear layer forms, and for moderate Reynolds numbers transition to turbulence takes place. Increased entrainment of high-speed fluid due to the turbulent mixing may result in reattachment of the separated region and formation of a laminar separation bubble. At higher incidence, the bubble breaks down, either separating completely or forming a longer bubble. In either case, the form drag increases, and the lift-curve's slope decreases. The ultimate goal of all this is to improve the airfoil's performance by increasing the lift-to-drag ratio. However, induced drag is caused by the lift generated on a lifting surface with a finite span. Moreover, more lift is generated at higher incidence, but form drag also increases at these angles.

All of the above points to potential conflicts as one tries to achieve a particular control goal only to affect another goal adversely. An ideal method of control that is simple, inexpensive to build and operate, and does not have any trade-off does not exist, and the skilled engineer has to make continuous compromises to achieve a particular design goal. Keeping this in mind, we now proceed to review selected control methods available to the designer of low-Reynolds-number lifting surfaces such as turbine blades and remotely piloted aircraft.

9.6 Separation Control

Fluid particles in a boundary layer are slowed down by wall friction. If the external potential flow is sufficiently retarded owing, for example, to the presence of an adverse pressure gradient, the momentum of those particles will be consumed by both the wall shear and the pressure gradient. At some point (or line), the viscous layer departs or breaks away from the bounding surface. The surface streamline nearest to the wall leaves the body at this point, and the boundary layer is said to separate. At separation, the rotational flow region next to the wall abruptly thickens, the normal velocity component increases, and the boundary-layer approximations are no longer valid. Because of the large energy losses associated with boundary-layer separation, the performance of a lifting surface is often controlled by the separation location. If separation is postponed, the pressure drag is decreased, and the circulation, and hence the lift at high angles of attack, is enhanced.

As discussed in the previous chapter, separation control methods include the modulation of pressure gradient via shaping, wall suction, surface cooling in gases, and surface heating in liquids. The first of these strategies is the simplest and is most suited for low-Reynolds-number situations. Streamlining can greatly reduce the steepness of the pressure rise. In contrast to turbulent flows, laminar boundary layers can only support very small adverse pressure gradients without separation. Transition on the upper surface of a lifting surface typically occurs at the first onset of adverse pressure gradient if the Reynolds number exceeds 10^6. The separation-resistant turbulent boundary layer that evolves in the pressure recovery region results in higher maximum lift at relatively large angle of stall. Depending on the severity of the initial adverse gradient, and hence on the airfoil shape, laminar separation may take place prior to transition. Regardless of whether or not the flow subsequently reattaches, the laminar separation leads to higher form drag and lower maximum lift. Delicate contouring of the airfoil near the minimum pressure point to lessen the severity of the adverse pressure gradient may be used to accomplish separation-free transition.

As an example of the effects of the lifting surface's shape on its performance, consider the lift curves for the three airfoil sections NACA 63_3–018, NACA 63–009, and NACA 64A006. These airfoils have maximum thicknesses of $0.18c$, $0.09c$, and $0.06c$, respectively, where c is the chord. The respective leading edge radii are $0.021c$, $0.006c$, and $0.003c$. Figure 9.6, adapted from the measurements by McCullough and Gault (1951), depicts C_L versus α curves for the three sections at a chord Reynolds

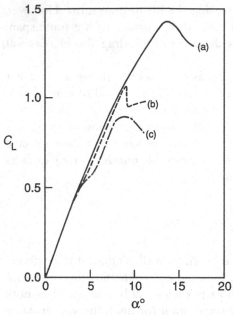

Figure 9.6: Lift curves for three airfoils at $Re_c = 5.8 \times 10^6$. (a) NACA 63_3–018; (b) NACA 63–009; (c) NACA 64A006 [after McCullough and Gault 1951].

number of 5.8×10^6. For the thick section, NACA 63_3–018, transition takes place near the minimum pressure point (Fitzgerald and Mueller 1990). Stalling in this case is of the trailing-edge type and is preceded by a gradual movement of the separation point of the turbulent boundary layer forward from the trailing edge as α increases. A laminar separation bubble is formed on the other two sections at small incidence. However, the NACA 63–009 section experiences a sudden leading-edge stall when the bubble bursts with no subsequent reattachment, whereas the NACA 64A006 section experiences a more gradual thin-airfoil stall (McCullough 1955; Crabtree 1957). In the latter case, the short bubble breaks down into a longer bubble at an angle of attack of $5°$, causing a slight discontinuity in the lift curve. Subsequent increase in α leads to a movement of the reattachment point toward the trailing edge. The maximum lift in this case is about 40% lower than that for the thick airfoil. The stall angle is also lower.

A second example is provided for an airfoil specifically designed for the low-Reynolds-number regime. The carefully contoured Eppler 61 has a maximum thickness of $0.056c$ and is highly cambered. Mueller and Burns (1982) reported lift, drag, and smoke visualization data for this airfoil section in the range of Reynolds numbers of 3×10^4–2×10^5. A sample of their lift and drag curves at three different speeds is depicted in Figure 9.7. At a negative angle of attack of about $\alpha = -3°$, the flow around the cambered airfoil separates at the leading edge on the lower surface without further reattachment. Zero lift is measured at this angle and is correlated with the appearance of smooth smokelines above and below the airfoil to form an uncambered, symmetrical shape, as shown in the top photograph in Figure 9.8. In this figure, the chord Reynolds number is $Re_c = 8.7 \times 10^4$, and three angles of attack are depicted: $\alpha = -3°$, $\alpha = 0°$, and $\alpha = 8°$. The results explain the deviation of the zero-lift angle as well as the shape of the lift curve from those predicted by thin airfoil theory. Strong Reynolds number effects are evident in both the lift and drag curves. At increasing angles of attack, the performance of the Eppler 61 airfoil is similar to that of the other thin airfoil depicted in Figure 9.6c, although the maximum lift coefficient is higher in the former case.

9.7 Transition Control via Shaping

Streamlining a body to prevent separation and reduce form drag is quite an old art, but the stabilization of a boundary layer by pushing the longitudinal location of the pressure minimum as far back as possible dates to the 1930s and led to the

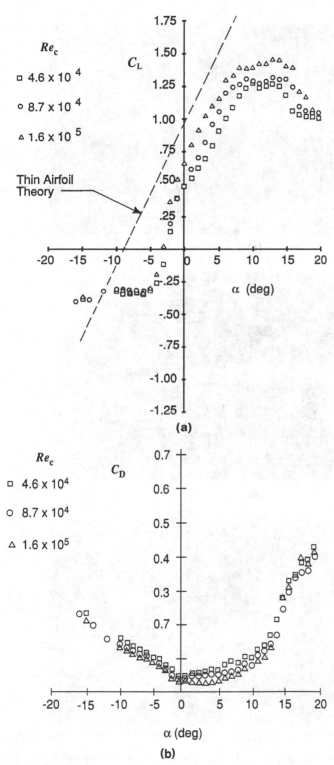

Figure 9.7: Reynolds number effects on lift (top curves) and drag (bottom curves) of the Eppler 61 airfoil [after Mueller and Burns 1982].

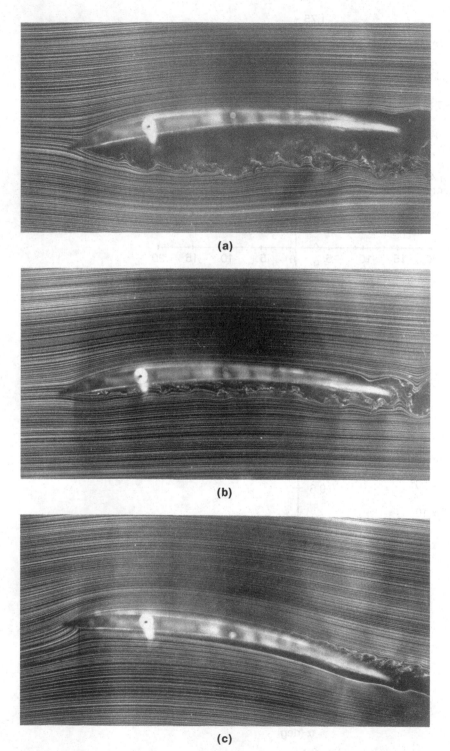

Figure 9.8: Smoke-wire flow visualization of the low-Reynolds-number Eppler 61 airfoil. $Re_c = 8.7 \times 10^4$. (a) $\alpha = -3°$; (b) $\alpha = 0°$; (c) $\alpha = 8°$ [after Mueller and Burns 1982].

successful development of the NACA 6-Series NLF airfoils. Newer, low-Reynolds-number lifting surfaces used in sailplanes, low-speed drones, and executive business jets have their maximum thickness point far aft of the leading edge. The recent success of the *Voyager's* 9-day, unrefueled flight around the world was due in part to a wing design employing natural laminar flow to approximately $0.5c$.

The favorable pressure gradient extends to the longitudinal location of the pressure minimum. Beyond this point, the adverse pressure gradient becomes steeper and steeper as the peak suction is moved further aft. For an airfoil, the desired shift in the point of minimum pressure can only be attained in a certain narrow range of angles of incidence. On the basis of the shape, angle of attack, Reynolds number, surface roughness, and other factors, the boundary layer either becomes turbulent shortly after the point of minimum pressure or separates first and then undergoes transition. One of the design goals of NLF is to maintain attached flow in the adverse pressure gradient region, and some method of separation control may have to be used there.

Subtle changes in the magnitude and extent of favorable pressure gradient, leading edge radius, and other shape variables can have pronounced effects on the airfoil's performance. As an example, consider the lift and drag characteristics of a conventional section (NACA 23015) and a laminar-flow section (NACA 66$_2$–215). Both airfoils are cambered, and both have maximum thickness of $0.15c$. However, the maximum thickness point is located at $0.3c$ for the conventional section and at $0.45c$ for the laminar-flow one. The leading-edge radius is $0.025c$ and $0.014c$ for the two respective airfoils. The point of minimum pressure, as computed for the basic symmetric section at zero incidence, is located at $0.15c$ for the conventional airfoil, whereas this point is pushed back to $0.6c$ for the laminar-flow airfoil. As seen from the lift curves depicted in Figure 9.9 for a chord Reynolds number of $Re_c = 9 \times 10^6$, the laminar-flow section has slightly higher lift at small angles of attack than the conventional section, but stalling occurs at lower incidence and the maximum lift is smaller for the laminar-flow airfoil. The lift-drag polars for the same two airfoils are shown in Figure 9.10. The sudden increase in drag for the laminar-flow section occurs at an angle of attack of about $\alpha = 1.5°$. This is caused by a forward movement of the minimum pressure point and corresponding accentuation of the adverse pressure gradient as the flow on the upper surface accelerates sharply to round the airfoil's sharp nose. The increased adverse pressure gradient leads to early transition and corresponding drag increase. The maximum lift-to-drag ratios for the conventional and laminar-flow airfoils are 125 and 102, respectively. However, the laminar-flow section is designed to cruise in the low-drag

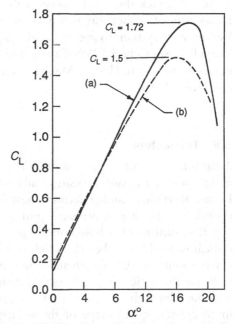

Figure 9.9: Lift curves for (a) conventional section (NACA 23015); and (b) laminar-flow section (NACA 66$_2$–215). Chord Reynolds number in both cases is $Re_c = 9 \times 10^6$ [after Abbott and von Doenhoff 1959].

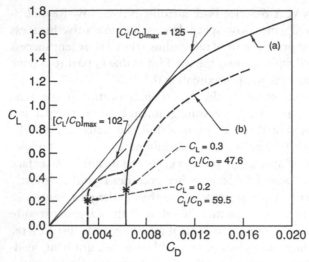

Figure 9.10: Lift-drag polars for the two airfoils depicted in Figure 9.9 [after Abbott and von Doenhoff 1959].

region ($\alpha < 1.5°$). At the design lift coefficients of 0.3 and 0.2 for the conventional and laminar-flow sections, the lift-to-drag ratios are 47.6 and 59.5, respectively.

Factors that limit the utility of NLF include crossflow instabilities and leading-edge contamination on swept wings, insect, and other particulate debris; high unit Reynolds numbers of conventional aircraft at lower cruise altitudes (not a limiting factor for low-Reynolds-number aircraft); and performance degradation at higher angles of attack due to the necessarily small leading-edge radius of NLF airfoils. Reductions of surface waviness and smoothness of modern production wings, special leading-edge systems to prevent insect impacts and ice accretion (Korkan, Cross, and Cornell 1986; Cebeci 1989; Zaman and Potapczuk 1989), higher cruise altitudes of newer airplanes, and higher Mach numbers all favor the application of NLF (Runyan and Steers 1980).

9.8 Turbulators

A turbulent boundary layer is more resistant to separation than a laminar one, and mostly for that reason transition advancement may be desired in some situations. In low-Reynolds-number terminology, the transition promoting devices are called turbulators. For a zero-pressure-gradient boundary layer, transition typically occurs at a Reynolds number based on distance from leading edge of the order of 10^6. The critical Reynolds number Re_{crit} below which perturbations of all wave numbers decay is about 6×10^4. To advance the transition Reynolds number, one may attempt to lower the critical Re, increase the growth rate of Tollmien–Schlichting waves, or introduce large disturbances that can cause bypass transition. The first two routes involve altering the shape of the velocity profile—making it less full—using wall motion, injection, adverse pressure gradient, surface heating in gases or cooling in liquids, or other strategies to increase the near-wall viscosity. The third route, exposing the boundary layer to large disturbances, is much simpler to implement though more difficult to analyze (Smith and Kaups 1968; Cebeci and Chang 1978; Nayfeh et al. 1986; Cebeci and Egan 1989).

Morkovin (1984) broadly classified the large disturbances that can cause by-pass transition into steady or unsteady ones originating in the freestream or at the body surface. The most common example is single, multiple, or distributed roughness elements placed on the wall. The mechanical roughness elements, in the form of serrations, strips, bumps, or ridges, are typically placed near the airfoil's leading edge. If the roughness characteristic length is large enough, the disturbance introduced is nonlinear, and bypass transition takes place. For a discrete three-dimensional roughness element of height-to-width ratio of one, Tani (1969) reported a transition Reynolds number of $Re_{\delta^*} \simeq 300$ for a roughness Reynolds number of $Re_\Upsilon \sim 10^3$. Here, Re_{δ^*} is based on the velocity outside the boundary layer and the displacement thickness ($Re_{\delta^*} \equiv U_o \delta^*/\nu$), and Re_Υ is based on the height of the roughness element Υ and the velocity in the undisturbed boundary layer at the height of the element ($Re_\Upsilon \equiv \overline{U}(\Upsilon)\, \Upsilon/\nu$). Note that the transition Reynolds number, Re_{δ^*}, indicated above is below the critical $Re_{\delta^*}|_{\text{crit}} = 420$ predicted from the linear stability theory. For a roughness Reynolds number of about 600, transition occurs at $Re_{\delta^*} \simeq 10^3$. For a smooth surface, transition typically takes place at $Re_{\delta^*} \simeq 2.6 \times 10^3$. An important consideration when designing a turbulator is to produce turbulence and suppress laminar separation without causing the boundary layer to become unnecessarily thick. A thick, turbulent, wall-bounded flow suffers more drag and is more susceptible to separation than a thin one. Consistent with this observation, available data (Figure 9.1) indicate that a rough airfoil has higher lift-to-drag ratio than a smooth one for $Re_c < 10^5$ but that this trait is reversed at higher Reynolds numbers.

For low-Reynolds-number airfoils, performance may be improved by reducing the size of the laminar separation bubble through the use of transition ramps (Eppler and Somers 1985), boundary layer trips (Davidson 1985; Van Ingen and Boermans 1986), or even pneumatic turbulators (Pfenninger and Vemuru 1990). Donovan and Selig (1989) provided extensive data using both methods for 40 airfoils in the Reynolds number range of $Re_c = 6 \times 10^4$–3×10^5. A long region of roughly constant adverse pressure gradient on the upper surface of a lifting surface (termed a bubble ramp) achieves a lower drag than the more conventional laminar-type velocity distribution in which the pressure initially remains approximately constant and then quickly recovers. Trips were also used in Donovan and Selig's experiments to shorten the separation bubble. A simple two-dimensional trip performed as well or better than zig-zag tape, hemisphere bumps, and normal blowing. Donovan and Selig concluded that an airfoil that performs poorly at low Reynolds numbers can be improved through the use of turbulators. Trips were less effective, however, at improving airfoils that normally had low drag. Other large disturbances that can lead to early transition include high turbulence levels in the freestream, external acoustic excitations, particulate contamination, and surface vibration. These are often termed environmental tripping. Transition can also be effected by detecting naturally occurring T–S waves and artificially introducing in-phase waves. Transition can be considerably advanced, on demand, using this wave superposition principle.

9.9 Parting Remarks

Available and contemplated flow control methods particularly suited for low-Reynolds-number lifting surfaces have been surveyed. The flow around these airfoils is dominantly affected by the formation of a separation bubble. The laminar separation makes the interrelation between transition, separation, lift, and drag controls particularly salient, presenting an additional degree of complexity. Several of these control techniques are suited for high- as well as low-speed airfoils.

Low-Reynolds-number airfoils span the range of 10^4–10^6. In this regime, laminar separation, transition, and reattachment may all occur within a short distance on the upper surface of an airfoil at incidence. The precise conditions for the occurrence of a laminar separation bubble depend on the local Reynolds number, pressure distribution, surface curvature, surface roughness, and freestream turbulence as well as other environmental factors.

To control the flow around a low-Reynolds-number airfoil to achieve improved performance, one must carefully consider potential conflicts in trying to achieve a particular control goal while inadvertently causing an adverse effect on another goal. A laminar boundary layer is less able to resist separation but is characterized by a very low skin-friction drag. However, if the flow separates, lift decreases and form drag increases substantially. In the low-Reynolds-number regime, laminar separation is often followed by transition to turbulence in the separated free-shear layer and subsequent reattachment to form a closed bubble. Bubble bursting at higher incidence leads to loss of lift and increased drag.

CHAPTER TEN

Drag Reduction

There does not exist a category of science to which one can give the name applied science. There are science and the applications of science, bound together as the fruit of the tree which bears it.
    (Louis Pasteur, 1822–1895)

Science becomes dangerous only when it imagines that it has reached its goal.
    (George Bernard Shaw, 1856–1950)

PROLOGUE

Drag is the force by which a fluid resists the relative motion of a solid. An equal and opposite reaction force acts on the body surface as a result of the fluid deformation, and drag is the component of this force parallel to the direction of the relative velocity vector. The fluid can be external or internal to the solid boundaries, and the solid surface can be rigid or compliant. Billions of gallons of fossil fuel are used annually to overcome the drag encountered by vehicles moving in air or water and the fluid resistance in gas, water, or oil pipelines. Flow control aims at minimizing this drag force, and the subject is explicitly or implicitly interwoven in every chapter of this book. Delaying laminar-to-turbulence transition (Chapter 6) is usually sought in order to benefit from the much lower skin friction associated with laminar boundary layers. Preventing separation (Chapter 8) means reducing the pressure drag, and when separation is provoked, as for example in delta wings, it is desirable to keep the associated drag penalty to a minimum. Compliant coatings (Chapter 7) are sought to reduce drag as well as to achieve other beneficial effects. Low-Reynolds-number vehicles (Chapter 9) notoriously suffer from low lift-to-drag ratio, and flow control attempts to remedy that. When higher rates of mass, momentum, or heat transfer are desired (Chapter 11), it is imperative that the concomitant drag penalty be kept to a minimum. The futuristic control systems envisioned in Chapter 14 deal for the most part with drag reduction. Reducing drag saves enormous resources spent to overcome it and is clearly advantageous to the environment, the economy, and the overall industrial competitiveness of any country. This chapter is a central, albeit brief, treatise on drag reduction that unites the disparate pieces of the puzzle discussed in more detail elsewhere in the book and ties a few loose ends not covered in other chapters.

10.1 Background

Whenever there is relative motion between a solid body and the fluid in which it is immersed, the body experiences a net force due to the action of the fluid. As depicted in Figure 10.1, this surface force can be decomposed into (1) a drag component parallel to the relative velocity vector, (2) a lift component perpendicular to the direction of motion, and (3) a side force resulting from any body asymmetries about the lift–drag (L–D) plane. For bodies that are symmetric about this plane, the side force, yawing moment, and rolling moment all vanish, leaving only lift, drag, and pitching moment acting on the body. If the body has a second plane of symmetry perpendicular to the L–D plane, the principal chord line of the body is the intersection of those two symmetry planes and is parallel to the freestream. In this case, both lift and pitching moment also vanish, and the body experiences only drag.

The total force acting on the body is equal and opposite to that on the fluid and is obtained by integrating the shear stress and pressure along the body surface

$$\mathbf{F} = \iint_A \tau_\mathrm{w}\, dA - \iint_A p\, d\mathbf{A}. \tag{10.1}$$

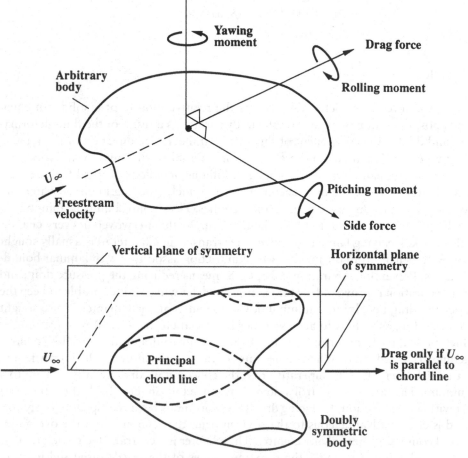

Figure 10.1: Forces and moments on a body immersed in a uniform flow.

This surface force is related to the fluid deformation and, in the case of a Newtonian fluid, the relation between the surface force per unit area (i.e., stress tensor) and rate of strain tensor is linear (see Section 2.2). In the limit of zero relative velocity between the body and fluid, the first term on the right-hand side of Eq. (10.1) vanishes, and the second term is the integral of the hydrostatic pressure around the body. This is the buoyancy force that Archimedes had shown over 2200 years ago to be equal and opposite to the weight of the displaced fluid.

The component of **F** parallel to the direction of motion is termed drag, and a body will move at a constant speed when the thrust generated by its propulsion system is equal to the drag force. The total drag consists of skin friction, equal to the streamwise component of the integral of all tangential stresses over the body surface, and pressure drag, equal to the streamwise component of the integral of all normal stresses. Flow separation is the major source of pressure drag—at least for blunt bodies—with additional contributions due to displacement effects of the boundary layer, wave resistance in a supersonic flow or at an air–water interface, drag induced by lift on a finite body, and virtual mass effects for time-dependent motions. The latter "pressure drag" exists even in inviscid fluids and occurs because, when the body velocity changes, the total kinetic energy of the fluid changes also. As depicted in Figure 10.2, the term form drag is used to denote the portion of the pressure drag that remains after subtracting out the drag due to lift. The sum of form drag and skin-friction drag sometimes is termed profile drag, which indicates the total drag that would exist had the aspect ratio been infinite (i.e., zero-induced drag).

At very low Reynolds numbers, say <1, viscous forces dominate, and pressure drag and skin-friction drag are about the same order of magnitude even when separation is absent. In this creeping flow regime, the drag is proportional to the relative velocity, the drag coefficient is inversely proportional to the Reynolds number, and the body shape has some but no major effect on the total drag.

At high Reynolds numbers, say greater than 1000, inertial effects dominate except for a thin layer near the body surface for which viscous forces and inertia are equally important. Drag for blunt bodies is large and is dominated by pressure drag caused by global separation, which usually occurs near the position of minimum pressure or maximum body thickness. The pressure drag is typically several orders of magnitude higher than the skin-friction drag. If Re is sufficiently high, the boundary layer is turbulent and is better able to resist separation longer, resulting in a narrower wake and lower total drag as compared with the laminar boundary layer case. The dimples on a golf ball, though increasing the skin-friction drag, lead to improved performance because the increased surface roughness trips the boundary layer and thus significantly lowers the dominant pressure drag.

For streamlined bodies at high Re, separation can usually be avoided all the way to the vicinity of the trailing edge. In that case, most of the drag is due to skin friction and, for that reason, higher total drag is observed when the boundary layer is turbulent. Still, the total drag is much smaller than that for blunt body having the same surface area. A streamlined wing at sufficiently high angle of attack is effectively a blunt body, and the resulting leading-edge separation is termed stall and leads to substantial increase in drag and loss of lift.

For all high-Reynolds-number cases, the total drag D can be estimated from a wake survey using the momentum integral equation for steady, incompressible flow

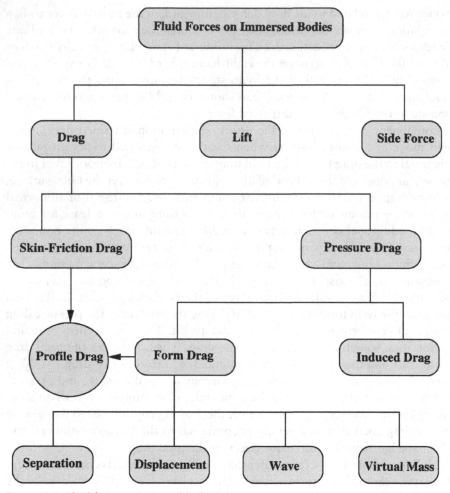

Figure 10.2: Fluid forces on immersed bodies.

with negligible body forces

$$D = \iint_A \rho U \left(U_\infty - U\right) dA \qquad (10.2)$$

The only restriction here is that the upstream and downstream planes at which the freestream velocity U_∞ and wake profile $U(y, z)$ are measured are far enough from the body such that the pressure in both planes is equal to the ambient pressure. At low Re, this condition is satisfied at distances that are too far from the body, at which location the velocity deficit $(U_\infty - U)$ would be too small to measure accurately.

The object of this chapter is to review available methods to reduce total drag. For a vehicle, reduced drag means longer range, reduced fuel cost and volume, higher payload, or increased speed. For pipes and channels, for which close to 100% of the drag is due to skin friction, drag reduction can result in improved throughput, reduced pumping power, or reduced duct size and cost.

10.2 Drag Reduction

Nature provides numerous instances in which drag reduction is essential for the survival of many species of air and marine animals (Bushnell and Moore 1991). Here, we cite instead two examples of the importance of minimizing the drag of man-made vehicles. At present, the annual fuel cost for all commercial airlines in the United States is about $10 billion—an expenditure that, thanks to declining oil prices adjusted for inflation, has not changed much in over a decade (Hefner 1988; Bushnell 1998). At subsonic cruising speeds, approximately half of the total drag of conventional takeoff and landing aircraft is due to skin friction. Hence, a reduction in skin-friction drag of 20% translates into an annual fuel saving of $1 billion. Not only is this a substantial sum, but many also believe that a return of the 1973 energy crisis is inevitable (Phillips 1979; Kannberg 1988). Although the world may have another 100-year supply of natural petroleum, it is estimated that the supply of the United States will be virtually exhausted early in the next century (Nagel, Alford, and Dugan 1975). Fuel conservation is certainly an important tool to ward off future shortages.

A stated goal of the United States National Aeronautics and Space Administration (NASA) is to improve the lift-to-drag ratio of the commercial fleet of aircraft by a factor of two during the next two decades. This will maintain the competitive edge that the U.S. aerospace industry now enjoys over European and Asian manufacturers. Doubling L/D would have to be done with a combination of clever drag reduction and high-lift devices. There is no shortage of ideas, but application to the real world is another matter altogether (Bushnell 1998; Perrier 1998). For example, to be acceptable to an airline, a drag-reducing device must not add too much weight to the aircraft because this causes a proportional drag increase, must not be too expensive because this increases the capital cost, and must not interfere with maintenance by either compromising safety or increasing operational cost.

The second example is from the military sector. The amount of propulsive power available for an underwater vehicle is limited by the volume allocated to its power plant and the efficiency of the various propulsive components. For these vehicles, about 90% of the total drag is due to skin friction. Accordingly, a reduction in skin-friction drag of 20% translates into an increase in speed of 6.8%. Although modest, this extra speed may be vital for the survival of a submarine being chased by another underwater vehicle.

Attempts to reduce drag go back to antiquity, as mentioned in Chapter 1. Streamlining and other control methods summarized in Chapter 8 can eliminate most of the pressure drag due to flow separation. Some form drag remains, however, even when the flow continues to be attached to the trailing edge. Owing to the displacement effects of the boundary layer, the pressure distribution around the body differs from the symmetric distribution predicted by potential flow theory. This remnant drag can be reduced by keeping the boundary layer as thin as possible.

For a blunt body, passive and active methods to reduce wake momentum deficit are available. For missiles, for example, the base drag component—with no jet flow at the base—can be as much as 50% of the total drag. In general, compared with two-dimensional bodies, the base drag penalty is lower for axisymmetric bodies because the vortex shedding process is much less intense in the latter case. Base drag is also generally lower at supersonic speeds inasmuch as compressibility effects tend to suppress vortex shedding. At moderate Reynolds numbers, Strykowski and Sreenivasan

(1990) have demonstrated that placing a tiny cylinder in just the right place along-side and well outside the wake of a much larger circular cylinder can significantly suppress vortex shedding from the mother cylinder and thus reduce its drag.

Boundary-layer tripping for advancing transition, trailing-edge splitter plate for disrupting the vortex formation process, base cavities, ventilated cavities, locked-vortex afterbodies, multistep afterbodies, and afterbodies employing nonaxisymmetric boat-tailing concepts are among the passive techniques to modify the flowfield around a bluff body. Viswanath (1995) provided a comprehensive review of these passive methods for subsonic and supersonic flows around two-dimensional and axisymmetric bodies. Among the active techniques to increase base pressure and thus to reduce pressure drag are transpiration and vibration. Continuous or pulsating base bleed is used to modify the flow in the separated region. The latter, with zero net mass addition, has been shown to be very effective when pulsating at twice the Kármán shedding frequency (Williams and Amato 1989).

Wave resistance and induced drag can also be reduced by geometric design. By sweeping the wings of a subsonic aircraft, drag divergence is delayed to higher Mach numbers, thus allowing the aircraft to fly at higher speeds without experiencing a sudden increase in drag. Additionally, the so-called area rule or coke-bottle effect typically leads to a factor of two reduction in wave drag at Mach number of one (Whitcomb 1956). For surface ships, for which typically half the drag is due to wave resistance, bow and stern bulbs can reduce the energy dissipated into waves at the air–water interface. An even simpler solution to reduce wave drag is to operate the ship well below the hull design speed, as is the case for supertankers.

The induced drag of an aircraft's wing, about 25% of the airplane's total drag at subsonic cruising speeds, is inversely proportional to its aspect ratio and, hence, a lifting surface is typically designed with as large an aspect ratio as permissible by structural considerations and desired degree of maneuverability. End plates or other vortex diffusers can also be used to reduce the induced drag further. In December 1986, the *Voyager* aircraft completed the first nonstop, nonrefueled flight around the world. The primary design goal was to use fuel sparingly to sustain the 9-day flight. Several innovations contributed to this success, including the use of lightweight graphite honeycomb composites; unique wing, propeller, and body design; and a high-specific-energy, air-cooled power plant. But, additionally, the unusually large wingspan used (33.8 m, including the winglets) contributed to a significant reduction in induced drag and proportional fuel saving. The *Daedalus* became in 1988 the first man-powered, heavier-than-air vehicle to sustain flight for a distance of 110 km. Here, too, a wing with a large aspect ratio and ultra-light-weight material contributed to the record-breaking flight.

Most of the current research efforts are directed toward reducing the skin-friction drag, and this topic will occupy the remainder of this chapter. According to Bushnell (1983), the leverage in this area of research is quite considerable and justifies the study of unusual or high-risk approaches on an exploratory basis.

10.3   Skin-Friction Reduction

As stated, skin friction accounts for about 50% and 90% of the total drag of, respectively, subsonic aircraft and underwater vehicles. Let us start by estimating typical

Reynolds numbers for the vehicles whose drag is to be minimized. A commercial air-craft traveling at a speed of 300 m/s would have a unit Reynolds number of 2×10^7/m at sea level and 1×10^7/m at an altitude of 10 km. Owing to the much smaller kinematic viscosity of water, an underwater vehicle moving at a modest speed of 10 m/s ($\approx$20 knots) would have the same unit Reynolds number of 1×10^7/m.

Three flow regimes can be identified for the purpose of reducing skin friction. First, if the flow is laminar, typically at Reynolds numbers based on distance from leading edge of $<10^6$, then methods of reducing the laminar shear stress are sought. Recall from Section 2.4 the streamwise momentum equation written at the wall

$$\rho v_w \frac{\partial u}{\partial y}\bigg|_{y=0} + \frac{\partial p}{\partial x}\bigg|_{y=0} - \frac{\partial \mu}{\partial y}\bigg|_{y=0} \frac{\partial u}{\partial y}\bigg|_{y=0} = \mu \frac{\partial^2 u}{\partial y^2}\bigg|_{y=0} \tag{10.3}$$

The right-hand side of this equation is the wall flux of spanwise vorticity. When this term is positive, the vorticity flux is negative, and the wall is then a sink of spanwise vorticity. The streamwise velocity profile in this case is inflectional and has lower slope at the wall, meaning lower skin friction. Therefore, any or a combination of the following techniques can be used to lower the laminar skin friction: upward wall motion, injection of fluid normal to the wall, adverse pressure gradient, wall heating in air, or wall cooling in water. Note that any of these methods will promote flow instability and separation. These tendencies have to be carefully considered when deciding how far to go with the attempt to lower C_f.

Two other techniques can be used to lower laminar skin friction. Narasimha and Ojha (1967) considered the higher-order effects of moderate longitudinal surface curvature. Their similarity solutions show a definite decrease in skin friction when the surface has convex curvature in all cases, including zero pressure-gradient. Narasimha and Ojha (1967) attributed the decrease in C_f to the tendency for the velocity in the potential flow region to decrease away from the surface. The second technique is used in rarefied gas flows. Appropriate surface preparation can be used to lower the tangential momentum accommodation coefficient and, thus, introduce a measurable slip velocity at the wall (Steinheil et al. 1977; Gampert, Homann, and Rieke 1980). We will return to the issue of slip velocity and accommodation coefficient in Chapter 13.

Secondly, in the range of Reynolds numbers from 1×10^6 to 4×10^7, active and passive methods to delay transition as much as possible are sought. These techniques were reviewed in Chapter 6 and can result in substantial savings. As shown in Figure 6.1, the skin-friction coefficient in the laminar flat-plate can be as much as an order of magnitude less than that in the turbulent case. Note, however, that all the stability modifiers discussed in Section 6.4 result in an increase in the skin friction over the unmodified Blasius layer. The object is, of course, to keep the penalty below the saving (i.e., the net drag will be above that of the flat-plate laminar boundary layer but well below the viscous drag in the flat-plate turbulent flow).

Thirdly, for $Re > 4 \times 10^7$, transition to turbulence cannot be delayed with any known practical method[1] without incurring a penalty that exceeds the savings. The task is then to reduce the skin-friction coefficient in a turbulent boundary layer. This

---

[1] The passive compliant coating that could indefinitely suppress Tollmien–Schlichting instabilities as discussed in Section 7.6.2 may be an exception to this statement, but as of this writing the very promising theoretical finding of Section 7.6.2 is yet to be validated experimentally.

topic will be covered in the following section. Relaminarization will be covered in Section 10.5, although achieving a net saving here is problematic at present. Futuristic control systems to reduce skin friction in turbulent wall-bounded flows will be discussed in Chapter 14.

For a vehicle with a unit Reynolds number of 1×10^7/m, the first regime exists in the first few centimeters and, hence, is of no great consequence. Transition delay is feasible on the wing and other appendages of an aircraft but not on its much longer fuselage. On the fuselage, where half of the skin-friction drag takes place, some alteration of the turbulence structure is sought. Short underwater bodies, such as torpedoes, are ideal targets for applying transition control methods. On the much longer submarine, these techniques are feasible on the first few meters of its surface. Beyond that, turbulence drag reduction or relaminarization is sought.

10.4   Reduction of C_f in Turbulent Flows

At very large Reynolds numbers, transition can no longer be postponed, and the boundary layer becomes turbulent and, thus, an excellent momentum conductor. Such a flow is less prone to separation but is characterized by large skin friction. Several techniques are available to reduce the turbulent skin-friction coefficient, but only very few are in actual use. The basic reason is that the majority of these methods are relatively new and are thus still in the research and development stage. The various drag-reduction methods discussed here will be grouped into four categories: techniques that reduce the near-wall momentum, methods involving the introduction of foreign substances, techniques involving the geometry, and recent innovations. Futuristic strategies are deferred to the last chapter of this book. Several excellent reviews of the various turbulent skin-friction reduction methods are available: Bushnell (1983), Bandyopadhyay (1986b), Wilkinson et al. (1988), Pollard (1997, 1998), and Blackwelder (1998). Bushnell and McGinley (1989) and Gad-el-Hak (1989) provided more general articles addressing turbulence control in wall flows to achieve a variety of desired goals. The former article in particular summarizes the known sensitivities of wall-bounded flows to various inputs having a first-order influence upon the flow structures. The books edited by Bushnell and Hefner (1990), Gad-el-Hak et al. (1998), and Meng (1998) offer many useful articles on drag reduction.

10.4.1   Reduction of Near-Wall Momentum

Methods of skin-friction drag reduction in turbulent boundary layers that rely on reducing the near-wall momentum are reviewed in this subsection. Recall the definition of local skin-friction coefficient

$$C_f \equiv \frac{\tau_w}{\frac{1}{2}\rho U_o^2} = \frac{2\nu}{U_o^2}\left[\frac{\partial \overline{U}}{\partial y}\right]_{y=0} = 2\left[\frac{u_\tau}{U_o}\right]^2, \tag{10.4}$$

where $\overline{U}$ is the time-averaged velocity component in the streamwise direction, U_o is the velocity outside the boundary layer, and u_τ is the friction velocity. Thus, one brute-force way of achieving drag reduction is to lower the mean-velocity gradient at the wall or make the curvature of the velocity profile there as positive as possible, as was done in the laminar case. Equation 10.3 is valid instantaneously and is therefore applicable to the turbulent case. The influence of wall transpiration, shaping, or heat transfer on the mean velocity profile is complicated by the additional effects of these modulations on the Reynolds stress term. However, these influences are qualitatively in the same direction as in the simpler laminar case. Thus, lower skin friction is achieved by driving the turbulent boundary layer toward separation. This is accomplished by injecting fluid normal to the wall, shaping to produce adverse pressure gradient, surface heating in air, surface cooling in water, or any other strategy designed to increase the fluid viscosity at the wall. These methods of control in general result in an increase in turbulence intensity (Wooldridge and Muzzy 1966).

Although in the reverse flow region downstream of the separation line the skin friction is negative, the increase in pressure drag is far more than the saving in skin-friction drag. The goal of these methods of control is to avoid actual separation, that is, lower C_f but not any lower than zero (the criterion for steady, two-dimensional separation). The papers by Stratford (1959a,b) provide useful discussion on the prediction of turbulent-boundary-layer separation and the concept of flow with continuously zero skin-friction throughout its region of pressure rise. By specifying that the turbulent boundary layer be just at the threshold of separation, without actually separating, at all positions in the pressure rise region, Stratford (1959b) experimentally verified that such a flow achieves a specified pressure rise in the shortest possible distance and with the least possible dissipation of energy. An airfoil that could utilize the Stratford's distribution immediately after transition from laminar to turbulent flow would be expected to have a very low drag (Liebeck 1978).

Two additional methods to reduce the near-wall momentum are tangential injection and ion wind. In the former, a low-momentum fluid is tangentially injected from a wall slot. For some distance downstream, the wall senses the lower injection velocity and not the actual freestream. Bushnell (1983) termed this situation a wall wake as opposed to a wall jet (used in high-lift devices to keep the flow attached). Ion wind, generated by an electrode in the wall and discharging either to space or to a ground on the wall further downstream, causes an increase in the normal velocity component near the wall. Because there is no net mass transfer through the surface, the near-wall longitudinal momentum is again reduced, and the boundary layer is driven toward separation (Malik, Weinstein, and Hussaini 1983). The source and amount of power to produce a measurable drag reduction are among the outstanding issues to be considered with this technique. Among the possibilities is the use of the natural buildup of static electricity on an aircraft that at present is dissipated at discharge points to avoid large-scale arcing.

When attempting to reduce drag by driving the boundary layer toward separation, a major concern is the flow behavior at off-design conditions. A slight increase in angle of attack, for example, can lead to separation and consequent large drag increase as well as loss of lift. High-performance airfoils with lift-to-drag ratio of over 100 utilize carefully controlled adverse pressure gradient to retard the near-wall fluid, but their performance deteriorates rapidly outside a narrow envelope (Carmichael 1974).

In the case of tangential or normal injection, the source of fluid to be injected is important. Large ram-drag penalty is associated with using freestream fluid, and reaching a break-even point is highly problematic. A lower loss source is fluid withdrawn from somewhere else to control transition or separation—for example, from the wings and empennage of an aircraft. This point will be discussed further in Section 10.6.

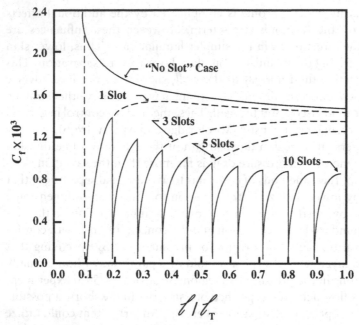

Figure 10.3: Computed skin-friction reduction due to tangential injection [after Howard et al. 1975].

Howard, Hefner, and Srokowski (1975) conducted finite-difference calculations to show the effect of single- and multislot injection on skin-friction drag of a fuselage shape typical of long-haul subsonic transport. A sample of their results is shown in Figure 10.3 for up to 10 tangential slots. The abscissa in the figure indicates the streamwise location of each slot, ℓ, normalized with the total length of the fuselage, ℓ_T. The beneficial effect of injection is most pronounced immediately downstream of the slot exit and diminishes with distance from the slot. Performance deterioration is clearly a result of the mixing between the low-momentum slot flow and the high-momentum boundary-layer flow. Note, however, that the asymptotic level of skin friction with multislots is lower than that in the unperturbed flow. Without taking into account the system penalties for collecting, ducting, and injecting the secondary flow, Howard et al. (1975) computed a reduction in total skin-friction drag of the order of 50%. They maintained that further optimization may provide even greater skin-friction reductions to help compensate for system losses.

As was the case when surface heating or cooling was considered for transition delay or separation postponement, this technique should be used for skin-friction reduction only when a heat source or sink is readily available. Rejected heat from the vehicle's power plant is a free source of energy, and cryofuel is an excellent heat sink that does not require energy expenditure.

10.4.2 Introduction of Foreign Substance

Turbulent skin-friction drag can be reduced by the addition of several foreign substances. Examples include long-chain molecules, surface-active agents and microbubbles in liquid flows, and small solid particles and fibers in either gases or liquids. In general, the addition of these substances leads to a suppression of the Reynolds stress production in the buffer zone that links the linear and the logarithmic portions of the mean velocity profile. Thus, the turbulent mixing is inhibited, and a consequent reduction in the viscous shear stress at the wall is achieved.

Drag reduction by solutions of macromolecules (molecular weight $>10^5$) is perhaps the more mature technology with well over 5000 research papers in existence (Virk 1975; White and Hemmings 1976; Hoyt 1979; Hendricks et al. 1989; Nadolink and Haigh 1995; Gyr and Bewersdorff 1995). Toms (1948) observed that the addition of a few parts per million of polymethyl methacrylate to a turbulent pipe flow of monochlorobenzene reduced the pressure drop substantially below that of the solvent alone at the same flow rate. Successful long-chain molecules for water flow include carboxymethyl cellulose, guar gum, copolymer of polyacrylamide and polyacrylic acid, and polyethylene oxide. Skin-friction drag reduction of up to 80% is possible in both external and internal flows with 100 ppm or less of the largest molecules.

The viscosity of these dilute solutions of polymers is typically 10% higher than that of the solvents, although most polymer solutions are detectably shear thinning at higher concentrations (Berman 1978). The polymer molecules at rest are in the form of spherical random coils in which each monomer is joined to the preceding one at a random angle. The onset of drag reduction is associated with the expansion of these coils outside the viscous sublayer. Hinch (1977) asserted that this unfolding explains the dramatic effects of few parts per million (by weight) on the drag. The effective hydrodynamic volume fraction shows large concentration once the polymer molecules are stretched. Therefore, it is the volume fraction of spheres just enclosing the separate polymers rather than the weight that is the appropriate measure of the added substance.

When the macromolecules are coiled, the viscosity of the dilute solution changes by a few percent, whereas if they are fully stretched the apparent viscosity increases by several orders of magnitude. Onset of drag reduction occurs when $D_M u_\tau / \nu = 0.015$, where D_M is the effective molecular diameter. This diameter is about two orders of magnitude less than the shortest dynamically significant length in a typical turbulent flow (Virk et al. 1967; Lumley 1969). Numerous experiments have indicated that the drag reduction is associated with a thickening of the buffer layer and a corresponding shift of the logarithmic part to accommodate the larger buffer zone (Reischman and Tiederman 1975). The effect of the additive may then be represented as a virtual slip at the wall (Virk et al. 1967).

The percentage drag reduction increases as the polymer concentration is increased and the buffer layer is thickened. This process tends to an asymptote when the buffer layer reaches the center of the pipe or the edge of the boundary layer. As the Reynolds number is increased at a given concentration, a second limit in drag reduction is reached when the macromolecules become fully stretched. Moreover, degradation or breaking of the molecules by the flow occurs at sufficiently high strain rates, with a consequent loss of the ability to reduce the drag (Berman and George 1974). Because of this latter point, care has to be exercised when mixing the macromolecules into the solvent as well as when delivering the solution into the flow.

By injecting the polymer in different regions of the boundary layer, Wells and Spangler (1967), Wu and Tulin (1972), and McComb and Rabie (1979) have shown that the additive must be in the wall region for drag reduction to occur. Furthermore, the channel-flow experiments of Tiederman et al. (1985) have clearly demonstrated that drag reduction is measured downstream of the location where the additives injected into the viscous sublayer begin to mix in significant quantities with the buffer region (i.e., $10 < u_\tau y/\nu < 100$. At streamwise locations where drag reduction does occur, the dimensionless spanwise streak spacing increases and the average bursting rate decreases, although the decrease in bursting frequency is larger than the corresponding increase in low-speed streak spacing. Tiederman et al. (1985) concluded that the polymers have a direct effect on the flow processes in the buffer region and that the linear sublayer appears to have a rather passive role in the interaction of the inner and outer portions of a turbulent wall layer.

Many mechanisms have been proposed to explain the experimental observations. Notable among these is the work of Lumley (1973, 1977), Landahl (1973, 1977), and the more recent quantitative theory by Ryskin (1987). According to Lumley's model, the macromolecules are expanded outside the viscous sublayer owing to the fluctuating strain rate. The resulting increase in effective viscosity damps only the small dissipative eddies. The suppression of small eddies in the buffer layer leads to increased scales, a delay in the reduction of the velocity profile slope, and consequent thickening of the wall region. In the viscous sublayer, the scales remain unchanged because the effective viscosity of the dilute solution in steady shear is only slightly affected. In the buffer zone, the scales of the dissipative and energy-containing eddies are roughly the same and, hence, the energy-containing eddies will also be suppressed, resulting in reduced momentum transport and reduced drag.

Lumley's more conservative general hypothesis is not incompatible with Landahl's specific, rather speculative mechanism (Lumley and Kubo 1984). The polymer suppresses the small eddies responsible for the inflectional profile and the secondary instability in the buffer zone. Kim et al. (1971) have shown that the greatest turbulent kinetic energy production for a Newtonian boundary layer occurs near the beginning of the buffer zone at $y \approx 12\nu/u_\tau$. A sharp maximum in the turbulent intensity, $\sqrt{\overline{u^2}}$, occurs at the same location. For a dilute solution of drag-reducing polymer, the peak of $\sqrt{\overline{u^2}}$ moves away from the wall and becomes much broader as compared with the solvent alone. Production of Reynolds stress in this important buffer zone is thus diminished.

In a recent article, Sreenivasan and White (1999) challenged the prevailing theories for polymer drag reduction. They maintained that the connection between the fluctuating strain rates and the large extensional viscosity is only circumstantial and that polymer coils can only be partially stretched in a random field of strain rates. If true, the elongational viscosity cannot be measurably higher than the solvent viscosity, and a different mechanism must therefore be responsible for the observed drag reduction. Sreenivasan and White (1999) argued that the elastic theory advanced by de Gennes (1990) is compatible with at least two experimental observations: the dependence of drag reduction onset on polymer concentration and the maximum drag-reduction asymptote. Basically, the elastic energy stored by the partially stretched polymers has an effect on the flow only when exceeding the turbulent energy. When the two energies become comparable, the elastic energy interferes with the usual turbulent cascade mechanism by not allowing it to proceed all the way to the Kolmogorov scales. Thus, the nonlinear action that generates small scales of turbulence will be terminated at some scale larger than the Kolmogorov scale, leading to increased buffer layer thickness and reduced drag. Sreenivasan and White (1999) carefully computed the order of magnitude of the expected flow behavior and successfully compared the results to existing experiments. Their results, albeit tentative, provide the framework needed for further targeted study—particularly at high Reynolds numbers.

The use of drag-reducing polymers in pipe lines is cost-effective in several applications. When it was discovered that the winter cooling of crude oil in the trans-Alaskan pipeline had been underestimated, polymer additives offered an inexpensive alternative to increasing the number or power of the pumping stations. Polymer has also been used in the the the Iraq–Turkey pipeline. Long-chain molecules are presently used in waste-disposal plants, fire-fighting equipment, and high-speed water-jet cutting. For surface ships or underwater vehicles practical considerations include the cost of the polymer and how, with what, and when to mix it as well as how to eject it into the boundary layer (i.e., using pumps or pressurized tanks), through slots or porous surfaces, and so forth. A very important additional consideration is the portion of the payload that has to be displaced to make room for the additive. Oil companies appear to have concluded that the use of polymer for supertankers is just at the break-even point economically. That assessment will undoubtedly change as fuel prices rise and perhaps as manufacturing costs of polymers drop. For submarines and torpedoes, where space is a prime consideration, it seems that the volume occupied by the additive is a more important obstacle to its routine use to date.

Other substances that can be added to a clear fluid to reduce skin-friction drag include surfactants (Savins 1967; Patterson, Zakin, and Rodriguez 1969; Zakin et al. 1971; Aslanov et al. 1980), microbubbles (McCormick and Bhattacharya 1973; Bogdevich and Malyuga 1976; Bogdevich et al. 1977; Madavan, Deutsch, and Merkle 1984, 1985; Merkle and Deutsch 1985, 1989), large length-to-diameter particles such as fibers, (Hoyt 1972; Lee et al. 1974; McComb and Chan 1979, 1981), and spherical particles (Soo and Trezek 1966; Pfeffer and Rosetti 1971; Radin 1974; Radin, Zakin, and Patterson 1975; Povkh, Bolonov, and Eidel'man 1979). Lumley (1977, 1978a) argued that the same mechanism discussed in conjunction with the polymer case governs the interaction of most of these substances with the turbulent boundary layer. The additives affect only the dissipative scales of the turbulence, thus increasing the scales of dissipation in the buffer layer. The energy-containing eddies are also suppressed and, consequently, the momentum transport is reduced.

A surface-active substance, such as soap, is that which, when dissolved in water or in an aqueous solution, reduces its surface tension. The majority of surfactants appear to reduce the turbulent skin-friction drag only in the presence of electrolytes. These additives are stable against mechanical destruction, which affords a certain advantage over polymer molecules that break down when the strain rate is sufficiently high. Although drag-reducing polymer flows undergo transition at a lower Reynolds number than Newtonian flows (Lumley and Kubo 1984), some recent experiments indicate that surfactants are very effective in delaying laminar-to-turbulence transition (A. J. Smits, private communication; Sabadell 1988).

The idea of placing a thin layer of gas between the wall and its water-boundary layer dates back to the last century. Substantial drag reductions are potentially possible because of the lower density of the gas. Unfortunately, the various instabilities associated with gas–liquid interface result in drag increase. Extremely small gas bubbles (microbubbles) injected through a porous wall or produced by electrolysis do not suffer from the stability problems of a gas film and result in skin-friction reduction as high as 80%. Maximum reduction is obtained when the gas volume fraction approaches the bubble-packing limit. Skin friction is reduced because of the substantially lower density and also because of the usual increase in bulk viscosity due to particles (Batchelor and Green 1972), which damps the small-scale motions in the buffer layer. Legner (1984) presented a simple phenomenological model to predict microbubble drag reduction in turbulent boundary layers. Application of the microbubbles technique for surface ships is quite feasible. Compressed atmospheric air is injected through a microporous skin over portions of the hull. This has the additional advantage of reducing fouling drag and fouling maintenance costs because water is kept away from the hull surface. The situation is different for underwater vehicles. There, a source of air is not readily available, and using electrolysis for bubble production will not yield net energy saving. Moreover, the presence of bubbles may adversely affect the flow-induced noise, which is an important consideration for naval vehicles.

Spherical or large length-to-diameter particles can be used in both air and water flows. For spherical particles, drag reduction of up to 50% is feasible for certain parameter values, although drag increase is also possible. Heavy but small particles may reduce the drag by inducing a stable density stratification near the wall, thus driving the turbulence toward relaminarization. Lighter particles that are large enough to interact with the smallest turbulence scales reduce the drag through the same mechanism as in the polymer case (i.e., suppressing the dissipative eddies and increasing the scale of dissipation).

McComb and Chan (1981) reported up to 80% drag reduction using the naturally occurring macrofiber chrysotile asbestos dispersed at a nominal concentration of 300 ppm by weight in a 0.5% aqueous solution of Aerosol OT. The individual fibrils of chrysotile asbestos have a mean diameter of 40 nm and a mean length-to-diameter ratio in an undegraded suspension in the range of 10^3–10^4. Like polymers, macrofibers readily break down (McComb and Chan 1979) and have to be carefully dispersed into the solvent and delivered to the boundary layer. In principle, fibers can also be used in gases, although their drag-reducing potential has yet to be demonstrated.

10.4.3 Methods Involving Geometry

Under this classification are some of the most recently researched techniques to reduce the turbulent skin-friction drag. These include large-eddy breakup devices (LEBUs), riblets, compliant surfaces, wavy walls, and other surface modifications. The book edited by Bushnell and Hefner (1990) offers comprehensive reviews of these and other drag-reduction strategies. With few exceptions, there is little theoretical basis for how these geometrical modifications affect the skin friction, and most of the present knowledge comes from experiments. Needless to say, the lack of analytical framework makes optimization for a given flow condition as well as extension to other flow regimes very tedious tasks.

The LEBUs are designed to sever, alter, or break up the large vortices that form the convoluted outer edge of a turbulent boundary layer. A typical arrangement consists of one or more splitter plates placed in tandem in the outer part of a turbulent boundary layer, as sketched in Figure 10.4. It is of course very easy to reduce substantially the skin friction in a flat-plate boundary layer by placing an obstacle above the surface. What is difficult is to ensure that the device's own skin-friction and pressure drag do not exceed the saving. The original screen fence device of Yajnik and Acharya (1978) and the various sized honeycombs used by Hefner, Weinstein, and Bushnell (1980) did not yield a net drag reduction. In low-Reynolds-number experiments, very thin elements placed parallel to a flat plate have a device total drag that is nearly equal to laminar skin friction. A net drag reduction of the order of 20% is feasible with

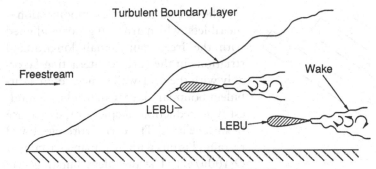

Figure 10.4: Sketch of a tandem arrangement of an LEBU.

two elements placed in tandem with a spacing of $\mathcal{O}$ [10δ] (Corke et al. 1980, 1981). These ribbons have typically a thickness and a chord of the order of 0.01δ and δ, respectively, and are placed at a distance from the wall of 0.8δ. Several experiments report a more modest drag reduction (Hefner, Anders, and Bushnell 1983) or even a drag increase, but it is believed that a slight angle of attack of the thin element can result in a laminar separation bubble and a consequent increase in the device pressure drag. Flat ribbons at small, positive angles of attack produce larger skin-friction reductions. This is consistent with the analytical result that a device producing positive lift away from the wall is more effective (Gebert 1988). In any case, a net drag reduction should be achievable, at least for devices having chord Reynolds numbers $<10^6$, if extra care is taken to polish and install the LEBU.

Net drag reduction has been documented even in the presence of favorable or adverse pressure gradient in the main flow (Bertelrud, Truong, and Avellan 1982; Plesniak and Nagib 1985). Additionally, Anders and Watson (1985) have demonstrated that an LEBU having an airfoil shape is nearly as effective in reducing the net drag as a flat ribbon. A wing is several orders of magnitude stiffer than a thin ribbon, which affords a certain advantage for extending the technique to field conditions. At flight Reynolds and Mach numbers, the possibility of paying for transitional or turbulent skin friction as well as wave drag on the device itself is real (Anders et al. 1984, 1988), and achieving net drag reduction in this circumstance is problematic at present (Anders 1990b). For an airplane, an LEBU is likely to take the form of a ring around the fuselage.

Two analytical attempts to explain the mechanisms involved with the LEBUs are noted in here. Basically, the LEBU acts as an airfoil on a gusty atmosphere, and a vortex unwinding mechanism is activated. Dowling's (1985) inviscid model indicates that the vorticity shed from the LEBU's trailing edge as a result of an incident line vortex convected past the device tends to cancel the effect of the incoming vortex and to reduce the velocity fluctuations near the wall. Atassi and Gebert (1987) and Gebert (1988) modeled the incoming turbulent rotational flow as two- or three-dimensional harmonic disturbances. They use the rapid distortion approximation and unsteady aerodynamic theory to compute the fluctuating velocity downstream of thin-plate and airfoil-shaped devices. The fluctuating normal velocity component is most effectively suppressed for a range of frequencies that scales with the freestream velocity and the device chord. This important result determines the optimum size of an LEBU by selecting this frequency range to correspond to that of the large-scale eddies in a given turbulent boundary layer.

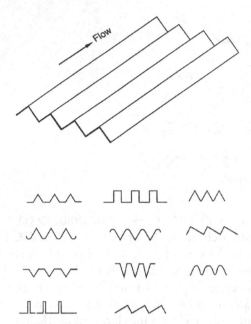

Figure 10.5: Fin shapes for longitudinally ribbed surface.

The second geometrical modification is the riblets, which are wall grooves aligned with the freestream. Small longitudinal striations in the surface interacting favorably with the near-wall structures in a turbulent boundary layer can produce a modest drag reduction in spite of the increase in surface area. The early work employed rectangular fins with height and spacing of $\mathcal{O}[100\nu/u_\tau]$. The turbulent bursting rate was reduced by about 20%, and a modest 4% net drag reduction was observed (Liu, Kline, and Johnston 1966). In a later refinement of this technique, Walsh and his colleagues at NASA–Langley examined the drag characteristics of longitudinally ribbed surfaces having a wide variety of fin shapes (see Figure 10.5) that included rectangular grooves, V-grooves, razor blade grooves, semicircular grooves, and alternating transverse curvature (Walsh and Weinstein 1978; Walsh 1980, 1982, 1983, 1990; Walsh and Lindemann 1984). A net drag reduction of 8% is obtained using V-groove geometry with sharp peak and either sharp or rounded valley. Optimum height and spacing of the symmetric grooves are about $15\nu/u_\tau$. Although these dimensions would be extremely small for the typical Reynolds numbers encountered on an airplane or a submarine (peak-to-valley height $\approx 35~\mu$m), such riblets need not be machined on the surface. Thin, low-specific-gravity plastic films with the correct geometry on one side and an adhesive on the other are presently available commercially, and existing vehicles could readily be retrofitted. In fact, these tapes were successfully tested at $Ma = 0.7$ on a T–33 airplane and on a Lear jet. The performance of the riblets in flight was similar to that observed in the laboratory (Anders et al. 1988). In water, riblets were employed on the rowing shell during the 1984 Summer Olympics by the United States rowing team. Similar riblets were also used on the submerged hull of the winner of the 1987 America's Cup yacht race, the *Stars and Stripes*, with apparent success.

Riblets are thought to restrain the movement of the near-wall longitudinal vortices and therefore maintain their coherence. The stabilized rolls or coherent structures present a barrier to the usual cascade to small scales and hence impede the rate of energy loss with a resulting drag reduction despite the increased surface area. Riblets are effective in the presence of moderate, adverse, and favorable pressure gradients. The loss of drag-reducing effectiveness as flow conditions vary off-design is gradual. Moreover, the percentage drag reduction slowly decreases to zero as the yaw angle between the flow and the grooves goes up to 30°. Surprisingly, drag does not increase at yaw angles >30°. Remaining practical problems include cost, weight penalty, particulate clogging, ultraviolet radiation, and film porosity and resistance to hydraulic fluids and fuel. Recent reviews of riblets were presented by Pollard (1997, 1998).

Curiously, fast sharks have a surface covering of dermal denticles with flow-aligned keels having the optimal riblet spacing as determined by Walsh (Bechert, Hoppe, and Reif 1985). As the shark grows, new keels are added onto the sides of the denticles without changing the keel-to-keel spacing. A clue to how riblets result in a net drag reduction despite the increase in wetted area is provided by the experiment of Hooshmand et al. (1983), who showed that, for an optimum V-grooved surface, C_f is reduced by 40% in the valleys and increased by 10% at the peaks. It appears that the riblets severely retard the flow in the valleys, which places a slip boundary condition upon the turbulence production process. Lumley and Kubo (1984) argued that the streamwise vortices in the near-wall region are forced to negotiate the sharp peaks of the riblets, causing increased losses. To stay in equilibrium, the eddies must increase their energy gain, and one way to do that is to grow larger (energy gain $\sim$(size)^2; loss $\sim$(size)). As was the case with polymers, the larger scales result in the secondary instability and the sharp change in profile slope occurring farther from the wall. The mean velocity is, thus, higher for the same friction velocity, and the coefficient of skin friction is reduced.

The third geometrical modification to reduce the skin friction is fixed or moving wavy walls. Kendall (1970) observed that the integrated skin friction in a turbulent boundary layer over rigid, sinusoidal waves can be as much as 25% below that over an equivalent flat surface. Kendall's waves were relatively shallow, having a height-to-wavelength ratio of $h/\lambda = 0.028$, with $\lambda = \mathcal{O}[\delta]$. Sigal (1971) reported a reduction in integrated skin friction of about 12% when using 10% larger amplitude waves ($h/\lambda = 0.031$) than that used by Kendall. The transverse waves provide alternating regions of longitudinal concave and convex curvatures along with alternating adverse and favorable pressure gradients. The coincidence of convex curvature and favorable pressure gradient promotes relaminarization as the wave crests are approached, and the near-wall momentum is reduced in regions having adverse pressure gradient. Accordingly, the integrated skin-friction drag is reduced. Unfortunately, viscous forces cause a downstream phase shift between the pressure distribution and the wave relative to the 180° out-of-phase relationship predicted from potential flow theory for a sinusoidal wave. This phase shift causes an additional pressure drag not present in the flat-surface case. The net result is that the total drag for the wavy surface increases despite the decrease in skin friction. Cary, Weinstein, and Bushnell (1980) argued that there may be two approaches to overcome this problem. First, because pressure drag varies as $(h/\lambda)^2$, further reduction of h/λ may yield a net drag reduction. Secondly, certain nonsinusoidal waves may have more favorable pressure and curvature effects on the viscous drag and a less skewed pressure distribution relative to the waves and therefore less pressure drag. Cary et al.'s numerical results indicate the soundness of the first approach. Net drag reduction of about 13% is attained for the optimum h/λ of 0.005. This result seems to be insensitive to the ratio of wave height to the viscous scale—at least for $hu_\tau/\nu < 33$. Experiments to test the second approach were only partially successful (Lin et al. 1983). Skewed waves with gradual, straight downstream-facing slope and steeper, sinusoidal upstream-facing surface have lower pressure drag than a symmetric sine wave. However, the asymmetric surface is not as effective in reducing the viscous drag with the result that the net drag reduction is only 1–2%, which is hardly worth the effort (Wilkinson et al. 1987, 1988).

Moving wavy walls are divided into two categories: driven, noninteractive walls and compliant surfaces. In the former technique, a flat or wavy wall is translated in the streamwise direction. The rectilinear wall motion essentially acts as a slip boundary condition and reduces the skin friction. Obviously the boundary layer could be completely eliminated when $U_{\text{wall}} = U_\infty$. A moving wavy wall can produce a thrust at high enough translational speed but causes an additional pressure drag at lower speed. In any case, this method of control is in general impractical and is used mainly to provide controlled experiments to determine what type of wave motion is required to achieve a given result, which is much the same as the laminar flow-control case.

Wall motion can also be generated by using a flexible coating with sufficiently low modulus of rigidity (Chapter 7). The idea is very appealing if a particular kind of fluid–solid interaction yields a net drag reduction. Covering a vehicle with a compliant coating is relatively simple; does not require modification of existing design; does not require slots, ducts, or internal equipment of any kind; and no energy expenditure is needed. The search for such a surface has been very elusive, however, despite reports of substantial drag reduction by some researchers. Irreproducibility seems to be an outstanding characteristic of the body of compliant coating experimental evidence (Bushnell et al. 1977; McMichael et al. 1980; Gad-el-Hak 1986a,b, 1987a, 1996a). The window of opportunity for a favorable coating response is rather narrow. For flow speeds below the transverse wave speed in the solid, no significant interaction between the fluid and the compliant surface takes place. Above the transverse wave speed, ample opportunity for interaction exists; however, hydroelastic instabilities appear, and the resulting large-amplitude surface waves increase the drag because of the roughness-like effects (Gad-el-Hak, Blackwelder, and Riley 1984; Gad-el-Hak and Ho 1986b; Gad-el-Hak 1996a). A second constraint concerns the density ratio of the fluid and the solid. The extremely low density of air makes finding a reasonable coating material highly unlikely, although in liquids the situation is different. Flexible surfaces that interact favorably with Tollmien–Schlichting waves in water boundary layers are readily available (Gad-el-Hak 1996a). For turbulent boundary layers, the task is to find a coating that responds to the flow, particularly to its preburst wall-pressure signature, with a surface motion that can alter, modify, or interrupt the sequence of events leading to a burst. The analytical work of Purshouse (1977) indicates that an anisotropic coating would be more likely to achieve a net drag reduction in a turbulent boundary layer. More discussion on these issues was presented in Chapter 7.

Other geometric modifications on the surface with a drag-reducing potential include fixed waves aligned with the flow (essentially large-scale riblets), micro air-bearings, compound or three-dimensional riblets, sieves, furry (wheatfield-type) surfaces, and sword-fish configuration. These techniques have been reviewed by Bushnell (1983) and more recently by Wilkinson et al. (1988). The last of these methods is the most straightforward. As a boundary layer thickens, the Reynolds number increases and the coefficient of skin friction decreases, with the result that the local skin friction is higher in the forward part of a vehicle and quite low in the aft end. Consider, for example, a 50-m-long aircraft with a unit Reynolds number of 10^7/m and ignore pressure gradient and other effects. Near the nose, the skin-friction coefficient is about 0.003 (Figure 6.1). Near the tail, $Re \approx 5 \times 10^8$ and $C_f \approx 0.0018$ (i.e., 40% lower). By substantially reducing the wetted area of the forebody—the sword-fish configuration—drag reduction may be attainable. This technique is particularly useful when combined with a convex aftbody.

10.4.4   Recent Innovations

Though against long-held believes that rough surfaces have more drag than their smooth counterparts, the riblet experience has apparently inspired Sirovich and Karlsson (1997) to search for other kind of roughness that may lead to skin-friction reduction. Their finding is one of those rare instances in which a better understanding of the physics of a turbulent flow led to a successful flow-control strategy rather than the more common reverse flow of ideas (Carpenter 1997). Sirovich and his colleagues believe—as do many researchers—that the streamwise vortices are the primary structures in the near-wall region. But, additionally, weaker oblique plane waves exist in this region propagating downstream and interacting with the streamwise vortices. Bursting events are the results of such interaction. Computer simulations have indicated that artificially randomizing the phases of the vortices and oblique waves leads to less bursting and consequent drag reduction. Sirovich and Karlsson (1997) tested this hypothesis in a fully developed turbulent channel flow at modest Reynolds numbers. Indeed surfaces covered with special randomized patterns of small V-shaped protuberances produced drag lower by up to 12% as compared to smooth surface. Regular patterns of the same roughness led to a rise in drag by 20% and corresponding increase in mixing.

There are two other drag-reducing strategies worth noting before closing this section. Jung, Mangiavacchi, and Akhavan (1992) and Akhavan, Jung, and Mangiavacchi (1992) reported a drag reduction of up to 40% in a direct numerical simulation of a turbulent channel flow when one of the duct walls undergoes high-frequency, spanwise oscillations. The maximum drag reduction is achieved when the oscillation period is $P^+ \equiv P u_\tau^2 / \nu = 100$. The rapid oscillations significantly damped the normal and tangential Reynolds stresses, whereas the peak location of each stress moved away from the wall and the buffer layer thickened. In a flat-plate boundary layer experiment, Laadhari, Skandaji, and Morel (1994) confirmed the turbulence suppression observed in the DNS data, but they did not measure the skin friction. In a more recent wind-tunnel experiment, Choi, DeBisschup, and Graham (1998) and Choi and Clayton (1998) did measure the skin friction using the Preston tube and near-wall slope of the mean velocity profile. They reported a 45% drag reduction at optimum oscillations. Choi and Graham (1998) extended the flat-plate work to show that circular-wall-oscillations in a turbulent pipe flow can lead to skin-friction reduction of up to 25%. The lower skin friction in both physical and numerical experiments has of course to be carefully balanced against the energy used to generate the wall oscillations.

The second strategy has recently been proposed by Schoppa and Hussain (1998). Using direct numerical simulations of turbulent channel flow, they indicated drag reduction of 20% when imposed counterrotating vortices are used and 50% when colliding, z-directed wall jets are employed. The former method can be realized experimentally using large-scale, passive vortex generators, whereas the latter—much like the wall oscillations above—is a predetermined active control method (see Figure 3.3) that does not require sensors or control logic but does require energy expenditure that has to be accounted for when computing the net saving. Schoppa and Hussain's strategy has yet to be validated using physical experiments.

10.5 Relaminarization

Several articles define and explain the reversion of a turbulent flow to the laminar state, and this phenomenon is variously also known as retransition, inverse or reverse transition, or relaminarization (Preston 1958; Patel and Head 1968; Bradshaw 1969). Narasimha and Sreenivasan (1979) provided a comprehensive review and analysis of the phenomenon. Instances of relaminarization may be found in a stably stratified atmosphere, spatially or temporally accelerated shear flows, coiled pipes, swirling flames, tip vortices behind finite lifting surfaces, far wakes of finite drag-producing bodies, air flow as it progresses from the trachea to the segmental bronchi in human lungs, and shear flows subjected to magnetic fields. Various authors use one or more of the following syndromes to diagnose the reversion of a turbulent flow: reduction in heat transfer or skin-friction coefficient, cessation of bursting events near the wall or entrainment of potential flow at the outer region of a boundary layer, reduction in high-frequency velocity fluctuations, breakdown of the log-law in the wall layer, spreading of intermittency to wall region, and overlapping of energy-containing and -dissipating eddy scales. In the quasi-laminar state, the velocity fluctuations are not necessarily zero, but rather their contribution to the dynamics of the mean flow becomes inconsequential. In other words, the turbulent fluctuations inherited from the previous history of the flow are no longer influencing the transport of mass, momentum, or energy. Narasimha and Sreenivasan (1979) adopted a pragmatic definition of reversion. They maintained that a flow has relaminarized if its subsequent development can be understood without recourse to any model for turbulent shear flow.

The most obvious mechanism for the occurrence of relaminarization is dissipation. When the Reynolds number goes down in a turbulent flow (e.g., by enlarging a duct or by branching a channel flow), the viscous dissipation may exceed the production of turbulent energy, and the flow may revert to a quasi-laminar state. This type of reversion tends to be rather slow. Turbulent energy can also be destroyed or absorbed by the work done against external forces such as buoyancy (e.g., in stably stratified fluids), centrifugal force (e.g., in convex boundary layers), or Coriolis force (e.g., in rotating channel flows). The governing parameter in these absorptive-type reversions is the Richardson number (Ri), and relaminarization proceeds rapidly once the critical value of Ri is exceeded. In both dissipative- and absorptive-type of reversion, the turbulence energy is decreased, but more significantly the velocity components that generate the crucial Reynolds stresses are "decorrelated." This explains the strong effects on turbulence of, say, a very mild positive curvature (So and Mellor 1973; Smits and Wood 1985). In this case, the amplitude as well as the phase of the different components of the fluctuating motion are affected by the additional strain rate imposed upon the flow—particularly away from the wall.

The third mechanism to effect relaminarization is observed in highly accelerated flows. Narasimha and Sreenivasan (1973) argued that, in the outer layer of such flows, the pressure forces dominate over the nearly frozen Reynolds stress field. The $\overline{uv}$ term remains at about the same level as the zero-pressure-gradient case over much of the boundary layer. Near the wall, however, dissipation dominates, and the flow is stabilized by the acceleration. The turbulence-reenergizing bursts diminish in frequency and may stop altogether, and a thin, new laminar boundary layer grows from the wall within the old boundary layer. According to Morkovin (1988), the turbulence in the outer layer is "starved," and the inner laminar layer is "buffeted" by the decaying, wakelike turbulence of the outer region. The skin friction and the heat transfer rate in the relaminarized inner layer is less than those expected in a corresponding turbulent flow.

Narasimha and Sreenivasan (1979) asserted that several different reverting flows can be considered in the light of the three archetypes summarized above. It is not uncommon, however, to effect relaminarization through a combination of mechanisms rather than a single one. In internal flows, relaminarization can be accomplished by gentle wall injection (to avoid separation) or by wall heating (for gases). In both these cases, the flow accelerates, thus activating the third mechanism above. Moreover, when a gas is heated, its kinematic viscosity is increased, resulting in a lower Reynolds number that may fall below the critical value, thus activating the first mechanism.

In external flows, Eq. 10.3 indicates that wall suction, surface heating in liquids, or surface cooling in gases may have the same effect on the near-wall region of a turbulent boundary layer as favorable pressure gradient. It appears that only the first of these stimuli has been experimentally demonstrated. Dutton (1960) observed that a suction coefficient $C_q \equiv |v_w|/U_o \approx 0.01$ is sufficient to relaminarize a boundary layer at $Re = \mathcal{O}[10^6]$. The initially turbulent flow approaches the laminar asymptotic state (Section 6.4) appropriate to the particular value of C_q used. Although the mean-velocity gradient near the wall is increased, the Reynolds shear stress is reduced at a faster rate, leading to a reduction in the turbulent energy production.

The question of immediate concern to this chapter is whether or not suction, or in fact any of the other reversion stimuli, is a viable drag reduction method for turbulent boundary layers. The suction rate necessary for relaminarization is too high to be applied profitably over an entire surface, although it may be feasible to apply massive suction through a spanwise slot (perhaps enough to ingest the entire mass flow in the boundary layer) followed by gentle suction (or any of the other stability modifiers discussed in Section 6.4) to prevent transition of the newly formed laminar sublayer. This issue will be discussed again in the next section.

Convex (or positive) curvature affects the boundary layer locally in much the same way as an LEBU. With a relatively large radius of curvature, of the order of $10\,\delta$, positive Reynolds stress is produced in the outer flow, and the wall shear is reduced. Downstream of the end of the curvature region, the flow relaxes very slowly to its unperturbed state. Wilkinson et al. (1988) suggested that a convex surface may be better suited for drag reduction than an LEBU device—particularly in supersonic flows in which wave drag penalty on the LEBU blade is rather large. Obviously, a vehicle surface cannot be all convex, and short regions of concave curvature on the way to the convex portions are unavoidable. Bushnell (1983) speculated that the limited extent of the concave regions may not allow formation of any lasting alterations to the turbulent structures. Net reductions of drag of the order of 20% appear to be obtainable using surface curvature. Bandyopadhyay (1990) provided a useful review of viscous drag reduction using convex curvature concepts. Little research is available on the feasibility of employing other relaminarization techniques to achieve drag reduction.

10.6 Synergism

Ideally, one searches for combinations of drag reduction methods for which the total favorable effect is greater than the sum. For example, Lee et al. (1974) have shown that the use of polymers and fibers together can produce much larger reduction in skin friction than either additive alone. A slightly different strategy, but perhaps no less beneficial, is to employ a second control method that reduces the drag penalty of a given technique while adding some saving of its own. An example of this is the use of fluid that is being withdrawn from a certain portion of a vehicle for injection into another location, thus avoiding the ram drag penalty associated with blowing freestream fluid. For an aircraft, tangential slot injection or normal distributed injection may be used to reduce the skin friction on portions of the fuselage, as discussed earlier. Possible sources of low-loss air include LFC suction from the wings and empennage, relaminarization slots on the fuselage, and active or passive bleed air for separation control.

Several possible combinations of drag reduction techniques were briefly mentioned elsewhere in this chapter. The sword-fish configuration (small wetted area) may be used in the forward part of a vehicle, where the local skin friction is high, followed by the rest of the body, where the boundary layer is thicker and the skin-friction coefficient is low. A nose spike followed by a short region of concave curvature and then a longer convex portion may be used in supersonic flows. Massive suction through a spanwise slot may be used to ingest the entire mass flow in a turbulent boundary layer followed by any of the stability modifiers discussed in Section 6.4 to maintain the growing laminar boundary layer downstream.

The suction rate necessary for establishing an asymptotic turbulent boundary layer independent of streamwise coordinate ($d\delta_\theta / dx = 0$) is much lower than the rate required for relaminarization but still not low enough to yield net drag reduction. For $Re = \mathcal{O}[10^6]$, Favre et al. (1966), Rotta (1970), and Verollet et al. (1972), among others, reported an asymptotic suction coefficient of $C_q \approx 0.003$. From Eq. 2.44 rewritten for a steady, incompressible, zero-pressure-gradient boundary layer on a flat plate, the corresponding skin-friction coefficient is $C_f = 2C_q = 0.006$, indicating higher skin friction than if no suction were applied. The problem is that fluid withdrawn through the wall has to come from outside the boundary layer where the streamwise momentum per unit mass is at the relatively high level of U_∞, as was discussed in Section 6.4 in connection with using suction for transition delay. To achieve a net drag reduction with suction, the process must be further optimized. The results of Eléna (1975, 1984) and more recently of Antonia et al. (1988) indicate that suction causes an appreciable stabilization of the low-speed streaks in the near-wall region. The maximum turbulence level at $y^+ \approx 12$ drops from 15 to 12% as C_q varies from 0 to 0.003. More dramatically, the tangential Reynolds stress near the wall drops by a factor of two for the same variation of C_q. The dissipation length scale near the wall increases by 40%, and the integral length scale by 25% with the suction.

Gad-el-Hak and Blackwelder (1987c, 1989) suggested that one possible means of optimizing the suction rate is to be able to identify where a low-speed streak is located and to apply a small amount of suction under it. Assuming that the production of turbulent energy is due to the instability of an inflectional $U(y)$ velocity profile, one needs to remove only enough fluid so that the inflectional nature of the profile is alleviated. An alternative technique that could reduce the Reynolds stress is to inject fluid selectively under the high-speed regions. The immediate effect would be to decrease the viscous shear at the wall, resulting in less drag. In addition, the velocity profiles in the spanwise direction, $U(z)$, would have a smaller shear, $\partial U / \partial z$, because the injection would create a more uniform flow. Because Swearingen and Blackwelder (1984) and Blackwelder and Swearingen (1990) have found that inflectional $U(z)$ profiles occur as often as inflection points are observed in $U(y)$ profiles, injection under the high-speed regions would decrease this shear and hence the resulting instability. Because the inflectional profiles are all inviscidly unstable with growth rates proportional to the shear, the resulting instabilities would be weakened by the suction or injection process. Gad-el-Hak and Blackwelder (1987c, 1989) proposed to combine suction with nonplanar surface modifications. Minute longitudinal roughness elements, if properly spaced in the spanwise direction, greatly reduce the spatial randomness of the low-speed streaks (Johansen and Smith 1986). By withdrawing the streaks forming near the peaks of the roughness elements, less suction should be required to achieve an asymptotic boundary layer. Experiments by Wilkinson and Lazos (1987) and Wilkinson (1988) combined suction or blowing with thin-element riblets. Although no net drag reduction has yet been attained in these experiments, the results indicate some advantage in combining suction with riblets, as proposed by the patent of Blackwelder and Gad-el-Hak (1990).

An LEBU influencing the outer structures of a turbulent boundary layer may be combined with drag reduction methods that mainly influence the near-wall region. Walsh and Lindemann (1984) showed that the performance of riblets in the presence of an LEBU is approximately additive and not quite a synergetic effect. Wilkinson et al. (1988) proposed using an LEBU device to reduce the free mixing between tangentially injected fluid and the boundary layer flow, thus extending the low skin-friction region. Bushnell (1983) suggested that the suction mass-flow rate required to achieve relaminarization of a turbulent boundary layer may be reduced by the use of an LEBU ahead of the suction region.

The LEBU model of Gebert (1988) indicates that the fluctuating normal velocity component is most effectively suppressed for a particular incoming harmonic disturbance. The modest drag reduction currently achieved using an LEBU device in a natural boundary layer may, therefore, be greatly enhanced by cyclically generating the large outer structures using the method developed by Gad-el-Hak and Blackwelder (1987a). In this case, the large eddies arrive at a given location at a precisely controlled period that matches the optimum frequency for the LEBU device. Alternatively, Goodman's (1985) technique to generate closely spaced turbulent spots in the spanwise and streamwise directions may be used for tripping a laminar boundary layer. An LEBU device placed downstream will encounter more uniformly paced large eddies, and its performance may be enhanced. Of course, in both schemes the penalty associated with artificially generating the large-eddy structures must be considered.

In a recent article, Fontaine et al. (1999) explored the drag reduction achieved by combining carbon dioxide microbubbles injection with homogenous dilute solutions of either drag-reducing polymer (polyethylene oxide) or surfactant (Aerosol OT). At low gas-injection rates, their results indicated that the combination of microbubbles injection with polymer additives can yield drag reduction levels exceeding those of the individual systems, but no synergism was observed. Additionally, reducing the bubble size through the addition of the surfactant did not favorably affect the characteristics of microbubbles drag reduction. Fontaine et al. (1999) speculated that, though the bubble size was reduced by as much as 23% in the presence of the surface-tension-reducing Aerosol OT, the diminished bubble diameter was still large compared with the turbulent scales in their water boundary layer.

Most of the combined drag reduction techniques discussed or proposed in this section are speculative owing to the scarcity of research in the area of synergism. Unguided by any theoretical framework and faced with the complexity of each of the methods used individually, one is tempted to stay away from the much more difficult situation in which two or more techniques are combined. However, the potential for achieving a total effect larger than the sum should stimulate more research for combinations of methods. In fact, in some situations one may not have much of a choice. Certain techniques such as suction may not yield net drag reduction and must, therefore, be combined with something else to achieve the desired goal. Other methods such as riblets yield such modest drag reductions (less than 10%) that it may not be worth the effort to implement the individual technique.

10.7 Parting Remarks

Drag reduction is a subject that is explicitly or implicitly interwoven in every chapter of this book. The practical benefits of achieving even a modest reduction in fluid resistance to moving bodies are enormous. There is no shortage of ideas on how to reduce the drag. Many challenges remain, however, when attempting to apply laboratory curiosities to real-life situations. For airplanes and submarines, streamlining has led to significant reduction in the pressure drag component attributed to separation. That is not the case for most land vehicles which, despite substantial improvement during the last few decades, are still more or less blunt bodies. Natural laminar flow (NLF) and laminar flow control (LFC) make it possible to have aircraft wings with substantial laminar regions and thus to benefit from the lower skin friction associated with this flow regime. Several techniques like polymer injection, LEBUs, and riblets can reduce the skin friction in turbulent boundary layers, but there are certain obstacles for the widespread use of any of these methods. Polymers, in spite of their very impressive percentage drag reduction and their successful application to internal flows, are not sufficiently inexpensive and require large volume for their storage onboard ships and submarines. Riblets lead to very modest drag reduction, and LEBUs are not effective at high Reynolds numbers. Newer strategies such as the random roughness of Sirovich and Karlsson (1997) or the large-scale vortex generators of Schoppa and Hussain (1998) still require more validation—particularly at field Reynolds numbers. The grand problem remains to find a practical, cost-effective method to reduce the turbulent skin-friction drag encountered on the long fuselage of commercial aircraft or of water vessels. We will return to this topic in Chapter 14.

CHAPTER ELEVEN

Mixing Enhancement

*Now I think hydrodynamics is to be the root of all physical science, and is
at present second to none in the beauty of its mathematics.*
(William Thomson (Lord Kelvin), 1824–1907)

*I pass with relief from the tossing sea of Cause and Theory to the firm
ground of Result and Fact.*
(Sir Winston Leonard Spencer Churchill, 1874–1965,
in The Malakand Field Force*)*

PROLOGUE

This chapter deals with the enhancement of rates of mass, momentum, and heat
transfer. This is desired to improve the performance of mixers of nonreacting species;
chemical reactors, including combustors; lifting surfaces; heat exchangers, including
those in HVAC systems; and numerous other man-made devices. In nature, the efficient turbulence mixing is responsible for the nearly uniform distribution of oxygen,
carbon dioxide, and internal energy in the Earth's atmosphere and oceans. Without
that more-or-less uniform distribution, seasonal and latitudinal temperature changes
would be even more extreme. Oxygen would be more concentrated in the equatorial
rain forests, and carbon dioxide would be more concentrated in industrial and urban centers. Under such circumstances, city dwellers at least would not survive long
on this third planet from the sun. In plants and animals, the efficient transport of
mass and energy through limited spaces is also vital for their survival. The subject of
mixing enhancement is very broad, is important to entire industries, and is covered
extensively in many journals and books. The coverage here by necessity will be rather
brief, focusing mainly on enhancing convective mass, momentum, and heat transport
in both laminar and turbulent flows.

Flow control can be used to augment mixing in both free-shear and wall-bounded
flows by increasing the effective area through which transport takes place, by setting
off resonant flow instabilities, by advancing laminar-to-turbulence transition, and by
enhancing the turbulence once the shear flow is already turbulent. In cases in which
early transition is not possible (or desired), as for example when the Reynolds number is not high enough because of small density, small length-scale, small velocity,
or large viscosity, chaotic advection can be effected in laminar flows to enhance the

convective rates of mass and heat transfer. The effective generation of secondary flows, recirculation zones, or flow unsteadiness is an additional tool to enhance mixing in both laminar and turbulent flows. Successful examples of this strategy include the use of vortex generators on airplane wings and coiled tubes in heat exchangers. There is a drag penalty associated with increasing the surface area as well as with all the other methods used to enhance mixing, and attempts must be made to minimize this penalty.

11.1   Introduction

According to Ottino (1990), the Reynolds number in problems in which mixing is important varies by an amazing 40 orders of magnitude. Over periods measured in millions of years, mixing in the Earth mantle may take place at Reynolds numbers as low as 10^{-20}, and for mixing in the interior of the stars the Reynolds number may reach 10^{20}; the Reynolds number of man-made devices is somewhere between these two extremes. The flow regimes may be creeping, laminar or turbulent, diffusion dominated or advection dominated, subsonic or hypersonic, and so forth. The fluids involved may be a single fluid or many, miscible or immiscible, Newtonian or non-Newtonian, one-phase or multiphase, and so on.

In accordance with the user group, the mass transport phenomenon is variously termed mixing, stirring, blending, fusing, mingling, coalescing, or agitating. In this chapter, mixing is used generically to denote the transport of mass, momentum, or energy. There is no monolithic solution to the mixing problem. It is obvious, for example, that exploiting turbulence to enhance mixing is possible only in some problems but impractical in many others.

At times, we may desire to enhance the rate of mass or heat transfer without unduly increasing the rate of momentum transfer—the latter translating directly into a skin-friction drag penalty. The decoupling between momentum transfer on the one hand and heat and mass transfer on the other is not always easy to accomplish of course, because the tools used to increase the rate of convective transport of one quantity typically increase the transport rate of all others. At other times, we may purposefully enhance the rate of momentum transfer, for example, to prevent boundary-layer separation. In this case, the concomitant increase in skin-friction drag is quite acceptable as compared with the alternative: a much higher increase in pressure drag had the flow separated.

Given unlimited time and space, desired transport of chemical species, momentum, or energy can eventually be achieved solely by molecular diffusion processes. In the combustion chamber of a jet engine, however, time and space are both rare commodities. Mixing the fuel with its rapidly moving oxidant must take place in the shortest distance possible and must also be nearly complete to ensure higher thermodynamic efficiency, less pollution, and so forth. Diffusion-dominated mixing—although good enough for the laminar flame of a candle—is clearly too slow for the jet engine and many other applications.

This chapter focuses on mixing enhancement in advection-dominated (i.e., high-Reynolds-number) flows. Note, however, that, regardless of the global Reynolds number (or more appropriately for mass and heat transfer the Péclet number) regime, ultimately molecular diffusion dominates at sufficiently small-scale, high-gradient regions. Mixing of species by advection is related to the stretching and folding of material surfaces in three-dimensional flowfields or material lines in two-dimensional ones. This is the mechanism by which distant fluid particles eventually come in contact with one another, that is, become mixed. The enhanced stretching and folding processes can result in greater surface areas and higher gradients available for the molecular diffusion mechanism to effectively finish the job by smearing and ultimately eliminating any remaining discontinuities or gradients in velocity, temperature, species concentration, and so forth. Provided that the flowfield is already known and that the advection dominates, nearly complete description of the mixing processes can be obtained purely kinematically down to the scales at which molecular diffusion becomes appreciable. The book by Ottino (1989a) emphasizes this point of view and provides a comprehensive treatment of the kinematics of mixing.

For some applications, the efficient mixing or transport mechanism of turbulence may be feasible as well as desirable. For example, a turbulent boundary layer in general is more resistant to separation than a laminar one, turbulence is used to homogenize fluid mixtures and to accelerate chemical reactions, and turbulent heat exchangers are much more effective than laminar ones.

For other applications, the Reynolds number may be too low, and advancing laminar-to-turbulence transition may be impractical. Mixing the highly viscous ingredients needed to make chocolate is such an application. In this case, it does not make much economic sense to increase the flow speed or scale sufficiently to promote turbulence. Instead, newly advanced techniques to effect chaotic mixing in laminar flows may be used to improve the efficiency of the mass transfer processes involved. Increasing the surface area using external and internal fins and inducing appropriate secondary flows are additional strategies to improve the rates of mass, momentum, and heat transfer in both laminar and turbulent flows.

11.2 Mixing

Stirring or mixing is basically a process that involves a reduction of length scales (i.e., thinning of material volumes) accomplished by stretching and folding of material surfaces. This archetypal mixing process is not unlike that found in the interior of stars or that used in making puff pastries. The primary transport mechanism associated with stretching and folding is macroscopic fluid motion termed advection. Microscopic fluid motions, on the other hand, are responsible for the molecular diffusion processes. The rate of transport of mass, momentum, or heat by molecular diffusion depends on (1) the surface area, (2) the appropriate diffusion coefficient (e.g., binary mass diffusion coefficient when blending two chemical species, viscosity coefficient when transporting momentum or vorticity, and thermal conductivity when conducting heat), and (3) the gradient of the transported quantity. For simple fluids like air and water under most circumstances, the transport coefficients are scalars, and the respective diffusion relations are linear but can be nonlinear as well as quite involved for complex fluids and flows. For example, the heat conducted per unit time per unit area in an isotropic, Fourier fluid is related linearly to the temperature gradient

$$\mathbf{q} = -\kappa_{\mathrm{f}} \nabla T \qquad (11.1)$$

where $\mathbf{q}$ is the heat flux due to molecular diffusion, T is the temperature field, and the thermal conductivity κ_f is a scalar that in general depends weakly on temperature and pressure.[1]

The diffusion coefficients are physical (thermodynamic) properties of the fluid(s) involved and, short of modest changes with pressure and temperature, not much can be done about them. Increasing the surface area available for the diffusion process results in proportional increase in the rate of transport. Similarly, increasing the gradient of the transported quantity proportionally enhances the diffusion process. That is where mixing by advection comes into play. If done right, the stretching and folding of material surfaces can dramatically increase both the area and the gradient available for the molecular diffusion to become pronounced. Secondary flow, turbulence, and chaotic advection can all lead to mixing enhancement. As always in flow control, the challenge is to accomplish the task with a strategy that is simple, is inexpensive to construct and operate, and has the least adverse effect on other flow considerations.

We start by recalling the important nondimensional parameters for the problem. The Reynolds number is a characteristic ratio of advective momentum flux to diffusive momentum flux. Thus,

$$Re \equiv \frac{v_o L}{v} \tag{11.2}$$

where v_o and L are respectively the convective velocity and length scales of the flow, and v is the fluid kinematic viscosity. The heat transfer Péclet number is defined as the ratio of advective heat transfer to conduction

$$Pe \equiv RePr = \frac{v_o L}{v_T} \tag{11.3}$$

where Pr is the Prandtl number ($\equiv v/v_T$), and v_T is the fluid thermal diffusivity ($\equiv \kappa_f/\rho c_p$). Similarly, the mass transfer Péclet number is defined as the ratio of advective mass transfer to mass diffusion

$$Pe \equiv ReSc = \frac{v_o L}{\mathcal{D}_{AB}} \tag{11.4}$$

where Sc is the Schmidt number ($\equiv v/\mathcal{D}_{AB}$), and $\mathcal{D}_{AB}$ is the binary mass diffusion coefficient between chemical species A and B. The focus in this chapter is on mixing by advection. Therefore, the Reynolds number and both Péclet numbers are assumed to be large.

Finally, the Nusselt number is equal to the dimensionless temperature gradient at the wall and provides a measure of the convective heat transfer occurring at the surface

$$Nu \equiv \frac{hL}{\kappa_f} = \left.\frac{\partial T^*}{\partial y^*}\right|_{y^*=0} \tag{11.5}$$

where h is the convection heat transfer coefficient, κ_f is the fluid thermal conductivity, and the superscript $*$ denotes dimensionless quantity. Similarly, the Sherwood

---

[1] Note that once κ_f varies with T, the heat flux-temperature relation (11.1) becomes nonlinear.

number is equal to the dimensionless concentration gradient at the wall and provides a measure of the convective mass transfer occurring at the surface

$$Sh \equiv \frac{h_m L}{\mathcal{D}_{AB}} = \frac{\partial C_A^*}{\partial y^*}\bigg|_{y^*=0} \tag{11.6}$$

where h_m is the convection mass transfer coefficient, and C_A is the concentration (or mass fraction) of species A.

11.2.1 Kinematical Description

To analyze advection-dominated mixing, we may simply follow the fluid surfaces as they stretch and fold by the rate-of-strain field. In principle, this kinematical description of the process requires only knowledge of the velocity field and of the initial distribution of the particular transported quantity. At some scale of the motion at which molecular diffusion becomes important, the pure kinematical description is inadequate, and more involved analysis is needed. As a background, we first elaborate on the difference between a passive scalar and a passively transported tracer. Although the literature on these two seemingly simple concepts is sufficiently muddled, we will argue below that a passive scalar is not necessarily a passively transported tracer. As maintained by Hassan Aref (private communication), there are different degrees of 'passivity.'

A passive scalar, or passive contaminant, is one that does not affect the dynamics of the flow in which it resides. In this case, the conservation equation for that scalar is decoupled from the continuity and momentum equations describing the flow dynamics. For example, temperature is considered a passive contaminant if the flow is incompressible and its density and viscosity are both independent of temperature. In this case, the energy equation is decoupled from the Navier–Stokes equations. Provided that the thermal conductivity is constant, the decoupled energy equation is linear and can usually readily be integrated *after* the velocity field is computed from the continuity and momentum equations. The exception is turbulent flowfields, which—short of direct numerical simulations—are known only probabilistically, and nonintegrable laminar flows, that is, those admitting chaotic solutions to the energy equation despite a well-behaved velocity field (see Section 11.6).

A tracer that follows fluid particles—a passively transported tracer—is one for which diffusion is negligible and no source terms exist in its conservation equation. For example, temperature T is passively transported or simply advected by the flow and described by

$$\frac{DT}{Dt} = \frac{\partial T}{\partial t} + \mathbf{u} \cdot \nabla T = 0 \tag{11.7}$$

if both conduction (molecular diffusion of heat) and dissipation are neglected in the energy equation. The former requires an infinite Péclet number, and the latter requires the ratio of Eckert number to Reynolds number to be vanishingly small (see Eq. (2.25)).

It is clear then that a passive scalar is not necessarily a passively transported tracer, and vice versa. For example, the low-speed flow of a low-Prandtl-number fluid (such as mercury) can be described by a decoupled energy equation provided that the temperature differences involved are not too high. In other words, the temperature in this case can be treated as a passive contaminant. Heat conduction is important, however, and the energy diffuses appreciably as it is advected by the flow and cannot therefore be considered a passively transported tracer. For a reverse example, consider the high-speed flow of a gas away from solid surfaces. For such a compressible, inviscid, nonconducting flow, the energy equation is coupled to the momentum and continuity equations, and the temperature field does affect the flow dynamics, that is, T can no longer be considered a passive scalar. But Eq. (11.7) still holds, albeit with an unknown velocity field, and T is a passively transported tracer constant along any particular particle path.

Similar arguments can be made if species concentration replaces temperature. In this case, the mass transfer Péclet number determines the importance of advection relative to molecular diffusion, and chemical reactions, if any, provide the source terms in the species conservation equation. Only a nonreacting, nondiffusing species can be considered a passively transported tracer.

Within the preceding approximation, the passively transported tracer T remains constant along any particular particle path, and the Lagrangian equivalent to the Eulerian equation (11.7) reads

$$\frac{d\mathbf{X}}{dt} = \mathbf{u}(\mathbf{x}, t) \tag{11.8}$$

where $\mathbf{X}$ is the instantaneous position of a fluid particle[2] characterized by its initial position $\mathbf{X}_0$ at a reference time say $t = 0$, and $\mathbf{u}$ is the Eulerian velocity field, which is presumably a known function of position and time determined from integrating the continuity and momentum equations. Equation (11.8) is a dynamical system describing the particle paths. For time-dependent flows, the particle paths, streamlines, and streaklines may not coincide, and the dynamical system (11.8) is nonautonomous.

To compute the spatial distribution of a passively transported tracer at the present time t, Eq. (11.8) is integrated backwards, in the direction of negative time, to determine the initial positions of all fluid particles whose present positions are of interest. The value of T at any present fluid particle position is then simply equated to the (presumably known) tracer value at the initial particle position at $t = 0$. In accordance with the velocity field, Eq. (11.8) can be linear or nonlinear and can be either integrable or nonintegrable. Chaotic (i.e., nonintegrable) solutions can be expected (but not guaranteed) if the number of degrees of freedom of the dynamical system is at least three and if Eq. (11.8) is nonlinear. Two-dimensional, unsteady flows and three-dimensional steady flows can thus exhibit chaotic behavior of particle trajectories even if the velocity field itself is nonchaotic.

As a result of chaotic particle motions, enhanced mixing and transport take place because a group of adjacent particles in chaotic motion will tend to deviate exponentially from one another, and this leads to substantial mixing with the surrounding fluid in a much shorter time than would be expected for regular motion. The resulting tremendous stretching of finite fluid regions paves the way for an enhanced activity by diffusion due to the greater surface area and gradient available for the molecular processes. We will return to the issue of chaotic advection in Section 11.6.

[2] Note that $\mathbf{x}$ and $\mathbf{X}$ are the same quantity: position in physical space. The former is the spatial position—an independent variable—in the Eulerian frame. The latter symbol is the material point position field—a dependent variable—in the Lagrangian frame.

11.2.2 Enhancement Strategies

To close this section on mixing, the following list categorizes the variety of strategies available to enhance the transport of mass, momentum, and heat:

- Increase the surface area of the solid using external and internal fins.
- Set off resonant flow instabilities.
- Advance laminar-to-turbulence transition.
- Enhance turbulence Reynolds stresses.
- Induce appropriate secondary flows, recirculation zones, or flow unsteadiness in both laminar and turbulent regimes.
- Effect chaotic advection in laminar flows.

The first item on this list deals with the *solid* surface area available for the transport of mass, momentum, and energy. All other items exploit the augmented stretching and folding of material surfaces to increase the *fluid* surface areas and gradients available for the molecular diffusion mechanism to finish the mixing task effectively.

11.3 Early Transition

Turbulence transports mass, momentum, and energy with equal zeal. In fact, the ability to disperse and mix is the most prominent syndrome of turbulent flows. Thus, as compared with laminar flows, turbulent boundary layers are thicker, are more resistant to separation, have higher skin friction, and are characterized by higher coefficients of convective mass and heat transfer. Mixing in free-shear flows is also dramatically enhanced by early transition to turbulence, an occurrence that is readily achieved at relatively low Reynolds numbers.

An important characteristic of turbulent flows is the random motion ability to transport or mix mass, momentum, and energy. The respective rates of transport are typically several orders of magnitude greater than the corresponding fluxes owing to molecular diffusion. The turbulence Reynolds number, defined using a characteristic turbulence velocity u (such as the rms of the streamwise velocity fluctuations) and a length scale ℓ (such as the characteristic size of large eddies),

$$Re \equiv \frac{u\ell}{\nu} \tag{11.9}$$

gives approximately the ratio of rate of momentum transfer due to turbulence to that due to molecular motion (i.e., due to viscous transport). Looking at it from a different perspective, in a problem with an imposed length scale, Re^{-1} is the ratio of a turbulence time scale to a molecular diffusion time scale that would prevail in the absence of turbulence. Turbulence is almost always a high-Reynolds-number phenomenon, and hence the turbulence diffusion is much faster than the molecular one. For example, the warmth of a hot-water radiator is felt throughout a typical-sized room in minutes rather than hours or even days. This enhanced transport, as compared with molecular diffusion, is caused by the weak natural-convection turbulent currents that are driven by the temperature and density gradients in the vicinity of the radiator.

The Reynolds number above should not be confused with that in Eq. (11.2). Consider, for example, a high-Reynolds-number, boundary-layer-type flow. In this case, the Reynolds number in Eq. (11.2) is related to the ratio of the convective length scale along the flow to the much smaller diffusive length scale across the flow. The turbulence Reynolds number in Eq. (11.9), on the other hand, reflects the ratio of the cross-stream length scale in a turbulent boundary layer to the cross-stream length scale in a laminar flow having the same $v_0 L/v$. The tangential Reynolds stress ($-\rho\overline{uv}$) very efficiently drives the cross-stream growth in the turbulent flow case, whereas the generally smaller viscous stress ($\mu\partial\overline{U}/\partial y$) is responsible for the more modest laminar boundary layer growth. Having a larger Re in Eq. (11.2) than that in Eq. (11.9) merely indicates that the streamwise length scale is larger than the cross-stream one even in the turbulent flow case (i.e., the boundary layer is still relatively thin).

Similar arguments can be made for heat and mass transfer. The turbulence Péclet number Pe defined using the thermal diffusivity v_T or the binary mass diffusivity $\mathcal{D}_{AB}$,

$$Pe \equiv \frac{u\ell}{v_T} \tag{11.10}$$

$$Pe \equiv \frac{u\ell}{\mathcal{D}_{AB}} \tag{11.11}$$

compares respectively the heat transfer rate or the mass transfer rate due to turbulence to that due to molecular diffusion.

11.3.1 Wall-Bounded Flows

Advancing transition for wall-bounded flows is discussed first, and the topic of free-shear flows is deferred to the following subsection. As a means for preventing flow separation, particularly for low-Reynolds-number lifting surfaces, early transition in wall-bounded flows was considered in Chapters 8 and 9. Methods to advance boundary-layer transition are of course the mirror image of the transition-delaying strategies discussed in Chapter 6.

Recall that in low-Reynolds-number terminology the transition-promoting devices are called turbulators. For a zero-pressure-gradient boundary layer, transition typically occurs at a Reynolds number based on distance from leading edge of the order of 10^6. The critical Reynolds number Re_{crit} below which perturbations of all wave numbers decay is about 6×10^4. To advance the transition Reynolds number, one may attempt to lower the critical Re, increase the growth rate of Tollmien–Schlichting waves, or introduce large disturbances that can cause bypass transition. The first two routes involve altering the shape of the velocity profile—making it less full—using wall motion, injection, adverse pressure gradient, surface heating in gases or cooling in liquids, or other strategies to increase the near-wall viscosity. The third route, exposing the boundary layer to large disturbances, is much simpler to implement though more difficult to analyze.

Morkovin (1984) broadly classified the large disturbances that can cause bypass transition into steady or unsteady ones originating in the freestream or at the body surface. The most common example is single, multiple, or distributed roughness elements placed on the wall. For lifting surfaces, for example, the mechanical roughness elements, in the form of serrations, strips, bumps, or ridges, are typically placed near the airfoil's leading edge. If the roughness characteristic length is large enough, the disturbance introduced is nonlinear, and bypass transition takes place. For a discrete three-dimensional roughness element of height-to-width ratio of one placed on a flat plate, Tani (1969) reported a transition Reynolds number of $Re_{\delta*} \simeq 300$ for a roughness Reynolds number of $Re_{\Upsilon} \simeq 10^3$. Here, $Re_{\delta*}$ is based on the velocity outside the boundary layer and the displacement thickness, and Re_{Υ} is based on the height of the roughness element Υ and the velocity in the undisturbed boundary layer at the height of the element. Note that the transition Reynolds number indicated above is below the critical $Re_{\delta*}|_{crit} = 420$ predicted from the linear stability theory for a zero-pressure-gradient flat plate. For a roughness Reynolds number of about 600, transition occurs at $Re_{\delta*} \simeq 10^3$. For a smooth surface, in contrast, transition typically takes place at $Re_{\delta*} \simeq 2.6 \times 10^3$.

Other large disturbances that can lead to early transition include high turbulence levels in the freestream, external acoustic excitations, particulate contamination, and surface vibration. These are often termed environmental tripping. Transition can also be effected by detecting naturally occurring T–S waves and artificially introducing in-phase waves. Transition can be considerably advanced, on demand, using this wave superposition principle.

Early transition can also be achieved by exploiting other routes to turbulence such as Taylor–Görtler or crossflow vortices. For example, a very mild negative curvature of $(0.003/\delta^*)$ results in the generation of strong streamwise vortices. In this case, the transition Reynolds number is lowered from $Re_{\delta*} \simeq 2600$ for the flat-plate case to $Re_{\delta*} \simeq 700$ for the concave surface. For high-Mach-number flows, the general decay in spatial amplification rate of T–S waves makes conventional tripping more difficult as the Mach number increases. For these flows, trips that generate oblique vorticity waves of appropriate wavelength to cause resonant growth of three-dimensional modes may be most effective to advance the transition location. For example, on a sharp cone at $Ma = 3.5$, Corke and Cavalieri (1997) and Corke et el. (1999) used an array of plasma (corona-discharge) electrodes to excite oblique vorticity-mode pairs and successfully advanced the transition of the supersonic boundary layer.

11.3.2 Free-Shear Flows

As discussed in Chapter 3, free-shear flows, such as jets, wakes, and mixing layers, are characterized by inflectional mean-velocity profiles and are therefore susceptible to inviscid instabilities. Viscosity is only a damping influence in this case, and the prime instability mechanism is vortical induction. Control goals for such flows include transition delay or advancement, mixing enhancement, and noise suppression. Free-shear flows and separated boundary layers are intrinsically unstable and lend themselves more readily to manipulation.

Free-shear flows originate from some kind of surface upstream, be it a nozzle, a moving body, or a splitter plate, and flow-control devices can therefore be placed on the corresponding walls, albeit far from the fully developed regions. Examples of such control include changing the geometry of a jet exit from circular to elliptic (Gutmark and Ho 1986), using periodic suction or injection in the lee side of a blunt body to affect its wake (Williams and Amato 1989), and vibrating the splitter plate of a mixing layer (Fiedler et al. 1988). These and other techniques are extensively reviewed by Fiedler and Fernholz (1990), who offer a comprehensive list of relevant references, and more recently by Gutmark et al. (1995), Viswanath (1995), and Fiedler (1998).

Because of the nature of their instabilities, free-shear flows undergo transition at extremely low Reynolds numbers as compared with wall-bounded flows. The critical Reynolds number for jets, wakes, and mixing layers is typically $Re_{\text{crit}} = \mathcal{O}[10]$. Many techniques are available to delay laminar-to-turbulence transition for both kinds of flows, but none would do that to indefinitely high Reynolds numbers. Therefore, for Reynolds numbers beyond a reasonable limit, one should not attempt to prevent transition but rather deal with the ensuing turbulence. Of course early transition to turbulence can be advantageous in some circumstances, for example to achieve separation delay, enhanced mixing or augmented heat transfer. The task of advancing transition is generally simpler than trying to delay it and is more relevant to the present chapter. Unless the fluid is very viscous or the speed and length scales are extremely small, jets, wakes, and mixing layers will readily undergo laminar-to-turbulence transition close to their source.

In addition to grouping the different kinds of hydrodynamic instabilities as inviscid or viscous, one could also classify them as convective or absolute based on the linear response of the system to an initial localized impulse (Huerre and Monkewitz 1990). A flow is convectively unstable if, at any fixed location, this response eventually decays in time; in other words, if all growing disturbances convect downstream from their source. Convective instabilities occur when there is no mechanism for upstream disturbance propagation, as, for example, in the case of rigid-wall boundary layers. If the disturbance is removed, then perturbation propagates downstream and the flow relaxes to an undisturbed state. Suppression or enhancement of convective instabilities is particularly effective when applied near the point where the perturbations originate.

If any of the growing disturbances has zero group velocity, the flow is absolutely unstable. This means that the local system response to an initial impulse grows in time. Absolute instabilities occur when a mechanism exists for upstream disturbance propagation, as, for example, in the separated flow over a backward-facing step where the flow recirculation provides such mechanism. In this case, some of the growing disturbances can travel back upstream and continually disrupt the flow even after the initial disturbance is neutralized. In some flows (e.g., two-dimensional blunt-body wakes), certain regions are absolutely unstable whereas others are convectively unstable. The upstream addition of acoustic or electric feedback can change a convectively unstable flow to an absolutely unstable one, and self-excited flow oscillations can thus be generated (e.g., Reisenthel, Xiong, and Nagib 1991). In any case, identifying the character of flow instability facilitates its effective control (i.e., suppressing or amplifying the perturbation as needed).

11.4 Turbulence Enhancement

As emphasized earlier, an important characteristic of turbulent flows is the random-motion ability to transport or mix mass, momentum, and energy. The respective rates of transport are typically several orders of magnitude greater than the corresponding fluxes due to molecular diffusion. But can one improve on that impressive feat? The mass, energy, and momentum transport in a turbulent flow are accomplished via strong correlations between the corresponding fluctuating quantities (e.g., $\overline{c_A v}$, $\overline{\theta v}$, $\overline{uv}$, where c_A is the concentration fluctuations of species A, θ is the temperature fluctuations, and u and v are respectively the fluctuating streamwise and cross-stream velocity components). It is obvious, then, that to enhance turbulence transport one has to strengthen the aforementioned correlations. This can be achieved by effecting appropriate secondary flows, recirculation zones, or flow unsteadiness. We offer here a few examples of passive and active control methods to accomplish this, first for boundary layers and then for free-shear flows.

11.4.1 Turbulent Wall-Bounded Flows

To enhance turbulence in wall-bounded flows, one may reverse the stimuli discussed in Section 10.5 in connection with relaminarization by using injection instead of suction in external flows, decelerating flows instead of accelerating ones, negative curvature (concave) instead of positive curvature (convex), unstable stratification instead of a stable one, and so on. The reverse stimuli will lead to an increase in turbulence production and enhanced mass, momentum, and heat transfer. Roughness will also enhance the turbulence, but its associated drag penalty must be carefully considered. Other devices to enhance the turbulence mixing include vane-type vortex generators, which draw energy from the external flow, or Wheeler-type or Kuethe-type generators, which are fully submerged within the boundary layer and presumably have less associated drag penalty (Rao and Kariya 1988). These and other passive and active strategies to energize the near-wall turbulence and augment the cross-stream momentum transport via Reynolds stresses were extensively discussed in Section 8.8 in connection with separation delay.

The turbulence levels in the near-wall region can be modestly augmented by freestream excitation. In a unique experiment, Bandyopadhyay (1988) used a pitching airfoil located outside a turbulent boundary layer to resonate the viscosity-dominated unsteady flow in a row of minute transverse square cavities at the wall. In a narrow band of pitching frequencies, the near-wall turbulence levels increased by as much as 6%, whereas the rest of the boundary layer seemed transparent to the external disturbances. This result is particularly significant because it demonstrates that near-wall events may be controlled from the freestream and, therefore, that a dynamic coupling between disparate scales may exist.

11.4.2 Turbulent Free-Shear Flows

Turbulent free-shear flows, like their wall-bounded counterparts, are dominated by large-scale coherent structures that, in turn, are filled with incoherent motions of smaller scales. The ratio of the largest to the smallest scales varies as $Re^{(3/4)}$, and the large eddies control the most important dynamics of the flow. To strengthen the correlations described above, most turbulence-enhancement strategies for free-shear flows center around controlling the processes that produce and sustain the large eddies. The goal is to achieve an order one effect using an order ε perturbation. The control is best carried out in the early stages of development because the growth of shear layers is rather sensitive to their initial conditions.

As examples of targeting large-scale structures, we first focus on flow in turbulent jets. The dominant structures in axisymmetric jets are toroidal rings in the near field and helical structures in the far field. Superposition of a plane axial acoustic perturbation having a frequency that falls within the sensitive range of the basic flow locks in the initial formation of vortex rings to the excitation frequency (Zaman and Hussain 1980). This causes the subsequent growth and spreading of the jet by vortex pairing to be augmented. Other strategies to enhance mixing in jets include (1) using an elliptical orifice, thus producing elliptical vortex structures that exhibit large major–minor axis oscillations caused by a curvature-dependent self-induction and interaction between azimuthal and axial vortices (Ho and Gutmark 1987; Hussain and Husain 1989; Husain and Hussain 1991, 1993; Gutmark and Grinstein 1999); (2) using suction at the jet lip to produce local regions of backflow and large-amplitude, self-excited oscillations associated with the resulting absolute instability (Strykowski and Niccum 1991); and (3) forcing the jet with multiple frequencies and modes to make the near-field ring vortices slightly eccentric (Reynolds 1996).

In this latter strategy, the rings act to tilt one another, and thus they can be made to move in different directions. By adding helical perturbations to axial ones, the vortex rings leave the nozzle lip eccentrically, and for an axial-to-helical frequency ratio of 2, the jet can be split into two distinct branches that emerge at an angle to the jet axis—the bifurcating jet of Juvet and Reynolds (1989). If the ratio of frequencies is incommensurate but in the range of 1.7–3.5, the vortex rings are sent in all directions and the jet explodes into a shower of vortex rings—the blooming jet of Lee and Reynolds (1985). As compared with the spreading angle of about 4° for an unperturbed axisymmetric turbulent jet, the spreading angle of a blooming jet can be as much as 75°, which is achieved at $Re_D = 5000$ and an axial–orbital frequency ratio of 2.3. Note, however, that as the vortex rings wander away from the primary jet axis, the blooming jet does not remain centered. This prompted some (e.g., Husain and Hussain 1983) to term the enhanced mixing as apparent and not real. With increased intensity of the required acoustic perturbations, these remarkable phenomena have been extended to jet Reynolds numbers as high as 10^5.

The ability of reactants to come together is a critical factor in determining the rate at which the product of a chemical reaction is formed. This is particularly true in reactions in which the reactive and diffusive time scales are longer than the advective time scale. In this case, the overall reaction yield is very sensitive to fluid mixing. For fast chemical reactions, mixing is relatively slow, desired reactions are slowed, and undesired ones are often enhanced. Product selectivity is reduced, and excessive production of waste results. A notorious example is the highly undesirable formation of nitric oxide (NO) and nitrogen dioxide (NO_2) when oxygen combines with atmospheric nitrogen rather than with the hydrocarbon fuel in internal combustion engines. Both NO_x compounds are considered toxic, and NO is related to the formation of photochemical smog. Improvements of combustion processes not only decrease fuel consumption but also reduce pollutant generation (Pope 1996). For example, simply imparting swirl to the incoming air in jet-engine combustors improves fuel economy and pollution emission. The same technique is now used in high-efficiency home oil burners. Active control strategies to enhance mixing in combustion applications are reviewed by Haile et al. (1998).

Lastly, methods to enhance mixing in compressible free-shear flows are reviewed by Gutmark et al. (1995). There the challenge stems from the inherently low growth rates of supersonic and hypersonic shear layers.

11.5 Heat Transfer

The size, performance, and efficiency of many engineering systems are often limited by the ability to transfer heat to or from the system. Consider the supercomputer, for which the speed of computing is limited by the distance through which the electrons must travel at nearly the speed of light. Electronic chips and other components must, therefore, be miniaturized and packaged tightly. Tremendous amounts of heat must be removed from a rather modest surface area. The central processing unit (CPU) of a present-generation supercomputer is not much larger than, say, a refrigerator. To limit the temperature of the CRAY–2 to a tolerable range, its entire CPU is immersed in a dielectric liquid coolant (trade name FC–75). This is obviously an expensive, but perhaps unavoidable, approach to achieving the desired goal.

If a boundary-layer-type flow is considered and radiation is neglected, the surface heat loss per unit time due to the macroscopic motion of a fluid is given by Newton's law of cooling

$$q_s = -\kappa_f A \frac{\partial T}{\partial y}\bigg|_{y=0^+} = h A (T_s - T_o) \tag{11.12}$$

where κ_f is the thermal conductivity of the fluid, A is the surface area, h is the convection heat-transfer coefficient that depends on the nature of the boundary layer and fluid motion as well as a host of fluid thermodynamic and transport properties, T_s is the surface temperature, and T_o is the fluid temperature at the edge of the shear layer. It is clear, then, that to maximize heat removal by convection, the surface area has to be maximized using fins or inserts, for example; the convection coefficient has to be maximized via secondary-flow generation, chaotic advection, early transition, or turbulence enhancement; or the temperature difference between the solid surface and the outer flow has to be maximized or a combination of these techniques needs to be used. Identical arguments can be made regarding convective mass transfer. There species concentration replaces temperature, and mass diffusion coefficient replaces thermal conductivity.

Convective heat transfer enhancement methods were reviewed by Bergles and Morton (1965), Bergles (1978, 1981, 1985), Webb et al. (1981), Bergles and Webb (1985), Nakayama (1986), Žukauskas (1986), Webb (1987, 1993), and Chang and Sen (1995). The *Journal of Enhanced Heat Transfer* is devoted to the topic and provides up-to-date references. The simplest scheme involves the introduction of distributed roughness on the heat-transfer surfaces. This would destabilize the wall region and intensify the mixing process, leading to an increase in the convection-heat-transfer coefficient. Patera (1986) and Patera and Mikić (1986) introduced the concept of resonant heat transfer enhancement based on excitation of shear-layer instabilities in internal separated flows. In a two-dimensionally periodically grooved channel, Tollmien–Schlichting-like waves forced by the Kelvin–Helmholtz shear layer instabilities at the groove's edge become unstable and take the form of self-sustained oscillations for Reynolds numbers greater than a critical value. For lower *Re*, oscillatory perturbations result in subcritical resonant excitation. For both laminar and turbulent flows, the resulting large-scale motions lead to significant lateral mixing and correspondingly dramatic enhancement of convection heat transfer coefficient. Other devices to enhance convective heat transfer in pipes include coil-spring wire inserts, twisted tape inserts, and longitudinal and helical ribs. These artifacts introduce additional no-slip surfaces as well as increase the momentum flux in a direction transverse to the flow, leading to significant pressure-drop penalty.

The secondary motion in the transverse plane of a coiled tube provides higher heat transfer rates compared with straight tubes. The same geometry can of course

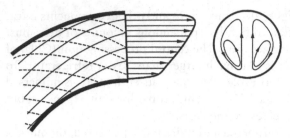

Figure 11.1: Secondary flow in a curved segment of a pipe.

be used to enhance the rate of mass transfer in chemical reactors, for example. The Dean number is defined as

$$De \equiv \frac{Re}{2}\sqrt{\frac{a}{R}} \qquad\qquad (11.13)$$

where Re is the Reynolds number based on the bulk velocity and pipe diameter, a is the radius of the cross section, and R is the radius of curvature. This dimensionless parameter measures the curvature effects compared with viscous effects. At a critical De, a secondary flow emerges in a curved pipe as a second stable laminar flow. Fluid particles near the flow axis have higher velocity and hence are acted upon by a larger centrifugal force than the slower particles in the vicinity of the wall. This results in a pair of streamwise vortices with flow directed outwards in the center and inwards (toward the center of curvature) near the wall, as indicated in Figure 11.1.

A material line initially perpendicular to the secondary upwelling is stretched the most, albeit linearly, as it is advected downstream by the primary flow.[3] The secondary flow causes faster or slower primary flow and corresponding higher or lower shear stress near the outside or inside of the bend. The coiling leads to an increase in the convection coefficient as well as the pressure drop. At higher Dean numbers, other instabilities develop, and the flow eventually becomes turbulent. There is no accompanying increase in surface area when a tube is coiled, and if the Reynolds number (or more precisely the Dean number) is not too high, the flow remains laminar.[4] The increase in Nusselt number or Sherwood number in this case is typically more than the increase in pressure drop, making coiled tubes more useful for practical heat exchangers and chemical reactors (Shah and Joshi 1987).

Forming even more twists in a coiled tube may generate tertiary flow that enhances the mixing even further. In a recent article, Knight (1999a) speculated that the unusually complex plumbing patterns of human arteries, respiratory ducts, and urinary tracts may have evolved for precisely this advantage. He then raised the intriguing possibility of improving on present heart bypasses—in which a section of blood vessel in the form of a planar arch is spliced into a coronary artery to divert blood around a blockage—using instead twisted, nonplanar bypasses. The resulting helical bypass, indicated schematically in Figure 11.2, may lead to improved mixing, smoothing out the undesired shear-stress extremes at the outside and inside of the

---

[3] Contrast this linear stretching to the exponential one associated with chaotic advection (Section 11.6).

[4] In fact, the centrifugal forces accompanying tight coiling may be sufficiently strong to relaminarize what would otherwise be a turbulent pipe flow. The stabilizing influence in this case is analogous to that in stably stratified flows.

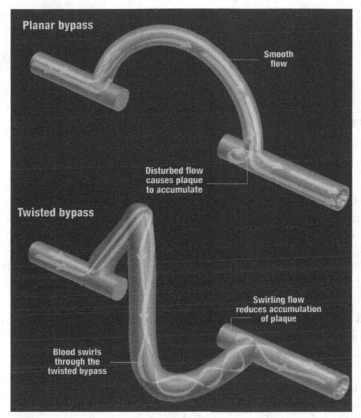

Figure 11.2: Conventional planar bypass and proposed twisted bypass graft.

bend and thus discouraging the accumulation of fat, proteins, and cholesterol at the junction to the large artery. The spiral bypass better resembles natural blood vessels and may lower the risk of two common modes of graft failure: (1) plaque accumulation at the "dead spot" in the blood stream where the graft meets the primary artery, and (2) excessive division of muscle cells in the arterial wall, which is an abnormality thought to be triggered or at least aggravated by the presence of high-shear-stress regions.

In microdevices (see Chapter 13), both radiative and convective heat loss or gain are enhanced by the huge surface-to-volume ratio. For a device with a characteristic length of 1 μm, this ratio is a million times that of a device having a length of 1 m. Consider a device having a characteristic length L_s. Use of the lumped capacitance method to compute the rate of convective heat transfer, for example, is justified if the Biot number ($\equiv h L_s / \kappa_s$, where h is the convection heat transfer coefficient and κ_s is the thermal conductivity of the solid) is less than 0.1. Small L_s implies small Biot number and a nearly uniform temperature within the solid. Within this approximation, the rate at which heat is lost by the solid to the surrounding fluid is given by

$$\rho_s L_s^3 c_s \frac{dT_s}{dt} = -h L_s^2 (T_s - T_\infty) \tag{11.14}$$

where ρ_s and c_s are respectively the density and specific heat of the solid, T_s is its (uniform) temperature, and T_∞ is the ambient fluid temperature. Solving Eq. (11.14) for constant convection coefficient is trivial, and the temperature of a hot surface drops exponentially with time from an initial temperature T_i,

$$\frac{T_s(t) - T_\infty}{T_i - T_\infty} = \exp\left[-\frac{t}{\mathcal{T}}\right] \tag{11.15}$$

where the time constant $\mathcal{T}$ is given by

$$\mathcal{T} = \frac{\rho_s L_s^3 c_s}{h L_s^2} \tag{11.16}$$

For small devices, the time it takes the solid to cool down is proportionally small. Clearly, the millionfold increase in surface-to-volume ratio implies a proportional increase in the rate at which heat escapes. Identical scaling arguments can be made regarding mass transfer.

11.6 Chaotic Mixing

Stirring by chaotic advection can be an attractive alternative for situations in which turbulence is either not possible or not desired. Mixing of highly viscous fluids or designers not willing to pay the turbulence drag penalty are two such circumstances. The idea here is to effect chaotic particle paths even for well-behaved velocity fields. A group of adjacent particles in chaotic motion will tend to deviate exponentially from each other, and this leads to substantial mixing with the surrounding fluid in a much shorter time than would be expected for regular motion. The resulting transport can be more effective than that in ordinary laminar flows—even those with superimposed secondary motions. The concept of chaotic advection has been demonstrated theoretically, experimentally, and numerically for mass and energy transport. Chaotic advection, or *Lagrangian chaos*, is possible for time-dependent, two-dimensional (or three-dimensional) flowfields as well as for steady three-dimensional flows.

Consider the diffusion equation for a passive scalar field $\theta(\mathbf{x}, t)$ having a constant diffusivity $\mathcal{D}$. In the Eulerian description, the partial differential equation that governs the diffusion process is linear because the velocity field is obtained independently from the continuity and momentum equations. The diffusion equation with no source terms reads

$$\frac{\partial \theta}{\partial t} + \mathbf{u} \cdot \nabla \theta = \mathcal{D} \nabla^2 \theta \tag{11.17}$$

where the advecting velocity field $\mathbf{u}(\mathbf{x}, t)$ is a prescribed function of the spatial coordinates $\mathbf{x}$ and time t. In a turbulent flow, $\mathbf{u}$ is a random vector field and therefore θ is also expected to be random. But the surprising result—if one considers the linearity of the governing equation—is that the scalar field θ can exhibit chaotic behavior even for simple laminar flows. This was demonstrated rather elegantly by Aref (1984), who coined the expression *chaotic advection*.

When the problem of a passively transported tracer is analyzed from a Lagrangian viewpoint, the governing equations may be nonlinear and nonintegrable. The resulting Lagrangian chaos is a spatial not a temporal one. Therefore, Eq. (11.17) is not the proper dynamical system to represent the problem, and its linearity or lack thereof is irrelevant to the chaotic behavior of θ. The velocity at a point may be constant or periodic, whereas the passively transported scalar may be aperiodic (i.e., chaotic) in space. Consider the trajectories of individual advected particles

$$\frac{dX}{dt} = u(x, y, z, t) \tag{11.18}$$

$$\frac{dY}{dt} = v(x, y, z, t) \tag{11.19}$$

$$\frac{dZ}{dt} = w(x, y, z, t) \tag{11.20}$$

The initial conditions are $\mathbf{X}|_{t=0} = \mathbf{X}_0$. These equations describe a finite-dimensional dynamical system and are the Lagrangian equivalent of Eq. (11.17) for the case of zero diffusivity. Chaotic particle paths lead to chaotic spatial distribution of the transported tracer.

The preceding equations are deterministic—meaning the forcing terms on the right-hand side are not random—for nonturbulent velocity fields. For steady, two-dimensional flow, the streamlines and particle paths coincide.[5] The equations for particle paths can in this case be written in terms of the streamfunction, $\psi(x, y)$,

$$\dot{X} = \frac{\partial \psi}{\partial y}, \quad \dot{Y} = -\frac{\partial \psi}{\partial x} \tag{11.21}$$

In the language of dynamical systems theory, the preceding are just Hamilton's canonical equations for one degree of freedom and hence are integrable when autonomous. Therefore, a regular velocity field leads to regular passive scalar distribution. On the other hand, if the two-dimensional flowfield is time-dependent, however innocuous, the nonautonomous Hamiltonian dynamical system can be nonintegrable, producing stochastic particle paths. Note that for plane flow, the phase-space flow of a Hamiltonian dynamical system corresponds to the configuration-space motion of advected particles. For three-dimensional flows, steady and unsteady laminar velocity fields can result in chaotic particle paths and therefore highly complex distribution of the passively transported tracer. In the steady case, a regular three-dimensional velocity field can produce chaotic particle paths and identically irregular streamlines—an amazing counterintuitive effect. These results provide a remarkable contrast between the Eulerian and Lagrangian representation of the same flow: regular Eulerian flowfields can, under the right circumstances, lead to highly irregular advection patterns.

Unlike the fractal dimension test, which can only be used in dissipative dynamical systems, the Lyapunov exponent provides a quantitative criterion for the presence of chaos in both conservative (Hamiltonian) and dissipative deterministic systems. A positive exponent implies chaotic dynamics and measures the sensitivity of the system to changes in its initial conditions. If the initial distance between two trajectories in phase space is d_0, at a small but later time the distance is on the average

$$d(t) = d_0 e^{\lambda t} \tag{11.22}$$

---

[5] As they also do in steady, three-dimensional flows.

where λ is the Lyapunov exponent. In the mixing problem, the Lyapunov exponent can be used to measure the rate at which an infinitesimal fluid volume is stretched. Each dimension of the flow has an associated exponent, and for a conservative, incompressible dynamical system, the sum of all exponents must be zero. For a nonchaotic system, stretching occurs at a linear rate, and thus all Lyapunov exponents must be zero. For a system that exhibits chaotic particle paths, the largest Lyapunov exponent is positive, denoting exponential separation of neighboring particles. The exponent must therefore be negative in at least one other dimension, implying a diminution of length scale in that direction.

If the two-dimensional or three-dimensional laminar flowfield is cleverly chosen, the dynamical system (Eqs. (11.18)–(11.20)) becomes nonintegrable with resulting stochastic response in the Lagrangian advection characteristics of a passively transported tracer. In this case, a group of adjacent particles will tend to deviate exponentially from one another, and this leads to substantial mixing with the surrounding fluid in a much shorter time than would be expected for regular motion. The resulting tremendous stretching of finite fluid regions paves the way for an enhanced activity by diffusion owing to the availability of greater surface area and gradient for the molecular processes. Despite the exponential divergence of neighboring states (in phase space), the stretching (i.e., divergence of neighboring trajectories) and folding (i.e., confinement to bounded space) mechanism is necessary to keep chaotic trajectories within a finite volume of phase space. The stretching and folding operation corresponds to what is called in dynamical systems theory a horseshoe map. Chaotic mixing can result from the breakup of homoclinic or hetroclinic loops into tangles. Those entanglements result from repeated intersections of the stable and unstable manifolds, giving rise to chaotic behavior. This causes a set of particles to be well mixed spatially as it is advected by even an innocuous flowfield.

The utility of chaotic advection has been demonstrated for mass as well as energy transport. Chaotic mixing can be effected even at extremely low Reynolds or Péclet numbers where turbulence is not a viable option, making the technique particularly useful for a range of applications in chemical engineering, low-speed flows, heat transfer, and materials processing. Because Lagrangian chaos can be achieved even at extremely low speeds, the enhanced scalar transport is not necessarily associated with enhanced momentum transport. This is a tremendous practical advantage. In contrast, scalar transport enhancement by turbulence is often at the expense of significant pressure drop, skin-friction increase, or both, and therefore power expenditure.

Herein we recall just a few examples from the recent deluge of demonstrations. The original analytical work in the field of chaotic advection is, as indicated earlier, attributable to Aref (1984), who also cited several earlier contributions by other researchers. Aref idealized an incompressible, inviscid fluid being stirred in a tank using point vortex agitator(s). Together with their images in the bounding contour, the agitators provide the source of unsteady potential flow. The motion is assumed wholly two-dimensional. Aref (1984) demonstrated by way of a numerical integration of the time-dependent version of Eq. (11.21) that chaotic particle motions are possible for certain unsteady movements of the stirring vortex. Two protocols were used to generate the successful unsteadiness: either a stirrer that jumps back and forth between two fixed positions or two stirrers at fixed positions that are run alternately for a given time interval. In either case, efficient stirring was achieved when chaotic particle paths were generated using the successful protocol.

Ottino (1989a,b; 1990) detailed the body of analytical and experimental work conducted by his research group. Chaotic mixing of miscible and immiscible fluids has been achieved in time-dependent creeping flows in two simple two-dimensional geometries. The first is the journal-bearing arrangement in which two eccentric cylinders counterrotate alternately for a fixed time period. The second geometry is a cavity flow in which the upper and lower walls are forced to move alternately—each parallel to itself but in opposite directions—using two independently driven timing belts. In both cases good mixing is achieved after just few periods for certain optimum movement protocols.

Again, for the creeping flow between two counterrotating eccentric cylinders, Ghosh et al. (1992) have computed the enhancement in the cross-stream heat (or mass) transport associated with either a steady recirculation region or chaotic advection. As compared with pure conduction (or molecular diffusion), the steady-state enhancement is proportional to the square root of a characteristic width of the recirculation region, but the effect diminishes as the Péclet number is lowered from its asymptotic infinite value. If time-periodic forcing is used to perturb the steady recirculation region, 100% time-averaged enhancement over pure diffusion is achieved at the optimum forcing frequency, leading to homoclinic entanglement of the recirculation region's bounding separatrices and hence to chaotic advection transport across those separatrices.

To avoid the practical difficulties of using time-periodic perturbations at a boundary, Acharya, Sen, and Chang (1992) investigated analytically, numerically, and experimentally the possibility of effecting chaotic advection in helical pipes by spatial periodicity in the downstream direction. Using as a base flow a conventional, fixed-axis coiled tube, they perturbed the customary secondary flow by periodic changes in the coiling. One such perturbation is shown in Figure 11.3, where each successive loop of the coil lies on a different, mutually perpendicular plane. Chaotic particle paths are observed when the switching length exceeds the critical value of one-half of the loop circumference. As compared with a conventional coiled tube using realistic fluids (i.e., finite Re, Pr, and Pe), the alternating-axis coiled heat exchanger leading to chaotic advection enhances the heat transfer by 6–8% with only a 1.5–2.5% pressure-drop penalty. Heat transfer enhancement by chaotic advection is reviewed by Chang and Sen (1995).

The same alternating-axis coiled tube shown in Figure 11.3 was investigated in an analytical study directed toward enhancing the rate of mass transfer. Sawyers et al. (1996) documented the benefits of chaotic advection on the overall yield in a slow, bimolecular chemical reaction. The two reactants are initially separated, and the flow is assumed laminar and steady with high mass Péclet number. In a straight tube, there is no flow in the transverse direction, and the interface separating the two reactants does not change as the flow progresses downstream. Mixing in this case is achieved solely by the very slow molecular diffusion process. Regular mixing produced by the secondary transverse flow in a helical coil linearly stretches the interface separating two reactants, leading to some improvement in mixing. Chaotic mixing, by contrast, leads to an exponentially growing interface length that can be related to a positive Lyapunov exponent. The area-averaged product mass fraction increases proportionally in this case, leading to a much higher yield of the reaction. Numerical solutions show that the enhancement by chaotic mixing also exists for fast reactions.

(a)

(b)

Figure 11.3: Conventional coiled tube and alternating-axis tube.

In a different kind of three-dimensional geometry, Sawyers, Sen, and Chang (1998) computationally investigated the heat transfer enhancement in the laminar flow inside corrugated channels. They compared the regular mixing in two-dimensional sinusoidal corrugations (with primary flow perpendicular to the corrugations) with the chaotic mixing resulting when the corrugations are sinusoidal in two orthogonal directions. The conventional geometry and the *egg-carton* one are

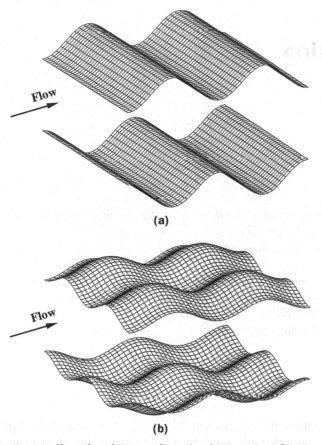

Figure 11.4: Channels with (a) two-dimensional corrugations; (b) egg-carton corrugations.

schematically depicted in Figure 11.4. As compared with flat-plate heat exchangers, the convection-heat-transfer coefficient is higher for the two-dimensional corrugations because of the presence of recirculation zones. The three-dimensional corrugations cause the boundary of those zones to be broken, thus allowing fluid particles to cross between the recirculation regions and the main flow. This allows heat transfer across what was a barrier to advection in the two-dimensional corrugated channel. The resulting enhanced mixing reduces the fluid temperature nonuniformities, thereby steepening the temperature gradient near the boundaries and increasing the heat transfer between the fluid and the channel walls.

Finally, for readers desiring either a gentle primer or more detailed treatment, the following books offer a broad range of sophistication and readability on the topics of nonlinear dynamical systems theory, chaos and chaotic advection: Gleick (1987), Ottino (1989a), Baker and Gollub (1990), Kim and Stringer (1992), Moon (1992), Ott (1993), Aref and Naschie (1995), and Nayfeh and Balachandran (1995).

CHAPTER TWELVE

Noise Reduction

The root of the matter is that the greatest stimulus of scientific discovery are its practical applications.
    (Lewis Fry Richardson, 1881–1953)

However far modern science and technics have fallen short of their inherent possibilities, they have taught mankind at least one lesson: Nothing is impossible.
    (Lewis Mumford, 1895–1990)

PROLOGUE

This chapter is concerned with noise suppression. Noise in the context of this book is undesired sound—particularly that generated by a fluid flow. Although tangential to the primary topics of this monograph, noise control is an important and broad topic and deserves, and is addressed in, its own books. Hence, the treatment here is rather cursory and consists simply of introducing the topic and relating noise control to other flow control topics discussed elsewhere in this book. Particularly for cold subsonic flows, small-scale turbulence fluctuations and unsteady flow oscillations, either in free-shear modes or interacting with solid surfaces, provide the primary sources for the flow-induced sound energy. For hot supersonic flows, the interaction of the turbulence large eddies with the flow is the dominant noise source. In either case, therefore, controlling the flow modulates the sound field favorably or adversely. The goal is of course to reduce noise pollution either for human comfort, for military advantage, or for the prevention of violent structural vibrations. Both passive and active control strategies can be employed to suppress flow-induced noise. This goal can be realized by reducing the vibrations of solid surfaces, eliminating or suppressing turbulence, or cleverly manipulating certain flow instabilities and coherent structures.

12.1 Introduction

Sound ordinarily refers to audible pressure fluctuations in the ambient air. However, sound can propagate as well in liquids and solids, and its frequencies can be lower (infrasonic waves below 20 Hz) or higher (ultrasonic waves above 20 kHz) than those that can stimulate the human ear and brain to the sensation of hearing. Sound can be a pleasant one like that emanating from a musical instrument in the hands of a talented Homo sapiens. It can also be an unpleasant noise such as that radiated from a jackhammer or an angry companion.

Sound can be generated by the vibrations of solid surfaces such as the strings of violins and similar musical instruments or the diaphragms of loudspeakers. It can also be generated by flow oscillations either directly or, more effectively, as a result of interactions with solid surfaces. Unsteady addition or removal of heat can also generate sound such as in singing flames. Noise is undesired sound, and the kind generated as a result of fluid flow is of primary interest in this book. In 1952, Lighthill focused on the sources of sound in the absence of vibrating surfaces and termed the genre "sound generated aerodynamically." The field is now known as aeroacoustics or, in water applications, hydroacoustics.

Audible sound levels vary over an enormous range, and therefore a relative logarithmic scale is typically used to express the sound's power, level of pressure fluctuations or intensity (energy flux or product of pressure and velocity perturbations). Sound measured in decibels (dB) is computed from the relation: $10 \log (X/Y)$, where X is either the sound power, the mean-square pressure fluctuations, or the mean sound intensity, and Y is a corresponding reference value typically related to the threshold of hearing. More subjectively, sound can also be measured in phons, where a loudness level of N phons is judged by the average ear to be as loud as a pure tone of frequency 1 kHz at a sound-pressure level of N dB. There is also the sone scale, which is a linear measure of loudness that is normalized so that 1 sone is a sound whose loudness level is 40 phons. A sound of 10 sones is 10 times as loud as a sound of 1 sone, and the audible sounds lie in a range of $\approx$0–100 sones. Finally, there is the perceived noise decibel (PNdB), the A-weighted sound level (dBA), B-weighted, C-weighted,....

The envelope for man-made machines is continuously being pushed toward both the miniature and the giant. As these machines become larger and more powerful, their radiated, often unpleasant sound levels typically increase. Noise pollution is an undesired byproduct of the present technological age, and efforts to minimize it must always be considered in any eventual product. To place machine noise in perspective, a human whisper involves acoustic power of roughly 10^{-10} watts (or 20 dB on the relative logarithmic scale), a shout $\approx 10^{-5}$ w (70 dB), a jackhammer ≈ 1 w (120 dB), a jet engine at takeoff $\approx 10^5$ w (170 dB), and a sizable rocket at launch $\approx 10^7$ w (190 dB). Therefore, a single modern jet plane generates at takeoff more noise than the combined shouting power of the entire world's population. For humans, unprotected exposure to sound levels above 100 dB for more than 15 minutes can cause hearing damage, and permanent hearing loss results when exposed to sound levels above 110 dB. The threshold of pain is between 130 and 140 dB.

For some machines, like supersonic transport, noise control becomes one of the limiting technologies. Out of concern for both air and noise pollution, the development of commercial supersonic transport (SST) in the United States was summarily halted in 1971 when the Congress voted to stop all further Federal funding for the faster-than-sound aircraft. Though the Europeans and the Soviets proceeded, with mixed success, to develop their own versions of SST, the United States has not yet recovered from that 30-year-old debacle. Noise control for jet engines remains a formidable barrier to the successful development of supersonic as well as hypersonic commercial aircraft.

Ultrasound is now used routinely in medicine to image tissues transparent to X-rays and in nuclear power plants to test for cracks and structural integrity. More relevant to this book, acoustic energy can be targeted toward certain flow regions to control transition (Chapter 6) or separation (Chapters 8 and 9). But the focus in this chapter is on attempting to reduce undesired flow-induced noise, not on the use of sound to control the flow or to perform other useful tasks. As we attempt to achieve other flow-control goals, however, the sound field may be affected favorably or adversely, and any unacceptable trade-off has to be scrutinized. Noise suppression may of course be the primary control goal, and the effect of that on drag, lift, mixing, and so forth, also has to be considered.

The main objective of this chapter is to introduce the topic of flow-induced noise and to describe a variety of passive and active strategies to reduce noise pollution either for human comfort, for military advantage, or for preventing violent structural vibrations. Understanding the sources of noise is an essential step toward devising ways to suppress the noise, and the chapter will start with a brief discussion of the fundamentals of acoustics and sources of sound.

The following sections merely scratch the surface of the important field of noise control. Readers who desire more information can consult the numerous journal articles available in the field appearing in, for example, the *Journal of Sound and Vibration*, *Journal of the Acoustical Society of America*, *Archives of Acoustics Quarterly*, and *Journal of Vibration and Acoustics*. A more gentle indoctrination in the topic can be had, however, by reading the books by, among others, Dowling and Ffowcs Williams (1983), Blake (1986), Bolton (1988), Wilson (1989), Foreman (1990), Tokhi and Leitch (1992), Atassi (1993), and Fung (1994). The proceedings and the two periodicals *Noise/News* and *Noise Control Engineering Journal* published by the Institute of Noise Control Engineering of the United States of America provide a wealth of information for the practically oriented. More fundamentally, the *Annual Review of Fluid Mechanics* has published 11 articles related to aero- and hydroacoustics in its first 30 volumes. Finally, the pioneering articles on sound generated aerodynamically authored by Sir James Lighthill can all conveniently be found in volume III of his collected work (Hussaini 1997).

12.2 Fundamentals of Acoustics

Sound is pressure perturbations that are transmitted in a compressible medium as successive compressive and rarefaction waves. Sound waves are longitudinal, mechanical waves; in other words, the material particles transmitting such a wave oscillate in the direction of propagation of the wave itself, and the waves carry energy and thus require a source of some sort. The elastic nature of the medium causes the vanishingly weak pressure pulses to propagate while maintaining quasi-thermodynamic equilibrium adiabatically (i.e., isentropically) and essentially with no viscous losses. Sound can, of course, propagate, reflect, refract, scatter and dissipate in gases, liquids and solids, but propagation through fluids is the primary concern here.

Acoustic disturbances are nearly always small, which allows linearization of the governing equations.[1] For example, the root-mean-square of the pressure perturbations at the threshold of hearing is 2×10^{-10} atmospheres, and at the threshold of pain is 2×10^{-3} atm. Consider a still fluid having uniform pressure p_0 and density ρ_0. Sound perturbs this field such that

$$p(\mathbf{x}, t) = p_0 + p'(\mathbf{x}, t) \tag{12.1}$$

$$\rho(\mathbf{x}, t) = \rho_0 + \rho'(\mathbf{x}, t) \tag{12.2}$$

$$u_k(\mathbf{x}, t) = 0 + v_k(\mathbf{x}, t) \tag{12.3}$$

where $v_k(\mathbf{x}, t)$ is the minute fluid particle velocity due to the passage of sound waves.

It can easily be shown (e.g., Moran and Shapiro 1995) that the rate of propagation of an infinitesimal pressure pulse, the so-called speed of sound, in an otherwise still substance is given by

$$a_0^2 = \frac{\gamma_0}{\rho_0 \alpha_0} = \left(\frac{\partial p}{\partial \rho} \right)_s \approx \frac{p'}{\rho'} \tag{12.4}$$

where γ_0 is the ratio of specific heats, and α_0 is the isothermal compressibility coefficient (defined in Chapter 2), both of which are computed at the uniform ambient conditions. The derivative $\partial p / \partial \rho$ is computed at constant entropy. Because the entropy is constant, the pressure and all other thermodynamic properties can be considered as a function of a single variable; for example, $p = p(\rho)$ only. In other words, sound is a disturbance in which the pressure can be determined from a knowledge of the density alone. The sound speed is itself a thermodynamic property whose value depends on the state of the medium through which sound propagates. For an ideal gas, $\alpha_0 = 1/p_0$ and therefore $a_0 = \sqrt{\gamma_0 R T_0}$, and the sound speed depends only on the gas type and its temperature. Owing to the much higher pressures required for liquids to change their volumes, the speed of sound in liquids is generally higher than in gases and higher yet than in solids. For a bubbly liquid, however, the speed of sound can be much lower than in either the gas or liquid alone.

We now restrict the discussion to fluids only. To develop the equation governing the propagation of sound waves, we start with the exact conservation relations for a continuum fluid (Chapter 2). In a Eulerian frame, the continuity equation in differential form reads

$$\frac{\partial \rho}{\partial t} + \frac{\partial}{\partial x_k} (\rho u_k) = 0 \tag{12.5}$$

and Newton's second law reads

$$\frac{\partial}{\partial t} (\rho u_k) + \frac{\partial}{\partial x_i} (\rho u_i u_k) = \frac{\partial \Sigma_{ik}}{\partial x_i} + \rho g_k \tag{12.6}$$

[1] Each of the simplifying features implemented here can of course be violated under certain unusual circumstances. For example, viscous effects become important when sound propagates over very long distances, and sound levels of 194 dB generate a not-too-small pressure fluctuations of 1 atm. There is also the issue of sound generated aerodynamically being strong enough to back-react with the fluid motions producing it. Nonlinear acoustics and other unusual effects are beyond the scope of this chapter.

where ρ is the fluid density, u_k is an instantaneous velocity component, Σ_{ik} is the stress tensor (surface force per unit area), and g_k is the body force per unit mass. The stress tensor can be split into two parts, hydrostatic pressure and normal and tangential viscous stresses

$$\Sigma_{ik} = -p\,\delta_{ik} + \tau_{ik} \tag{12.7}$$

For the weak acoustic perturbations, we linearize Eqs. (12.5) and (12.6) and neglect viscous and gravity forces.[2] The resulting mass and momentum conservation equations read, respectively,

$$\frac{\partial \rho'}{\partial t} + \rho_0 \frac{\partial v_k}{\partial x_k} = 0 \tag{12.8}$$

$$\rho_0 \frac{\partial v_k}{\partial t} + \frac{\partial p'}{\partial x_k} = 0 \tag{12.9}$$

If the velocity perturbation is eliminated by taking $\partial/\partial x_k$ of Eq. (12.9) and the result is subtracted from $\partial/\partial t$ of Eq. (12.8),

$$\frac{\partial^2 \rho'}{\partial t^2} - \frac{\partial^2 p'}{\partial x_k \partial x_k} = 0 \tag{12.10}$$

and Eq. (12.4) is used to eliminate the density and to yield finally the wave equation for pressure perturbations

$$\frac{\partial^2 p'}{\partial t^2} - a_0^2 \frac{\partial^2 p'}{\partial x_k \partial x_k} = 0 \tag{12.11}$$

This is a hyperbolic partial differential equation that describes the propagation—at the constant speed of sound a_0—of pressure perturbations in an otherwise still fluid. Identical equations can be written for all the other small quantities describing the acoustic field; for example, density perturbations ($\rho' = p'/a_0^2$), velocity $\mathbf{v}$, and velocity potential ϕ ($p' = -\rho_0 \frac{\partial \phi}{\partial t}$).[3]

12.3 Sources of Sound

Sound is small pressure perturbations organized into waves—sound waves. If we insist that sound must not anticipate its cause or in other words that information about the current source activity should not be contained in past waves that anticipate the present,[4] the homogenous equation (12.11) in the absence of boundaries has

---

[2] The relevant Reynolds number is $2\pi a_0 \lambda/\nu$, and is of the order of 10^8 in air at the most audible frequency but is much higher in water. Hence, viscous forces can be safely neglected for sound propagation over distances much less than $2\pi a_0 \lambda/\nu$ wavelengths. The relevant Froude number square is $a_0^2/g\lambda$, and therefore gravity is neglected for sound waves with wavelength $\lambda \ll 12$ km (in air) or 200 km (in water), which is a condition that is readily satisfied for most waves.

[3] For one-dimensional sound waves, the particle velocity is given by $v = \pm p'/\rho_0 a_0$, and v satisfies the one-dimensional version of the wave equation.

[4] This important principle is called the *causality condition* and in simple fields is indistinguishable from the *radiation condition*, which states that in open space, only outward-traveling waves can have any real existence.

only one solution: $p'(\mathbf{x}, t) = 0$ everywhere. In other words, complete silence is the only homogeneous sound field in unbounded space. Obviously, sound must derive its energy from a source of some sort. Mathematically, a nonzero source term on the right-hand side of Eq. (12.11) makes the differential equation nonhomogenous and can force nontrivial solutions. But what are some of those sources? A sound source is a forcing term in the wave equation and occurs when the criteria leading to the homogenous equation are violated in some way. Having a source term in either the continuity or momentum equation, breaking down the state equation $p = p(\rho)$, having a body in relative motion with a fluid, or having velocity fluctuations sufficiently strong to break down the linearization criterion all violate the essence of the homogenous wave equation: an expression of the conservation laws in a weakly disturbed homogenous material in which $p = p(\rho)$ only.

Vibrating solid surfaces, machinery containing rotating blades, unsteady free-shear or wall-bounded flows, and unsteady expansion or contraction of fluid due, for example, to unsteady heating or cooling are all possible sources of sound. Determining precisely the source(s) of a particular sound is not, however, an easy problem—least of all because a particular source provides a unique sound field, but the converse is certainly not true (Dowling and Ffowcs Williams 1983). Analyzing a sound field with the most sophisticated instruments cannot tell us the precise nature of its source; this problem has no unique solution!

What are the mechanisms that cause the kinetic energy associated with the motion of a fluid or a solid to be converted to acoustic energy? Having a lot of pressure fluctuations in a low-speed, free-shear turbulent flow, for example, does not necessarily result in a strong sound field. In fact, a minuscule portion of the kinetic energy associated with the turbulence fluctuations is converted to acoustic energy and radiated as sound. Lighthill (1962) termed the pressure fluctuations in an incompressible turbulent flow pseudosound. These fluctuations are induced directly by fluctuating vortical motions in the flow proper. Though we can "hear" pseudosound via a suitably placed microphone, stethoscope, or even our own ears, these pressure fluctuations lack the essential quality of propagation at or around the characteristic sound speed. The turbulent eddies are convected with the flow itself, and within them the pressure fluctuations are balanced, for the most part, by fluctuations in fluid acceleration. Without the benefit of a surface that may vibrate in reaction to the unsteady pressure forces, a very small percentage of the pressure fluctuations radiates as pressure waves, satisfies the wave equation, and is responsible for what is termed aerodynamic sound—real sound. The radiated sound falls off as the inverse first power of the distance, and thus in the far field it dominates over the pseudosound, which falls off at least as the inverse square of the distance. But how does one estimate the true sound as a small percentage of the pseudosound? This and other questions posed in this section were not answered satisfactorily until the pioneering work of Lighthill published in 1952 in which he introduced a first-principle theory for sound generated aerodynamically.

Having an unsteady source of mass (ρ) that forces the amount of matter in a fixed region in space to fluctuate can generate sound effectively. This is the loudly efficient mechanism by which puffs of steam or compressed air generate sound in a fluidic siren. Less effectively, a bell's surface, when struck, generates sound by adding "puffs of momentum" to the adjacent air. One can think of the siren as being replaced by a hypothetical distribution of point mass singularities, or monopole sound sources. The vibrating surface introduces no new fluid and is equivalent to a distribution of dipoles, forcing the rates of momentum (i.e., mass flux ρu_k) in a fixed region in space to fluctuate. A point dipole involves a neighboring pair of equal point monopoles with opposite signs. Though the instantaneous sum of the two sources is zero, the net-sound field of a dipole is nonzero only because the sources at infinitesimally different positions radiate sound to the point $(\mathbf{x}, t)$ at slightly different retarded time. A dipole has both a magnitude and direction, and the destructive interference or cancellation involved makes it a less efficient radiator of sound than the omnidirectional monopole. The final sound source in this hierarchy is a distribution of quadrupoles that forces the rates of momentum flux $(\rho u_i u_k)$ across fixed surfaces to fluctuate. A quadrupole consists of two positive monopoles and two negative ones and thus it introduces neither new mass nor new momentum. No external force acts on the turbulent fluid because internal actions and reactions balance. In this case, the total dipole strength is zero, and the effects of different hypothetical dipoles nearly, but not quite, cancel out at large distances. A quadrupole is the least efficient source of the triad because it involves a double tendency for source elements to cancel and is the pathetically ineffective mechanism by which sound is generated in subsonic, free-shear turbulent flows.

To answer the question of how sound is generated aerodynamically, Lighthill (1952) considered the flowfield to occupy a small part of a very large volume of fluid of which the remainder is at rest (Figure 12.1). The portion that contains the intense velocity fluctuations provides the source region outside of which the wave equation (12.11) is valid. To illustrate Lighthill's acoustic analogy, we seek the difference between the exact statements of the natural laws—valid in the flow region—and their acoustical approximations—valid in the *far field* or the still fluid region. Then the equations governing the fluctuations of density in the real fluid are compared with those that would be appropriate to a uniform acoustic medium at rest. The difference between the two sets of equations is considered as a hypothetical distribution of sound sources.

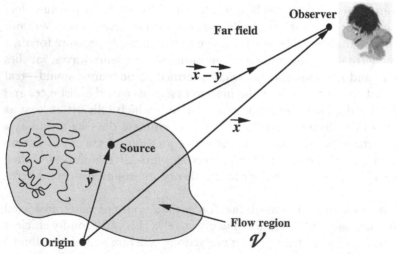

Figure 12.1: Sound emitted at $\mathbf{y}$ is heard by the observer at $\mathbf{x}$ at a time $\frac{|\mathbf{x} - \mathbf{y}|}{a_0}$ after emission.

We start with the exact expressions for the conservation of mass and momentum, the latter without body forces. Taking $\partial/\partial x_k$ of Eq. (12.6) and subtracting the result from $\partial/\partial t$ of Eq. (12.5), we get

$$\frac{\partial^2 \rho}{\partial t^2} = \frac{\partial^2 \rho'}{\partial t^2} = \frac{\partial^2}{\partial x_i \partial x_k}(\rho u_i u_k - \Sigma_{ik}) \tag{12.12}$$

Subtracting

$$a_o^2 \frac{\partial^2 \rho'}{\partial x_k \partial x_k} = \frac{\partial^2}{\partial x_i \partial x_k}\left(a_o^2 \rho' \delta_{ik}\right) \tag{12.13}$$

from both sides of Eq. (12.12), we reach the desired inhomogenous equation for sound

$$\frac{\partial^2 \rho'}{\partial t^2} - a_o^2 \frac{\partial^2 \rho'}{\partial x_k \partial x_k} = \frac{\partial^2 T_{ik}}{\partial x_i \partial x_k} \tag{12.14}$$

where $T_{ik} = \rho u_i u_k - \tau_{ik} + p\delta_{ik} - a_o^2 \rho' \delta_{ik}$, the Lighthill stress tensor, and we have used Eq. (12.7).

In effect, we have rearranged the governing equations to isolate a linear wave operator, the remaining terms being treated as a source that drives the wave motion. The source field for the acoustic waves is a quadrupole distribution whose instantaneous strength per unit volume is the tensor T_{ik}. The double divergence structure of the right-hand side of Eq. (12.14) indicates a double tendency for destructive interference and reveals the quadrupole nature of the source term. Contributions to the instantaneous Lighthill stress tensor come from normal and tangential Reynolds stresses, normal and tangential viscous stresses, and the difference between the actual hydrostatic pressure in the active flow region and that which would prevail in a uniform acoustic medium at rest. In the sound field proper, $p' \approx a_o^2 \rho'$, the weak flow is inviscid and the Reynolds stress is second order in the small-velocity fluctuations. In other words, $T_{ik} \approx 0$ outside the active flow region. Inside though, strong heating or cooling, pressure forces on a rigid body moving relative to the flow, viscous stresses, and turbulent fluctuations can all be sources of aerodynamic sound. The density fluctuations in the real flow are exactly those that would occur in a uniform acoustic medium subject to an external stress system given by the difference between the effective stresses in the real flow and the stresses in the uniform acoustic medium at rest.

Consider an observer located in the far field at $\mathbf{x}$, the solution of Eq. (12.14) reads

$$\rho'(\mathbf{x}, t) = \frac{\partial^2}{\partial x_i \partial x_k} \iiint_V \frac{T_{ik}\left(\mathbf{y}, t - \frac{|\mathbf{x}-\mathbf{y}|}{a_o}\right)}{4\pi a_o^2 |\mathbf{x} - \mathbf{y}|} \, d\mathcal{V} \tag{12.15}$$

The integration is carried out for all sources located in the flow region at $\mathbf{y}$. Every quadrupole element at $\mathbf{y}$ generates a field that travels out at the speed of sound, reaching the observer at $\mathbf{x}$ at a time $|\mathbf{x} - \mathbf{y}|/a_o$ later (Figure 12.1). The effect of each source element falls off inversely as the distance traveled $|\mathbf{x} - \mathbf{y}|$.

The integration above can in principle be carried out if the Lighthill stress tensor T_{ik} is known. For incompressible turbulent flows, this tensor is dominated by the instantaneous Reynolds stress $\rho\, u_i u_k$, a random quantity that in general is unknown analytically and is very difficult to measure. Therefore, heuristic models, based on a combination of experimental data, computations, and theoretical concepts, or simple dimensional analysis is used to estimate the sound field. For example, Lighthill (1952, 1954) estimated the mean-square value of the radiated sound field from a cold subsonic jet to be

$$\overline{\rho'^2} \sim \rho_o^2\, Ma^8 \frac{D^2}{|\mathbf{x}|^2}, \quad Ma \ll 1 \tag{12.16}$$

where Ma is the Mach number based on the jet-exit velocity; the ambient sound speed, D, is its diameter; and $|\mathbf{x}|$ is the distance between the jet exit and the observer. The small Mach number makes it possible to assume a compact source ($D \sim \lambda\, Ma \ll \lambda$, where λ is the characteristic wavelength of the generated sound) and greatly simplifies the dimensional analysis. Basically, for a compact source, the variation of retarded time as y ranges over all the source volume in the integral (12.15) is ignored, in effect making the source distribution equivalent to a point quadrupole.

Mercifully, less than 0.01% of the turbulence kinetic energy is converted to acoustic energy, and this percentage diminishes as the Mach number drops further from unity.[5] At low Mach numbers, jets as well as all compact sources are acoustically very inefficient, and $\overline{\rho'^2} \sim Ma^8$. The presence of a rigid surface, of much higher density than the surrounding fluid,[6] in the flow proper is equivalent to a distribution of dipoles and increases the sound-generation efficiency so that $\overline{\rho'^2} \sim Ma^6$. For flexible bodies capable of dilating, the resulting monopole singularities lead to an even more efficient sound generation, and $\overline{\rho'^2} \sim Ma^4$. And finally, for compressible turbulent jets, the source region can no longer be considered compact, the structure of the sound field is dominated by retarded time variations over the source region, and the number of eddies that can be heard at any one time increases linearly with the Mach number. In this case, the mean-square value of the radiated sound field is estimated to be a much higher $\overline{\rho'^2} \sim Ma^3$. These dimensional analysis results are well confirmed with experiments.

The Lighthill acoustic analogy provides reasonable predictions for subsonic free-shear flows and for noise mechanisms associated with the nearly steady aerodynamic loading of turbomachinery blades. The alternative stochastic wave model more recently developed by Tam (1995) allows direct calculations of sound radiation without recourse to the acoustic analogy. The semiempirical stochastic model theory has been successful in predicting jet noise when Mach waves are radiated (as a result of the supersonic convection speeds of large-eddy structures) or when broadband shock-associated noise results from the passage of large-scale structures through the shock-cell pattern in imperfectly expanded supersonic jets. We will return to supersonic jet noise mechanisms in the next section where, armed with knowledge of aerodynamic sound sources, we discuss methods to suppress flow-induced noise for both low-speed and high-speed flows.

---

[5] The acoustic efficiency η is the ratio of acoustic power radiated to jet power delivered. For subsonic jets, $\eta \sim 10^{-4} Ma^5$.

[6] Nekton, having about the same density as their surrounding fluid, swim in water outstandingly more quietly than the heavier-than-air avian species can fly in air.

12.4 Noise Control

Noise suppression involves the reduction of noise sources' efficiency and ability to convert kinetic energy to sound power, the interruption of sound transmission, the accelerated dissipation of acoustic energy into heat, or the active cancellation of sound waves using out-of-phase waves. These strategies can be implemented by a variety of passive, active, and reactive devices ranging in complexity from simple ear plugs to sophisticated antisound systems.

Unhealthy noise pollution is easier to quantify than what is considered a mere annoyance. Unprotected exposure to sound levels above 110 dB can result in permanent hearing loss. Sound that merely irritates, on the other hand, is obviously a subjective matter at any level. For most people, save teenagers, sound levels above 90 dB are uncomfortable and distracting. Noise containing discernible pure tones is, however, more likely to invoke a negative response than broadband noise of the same level. The screech tones radiating from imperfectly expanded supersonic jets can be maddening.

Noise control, a relatively young field of research, is required to maintain an acceptable or even pleasant environment, to avoid detection of certain aquatic vessels, to facilitate the proper operation of sonars on underwater vehicles, and finally to prevent violent structural vibrations. Heckl (1988) broadly classified noise suppression methods into four categories:

- **Commonsense application:** speed reduction, load reduction, increase of mass, general purpose enclosures, mufflers, screens, earmuffs.
- **Add-on measures:** multiple walls and floors, including flanking transmission, tuned absorbers and vibration coatings, elastic foundations, impedance mismatches in the propagation paths, and shift of resonances away from exciting frequencies.
- **Changes at the source:** reduction of time derivatives, for example, in hammers, punch presses, exhausts, pumps, and valves; interruption of feedback loops or other instabilities (wheel squeal caused by stick-slip, self-excitation in burners, whistles); reduction of roughness and free clearances (rolling noise, piston slap).
- **Active methods of control:** sound reflection and absorption in waveguides (ducts, beams), helmets with active noise control, interruption of feedback loops or other instabilities such as in burners, compensation of periodic excitation.

In the next six subsections, we will very briefly discuss several aspects of the art and science of noise suppression. Once again, we will merely scratch the surface, and the interested reader is encouraged to consult the books and journals listed at the end of Section 12.1.

12.4.1 Relation to Other Flow-Control Goals

We have established in Section 12.3 that turbulence is a major source of flow-induced noise—at least for cold subsonic flows. It is logical, therefore, that turbulence suppression, particularly at the noise-inducing scales, reduces the levels of radiated sound. Delaying laminar-to-turbulence transition, suppressing turbulence—particularly its tangential Reynolds stresses—or, in the extreme, relaminarization can each help to reduce noise levels. Controlling the mixing processes or targeting particular coherent structures can also modulate the sound field favorably or adversely. Strategies to achieve any of these goals are thoroughly discussed throughout this book. If turbulence is to be enhanced, for example, to prevent flow separation, the adverse effect of that control on noise levels has to be scrutinized. Conversely, if noise is to be controlled by suppressing turbulence, the trade-off in terms of drag or lift penalty has to be considered carefully. Occasionally, suppressing noise simultaneously leads to additional desired flow traits. For example, delaying laminar-to-turbulence transition leads to lower noise levels as well as lower skin-friction darg. As a second example, using antisound to reduce combustion-generated noise can also eliminate undesired flow instabilities. Lastly, for subsonic turbulent flows, weakening the fine-scale structures is beneficial to noise reduction, whereas breaking up the large eddies in supersonic flows leads to lower noise intensities.

As seen in Section 12.3, turbulence generates sound equivalently to a quadrupole source distribution. Organized structures in a turbulent flow play an important role in the source process and, therefore, controlling these eddies may be the key to noise suppression (Ffowcs Williams 1977). At the low Mach number typically encountered in underwater applications ($\mathcal{O}[0.01]$), unbounded turbulent flows have an extremely low acoustic efficiency according to Lighthill's (1952) theory. The fine-scale eddies are responsible for most of the radiated sound. The large differences between the noise levels predicted by the quadrupole theory and those actually generated by a turbulent boundary layer are often explained via acoustic scattering processes (Crighton 1984). Small rigid bodies like ribs, rivet heads, edges and angles, gas bubbles, or sharp edges (Blake and Gershfeld 1989) can destroy the cancellation between opposing elements of the quadrupole and may lead to the more efficient radiation of a dipole or even a monopole. Thus, to reduce noise, these efficient sound generators should be avoided as much as practical.

A navy's ability to conceal its submarines or to have reliable underwater communications is a crucial military advantage. In contrast to air, water transmits sound waves far better than it does electromagnetic waves like visible light or radio waves. Therefore, sound is used extensively for underwater communications. Sonar in water plays a similar role to radar in air. For water vessels, sonars are used for navigation and detection. There are at least two military advantages to suppressing the noise sources of aquatic vessels. First, noise radiated from a submarine, say, as a result of its hull vibrations, its power plant, its propulsion system, or any other sound source makes the vehicle less quiet and hence more susceptible to detection by enemy sonars. Second, if a sonar's transmission and receiving elements are buried beneath a turbulent boundary layer, the hydrophones receive the pseudosound resulting from the pressure fluctuations in the turbulent flow along with any legitimate incoming signal. The signal-to-noise ratio is lowered in this case, and the sonar's ability to detect enemy vessels and incoming torpedoes as well as to receive friendly communications is hampered as a consequence.

Mostly because of the latter problem, sonars are typically placed at the nose of submarines where the boundary layer is laminar and the pseudosound minimal. Search rates of these listening devices can be drastically improved by placing additional hydrophones farther downstream. However, the turbulence-induced noise is sufficient to interfere with the proper operations of these devices, and noise must therefore be suppressed before the search rate can be improved. A second example of the undesired side effects of turbulence comes from the aeronautics field. On commercial aircraft, the turbulent boundary layer outside the fuselage contributes significantly to cabin noise. Techniques to reduce flow-induced noise are therefore sought. Methods to delay transition (Chapter 6) or to relaminarize an already turbulent boundary layer (Chapter 10) should be useful to suppress the turbulence-induced noise in both of the preceding examples. It is feasible to delay transition on both vehicles, where the unit Re is about 10^7/m, for the first few meters. Beyond that, relaminarization or at least partial suppression of turbulence is the only option. As indicated in Section 10.5, achieving relaminarization over the entire length of an aircraft or a submarine may require large energy expenditure and is, therefore, not very practical. However, it may not be necessary to relaminarize an entire vehicle. Instead, it may be more feasible to target selected portions or even selected coherent structures of the boundary layer where, for example, a hydrophone is to be placed.

12.4.2   Passive Noise Control

Passive noise control devices include shields, rigid and compliant walls, mufflers, silencers, resonators, and absorbent materials. These can be placed around a noisy machine or used along an exhaust pipe. They work by reducing the acoustic energy transmitted to the environment either by reflecting it back or by dissipating part of the energy into heat. Radiated sound is also reduced when vibrations are suppressed using, for example, a properly designed foundation to isolate a machine from a concrete floor or an energy-absorbent viscoelastic liner to wrap around an entire machine or vehicle.

When a solid wall interrupts sound propagation in air, part of the acoustic energy is reflected, part is absorbed, and part is transmitted. The sound pressure incident on the wall causes the solid to vibrate, and the fraction of incident sound energy transmitted depends on the relative impedance (product of density and sound speed) of the wall and fluid. A wall constructed of dense material tends to be most effective in reducing noise transmission; the major part of the incident acoustic energy is either reflected or dissipated into heat owing to impedance mismatch, absorption, and damping. Generally, passive sound suppression devices work best to shield from, and to discourage, the transmission of high-frequency (short-wavelength) sound waves.

To illustrate this last statement, consider a planar sound wave incident perpendicular to a rigid plane wall of mass per unit area m. The fluid on either side of the wall has the same density ρ_0 and sound speed a_0, and the incident sound radian frequency is ω. The wall material has higher density and sound speed than the fluid. The aim is to compute the portion of the incident acoustic energy that would be reflected and the portion that would be transmitted through to the other side of the wall. The boundary conditions are continuity of the fluid and solid particle velocities on either side of the wall and a pressure difference across the wall that provides the force necessary to accelerate a unit area of the wall material. The particle velocity condition on the incident side of the wall immediately implies that both the transmitted and reflected waves have the same frequency as the incident wave but not the same wavelength. The velocity and pressure boundary conditions can easily be used together with the conventional relations for one-dimensional waves between pressure, particle velocity, and acoustic energy to give the following expression for the energy transmission coefficient ETC (see, for example, Dowling and Ffowcs Williams 1983):

$$\text{Energy transmission coefficient} \equiv \frac{\text{Energy transmitted through the surface}}{\text{Energy of the incident beam}}$$

$$ETC = \frac{4\rho_0^2 a_0^2}{4\rho_0^2 a_0^2 + \omega^2 m^2} \tag{12.17}$$

It is clear that very little high-frequency sound is transmitted through solid walls; most of it is reflected. In contrast, low-frequency waves travel through walls with very little attenuation. The proper dimensionless parameter is

$$\frac{\omega m}{\rho_0 a_0} \gg 1 \implies \text{sound waves are mostly reflected}$$

$$\frac{\omega m}{\rho_0 a_0} \ll 1 \implies \text{sound waves are mostly transmitted}$$

Low-frequency sound gets through all but the most massive walls.

Sound-absorbent coatings are made either of porous material or perforated material and are often used to dissipate acoustic energy. In the case of porous material like foam or expanded polystyrene, the passage of sound waves causes the air inside the pores to vibrate. The extremely narrow channels cause a large resistance to the fluid motion and the dissipation of sound energy into heat. Additional energy losses result from the heat exchange between the heated, compressed air or the cooled, rarefied air inside the pores and the solid skeleton. Perforated materials typically consist of a porous surface glued to a honeycomb structure and are designed to create beer-bottle-like geometries (i.e., numerous Helmholtz resonators). There, resonant-frequency pressure perturbations at the neck produce large velocities in the resonator. Again, large velocity in a narrow passage results in considerable viscous dissipation and reduction of acoustic energy. The depth of the perforations and the volume of the backing cavities can be optimized to resonate in the middle of the frequency range, where sound attenuation is desired.

Mufflers are typically used to quiet exhaust pipes in internal combustion engines. A muffler usually consists of inlet and outlet pipes of equal diameters and an expansion chamber between them. The area ratio and the length of the expansion chamber determine the muffler transmission loss, and the chamber length is tuned so that maximum attenuation occurs at the dominant frequencies of the incoming sound. Silencers are similar devices used in ducts or fluid intakes in industrial applications. The sound energy in mufflers and silencers is conserved. Any reduction in the transmitted wave energy must be coupled to a corresponding increase in the reflected energy. Both passive devices typically employ a combination of sound-absorbent materials designed to convert the high-frequency component of the noise into mechanical vibration and heat and destructive interference between the transmitted and reflected waves in the expansion chamber for the lower range of frequencies. The sound-absorbent material results in a net reduction in acoustic energy. Mufflers and silencers can also utilize Helmholtz resonators attached to the side of the duct and optimized to suppress noise in a narrow range of frequencies.

12.4.3 Compliant Coatings

Fluid–structure interaction is particularly important in underwater noise radiation owing to the relatively high density and bulk modulus of water. In contrast to a rigid wall, a compliant wall is sufficiently *soft* to deform elastically as a result of the pressure and viscous forces of a moving fluid. From a fundamental viewpoint, a rich variety of fluid–structure interactions exist when a fluid flows over a surface that can comply with the flow. The transverse wave speed in the solid (square root of the ratio of the wall modulus of rigidity and its density) relative to a characteristic flow speed determines the degree of compliance of the wall; a value below one indicates a compliant wall. Not surprisingly, instability modes proliferate when two wave-bearing media are coupled. Some waves are flow-based, some are wall-based, and some are a result of the coalescence of both kind of waves. A compliant coating can in principle be designed to achieve desired flow traits such as transition delay, noise suppression, and lower skin-friction drag in turbulent flow. The topic is thoroughly discussed in Chapter 7.

A compliant coating that causes transition delay or reduction in turbulence skin friction will also lead to attenuation of sound radiated by the boundary layer. This is because the wall-pressure fluctuations are dependent on turbulence levels, which in turn are related to the wall shear stress. The reverse is not necessarily true (i.e., a coating may attenuate the flow noise without affecting the hydrodynamic drag). In fact, the technology exists today for manufacturing energy-absorbent compliant liners for sound absorption, vibration reduction, and noise shielding, whereas the search for drag-reducing coating—particularly for turbulent flows—has thus far eluded researchers for about 50 years (Riley et al. 1988; Gad-el-Hak 1996a). As mentioned earlier, a submarine hull vibrates as a result of the surface forces in the surrounding flow or as a result of vibrations in other parts of the vessel such as its power plant. The hull vibrations are an effective source of radiated sound and can readily be detected by sonars far away. A properly designed compliant coating can act as an effective attenuator of hull vibrations and therefore reduce the likelihood of detection.

Flow noise is influenced by surface flexibility through two distinct mechanisms: either the surface acts as a sounding board excited by the turbulent pressure field or the surface compliance induces a change in the turbulence structure and, hence, modifies the pressure fluctuations (Ffowcs Williams 1965; Purshouse 1976; Dowling 1983, 1986). As mentioned earlier, flow-induced noise is currently considered the limiting performance factor for sonar systems placed on surface ships, submarines, and towed arrays. Major advances in the reduction of "self-noise" have been achieved by exploiting the first mechanism above, and further reductions may be possible if the nature of the turbulent boundary layer and its wall-pressure fluctuations can be altered. Von Winkle (1961) and Barger and von Winkle (1961) reported pressure fluctuation measurements on a streamlined body of revolution free-falling in a water tank. Flush-mounted hydrophones fabricated from lead-zirconate titanite were used to measure the instantaneous pressure. Their experiment indicated a dramatic reduction in the pressure coefficient when the body was covered with a Kramer-type compliant coating (see Chapter 7). This result may, however, have been due to transition delay caused by the flexible surface and not the changes in the turbulence structure.

12.4.4 Jet Noise

The quest to power commercial aircraft with ever more powerful jet engines has, in large measure, influenced noise suppression research for the last half century. Inspecting Eq. (12.16) reveals the strong dependence of the noise levels from cold subsonic jets on the Mach number and a weaker dependence on the jet diameter. This leads to what is by far the most powerful strategy to reduce jet noise: by reducing the jet velocity with a reciprocal increase in diameter to maintain a given thrust, the resulting noise reduction is proportional to the sixth power of velocity. The acoustic efficiency is the ratio of acoustic power radiated to jet power delivered and varies as Ma^5. By means of the downward trend in Mach number, jet engines of ever larger diameter have permitted an increase in engine power to be delivered at community-acceptable noise levels. Even at a prescribed thrust, there are of course weight and drag penalties when larger engines are utilized, and these have to be weighed carefully against the benefit (or necessity) of lower noise levels.

Turbojet engines with bypass reconcile the need to achieve high Carnot efficiency while maintaining a good Froude efficiency, so to speak. Jet engines were designed as early as the 1950s with a 40% bypass ratio. Part of the energy of the gas coming out of the combustion chamber was used to impel a surrounding additional 40% of the air mass, which bypassed the combustion process altogether. For a given thrust and engine efficiency, the sound energy was reduced at this bypass ratio by a factor of 5 (7 dB). For even more gain, higher bypass ratios are sought in modern designs.

Though they work in practice, there are some questions about the exact reasons why high-bypass-ratio engines have less noise than conventional engines. There is a reduction in the global Mach number of the jet and a concomitant reduction in radiated noise. Additionally, bypass air reduces the shear magnitude between the high-speed primary air and the freestream air. This in turn reduces the turbulence energy production and thus radiated sound. Because of the difference in temperatures of the hot primary air, the cooler bypass air, and the cold freestream, there is, however, a strong shear in Mach number (but not velocity) between the primary and bypass air and a strong shear in velocity (but not Mach number) between the bypass and freestream air. It appears that the Mach number is more critical to the sound generation, but the detailed measurements have not been done yet to pinpoint the sources of the sound. Therefore, at present the improvements in sound levels are semiempirical without knowledge of the precise causes.

There are two other techniques to reduce noise from subsonic jets, often utilized simultaneously. Neither strategy is suited for high-bypass-ratio engines. The use of either a multinozzle noise suppressor (e.g., Pratt & Whitney engines powering some Boeing 707s) or deeply corrugated jet exit (e.g., Rolls–Royce engines powering some Boeing 707s) benefits from both the so-called shielding principle and a reduction of the quadrupole intensity. The shielding principle (Lighthill 1962) states that if an observer is listening to a jet noise and is placed near the peak of its directional distribution, then the presence of another jet behind the first, shielded from his or her line of direct observation, does not significantly increase the noise heard. The reason is the scattering of the far jet noise by the near jet flow, which causes a directional redistribution of the noise from the far jet and shifts its direction of peak intensity. In effect, the far jet sound is heard at a directional average, not at a peak value. In addition to the shielding effect, both the multinozzle jet and corrugated-exit jet result in a shorter extent of the mixing layer, bringing the adjustment region where turbulence intensities begin to decrease closer to the orifice, and a reduced relative shear between the jet and the ambient air, which suppresses the turbulence intensities in the mixing region. These advantages effectively reduce the quadrupole intensity and thus the radiating sound. The total noise reduction in both the Pratt & Whitney and Rolls–Royce engines is around 10 dB, whereas the weight, thrust, and drag penalties resulting from the special noise-suppression designs are each around 0.7%. These penalties reduce the aircraft height above a typical community-noise measuring point, adversely affecting the radiated noise by less than 1 dB.

Both fine-scale and large-scale coherent structures in turbulent flows are capable of generating noise. However, the relative importance of the noise they produce depends on the jet Mach number and fluid temperature. Noise from cold subsonic jets is caused primarily by the small-scale structures. Unless the jet is very hot,[7] the turbulence convection Mach number, relative to the ambient sound speed, is below one, and the large eddies are ineffective noise generators. In contrast to subsonic jets, large-scale structures are the primary source of noise in hot supersonic jets. In this case, the large turbulence structures convect downstream at supersonic Mach number relative to the ambient sound speed. The supersonic eddies are capable of producing intense Mach wave radiation, which easily predominates over the noise from fine-scale turbulence. The resulting broadband sound is called turbulence mixing noise, and it radiates in the downstream direction. Tam's (1995) stochastic instability wave model treats the large coherent structures as being statistically equivalent to the instability waves of the jet. These waves in turn are considered as a wavy wall—having the same wavelength and wave speed as the instability wave—above which the supersonic flow convects and generates intense Mach (sound) waves that propagate in a preferred downstream direction. The Mach angle can be computed readily from the phase velocity c_p of the most amplified instability wave and the ambient sound speed a_0: $\theta = \sec^{-1} Ma_c = \sec^{-1} \frac{c_p}{a_0}$, where Ma_c is the convective Math number.

Most supersonic jets are imperfectly expanded, and this leads to the formation of a pattern of oblique shocks or expansion fans and two additional mechanisms for noise generation. For an underexpanded jet, an expansion fan is initiated at the nozzle lip to allow the gradual reduction of static pressure to ambient conditions. For an overexpanded jet, an oblique shock is formed at the lip to allow the abrupt increase in static pressure to match that outside the jet. Once formed, the expansion fan or shock wave is convected downstream until impinged on the mixing layer on the other side. The subsonic flow outside the jet cannot support either a shock or an expansion fan, and hence both have to reflect back into the jet plume. This process is repeated many times, forming a quasi-periodic pattern of shocks or expansions fans. The shock-cell pattern may be regarded as disturbances trapped inside the jet by the mixing layer at the peripheral. The jet flow behaves as a waveguide to the disturbances forming the shock pattern, which in turn leads to the two shock-associated sources of noise described below.

Both broadband shock noise and discrete screech tones result from the passage of large-scale structures through the shock-cell pattern. In contrast to the turbulence mixing noise, the two shock-associated sounds radiate primarily in the upstream

[7] The jet Mach number based on the local sound speed in the hot fluid is lower than the acoustically more relevant Mach number based on the sound speed in the cold ambient fluid.

direction. The broadband shock-associated noise is produced as a result of the constructive scattering of the large eddies by the shock cells in the jet plume. Because the quasi-periodic shock pattern can be conceived as made up of a superposition of waveguide modes of different wavelengths, each mode scatters sound in a preferred (upstream) direction. The screech noise is due to an acoustic feedback phenomenon. Here acoustic disturbances impinging on the nozzle lip—where the mixing layer is thin and receptive to external excitation—excite the intrinsic instability modes of the jet. The resulting waves convect downstream, gain strength, and interact with the aforementioned shock-cell pattern. The unsteady interaction generates acoustic waves that propagate upstream outside the jet and upon reaching the lip region close the feedback loop. The resulting sound is intense at a particular fundamental frequency, the screech tone frequency, and can sometimes be heard up to the fourth or fifth harmonic.

Noise radiated from supersonic jet engines cannot be reduced by simply increasing their size and thus reducing their Mach number, as done in subsonic jets. Extrawide supersonic engines would generate unacceptable shock-wave drag as well as supersonic-boom emissions. Systems to suppress noise in supersonic jets are generally more complex than their subsonic counterparts, making noise control a formidable barrier to the successful development of supersonic as well as hypersonic commercial aircraft. Improved mixing in supersonic jets breaks up their large-scale structures and has invariably the beneficial side effect of substantial noise reduction. Enhanced mixing devices include ejectors, taps, acoustically treated mixers and ejectors, and unconventional nozzle geometries. Rectangular and triangular jets radiate less noise than circular jets at high subsonic and supersonic speeds (Ahuja, Hayden, and Entrekin 1993; Rice and Raman 1993). Jet engine exhausts are *scalloped* to reduce the scale of their coherent structures and hence reduce the radiated noise. An active strategy to accomplish a similar feat involves forcing the jet acoustically to cause the natural train of unequally spaced vortex rings to become more uniform. The formation of larger scales is thus delayed, leading to a reduction of the jet spreading rate as well as radiated noise (Hussain and Hasan 1985). To reduce the shock-associated broadband and screech noise, researchers have attempted to produce shock-free supersonic plumes with nonaxisymmetric distributions of exit-plane momentum thickness or to create jets that are excited by oscillating control devices or by feedback of time-dependent disturbances within the initial jet column.

12.4.5 Turbomachinery Blades

The noise radiated from fans, propellers, turbofans, propfans, turbine blades, and the like is dominated by a series of pure tones whose fundamental is the basic rotational frequency multiplied by the number of blades. Any fluctuations generated by the turbulent flow itself are merely the high-frequency noise background to the pure tones and their harmonics. The turbomachine thrust and torque result from the blades' periodic forces acting on the fluid. These forces are acoustically equivalent to a distribution of dipoles. The turbomachine disk diameter is typically comparable to the generated acoustic wavelength, and thus the sound source cannot be treated as compact, thus complicating the use of the acoustic analogy to compute the radiated noise.

Upstream nonuniformities are caused by atmospheric gusts, inlet distortions, blade tip vortices, rotor-stator wake interactions, interactions between blade rows, and so forth. In the blade frame of reference, these nonuniformities appear as convected vortical waves and induce unsteady forces on the blade surfaces, causing forced vibrations and radiation of additional broadband as well as harmonic noise components. If the frequency of this aerodynamic excitation matches the turbomachine natural frequency, the forced vibration can be destructive. The high-frequency, broadband noise due to vortex shedding and lift fluctuations on high-speed blades is even more problematic to compute as compared with the pure tones. In hydrodynamic applications, the length scale associated with blade-row interactions is small compared with the wavelength of the radiated sound, and the source can therefore be treated as acoustically compact, making the use of the Lighthill's analogy convenient.

As in jets, here also the objective is to reduce the noise levels without compromising the machine performance measured, for example, in terms of thrust per unit of airflow (specific thrust) or thrust per unit of engine weight. Making the turbomachine larger while reducing the rotational frequency of its blades does lead to a reduction of radiated noise but adds weight, space, and drag penalties. Noise related to the fluctuating forces on the blades can be suppressed by reducing those fluctuations or by attenuating the radiated sound through the engine intake. McCune and Kerrebrock (1973) maintained that the perceived noise can be reduced by (a) reducing the radiated power, (b) shifting the radiated power to a frequency at which the ear is less sensitive or where attenuation is more rapid, and (c) changing the directivity pattern. They provided useful analysis of the performance penalties associated with noise reduction in turbomachines.

12.4.6 Antisound

In addition to passive devices, noise control can also be accomplished actively or reactively. Both of these strategies require energy expenditure. An example of predetermined active control is the forced oscillations applied to a jet exit to equalize the spacing between successive vortex ring instabilities (see Section 12.4.4). A second example of active control is the beaming of a continuous broadband noise in an office space to mask other intrusive sounds such as conversation in an adjacent cubicle. The result of masking is an *increase* in the noise level, but the subjective effect may be a reduction in annoyance. Reactive sound control is often termed antisound and is designed to reduce the noise levels, but the reactive system is quite a bit more involved than a mere predetermined active control system.

Reactive noise suppression systems are based on generating sound by auxiliary source with such an amplitude and phase that, in the region of interest, the sound wave interference from the original and auxiliary source results in considerable reduction of the noise levels. What makes these systems possible is, of course, the linearity of the governing equations and the consequent validity of the principle of superposition. Though the idea can perhaps be traced to the 1950s, significant advances in antisound were not made until the 1980s by, especially, Ffowcs Williams (1984). In principle, it is possible to cancel an existing sound field by introducing a new sound field of equal amplitude and opposite phase. The elements of an antisound system consist of sensors to measure the existing field, acoustic sources to produce the antisound, and microprocessor-based controllers that utilize inputs from the sensors to determine the appropriate output signals. These systems work best to cancel low-frequency sound. For higher-frequency sound, additional acoustic modes can propagate as well as disperse, complicating the application of the antisound concept.

Fortunately, passive devices work well at high frequencies, and a combined system may be able to suppress most of the noise.

The earliest application of antisound involved the reduction of fan noise in ducts. We cite here a few more recent examples. Rotating stall instability in axial compressors has been successfully controlled by active methods. Delay of surge in gas turbine engines has been demonstrated, allowing a significant increase in power beyond the surge boundary for uncontrolled engines. The fluttering of a flexible wing in a wind tunnel has been successfully stabilized by activating a wall-mounted loudspeaker. Other existing or potential applications include reduction of cabinet noise in aircraft and automobiles, combustion chambers in jet engines, electric power transformers, diesel engines, electric motors, pumps, compressors, and other turbomachinery. The pilots of the *Voyager* aircraft, which completed the first nonstop, nonrefueled flight around the world in December 1986, used an active-noise-control communications headset system.

The older antisound systems with fixed gain in the feedback loop could not achieve high noise reduction. However, with the recent availability of adaptive filter systems, much more impressive suppression of noise is feasible (Tichy, Warnaka, and Poole 1984). These systems are capable of adjusting the feedback loop for the magnitude and phase relationship of the spectral components and can quickly compensate for the sound path changes. These reactive-control devices seem to work best for low-frequency sound, for which passive silencers are relatively ineffective, and when the source is localized and accessible. Local control is obviously easier than global. Moreover, effective control is achieved when the system response within the frequency band of interest is dominated by relatively few modes.

Combustion chambers often act as acoustic resonators, and attempts to increase their density of energy release can lead to undesired combustion instabilities that can be violent enough to quench the flame or even to cause structural damage. When low-energy antisound systems were applied to combustion chambers to reduce the noise generated by the combustion instabilities, both the sound field and those buzz instabilities were suppressed, dramatically improving the chamber performance. It is conceivable that properly designed antisound systems can repeat this success in other fluid-flow situations in which instabilities contribute to the radiated sound as well as limit performance. Review papers on active control of noise are available by Warnaka (1982); Ericksson et al. (1988); Van Laere and Sas (1988); and Warner, Waters, and Bernhard (1988). Antisound techniques seem to have gone from the laboratory to application and commercialization in a remarkably short time.

CHAPTER THIRTEEN

Microelectromechanical Systems

> *If nature were not beautiful, it would not be worth studying it. And life would not be worth living.*
> (Jules Henry Poincaré, 1854–1912)

> *Everything should be made as simple as possible, but not simpler.*
> (Albert Einstein, 1879–1955)

PROLOGUE

This chapter provides background material on an essential element of the targeted flow-control strategy to be detailed in Chapter 14. Manufacturing processes that can create extremely small machines have been developed in recent years. Microelectromechanical systems (MEMS) refer to devices that have characteristic length of less than 1 mm but more than 1 μm, that combine electrical and mechanical components, and that are fabricated using integrated circuit batch-processing techniques. Electrostatic, magnetic, electromagnetic, pneumatic, and thermal actuators, motors, valves, gears, cantilevers, diaphragms, and tweezers of less than 100-μm size have been fabricated. These have been used as sensors for pressure, temperature, mass flow, velocity, and sound; as actuators for linear and angular motions; and as simple components for complex systems such as microheat engines and microheat pumps. Many of these microsensors and microactuators are potentially very useful for futuristic distributed control systems that are particularly suited for turbulent flows, as will be discussed in the next chapter. In the present chapter, we review the status of our understanding of fluid-flow phenomena particular to microdevices. In terms of applications, this and the following chapter emphasize the use of MEMS as sensors and actuators for flow diagnosis and control.

13.1 Introduction

How many times when you are working on something frustratingly tiny, like your wife's wrist watch, have you said to yourself, "If I could only train an ant to do this!" What I would like to suggest is the possibility of training an ant to train a mite to do this. What are the possibilities of small but movable machines? They may or may not be useful, but they surely would be fun to make.

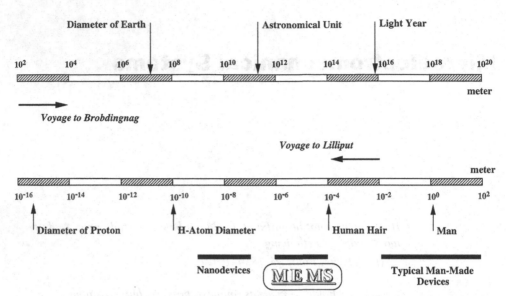

Figure 13.1: The scale of things in meters. Lower scale continues in the upper bar from left to right.

(From the talk "There's Plenty of Room at the Bottom," delivered by Richard P. Feynman at the annual meeting of the American Physical Society, Pasadena, California, 29 December 1959.)

Toolmaking has always differentiated our species from all others on Earth. Aerodynamically correct wooden spears were carved by archaic Homo sapiens close to 400,000 years ago. Man builds things consistent with his size, typically in the range of two orders of magnitude larger or smaller than himself, as indicated in Figure 13.1. Though the extremes of length scale are outside the range of this figure, human beings, at slightly more than 10^0 m, amazingly fit right in the middle of the smallest subatomic particle, which is approximately 10^{-26} m, and the extent of the observable universe, which is $\approx 1.42 \times 10^{26}$ m (15 billion light years). An egocentric universe indeed! But humans have always striven to explore, build, and control the extremes of length and time scales. In the voyages to Lilliput and Brobdingnag of *Gulliver's Travels*, Jonathan Swift (1726) speculated on the remarkable possibilities that diminution or magnification of physical dimensions provide. The Great Pyramid of Khufu was originally 147 m high when completed around 2600 B.C., whereas the Empire State Building constructed in 1931 is presently—after the addition of a television antenna mast in 1950—449 m high. At the other end of the spectrum of man-made artifacts, a dime is slightly less than 2 cm in diameter. Watchmakers have practiced the art of miniaturization since the thirteenth century. The invention of the microscope in the seventeenth century opened the way for direct observation of microbes and plant and animal cells. Smaller things were man-made in the latter half of this century. The transistor—invented in 1947—in today's integrated circuits has a size[1] of 0.18 μm (180 nanometers) in production and approaches 10 nm in research laboratories using electron beams. But what about the miniaturization of mechanical parts—machines—envisioned by Feynman (1961) in his legendary speech quoted above?

---

[1] The smallest feature in a microchip is defined by its smallest linewidth, which in turn is related to the wavelength of light employed in the basic lithographic process that is used to create the chip.

Manufacturing processes that can create extremely small machines have been developed in recent years (Angell, Terry, and Barth 1983; Gabriel et al. 1988, 1992; O'Connor 1992; Gravesen, Branebjerg, and Jensen 1993; Bryzek, Peterson, and McCulley 1994; Gabriel 1995; Hogan 1996; Ho and Tai 1996, 1998; Tien 1997; Busch-Vishniac 1998; Amato 1998). Electrostatic, magnetic, electromagnetic, pneumatic and thermal actuators, motors, valves, gears, cantilevers, diaphragms, and tweezers of less than 100-μm size have been fabricated. These have been used as sensors for pressure, temperature, mass flow, velocity, and sound; as actuators for linear and angular motions; and as simple components for complex systems such as microheat engines and microheat pumps (Lipkin 1993; Garcia and Sniegowski 1993, 1995; Sniegowski and Garcia 1996; Epstein and Senturia 1997; Epstein et al. 1997). The technology is progressing at a rate that far exceeds that of our understanding of the unconventional physics involved in the operation as well as the manufacturing of those minute devices (Knight 1999b). The present chapter focuses on one aspect of such physics: fluid-flow phenomena associated with microscale devices. In terms of applications, this and the following chapter will emphasize the use of MEMS as sensors and actuators for flow diagnosis and control.

The term microelectromechanical systems (MEMS) refers to devices that have characteristic length of less than 1 mm but more than 1 μm, that combine electrical and mechanical components, and that are fabricated using integrated circuit batch-processing technologies. Current manufacturing techniques for MEMS include surface silicon micromachining; bulk silicon micromachining; lithography, electrodeposition, and plastic molding (or, in its original German, *lithographie galvanoformung abformung*, LIGA); and electrodischarge machining (EDM). As indicated in Figure 13.1, MEMS are more than four orders of magnitude larger than the diameter of the hydrogen atom but about four orders of magnitude smaller than the traditional man-made artifacts. Nanodevices (some say NEMS) further push the envelope of electromechanical miniaturization.

Despite Feynman's demurring regarding the usefulness of small machines, MEMS are finding increased applications in a variety of industrial and medical fields with a potential worldwide market in the billions of dollars. Accelerometers for automobile airbags, keyless entry systems, dense arrays of micromirrors for high-definition optical displays, scanning electron microscope tips to image single atoms, microheat exchangers for cooling of electronic circuits, reactors for separating biological cells, blood analyzers, and pressure sensors for catheter tips are but a few in current use. Microducts are used in infrared detectors, diode lasers, miniature gas chromatographs, and high-frequency fluidic control systems. Micropumps are used for ink-jet printing, environmental testing, and electronic cooling. Potential medical applications for small pumps include controlled delivery and monitoring of minute amounts of medication, manufacturing of nanoliters of chemicals, and development of an artificial pancreas. Several new journals are dedicated to the science and technology of MEMS, for example the *IEEE/ASME Journal of Microelectromechanical Systems*, *Journal of Micromechanics and Microengineering*, and *Microscale Thermophysical Engineering*.

Not all MEMS devices involve fluid flows, but this chapter will focus on the ones that do. Microducts, micropumps, microturbines, and microvalves are examples of small devices involving the flow of liquids and gases. Microelectromechanical systems can also be related to fluid flows in an indirect way. The availability of inexpensive, batch-processing-produced microsensors and microactuators provides opportunities for targeting small-scale coherent structures in macroscopic turbulent shear flows. Flow control using MEMS promises a quantum leap in control system performance. This chapter will cover both the direct and indirect aspects of microdevices and fluid flows. Section 2 addresses the question of modeling fluid flows in microdevices, and Section 3 gives a brief overview of typical applications of MEMS in the field of fluid mechanics. The papers by Gad-el-Hak (1999) and Löfdahl and Gad-el-Hak (1999) provide more detail on MEMS applications in turbulence and flow control. In this book, Chapter 14 details reactive flow control strategies particularly suited for taming turbulence.

13.2   Flow Physics

13.2.1   Fluid Mechanics Issues

The rapid progress in fabricating and utilizing microelectromechanical systems during the last decade has not been matched by corresponding advances in our understanding of the unconventional physics involved in the operation and manufacture of small devices. Providing such understanding is crucial to designing, optimizing, fabricating, and operating improved MEMS devices.

Fluid flows in small devices differ from those in macroscopic machines. The operation of MEMS-based ducts, nozzles, valves, bearings, turbomachines, and so forth cannot always be predicted from conventional flow models such as the Navier–Stokes equations with a no-slip boundary condition at a fluid–solid interface, as routinely and successfully applied for larger flow devices. Many questions have been raised when the results of experiments with microdevices could not be explained via traditional flow modeling. The pressure gradient in a long microduct was observed to be nonconstant, and the measured flowrate was higher than that predicted from the conventional continuum flow model. Load capacities of microbearings were diminished, and electric currents needed to move micromotors were extraordinarily high. The dynamic response of micromachined accelerometers operating at atmospheric conditions was observed to be overdamped.

In the early stages of development of this exciting new field, the objective was to build MEMS devices as productively as possible. Microsensors were reading something, but not many researchers seemed to know exactly what. Microactuators were moving, but conventional modeling could not precisely predict their motion. After a decade of unprecedented progress in MEMS technology, perhaps the time is now ripe to take stock, slow down a bit, and answer the many questions that have arisen. The ultimate aim of this long-term exercise is to achieve rational-design capability for useful microdevices and to be able to characterize the operations of microsensors and microactuators definitively and with as little empiricism as possible.

In dealing with fluid flow through microdevices, one is faced with the question of which model to use, which boundary condition to apply, and how to proceed to obtain solutions to the problem at hand. Obviously, surface effects dominate in small devices. The surface-to-volume ratio for a machine with a characteristic length of 1 m is 1 m^{-1}, whereas that for a MEMS device having a size of 1 μm is 10^6 m^{-1}. The millionfold increase in surface area relative to the mass of the minute device substantially affects the transport of mass, momentum, and energy through the surface. The small length scale of microdevices may invalidate the continuum approximation altogether. Slip flow, thermal creep, rarefaction, viscous dissipation, compressibility, intermolecular forces, and other unconventional effects may have to be taken into account, preferably using only first principles such as conservation of mass, Newton's second law, conservation of energy, and so forth.

In this section, we discuss continuum as well as molecular-based flow models and the choices to be made. Computing typical Reynolds, Mach, and Knudsen numbers for the flow through a particular device is a good start to characterize the flow. For gases, microfluid mechanics has been studied by incorporating slip boundary conditions, thermal creep, viscous dissipation, and compressibility effects into the continuum equations of motion. Molecular-based models have also been attempted for certain ranges of the operating parameters. Use is made of the well-developed kinetic theory of gases, embodied in the Boltzmann equation, and direct simulation methods such as Monte Carlo. Microfluid mechanics of liquids is more complicated. The molecules are much more closely packed at normal pressures and temperatures, and the attractive or cohesive potential between the liquid molecules as well as between the liquid and solid ones plays a dominant role if the characteristic length of the flow is sufficiently small. In cases in which the traditional continuum model fails to provide accurate predictions or postdictions, expensive molecular dynamics simulations seem to be the only first-principle approach available to characterize liquid flows in microdevices rationally. Such simulations are not yet feasible for realistic flow extent or number of molecules. As a consequence, the microfluid mechanics of liquids is much less developed than that for gases.

13.2.2 Fluid Modeling

There are basically two ways of modeling a flowfield: either as the fluid really is, a collection of molecules, or as a continuum in which the matter is assumed continuous and indefinitely divisible. The former modeling is subdivided into deterministic methods and probabilistic ones, whereas in the latter approach the velocity, density, pressure, and so forth are defined at every point in space and time and conservation of mass, energy, and momentum leads to a set of nonlinear partial differential equations (Euler, Navier–Stokes, Burnett, etc.). Fluid-modeling classification is depicted schematically in Figure 13.2.

The continuum model, embodied in the Navier–Stokes equations, is applicable to numerous flow situations. The model ignores the molecular nature of gases and liquids and regards the fluid as a continuous medium describable in terms of the spatial and temporal variations of density, velocity, pressure, temperature, and other macroscopic flow quantities. For dilute gas flows near equilibrium, the Navier–Stokes equations are derivable from the molecularly based Boltzmann equation but can also be derived independently for both liquids and gases. In the case of direct derivation, some empiricism is necessary to close the resulting indeterminate set of equations. The continuum model is easier to handle mathematically (and is also more familiar to most fluid dynamists) than the alternative molecular models. Continuum models should therefore be used as long as they are applicable. Thus, careful considerations of the validity of the Navier–Stokes equations and the like are in order.

Basically, the continuum model leads to fairly accurate predictions as long as local properties such as density and velocity can be defined as averages over elements large compared with the microscopic structure of the fluid but small enough in comparison with the scale of the macroscopic phenomena to permit the use of differential

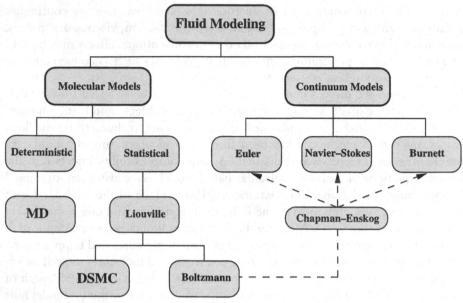

Figure 13.2: Molecular and continuum flow models.

calculus to describe them. Additionally, the flow must not be too far from thermodynamic equilibrium. The former condition is almost always satisfied, but it is the latter that usually restricts the validity of the continuum equations. As will be seen in Section 13.2.3, the continuum flow equations do not form a determinate set. The shear stress and heat flux must be expressed in terms of lower-order macroscopic quantities such as velocity and temperature, and the simplest (i.e., linear) relations are valid only when the flow is near thermodynamic equilibrium. Worse yet, the traditional no-slip boundary condition at a solid-fluid interface breaks down even before the linear stress–strain relation becomes invalid.

To be more specific, we temporarily restrict the discussion to gases for which the concept of mean free path is well defined. Liquids are more problematic, and we defer their discussion to Section 13.2.7. For gases, the mean free path $\mathcal{L}$ is the average distance traveled by molecules between collisions. For an ideal gas modeled as rigid spheres, the mean free path is related to temperature T and pressure p as follows:

$$\mathcal{L} = \frac{1}{\sqrt{2}\pi n \sigma^2} = \frac{kT}{\sqrt{2}\pi p \sigma^2} \tag{13.1}$$

where n is the number density (number of molecules per unit volume), σ is the molecular diameter, and k is the Boltzmann constant.

The continuum model is valid when $\mathcal{L}$ is much smaller than a characteristic flow dimension L. As this condition is violated, the flow is no longer near equilibrium, and the linear relation between stress and rate of strain and the no-slip velocity condition are no longer valid. Similarly, the linear relation between heat flux and temperature gradient and the no-jump temperature condition at a solid-fluid interface are no longer accurate when $\mathcal{L}$ is not much smaller than L.

The length scale L can be some overall dimension of the flow, but a more precise choice is the scale of the gradient of a macroscopic quantity, as, for example, the density ρ, as defined by

$$L = \frac{\rho}{\left|\frac{\partial \rho}{\partial y}\right|} \tag{13.2}$$

The ratio between the mean free path and the characteristic length is known as the Knudsen number

$$Kn = \frac{\mathcal{L}}{L} \tag{13.3}$$

and generally the traditional continuum approach is valid, albeit with modified boundary conditions, as long as $Kn < 0.1$.

There are two more important dimensionless parameters in fluid mechanics, and the Knudsen number can be expressed in terms of these two. The Reynolds number is the ratio of inertial forces to viscous ones

$$Re = \frac{v_0 L}{\nu} \tag{13.4}$$

where v_0 is a characteristic velocity, and ν is the kinematic viscosity of the fluid. The Mach number is the ratio of flow velocity to the speed of sound

$$Ma = \frac{v_0}{a_0} \tag{13.5}$$

The Mach number is a dynamic measure of fluid compressibility and may be considered as the ratio of inertial forces to elastic ones. From the kinetic theory of gases, the mean free path is related to the viscosity as follows:

$$\nu = \frac{\mu}{\rho} = \frac{1}{2}\mathcal{L}\bar{v}_m \tag{13.6}$$

where μ is the dynamic viscosity, and $\bar{v}_m$ is the mean molecular speed, which is somewhat higher than the sound speed a_0,

$$\bar{v}_m = \sqrt{\frac{8}{\pi \gamma}} a_0 \tag{13.7}$$

where γ is the specific heat ratio (i.e., the isentropic exponent). Combining Eqs. (13.3)–(13.7), we reach the required relation

$$Kn = \sqrt{\frac{\pi \gamma}{2}} \frac{Ma}{Re} \tag{13.8}$$

In boundary layers, the relevant length scale is the shear-layer thickness δ, and for laminar flows

$$\frac{\delta}{L} \sim \frac{1}{\sqrt{Re}} \tag{13.9}$$

$$Kn \sim \frac{Ma}{Re_\delta} \sim \frac{Ma}{\sqrt{Re}} \tag{13.10}$$

where Re_δ is the Reynolds number based on the freestream velocity v_o and the boundary layer thickness δ, and Re is based on v_o and the streamwise length scale L.

Rarefied gas flows are in general encountered in flows in small geometries such as MEMS devices and in low-pressure applications such as high-altitude flying and high-vacuum gadgets. The local value of the Knudsen number in a particular flow determines the degree of rarefaction and the degree of validity of the continuum model. The different Knudsen number regimes are determined empirically and are therefore only approximate for a particular flow geometry. The pioneering experiments in rarefied gas dynamics were conducted by Knudsen in 1909. In the limit of zero Knudsen number, the transport terms in the continuum momentum and energy equations are negligible, and the Navier–Stokes equations then reduce to the inviscid Euler equations. Both heat conduction and viscous diffusion and dissipation are negligible, and the flow is then approximately isentropic (i.e., adiabatic and reversible) from the continuum viewpoint, whereas the equivalent molecular viewpoint is that the velocity distribution function is everywhere of the local equilibrium or Maxwellian form. As Kn increases, rarefaction effects become more important, and eventually the continuum approach breaks down altogether. The different Knudsen number regimes are depicted in Figure 13.3 and can be summarized as follows:

$$\text{Euler equations (neglect molecular diffusion): } Kn \to 0 \; (Re \to \infty)$$
$$\text{Navier–Stokes equations with no-slip boundary conditions: } Kn \le 10^{-3}$$
$$\text{Navier–Stokes equations with slip boundary conditions: } 10^{-3} \le Kn \le 10^{-1}$$
$$\text{Transition regime: } 10^{-1} \le Kn \le 10$$
$$\text{Free-molecule flow: } Kn > 10$$

We will return to these regimes in the following subsections.

As an example, consider air at standard temperature ($T = 288$ K) and pressure ($p = 1.01 \times 10^5$ N/m^2). A cube 1 μm to the side contains 2.54×10^7 molecules separated by an average distance of 0.0034 μm. The gas is considered dilute if the ratio of this distance to the molecular diameter exceeds 7, and in the present example this ratio is 9, which barely satisfies the dilute gas assumption. The mean free path computed from Eq. (13.1) is $\mathcal{L} = 0.065 \mu$m. A microdevice with characteristic length of 1 μm would have $Kn = 0.065$, which is in the slip-flow regime. At lower

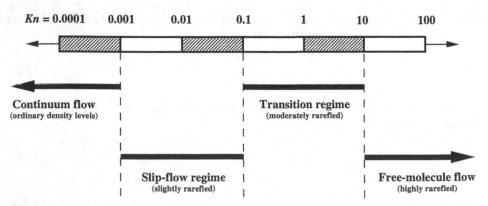

Figure 13.3: Knudsen number regimes.

pressures, the Knudsen number increases. For example, if the pressure is 0.1 atm and the temperature remains the same, $Kn = 0.65$ for the same 1-μm device, and the flow is then in the transition regime. There would still be over 2 million molecules in the same one-micron cube, and the average distance between them would be 0.0074 μm. The same device at 100 km altitude would have $Kn = 3 \times 10^4$, which is well into the free-molecule flow regime. The Knudsen number for the flow of a light gas like helium is about three times larger than that for air flow at otherwise the same conditions.

Consider a long microchannel for which the entrance pressure is atmospheric and the exit conditions are near vacuum. As air goes down the duct, the pressure and density decrease, whereas the velocity, Mach number, and Knudsen number increase. The pressure drops to overcome viscous forces in the channel. If isothermal conditions prevail,[2] density also drops, and conservation of mass requires the flow to accelerate down the constant-area tube. The fluid acceleration in turn affects the pressure gradient, resulting in a nonlinear pressure drop along the channel. The Mach number increases down the tube, limited only by the choked-flow condition ($Ma = 1$). Additionally, the normal component of velocity is no longer zero. With lower density, the mean free path increases, and Kn correspondingly increases. All flow regimes depicted in Figure 13.3 may occur in the same tube: continuum with no-slip boundary conditions, slip-flow regime, transition regime, and free-molecule flow. The air flow may also change from incompressible to compressible as it moves down the microduct. A similar scenario may take place if the entrance pressure is, say, 5 atm, while the exit is atmospheric. This deceivingly simple duct flow may in fact manifest every single complexity discussed in this section.

In the following six subsections, we discuss in turn the Navier–Stokes equations, compressibility effects, boundary conditions, molecular-based models, liquid flows, and surface phenomena.

13.2.3 Continuum Model

We recall in this subsection the traditional conservation relations in fluid mechanics. These equations were derived in Chapter 2. Here, we reemphasize the precise assumptions needed to obtain a particular form of those equations. A continuum fluid implies that the derivatives of all the dependent variables exist in some reasonable sense. In other words, local properties such as density and velocity are defined as averages over elements large compared with the microscopic structure of the fluid but small enough in comparison with the scale of the macroscopic phenomena to permit the use of differential calculus to describe them. As mentioned earlier, such conditions are almost always met. For such fluids, and on the assumption that the laws of nonrelativistic mechanics hold, the conservation of mass, momentum, and energy can be expressed at every point in space and time as a set of partial differential equations as follows:

$$\frac{\partial \rho}{\partial t} + \frac{\partial}{\partial x_k}(\rho u_k) = 0 \tag{13.11}$$

---

[2] More likely the flow will be somewhere in between isothermal and adiabatic, Fanno flow. In that case both density and temperature decrease downstream, the former not as fast as in the isothermal case. None of that changes the qualitative arguments made in the example.

$$\rho\left(\frac{\partial u_i}{\partial t} + u_k\frac{\partial u_i}{\partial x_k}\right) = \frac{\partial \Sigma_{ki}}{\partial x_k} + \rho g_i \tag{13.12}$$

$$\rho\left(\frac{\partial e}{\partial t} + u_k\frac{\partial e}{\partial x_k}\right) = -\frac{\partial q_k}{\partial x_k} + \Sigma_{ki}\frac{\partial u_i}{\partial x_k} \tag{13.13}$$

where ρ is the fluid density, u_k is an instantaneous velocity component (u, v, w), Σ_{ki} is the second-order stress tensor (surface force per unit area), g_i is the body force per unit mass, e is the internal energy, and q_k is the sum of heat flux vectors due to conduction and radiation. The independent variables are time t and the three spatial coordinates x_1, x_2, and x_3 or (x, y, z).

Equations (13.11), (13.12), and (13.13) constitute 5 differential equations for the 17 unknowns ρ, u_i, Σ_{ki}, e, and q_k. Absent any body couples, the stress tensor is symmetric, having only 6 independent components, which reduces the number of unknowns to 14. Obviously, the continuum flow equations do not form a determinate set. To close the conservation equations, the relation between the stress tensor and deformation rate, the relation between the heat flux vector, and the temperature field and appropriate equations of state relating the different thermodynamic properties are needed. The stress–rate of strain relation and the heat flux–temperature gradient relation are approximately linear if the flow is not too far from thermodynamic equilibrium. This is a phenomenological result but can be rigorously derived from the Boltzmann equation for a dilute gas if it is assumed that the flow is near equilibrium (see Section 13.2.6). For a Newtonian, isotropic, Fourier, ideal gas, for example, these relations read

$$\Sigma_{ki} = -p\delta_{ki} + \mu\left(\frac{\partial u_i}{\partial x_k} + \frac{\partial u_k}{\partial x_i}\right) + \lambda\left(\frac{\partial u_j}{\partial x_j}\right)\delta_{ki} \tag{13.14}$$

$$q_i = -\kappa\frac{\partial T}{\partial x_i} + \text{Heat flux due to radiation} \tag{13.15}$$

$$de = c_v\,dT \quad \text{and} \quad p = \rho\mathcal{R}T \tag{13.16}$$

where p is the thermodynamic pressure, μ and λ are the first and second coefficients of viscosity, respectively, δ_{ki} is the unit second-order tensor (Kronecker delta), κ is the thermal conductivity, T is the temperature field, c_v is the specific heat at constant volume, and $\mathcal{R}$ is the gas constant, which is given by the Boltzmann constant divided by the mass of an individual molecule $(k = m\mathcal{R})$. The Stokes hypothesis relates the first and second coefficients of viscosity, thus $\lambda + \frac{2}{3}\mu = 0$, although the validity of this assumption for other than dilute, monatomic gases has occasionally been questioned (Gad-el-Hak 1995). With the preceding constitutive relations and radiative heat transfer neglected, Eqs. (13.11), (13.12), and (13.13) respectively read

$$\frac{\partial \rho}{\partial t} + \frac{\partial}{\partial x_k}(\rho u_k) = 0 \tag{13.17}$$

$$\rho\left(\frac{\partial u_i}{\partial t} + u_k\frac{\partial u_i}{\partial x_k}\right) = -\frac{\partial p}{\partial x_i} + \rho g_i + \frac{\partial}{\partial x_k}\left[\mu\left(\frac{\partial u_i}{\partial x_k} + \frac{\partial u_k}{\partial x_i}\right) + \delta_{ki}\lambda\frac{\partial u_j}{\partial x_j}\right] \tag{13.18}$$

$$\rho c_v\left(\frac{\partial T}{\partial t} + u_k\frac{\partial T}{\partial x_k}\right) = \frac{\partial}{\partial x_k}\left(\kappa\frac{\partial T}{\partial x_k}\right) - p\frac{\partial u_k}{\partial x_k} + \phi \tag{13.19}$$

The three components of the vector equation (13.18) are the Navier–Stokes equations expressing the conservation of momentum for a Newtonian fluid. In the thermal energy equation (13.19), ϕ is the always positive dissipation function expressing the irreversible conversion of mechanical energy to internal energy as a result of the deformation of a fluid element. The second term on the right-hand side of Eq. (13.19) is the reversible work done (per unit time) by the pressure as the volume of a fluid material element changes. For a Newtonian, isotropic fluid, the viscous dissipation rate is given by

$$\phi = \frac{1}{2}\mu\left(\frac{\partial u_i}{\partial x_k} + \frac{\partial u_k}{\partial x_i}\right)^2 + \lambda\left(\frac{\partial u_j}{\partial x_j}\right)^2 \tag{13.20}$$

There are now six unknowns, ρ, u_i, p, and T, and the five coupled equations (13.17), (13.18), and (13.19) plus the equation of state relating pressure, density, and temperature. These six equations together with a sufficient number of initial and boundary conditions constitute a well-posed, albeit formidable, problem. The system of equations (13.17)–(13.19) is an excellent model for the laminar or turbulent flow of most fluids such as air and water under many circumstances, including high-speed gas flows for which the shock waves are thick relative to the mean free path of the molecules.

Considerable simplification is achieved if the flow is assumed incompressible, which is usually a reasonable assumption provided that the characteristic flow speed is less than 0.3 of the speed of sound. The incompressibility assumption is readily satisfied for almost all liquid flows and many gas flows. In such cases, the density is assumed either a constant or a given function of temperature (or species concentration). The governing equations for such flows are

$$\frac{\partial u_k}{\partial x_k} = 0 \tag{13.21}$$

$$\rho\left(\frac{\partial u_i}{\partial t} + u_k\frac{\partial u_i}{\partial x_k}\right) = -\frac{\partial p}{\partial x_i} + \frac{\partial}{\partial x_k}\left[\mu\left(\frac{\partial u_i}{\partial x_k} + \frac{\partial u_k}{\partial x_i}\right)\right] + \rho g_i \tag{13.22}$$

$$\rho c_p\left(\frac{\partial T}{\partial t} + u_k\frac{\partial T}{\partial x_k}\right) = \frac{\partial}{\partial x_k}\left(\kappa\frac{\partial T}{\partial x_k}\right) + \phi_{\text{incomp}} \tag{13.23}$$

where ϕ_{incomp} is the incompressible limit of Eq. (13.20). These are now five equations for the five dependent variables u_i, p, and T. As noted in Section 2.2, the left-hand side of Eq. (13.23) has the specific heat at constant pressure c_p and not c_v.

For both the compressible and the incompressible equations of motion, the transport terms are neglected away from solid walls in the limit of infinite Reynolds number ($Kn \to 0$). The fluid is then approximated as inviscid and nonconducting, and the corresponding equations read (for the compressible case)

$$\frac{\partial \rho}{\partial t} + \frac{\partial}{\partial x_k}(\rho u_k) = 0 \tag{13.24}$$

$$\rho\left(\frac{\partial u_i}{\partial t} + u_k\frac{\partial u_i}{\partial x_k}\right) = -\frac{\partial p}{\partial x_i} + \rho g_i \tag{13.25}$$

$$\rho c_v\left(\frac{\partial T}{\partial t} + u_k\frac{\partial T}{\partial x_k}\right) = -p\frac{\partial u_k}{\partial x_k} \tag{13.26}$$

The Euler equation (13.25) can be integrated along a streamline, and the resulting Bernoulli's equation provides a direct relation between the velocity and pressure.

13.2.4 Compressibility

The issue of whether to consider the continuum flow compressible or incompressible seems to be rather straightforward but is in fact full of potential pitfalls. This issue was covered in Chapter 2, but portions of that material are repeated here because of their importance to MEMS flows and to provide continuity between the different subsections in this section. If the local Mach number is less than 0.3, then the flow of a compressible fluid like air can—according to the conventional wisdom—be treated as incompressible. But the well-known $Ma < 0.3$ criterion is only a necessary but not a sufficient one to allow treatment of the flow as approximately incompressible. In other words, there are situations in which the Mach number can be exceedingly small while the flow is compressible. As is well documented in heat transfer textbooks, strong wall heating or cooling may cause the density to change sufficiently and the incompressible approximation to break down, even at low speeds. Less known is the situation encountered in some microdevices in which the pressure may strongly change because of viscous effects even though the speeds may not be high enough for the Mach number to go above the traditional threshold of 0.3. Corresponding to the pressure changes are strong density changes that must be taken into account when writing the continuum equations of motion. In this section, we systematically explain all situations relevant to MEMS for which compressibility effects must be considered.

Let us rewrite the full continuity equation (13.11) as follows:

$$\frac{\mathrm{D}\rho}{\mathrm{D}t} + \rho \frac{\partial u_k}{\partial x_k} = 0 \tag{13.27}$$

where $\frac{\mathrm{D}}{\mathrm{D}t}$ is the substantial derivative ($\frac{\partial}{\partial t} + u_k \frac{\partial}{\partial x_k}$) expressing changes following a fluid element. The proper criterion for the incompressible approximation to hold is that ($\frac{1}{\rho}\frac{\mathrm{D}\rho}{\mathrm{D}t}$) is vanishingly small. In other words, if density changes following a fluid particle are small, the flow is approximately incompressible. Density may change arbitrarily from one particle to another without violating the incompressible flow assumption. This is the case, for example, in the stratified atmosphere and ocean, where the variable-density–temperature–salinity flow is often treated as incompressible.

From the state principle of thermodynamics, we can express the density changes of a simple system in terms of changes in pressure and temperature by

$$\rho = \rho(p, T). \tag{13.28}$$

From the chain rule of calculus,

$$\frac{1}{\rho}\frac{\mathrm{D}\rho}{\mathrm{D}t} = \alpha \frac{\mathrm{D}p}{\mathrm{D}t} - \beta \frac{\mathrm{D}T}{\mathrm{D}t} \tag{13.29}$$

where α and β are, respectively, the isothermal compressibility coefficient and the bulk expansion coefficient, both previously defined in Section 2.3. The flow must be treated as compressible if pressure and temperature changes are sufficiently strong. Equation (13.29) must of course be properly nondimensionalized before it is decided whether a term is large or small. Here, we follow closely the procedure detailed in Panton (1996).

Consider first the case of adiabatic walls. Density is normalized with a reference value ρ_0, velocities with a reference speed v_0, spatial coordinates and time with, respectively, L and L/v_0, and the isothermal compressibility coefficient and bulk expansion coefficient with reference values α_0 and β_0, respectively. The pressure is nondimensionalized with the inertial pressure scale $\rho_0 v_0^2$. This scale is twice the dynamic pressure, that is, the pressure change as an inviscid fluid moving at the reference speed is brought to rest.

Temperature changes for the case of adiabatic walls result from the irreversible conversion of mechanical energy into internal energy via viscous dissipation. Temperature is therefore nondimensionalized as follows:

$$T^* = \frac{T - T_0}{\left(\frac{\mu_0 v_0^2}{\kappa_0}\right)} = \frac{T - T_0}{Pr\left(\frac{v_0^2}{c_{p_0}}\right)} \tag{13.30}$$

where T_0 is a reference temperature, μ_0, κ_0, and c_{p_0} are, respectively, reference viscosity, thermal conductivity, and specific heat at constant pressure, and Pr is the reference Prandtl number, $(\mu_0 c_{p_0})/\kappa_0$.

In the present formulation, the scaling used for pressure is based on the Bernoulli's equation and therefore neglects viscous effects. This particular scaling guarantees that the pressure term in the momentum equation will be of the same order as the inertia term. The temperature scaling assumes that the conduction, convection, and dissipation terms in the energy equation have the same order of magnitude. The resulting dimensionless form of Eq. (13.29) reads

$$\frac{1}{\rho^*}\frac{D\rho^*}{Dt^*} = \gamma_0 Ma^2 \left\{ \alpha^* \frac{Dp^*}{Dt^*} - \frac{PrB\beta^*}{A}\frac{DT^*}{Dt^*} \right\} \tag{13.31}$$

where the superscript * indicates a nondimensional quantity, Ma is the reference Mach number, and A and B are dimensionless constants defined by $A \equiv \alpha_0 \rho_0 c_{p_0} T_0$, and $B \equiv \beta_0 T_0$. If the scaling is properly chosen, the terms having the * superscript in the right-hand side should be of order one, and the relative importance of such terms in the equations of motion is determined by the magnitude of the dimensionless parameter(s) appearing to their left (e.g., Ma, Pr, etc.). Therefore, as $Ma^2 \to 0$, temperature changes due to viscous dissipation are neglected (unless Pr is very large as, for example, in the case of highly viscous polymers and oils). Within the same order of approximation, all thermodynamic properties of the fluid are assumed constant.

Pressure changes are also neglected in the limit of zero Mach number. Hence, for $Ma < 0.3$ (i.e., $Ma^2 < 0.09$), density changes following a fluid particle can be neglected, and the flow can then be approximated as incompressible. However, there is a caveat in this argument. Pressure changes due to inertia can indeed be neglected at small Mach numbers, and this is consistent with the way we nondimensionalized the pressure term above. If, on the other hand, pressure changes are mostly due to viscous effects, as is the case for example in a long duct or a gas bearing, pressure changes may be significant even at low speeds (low Ma). In that case the term $\frac{Dp^*}{Dt^*}$ in Eq. (13.31) is no longer of order one and may be large regardless of the value of Ma.

Density then may change significantly, and the flow must be treated as compressible. Had pressure been nondimensionalized using the viscous scale ($\frac{\mu_o v_o}{L}$) instead of the inertial one ($\rho_o v_o^2$), the revised equation (13.31) would have Re^{-1} appearing explicitly in the first term in the right-hand side, thus accentuating the importance of this term when viscous forces dominate.

A similar result can be gleaned when the Mach number is interpreted as follows:

$$
Ma^2 = \frac{v_o^2}{a_o^2} = v_o^2 \frac{\partial \rho}{\partial p}\bigg|_s = \frac{\rho_o v_o^2}{\rho_o} \frac{\partial \rho}{\partial p}\bigg|_s
$$

$$
\approx \frac{\Delta p}{\rho_o} \frac{\Delta \rho}{\Delta p} = \frac{\Delta \rho}{\rho_o} \tag{13.32}
$$

where s is the entropy. Again, the preceding equation assumes that pressure changes are inviscid, and therefore small Mach number means negligible pressure and density changes. In a flow dominated by viscous effects—such as that inside a microduct—density changes may be significant even in the limit of zero Mach number.

Identical arguments can be made in the case of isothermal walls. Here strong temperature changes may be the result of wall heating or cooling, even if viscous dissipation is negligible. The proper temperature scale in this case is given in terms of the wall temperature T_w and the reference temperature T_o as follows

$$
\hat{T} = \frac{T - T_o}{T_w - T_o} \tag{13.33}
$$

where $\hat{T}$ is the new dimensionless temperature. The nondimensional form of Eq. (13.29) now reads

$$
\frac{1}{\rho^*} \frac{D\rho^*}{Dt^*} = \gamma_o Ma^2 \alpha^* \frac{Dp^*}{Dt^*} - \beta^* B\left(\frac{T_w - T_o}{T_o}\right) \frac{D\hat{T}}{Dt^*} \tag{13.34}
$$

Here we notice that the temperature term is different from that in Eq. (13.31), for Ma is no longer appearing in this term, and strong temperature changes, that is, large $(T_w - T_o)/T_o$, may cause strong density changes regardless of the value of the Mach number. Additionally, the thermodynamic properties of the fluid are not constant but depend on temperature, and as a result, the continuity, momentum, and energy equations are all coupled. The pressure term in Eq. (13.34), on the other hand, is exactly as it was in the adiabatic case, and the same arguments made before apply: the flow should be considered compressible if $Ma > 0.3$ or if pressure changes due to viscous forces are sufficiently large.

Experiments in gaseous microducts confirm the preceding arguments. For both low- and high-Mach-number flows, pressure gradients in long microchannels are nonconstant, consistent with the compressible flow equations. Such experiments were conducted by, among others, Prud'homme et al. (1986); Pfahler et al. (1991); van den Berg et al. (1993); Liu et al. (1993, 1995); Pong et al. (1994); Harley et al. (1995); Piekos and Breuer (1996); Arkilic (1997); and Arkilic, Schmidt, and Breuer (1995, 1997a,b). Sample results will be presented in the following subsection.

13.2.5 Boundary Conditions

The equations of motion described in Section 13.2.3 require a certain number of initial and boundary conditions for proper mathematical formulation of flow problems. In this subsection, we describe the boundary conditions at a fluid–solid interface. Boundary conditions in the inviscid flow theory pertain only to the velocity component normal to a solid surface. The highest spatial derivative of velocity in the inviscid equations of motion is first-order, and only one velocity boundary condition at the surface is admissible. The normal velocity component at a fluid–solid interface is specified, and no statement can be made regarding the tangential velocity component. The normal velocity condition simply states that a fluid-particle path cannot go through an impermeable wall. Real fluids are of course viscous, and the corresponding momentum equation has second-order derivatives of velocity, thus requiring an additional boundary condition on the velocity component tangential to a solid surface.

Traditionally, the no-slip condition at a fluid–solid interface is enforced in the momentum equation, and an analogous no-temperature-jump condition is applied in the energy equation. The notion underlying the no-slip and no-jump condition is that, within the fluid, there cannot be any finite discontinuities of velocity or temperature. Those would involve infinite velocity or temperature gradients and thus produce infinite viscous stress or heat flux that would destroy the discontinuity in infinitesimal time. The interaction between a fluid particle and a wall is similar to that between neighboring fluid particles, and therefore no discontinuities are allowed at the fluid–solid interface either. In other words, the fluid velocity must be zero relative to the surface, and the fluid temperature must equal to that of the surface. But strictly speaking these two boundary conditions are valid only if the fluid flow adjacent to the surface is in thermodynamic equilibrium. This requires an infinitely high frequency of collisions between the fluid and the solid surface. In practice, the no-slip and no-jump condition leads to fairly accurate predictions as long as $Kn < 0.001$ (for gases). Beyond that, the collision frequency is simply not high enough to ensure equilibrium, and a certain degree of tangential-velocity slip and temperature jump must be allowed. This is a case frequently encountered in MEMS flows, and we develop the appropriate relations in this subsection.

For both liquids and gases, the linear Navier boundary condition empirically relates the tangential velocity slip at the wall $\Delta u|_\mathrm{w}$ to the local shear

$$\Delta u|_\mathrm{w} = u_\mathrm{fluid} - u_\mathrm{wall} = L_\mathrm{s} \frac{\partial u}{\partial y}\bigg|_\mathrm{w} \tag{13.35}$$

where L_s is the constant slip length, and $\frac{\partial u}{\partial y}|_\mathrm{w}$ is the strain rate computed at the wall. In most practical situations, the slip length is so small that the no-slip condition holds. In MEMS applications, however, that may not be the case. Once again we defer the discussion of liquids to Section 13.2.7 and focus for now on gases.

On the assumption that isothermal conditions prevail, the preceding slip relation has been rigorously derived by Maxwell (1879) from considerations of the kinetic theory of dilute, monatomic gases. Gas molecules, modeled as rigid spheres, continuously strike and reflect from a solid surface just as they continuously collide with each other. For an idealized perfectly smooth wall,[3] the incident angle exactly equals the reflected angle, and the molecules conserve their tangential momentum and thus exert no shear on the wall. This is termed specular reflection and results in perfect slip at the wall. For an extremely rough wall, on the other hand, the molecules

---

[3] At the molecular scale.

reflect at some random angle uncorrelated with their entry angle. This perfectly diffuse reflection results in zero tangential momentum for the reflected fluid molecules to be balanced by a finite slip velocity in order to account for the shear stress transmitted to the wall. A force balance near the wall leads to the following expression for the slip velocity:

$$u_{gas} - u_{wall} = \mathcal{L} \frac{\partial u}{\partial y}\bigg|_{w} \qquad (13.36)$$

where $\mathcal{L}$ is the mean free path. The right-hand side can be considered as the first term in an infinite Taylor series, which is sufficient if the mean free path is *relatively* small enough. Equation (13.36) states that significant slip occurs only if the mean velocity of the molecules varies appreciably over a distance of one mean free path. This is the case, for example, in vacuum applications, flow in microdevices, or both. The number of collisions between the fluid molecules and the solid in those cases is not large enough for even an approximate flow equilibrium to be established. Furthermore, additional (nonlinear) terms in the Taylor series would be needed as $\mathcal{L}$ increases and the flow is further removed from the equilibrium state.

For real walls some molecules reflect diffusively and some reflect specularly. In other words, a portion of the momentum of the incident molecules is lost to the wall, and a (typically smaller) portion is retained by the reflected molecules. The tangential-momentum-accommodation coefficient σ_v is defined as the fraction of molecules reflected diffusively. This coefficient depends on the fluid, the solid, and the surface finish and has been determined experimentally to be between 0.2–0.8 (Thomas and Lord 1974; Seidl and Steinheil 1974; Porodnov et al. 1974; Arkilic et al. 1997b; Arkilic 1997), the lower limit being for exceptionally smooth surfaces, whereas the upper limit is typical of most practical surfaces. The final expression derived by Maxwell for an isothermal wall reads

$$u_{gas} - u_{wall} = \frac{2 - \sigma_v}{\sigma_v} \mathcal{L} \frac{\partial u}{\partial y}\bigg|_{w} \qquad (13.37)$$

For $\sigma_v = 0$, the slip velocity is unbounded, whereas for $\sigma_v = 1$, Eq. (13.37) reverts to Eq. (13.36).

Similar arguments were made for the temperature-jump boundary condition by von Smoluchowski (1898). For an ideal gas flow in the presence of wall-normal and tangential temperature gradients, the complete slip-flow and temperature-jump boundary conditions read

$$u_{gas} - u_{wall} = \frac{2 - \sigma_v}{\sigma_v} \frac{1}{\rho \sqrt{\frac{2 \mathcal{R} T_{gas}}{\pi}}} \tau_w + \frac{3}{4} \frac{Pr(\gamma - 1)}{\gamma \rho \mathcal{R} T_{gas}} (-q_x)_w$$

$$= \frac{2 - \sigma_v}{\sigma_v} \mathcal{L} \left(\frac{\partial u}{\partial y} \right)_w + \frac{3}{4} \frac{\mu}{\rho T_{gas}} \left(\frac{\partial T}{\partial x} \right)_w \qquad (13.38)$$

$$T_{gas} - T_{wall} = \frac{2 - \sigma_T}{\sigma_T} \left[\frac{2(\gamma - 1)}{(\gamma + 1)} \right] \frac{1}{\rho \mathcal{R} \sqrt{\frac{2 \mathcal{R} T_{gas}}{\pi}}} (-q_y)_w$$

$$= \frac{2 - \sigma_T}{\sigma_T} \left[\frac{2\gamma}{(\gamma + 1)} \right] \frac{\mathcal{L}}{Pr} \left(\frac{\partial T}{\partial y} \right)_w \qquad (13.39)$$

where x and y are the streamwise and normal coordinates, ρ and μ are respectively the fluid density and viscosity, $\mathcal{R}$ is the gas constant, T_{gas} is the temperature of the gas adjacent to the wall, T_{wall} is the wall temperature, τ_{w} is the shear stress at the wall, Pr is the Prandtl number, γ is the specific heat ratio, and $(q_x)_{\text{w}}$ and $(q_y)_{\text{w}}$ are, respectively, the tangential and normal heat flux at the wall.

The tangential-momentum-accommodation coefficient σ_v and the thermal-accommodation coefficient σ_T are given by, respectively,

$$\sigma_v = \frac{\tau_i - \tau_r}{\tau_i - \tau_w} \tag{13.40}$$

$$\sigma_T = \frac{dE_i - dE_r}{dE_i - dE_w} \tag{13.41}$$

where the subscripts i, r, and w stand for, respectively, incident, reflected, and solid wall conditions; τ is a tangential momentum flux; and dE is an energy flux.

The second term in the right-hand side of Eq. (13.38) is the *thermal creep*, which generates slip velocity in the fluid opposite to the direction of the tangential heat flux (i.e., flow in the direction of increasing temperature). At sufficiently high Knudsen numbers, streamwise temperature gradient in a conduit leads to a measurable pressure gradient along the tube. This may be the case in vacuum applications and MEMS devices. Thermal creep is the basis for the so-called Knudsen pump—a device with no moving parts—in which rarefied gas is hauled from one cold chamber to a hot one.[4] Clearly, such a pump performs best at high Knudsen numbers and is typically designed to operate in the free-molecule flow regime.

In dimensionless form, Eqs. (13.38) and (13.39), respectively, read

$$u^*_{\text{gas}} - u^*_{\text{wall}} = \frac{2 - \sigma_v}{\sigma_v} Kn \left(\frac{\partial u^*}{\partial y^*} \right)_w + \frac{3}{2\pi} \frac{(\gamma - 1)}{\gamma} \frac{Kn^2 \, Re}{Ec} \left(\frac{\partial T^*}{\partial x^*} \right)_w \tag{13.42}$$

$$T^*_{\text{gas}} - T^*_{\text{wall}} = \frac{2 - \sigma_T}{\sigma_T} \left[\frac{2\gamma}{(\gamma + 1)} \right] \frac{Kn}{Pr} \left(\frac{\partial T^*}{\partial y^*} \right)_w \tag{13.43}$$

where the superscript * indicates dimensionless quantity, Kn is the Knudsen number, Re is the Reynolds number, and Ec is the Eckert number defined by

$$Ec = \frac{v_o^2}{c_p \Delta T} = (\gamma - 1) \frac{T_o}{\Delta T} Ma^2 \tag{13.44}$$

where v_o is a reference velocity, $\Delta T = (T_{\text{gas}} - T_o)$, and T_o is a reference temperature. Note that very low values of σ_v and σ_T lead to substantial velocity slip and temperature jump even for flows with small Knudsen number.

The first term in the right-hand side of Eq. (13.42) is first-order in Knudsen number, whereas the thermal creep term is second-order, meaning that the creep phenomenon is potentially significant at large values of the Knudsen number. Equation (13.43) is first-order in Kn. Using Eqs. (13.8) and (13.44), the thermal

[4] The terminology *Knudsen pump* has been used by, for example, Vargo and Muntz (1996), but according to Loeb (1961), the original experiments demonstrating such pumps were carried out by Osborne Reynolds.

creep term in Eq. (13.42) can be rewritten in terms of ΔT and Reynolds number. Thus,

$$u^*_{gas} - u^*_{wall} = \frac{2 - \sigma_v}{\sigma_v} Kn \left(\frac{\partial u^*}{\partial y^*} \right)_w + \frac{3}{4} \frac{\Delta T}{T_o} \frac{1}{Re} \left(\frac{\partial T^*}{\partial x^*} \right)_w \tag{13.45}$$

It is clear that large temperature changes along the surface or low Reynolds numbers lead to significant thermal creep.

The continuum Navier–Stokes equations with no-slip or no-temperature jump boundary conditions are valid as long as the Knudsen number does not exceed 0.001. First-order slip or temperature-jump boundary conditions should be applied to the Navier–Stokes equations in the range of $0.001 < Kn < 0.1$. The transition regime spans the range of $0.1 < Kn < 10$, and second-order or higher slip or temperature-jump boundary conditions are applicable there. Note, however, that the Navier–Stokes equations are first-order accurate in Kn, as will be shown in Section 13.2.6, and are themselves not valid in the transition regime. Either higher-order continuum equations (e.g., Burnett equations) should be used there, or molecular modeling should be invoked, abandoning the continuum approach altogether.

For isothermal walls, Beskok (1994) derived a higher-order slip-velocity condition as follows:

$$u_{gas} - u_{wall} = \frac{2 - \sigma_v}{\sigma_v} \left[\mathcal{L} \left(\frac{\partial u}{\partial y} \right)_w + \frac{\mathcal{L}^2}{2!} \left(\frac{\partial^2 u}{\partial y^2} \right)_w + \frac{\mathcal{L}^3}{3!} \left(\frac{\partial^3 u}{\partial y^3} \right)_w + \cdots \right] \tag{13.46}$$

Attempts to implement the slip condition of Eq. (13.46) in numerical simulations are rather difficult. Second-order and higher derivatives of velocity cannot be computed accurately near the wall. On the basis of asymptotic analysis, Beskok (1996) and Beskok and Karniadakis (1994, 1999) proposed the following alternative higher-order boundary condition for the tangential velocity, including the thermal creep term:

$$u^*_{gas} - u^*_{wall} = \frac{2 - \sigma_v}{\sigma_v} \frac{Kn}{1 - bKn} \left(\frac{\partial u^*}{\partial y^*} \right)_w + \frac{3}{2\pi} \frac{(\gamma - 1)}{\gamma} \frac{Kn^2 Re}{Ec} \left(\frac{\partial T^*}{\partial x^*} \right)_w \tag{13.47}$$

where b is a high-order slip coefficient determined from the presumably known no-slip solution, thus avoiding the computational difficulties mentioned above. If this high-order slip coefficient is chosen as $b = u''_w / 2u'_w$, where the prime denotes derivative with respect to y and the velocity is computed from the no-slip Navier–Stokes equations, Eq. (13.47) becomes second-order accurate in Knudsen number. Beskok's procedure can be extended to third- and higher-orders for both the slip-velocity and thermal creep terms.

Similar arguments can be applied to the temperature-jump boundary condition, and the resulting Taylor series reads in dimensionless form (Beskok 1996)

$$T^*_{gas} - T^*_{wall} = \frac{2 - \sigma_T}{\sigma_T} \left[\frac{2\gamma}{(\gamma + 1)} \right] \frac{1}{Pr} \left[Kn \left(\frac{\partial T^*}{\partial y^*} \right)_w + \frac{Kn^2}{2!} \left(\frac{\partial^2 T^*}{\partial y^{*2}} \right)_w + \cdots \right] \tag{13.48}$$

Again, the difficulties associated with computing second- and higher-order derivatives

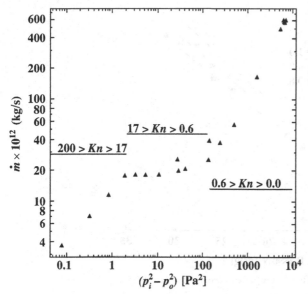

Figure 13.4: Variation of mass flow rate as a function of $(p_i^2 - p_o^2)$.
Original data acquired by S.A. Tison and plotted by Beskok et al.
(1996).

of temperature are alleviated by using an identical procedure to that utilized for the tangential velocity boundary condition.

Several experiments in low-pressure macroducts or in microducts confirm the necessity of applying the slip boundary condition at sufficiently large Knudsen numbers. Among them are those conducted by Knudsen (1909), Pfahler at al. (1991), Tison (1993), Liu et al. (1993, 1995), Pong et al. (1994), Arkilic et al. (1995), Harley et al. (1995), and Shih et al. (1995, 1996). The experiments are complemented by the numerical simulations carried out by Beskok (1994, 1996); Beskok and Karniadakis (1994, 1999); and Beskok, Karniadakis, and Trimmer (1996). Here we present selected examples of the experimental and numerical results.

Tison (1993) conducted pipe flow experiments at very low pressures. His pipe had a diameter of 2 mm and a length-to-diameter ratio of 200. Both inlet and outlet pressures were varied to yield Knudsen numbers in the range of $Kn = 0$–200. Figure 13.4 shows the variation of mass flow rate as a function of $(p_i^2 - p_o^2)$, where p_i is the inlet pressure and p_o is the outlet pressure.[5] The pressure drop in this rarefied pipe flow is nonlinear, which is characteristic of low-Reynolds-number, compressible flows. Three distinct flow regimes are identified: (1) the slip-flow regime, $0 < Kn < 0.6$; (2) transition regime, $0.6 < Kn < 17$, where the mass flow rate is almost constant as the pressure changes; and (3) free-molecule flow, $Kn > 17$. Note that the demarcation between these three regimes is slightly different from that mentioned in Section 13.2.2. As stated, the different Knudsen number regimes are determined empirically and are therefore only approximate for a particular flow geometry.

Shih et al. (1995) conducted their experiments in a microchannel using helium as a fluid. The inlet pressure varied, but the duct exit was atmospheric. Microsensors where fabricated in situ along their MEMS channel to measure the pressure.

[5] The original data in this figure were acquired by S.A. Tison and plotted by Beskok et al. (1996).

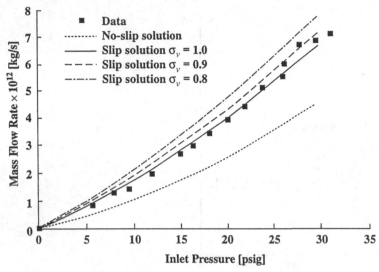

Figure 13.5: Mass flow rate versus inlet pressure in a microchannel [after Shih et al. 1995].

Figure 13.5 shows their measured mass flow rate versus the inlet pressure. The data are compared to the no-slip solution and the slip solution using three different values of the tangential-momentum-accommodation coefficient, 0.8, 0.9, and 1.0. The agreement is reasonable with the case $\sigma_\nu = 1.0$, indicating perhaps that the channel used by Shih et al. was quite rough on the molecular scale. In a second experiment (Shih et al. 1996), nitrous oxide was used as the fluid. The square of the pressure distribution along the channel is plotted in Figure 13.6 for five different inlet pressures. The experimental data (symbols) compare well with the theoretical predictions (solid lines). Again, the nonlinear pressure drop shown indicates that the gas flow is compressible.

Arkilic (1997) provided an elegant analysis of the compressible, rarefied flow in a microchannel. The results of his theory are compared with the experiments of Pong et al. (1994) in Figure 13.7. The dotted line is the incompressible flow

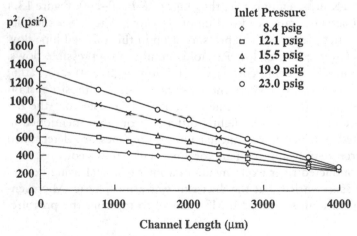

Figure 13.6: Pressure distribution of nitrous oxide in a microduct. Solid lines are theoretical predictions [after Shih et al. 1996].

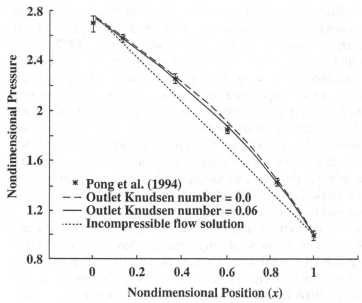

Figure 13.7: Pressure distribution in a long microchannel. The symbols are experimental data while the solid lines are different theoretical predictions [after Arkilic 1997].

solution, where the pressure is predicted to drop linearly with streamwise distance. The dashed line is the compressible flow solution that neglects rarefaction effects (assumes $Kn = 0$). Finally, the solid line is the theoretical result that takes into account both compressibility and rarefaction via slip-flow boundary condition computed at the exit Knudsen number of $Kn = 0.06$. That theory compares most favorably with the experimental data. In the compressible flow through the constant-area duct, density decreases, and thus velocity increases in the streamwise direction. As a result, the pressure distribution is nonlinear with negative curvature. A moderate Knudsen number (i.e., moderate slip) actually diminishes, albeit rather weakly, this curvature. Thus, compressibility and rarefaction effects lead to opposing trends, as pointed out by Beskok et al. (1996).

13.2.6 Molecular-Based Models

In the continuum models discussed in Section 13.2.3, the macroscopic fluid properties are the dependent variables, whereas the independent variables are the three spatial coordinates and time. The molecular models recognize the fluid as a myriad of discrete particles: molecules, atoms, ions, and electrons. The goal here is to determine the position, velocity, and state of all particles at all times. The molecular approach is either deterministic or probabilistic (refer to Figure 13.2). Provided that there is a sufficient number of microscopic particles within the smallest significant volume of a flow, the macroscopic properties at any location in the flow can then be computed from the discrete-particle information by a suitable averaging or weighted averaging process. This subsection discusses molecular-based models and their relation to the continuum models previously considered.

The most fundamental of the molecular models is a deterministic one. The motion of the molecules is governed by the laws of classical mechanics, although, at the expense of greatly complicating the problem, the laws of quantum mechanics can also be considered in special circumstances. The modern molecular dynamics computer simulations (MD) have been pioneered by Alder and Wainwright (1957, 1958, 1970) and reviewed by Ciccotti and Hoover (1986); Allen and Tildesley (1987); Haile (1993); and Koplik and Banavar (1995). The simulation begins with a set of N molecules in a region of space, each assigned a random velocity corresponding to a Boltzmann distribution at the temperature of interest. The interaction between the particles is prescribed typically in the form of a two-body potential energy, and the time evolution of the molecular positions is determined by integrating Newton's equations of motion. Because MD is based on the most basic set of equations, it is valid in principle for any flow extent and any range of parameters. The method is straightforward in principle, but there are two hurdles: (1) the choice of a proper and convenient potential for particular fluid and solid combinations and (2) the colossal computer resources required to simulate a reasonable flowfield extent.

For purists, the former difficulty is a sticky one. There is no totally rational methodology by which a convenient potential can be chosen. Part of the art of MD is to pick an appropriate potential and validate the simulation results with experiments or other analytical or computational results. A commonly used potential between two molecules is the generalized Lennard–Jones 6–12 potential to be used in Section 13.2.7 and further discussed in Section 13.2.8.

The second difficulty, and by far the most serious limitation of molecular dynamics simulations, is the number of molecules N that can realistically be modeled on a digital computer. Because the computation of an element of trajectory for any particular molecule requires consideration of *all* other molecules as potential collision partners, the amount of computation required by the MD method is proportional to N^2. Some saving in computer time can be achieved by cutting off the weak tail of the potential (see Figure 13.12) at, say, $r_c = 2.5\sigma$ and shifting the potential by a linear term in r so that the force goes smoothly to zero at the cutoff. As a result, only nearby molecules are treated as potential collision partners, and the computation time for N molecules no longer scales with N^2.

The state of the art of molecular dynamics simulations in the 1990s is such that with a few hours of CPU time, general-purpose supercomputers can handle around 10,000 molecules. At enormous expense, the fastest parallel machine available can simulate around 1 million particles. Because of the extreme diminution of molecular scales, these capabilities translate into regions of liquid flow of about 0.01 μm (100 Å) in linear size over time intervals of around 0.001 μs, which is just enough for continuum behavior to set in for simple molecules. To simulate 1 s of real time for complex molecular interactions (e.g., including vibration modes, reorientation of polymer molecules, collision of colloidal particles, etc.) requires unrealistic CPU time measured in thousands of years.

The MD simulations are highly inefficient for dilute gases, for their molecular interactions are infrequent. These simulations are more suited for dense gases and liquids. Clearly, molecular dynamics simulations are reserved for situations in which the continuum approach or the statistical methods are inadequate to compute important flow quantities from first principles. Slip boundary conditions for liquid flows in extremely small devices is such a case, as will be discussed in Section 13.2.7.

An alternative to the deterministic molecular dynamics is the statistical approach in which the goal is to compute the probability of finding a molecule at a particular position and state. If the appropriate conservation equation can be solved for the probability distribution, important statistical properties such as the mean number, momentum, or energy of the molecules within an element of volume can be computed from a simple weighted averaging. In a practical problem, it is such average quantities that concern us rather than the detail for every single molecule. Clearly, however, the accuracy of computing average quantities, via the statistical approach, improves as the number of molecules in the sampled volume increases. The kinetic theory of dilute gases is well advanced, but that for dense gases and liquids is much less so owing to the extreme complexity of having to include multiple collisions and intermolecular forces in the theoretical formulation. The statistical approach is well covered in books such as those by Kennard (1938); Hirschfelder, Curtiss, and Bird (1954); Schaaf and Chambré (1961); Vincenti and Kruger (1965); Kogan (1969); Chapman and Cowling (1970); Cercignani (1988); and Bird (1994) as well as in review articles such as those by Kogan (1973); Muntz (1989); and Oran, Oh, and Cybyk (1998).

In the statistical approach, the fraction of molecules in a given location and state is the sole dependent variable. The independent variables for monatomic molecules are time, the three spatial coordinates, and the three components of molecular velocity. These describe a six-dimensional phase space.[6] For diatomic or polyatomic molecules, the dimension of phase space is increased by the number of internal degrees of freedom. Orientation adds an extra dimension for molecules that are not spherically symmetric. Finally, for mixtures of gases, separate probability distribution functions are required for each species. Clearly, the complexity of the approach increases dramatically as the dimension of phase space increases. The simplest problems are, for example, those for steady, one-dimensional flow of a simple monatomic gas.

To simplify the problem we restrict the discussion here to monatomic gases having no internal degrees of freedom. Furthermore, the fluid is restricted to dilute gases and molecular chaos is assumed. The former restriction requires the average distance between molecules δ to be an order of magnitude larger than their diameter σ. That will almost guarantee that all collisions between molecules are binary collisions and thus enable one to avoid the complexity of modeling multiple encounters.[7] The molecular chaos restriction improves the accuracy of computing the macroscopic quantities from the microscopic information. In essence, the volume over which averages are computed has to have a sufficient number of molecules to reduce statistical errors. It can be shown that computing macroscopic flow properties by averaging over a number of molecules will result in statistical fluctuations with a standard deviation of approximately 0.1%, if one million molecules are used, and around 3% if one thousand molecules are used. The molecular chaos limit requires the length scale L for the averaging process to be at least 100 times the average distance between molecules (i.e., typical averaging over at least one million molecules).

Figure 13.8, adapted from Bird (1994), shows the limits of validity of the dilute gas approximation ($\delta/\sigma > 7$), the continuum approach ($Kn < 0.1$, as discussed

---

[6] The evolution equation of the probability distribution is considered; hence, time is the seventh independent variable.

[7] Dissociation and ionization phenomena involve triple collisions and therefore require separate treatment.

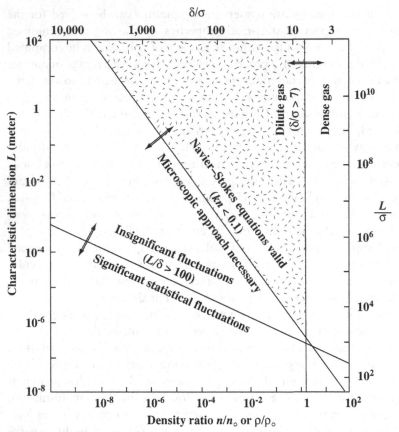

Figure 13.8: Effective limits of different flow models [after Bird 1994].

previously in Section 13.2.2), and the neglect of statistical fluctuations ($L/\delta > 100$). Using a molecular diameter of $\sigma = 4 \times 10^{-10}$ m as an example, the three limits are conveniently expressed as functions of the normalized gas density ρ/ρ_o or number density n/n_o, where the reference densities ρ_o and n_o are computed at standard conditions. All three limits are straight lines in the log–log plot of L versus ρ/ρ_o, as depicted in Figure 13.8. Note the shaded triangular wedge inside which both the Boltzmann and Navier–Stokes equations are valid. Additionally, the lines describing the three limits very nearly intersect at a single point. As a consequence, the continuum breakdown limit always lies between the dilute gas limit and the limit for molecular chaos. As density or characteristic dimension is reduced in a dilute gas, the Navier–Stokes model breaks down before the level of statistical fluctuations becomes significant. In a dense gas, on the other hand, significant fluctuations may be present even when the Navier–Stokes model is still valid.

The starting point in statistical mechanics is the Liouville equation, which expresses the conservation of the N-particle distribution function in $6N$-dimensional phase space,[8] where N is the number of particles under consideration. If only external

---

[8] Three positions and three velocities for *each* molecule of a monatomic gas with no internal degrees of freedom.

forces that do not depend on the velocity of the molecules,[9] the Liouville equation for a system of N mass points reads

$$\frac{\partial \mathcal{F}}{\partial t} + \sum_{k=1}^{N} \boldsymbol{\xi}_k \cdot \frac{\partial \mathcal{F}}{\partial \mathbf{x}_k} + \sum_{k=1}^{N} \mathbf{F}_k \cdot \frac{\partial \mathcal{F}}{\partial \boldsymbol{\xi}_k} = 0 \tag{13.49}$$

where $\mathcal{F}$ is the probability of finding a molecule at a particular point in phase space, t is time, $\boldsymbol{\xi}_k$ is the three-dimensional velocity vector for the kth molecule, $\mathbf{x}_k$ is the three-dimensional position vector for the kth molecule, and $\mathbf{F}$ is the external force vector. Note that the dot product in the preceding equation is carried out over each of the three components of the vectors $\boldsymbol{\xi}$, $\mathbf{x}$, and $\mathbf{F}$ and that the summation is over all molecules. Obviously such an equation is not tractable for a realistic number of particles.

A hierarchy of reduced distribution functions may be obtained by repeated integration of the Eq. (13.49). The final equation in the hierarchy is for the single particle distribution which also involves the two-particle distribution function. If molecular chaos is assumed, that final equation becomes a closed one (i.e., one equation in one unknown) and is known as the Boltzmann equation, the fundamental relation of the kinetic theory of gases. That final equation in the hierarchy is the only one that carries any hope of obtaining analytical solutions.

A simpler direct derivation of the Boltzmann equation is provided by Bird (1994). For monatomic gas molecules in binary collisions, the integro-differential Boltzmann equation reads

$$\frac{\partial(nf)}{\partial t} + \xi_j \frac{\partial(nf)}{\partial x_j} + F_j \frac{\partial(nf)}{\partial \xi_j} = J(f, f^*), \quad j = 1, 2, 3 \tag{13.50}$$

where nf is the product of the number density and the normalized velocity distribution function ($dn/n = f\,d\boldsymbol{\xi}$), x_j and ξ_j are respectively the coordinates and speeds of a molecule,[10] F_j is a known external force, and $J(f, f^*)$ is the nonlinear collision integral that describes the net effect of populating and depopulating collisions on the distribution function. The collision integral is the source of difficulty in obtaining analytical solutions to the Boltzmann equation and is given by

$$J(f, f^*) = \int_{-\infty}^{\infty} \int_{0}^{4\pi} n^2 (f^* f_1^* - f f_1) \xi_r \sigma \, d\Omega \, (d\boldsymbol{\xi})_1 \tag{13.51}$$

where the superscript * indicates postcollision values, f and f_1 represent two different molecules, ξ_r is the relative speed between two molecules, σ is the molecular cross-section, Ω is the solid angle, and $d\boldsymbol{\xi} = d\xi_1 \, d\xi_2 \, d\xi_3$.

Once a solution for f is obtained, macroscopic quantities such as density, velocity, temperature, and so forth, can be computed from the appropriate weighted integral of the distribution function. For example,

$$\rho = mn = m \int (nf) \, d\boldsymbol{\xi} \tag{13.52}$$

---

[9] This excludes Lorentz forces, for example.

[10] Constituting, together with time, the seven independent variables of the single-dependent-variable equation.

$$u_i = \int \xi_i\, f\, \mathrm{d}\boldsymbol{\xi} \tag{13.53}$$

$$\frac{3}{2}kT = \int \frac{1}{2}m\xi_i\xi_i\, f\, \mathrm{d}\boldsymbol{\xi} \tag{13.54}$$

If the Boltzmann equation is nondimensionalized with a characteristic length L and characteristic speed $[2(k/m)T]^{1/2}$, where k is the Boltzmann constant, m is the molecular mass, and T is temperature, the inverse Knudsen number appears explicitly in the right-hand side of the equation as follows:

$$\frac{\partial \hat{f}}{\partial \hat{t}} + \hat{\xi}_j \frac{\partial \hat{f}}{\partial \hat{x}_j} + \hat{F}_j \frac{\partial \hat{f}}{\partial \hat{\xi}_j} = \frac{1}{Kn}\hat{J}(\hat{f}, \hat{f}^*), \quad j = 1, 2, 3 \tag{13.55}$$

where the superscript ^ represents a dimensionless variable, and $\hat{f}$ is nondimensionalized using a reference number density n_o.

The five conservation equations for the transport of mass, momentum, and energy can be derived by multiplying the Boltzmann equation above by, respectively, the molecular mass, momentum, and energy and then integrating over all possible molecular velocities. Subject to the restrictions of dilute gas and molecular chaos stated earlier, the Boltzmann equation is valid for all ranges of Knudsen number from 0 to ∞. Analytical solutions to this equation for arbitrary geometries are difficult mostly because of the nonlinearity of the collision integral. Simple models of this integral have been proposed to facilitate analytical solutions; see, for example, Bhatnagar, Gross, and Krook (1954).

There are two important asymptotes to Eq. (13.55). First, as $Kn \to \infty$, molecular collisions become unimportant. This is the free-molecule flow regime depicted in Figure 13.3 for $Kn > 10$, where the only important collision is that between a gas molecule and the solid surface of an obstacle or a conduit. Analytical solutions are then possible for simple geometries, and numerical simulations for complicated geometries are straightforward once the surface-reflection characteristics are accurately modeled. Secondly, as $Kn \to 0$, collisions become important, and the flow approaches the continuum regime of conventional fluid dynamics. The second law specifies a tendency for thermodynamic systems to revert to the equilibrium state, smoothing out any discontinuities in macroscopic flow quantities. The number of molecular collisions in the limit $Kn \to 0$ is so large that the flow approaches the equilibrium state in a time short compared with the macroscopic time scale. For example, for air at standard conditions ($T = 288$ K; $p = 1$ atm), each molecule experiences, on the average, 10 collisions per nanosecond and travels 1 micron in the same time period. Such a molecule has already *forgotten* its previous state after 1 ns. In a particular flowfield, if the macroscopic quantities vary little over a distance of 1 μm or over a time interval of 1 ns, the flow of air at standard temperature and pressure is near equilibrium.

At $Kn = 0$, the velocity distribution function is everywhere of the local equilibrium or Maxwellian form

$$\hat{f}^{(0)} = \frac{n}{n_o}\pi^{-3/2}\exp[-(\hat{\boldsymbol{\xi}} - \hat{u})^2] \tag{13.56}$$

where $\hat{\boldsymbol{\xi}}$ and $\hat{u}$ are, respectively, the dimensionless speeds of a molecule and of the flow. In this Knudsen number limit, the velocity distribution of each element of the fluid instantaneously adjusts to the equilibrium thermodynamic state appropriate to the local macroscopic properties as this molecule moves through the flowfield. From the continuum viewpoint, the flow is isentropic, and heat conduction and viscous diffusion and dissipation vanish from the continuum conservation relations.

The Chapman–Enskog theory attempts to solve the Boltzmann equation by considering a small perturbation of $\hat{f}$ from the equilibrium Maxwellian form. For small Knudsen numbers, the distribution function can be expanded in terms of Kn in the form of a power series

$$\hat{f} = \hat{f}^{(0)} + Kn\,\hat{f}^{(1)} + Kn^2\,\hat{f}^{(2)} + \cdots \tag{13.57}$$

By substituting the preceding series in the Boltzmann equation (13.55) and equating terms of equal order, the following recurrent set of integral equations results:

$$\hat{J}\left(\hat{f}^{(0)},\, \hat{f}^{(0)}\right) = 0, \quad \hat{J}\left(\hat{f}^{(0)},\, \hat{f}^{(1)}\right) = \frac{\partial \hat{f}^{(0)}}{\partial \hat{t}} + \hat{\xi}_j \frac{\partial \hat{f}^{(0)}}{\partial \hat{x}_j} + \hat{F}_j \frac{\partial \hat{f}^{(0)}}{\partial \hat{\xi}_j}, \ldots \tag{13.58}$$

The first integral is nonlinear, and its solution is the local Maxwellian distribution, Eq. (13.56). The distribution functions $\hat{f}^{(1)}$, $\hat{f}^{(2)}$, and so forth each satisfy an inhomogeneous linear equation whose solution leads to the transport terms needed to close the continuum equations appropriate to the particular level of approximation. The continuum stress tensor and heat flux vector can be written in terms of the distribution function, which in turn can be specified in terms of the macroscopic velocity and temperature and their derivatives (Kogan 1973). The zeroth-order equation yields the Euler equations, the first-order equation results in the linear transport terms of the Navier–Stokes equations, the second-order equation gives the nonlinear transport terms of the Burnett equations, and so on. Keep in mind, however, that the Boltzmann equation as developed in this subsection is for a monatomic gas. This excludes the all-important air, which is composed largely of diatomic nitrogen and oxygen.

As discussed in Sections 13.2.2, 13.2.3, and 13.2.5, the Navier–Stokes equations can and should be used up to a Knudsen number of 0.1. Beyond that, the transition flow regime commences ($0.1 < Kn < 10$). In this flow regime, the molecular mean free path for a gas becomes significant relative to a characteristic distance for important flow-property changes to take place. The Burnett equations can be used to obtain analytical and numerical solutions for at least a portion of the transition regime for a monatomic gas, although their complexity has precluded realizing many results for realistic geometries (Agarwal, Yun, and Balakrishnan 1999). There is also some degree of uncertainty about the proper boundary conditions to use with the continuum Burnett equations, and experimental validations of the results have been very scarce. Additionally, as the gas flow further departs from equilibrium, the bulk viscosity ($= \lambda + \frac{2}{3}\mu$, where λ is the second coefficient of viscosity) is no longer zero, and the Stokes hypothesis no longer holds (see Gad-el-Hak 1995 for an interesting summary of the issue of bulk viscosity).

In the transition regime, the molecularly based Boltzmann equation cannot easily be solved either unless the nonlinear collision integral is simplified. Thus, clearly the transition regime is in dire need of alternative solution methods. The MD simulations, as mentioned earlier, are not suited for dilute gases. The best approach for the transition regime right now is the direct simulation Monte Carlo (DSMC) method developed by Bird (1963, 1965, 1976, 1978, 1994) and briefly described below. Some recent reviews of DSMC include those by Muntz (1989), Cheng (1993), Cheng and Emmanuel (1995), and Oran et al. (1998). The mechanics as well as the history of the DSMC approach and its ancestors are well described in the book by Bird (1994).

Unlike molecular dynamics simulations, DSMC is a statistical computational approach to solving rarefied gas problems. Both approaches treat the gas as discrete particles. Subject to the dilute gas and molecular chaos assumptions, the direct simulation Monte Carlo method is valid for all ranges of Knudsen number, although the method becomes quite expensive for $Kn < 0.1$. Fortunately, this is the continuum regime in which the Navier–Stokes equations can be used analytically or computationally. The DSMC method is therefore ideal for the transition regime ($0.1 < Kn < 10$), where the Boltzmann equation is difficult to solve. The Monte Carlo method is, like its name sake, a random number strategy based directly on the physics of the individual molecular interactions. The idea is to track a large number of randomly selected, statistically representative particles and to use their motions and interactions to modify their positions and states. The primary approximation of the DSMC method is to uncouple the molecular motions and the intermolecular collisions over small time intervals. A significant advantage of this approximation is that the amount of computation required is proportional to N, in contrast to N^2, for molecular dynamics simulations. In essence, particle motions are modeled deterministically, whereas collisions are treated probabilistically, each simulated molecule representing a large number of actual molecules. Typical computer runs of DSMC in the 1990s involve tens of millions of intermolecular collisions and fluid–solid interactions.

The DSMC computation is started from some initial condition and followed in small time steps that can be related to physical time. Colliding pairs of molecules in a small geometric cell in physical space are randomly selected after each computational time step. Complex physics such as radiation, chemical reactions, and species concentrations can be included in the simulations without the necessity of nonequilibrium thermodynamic assumptions that commonly afflict nonequilibrium continuum-flow calculations. Because DSMC is more computationally intensive than classical continuum simulations, it should therefore be used only when the continuum approach is not feasible.

The DSMC technique is explicit and time marching and therefore always produces unsteady flow simulations. For macroscopically steady flows, Monte Carlo simulation proceeds until a steady flow is established within a desired accuracy at sufficiently large time. The macroscopic flow quantities are then the time average of all values calculated after reaching the steady state. For macroscopically unsteady flows, ensemble averaging of many independent Monte Carlo simulations is carried out to obtain the final results within a prescribed statistical accuracy.

13.2.7 Liquid Flows

From the continuum point of view, liquids and gases are both fluids obeying the same equations of motion. For incompressible flows, for example, the Reynolds number is the primary dimensionless parameter that determines the nature of the flowfield. It is true that water, for example, has density and viscosity that are, respectively, three and two orders of magnitude higher than those for air, but if the Reynolds number and geometry are matched, liquid and gas flows should be identical.[11] For MEMS applications, however, we anticipate the possibility of nonequilibrium flow conditions and the consequent invalidity of the Navier–Stokes equations and the no-slip boundary conditions. Such circumstances can best be researched using the molecular approach. This was discussed for gases in Section 13.2.6, and the corresponding arguments for liquids will be given in this subsection. The literature on non-Newtonian fluids in general and polymers in particular is vast (for example, the bibliographic survey by Nadolink and Haigh, 1995, cites over 4900 references on polymer drag reduction alone) and provides a rich source of information on the molecular approach for liquid flows.

Solids, liquids, and gases are distinguished merely by the degree of proximity and the intensity of motions of their constituent molecules. In solids, the molecules are packed closely and confined, each hemmed in by its neighbors (Chapman and Cowling 1970). Only rarely would one solid molecule slip from its neighbors to join a new set. As the solid is heated, molecular motion becomes more violent, and a slight thermal expansion takes place. At a certain temperature that depends on ambient pressure, sufficiently intense motion of the molecules enables them to pass freely from one set of neighbors to another. The molecules are no longer confined but are nevertheless still closely packed, and the substance is now considered a liquid. Further heating of the matter eventually releases the molecules altogether, allowing them to break the bonds of their mutual attractions. Unlike solids and liquids, the resulting gas expands to fill any volume available to it.

Unlike solids, both liquids and gases cannot resist finite shear force without continuous deformation, that is, the definition of a fluid medium. In contrast to the reversible, elastic, static deformation of a solid, the continuous deformation of a fluid resulting from the application of a shear stress results in an irreversible work that eventually becomes random thermal motion of the molecules, that is, viscous dissipation. In a 1-μm cube, there are around 25 million molecules of air at standard temperature and pressure. The same cube would contain around 34 billion molecules of water. Thus, liquid flows are continuum even in extremely small devices through which gas flows would not. The average distance between molecules in the gas example is one order of magnitude higher than the diameter of its molecules, whereas that for the liquid phase approaches the molecular diameter. As a result, liquids are almost incompressible. Their isothermal compressibility coefficient α and bulk expansion coefficient β are much smaller compared with those for gases. For water, for example, a hundredfold increase in pressure leads to a less than 0.5% decrease in volume. Sound speeds through liquids are also high relative to those for gases, and as a result most liquid flows are incompressible.[12] The exception is propagation of ultra-high-frequency sound waves and cavitation phenomena.

The mechanism by which liquids transport mass, momentum, and energy must be very different from that for gases. In dilute gases, intermolecular forces play no role, and the molecules spend most of their time in free flight between brief collisions during which the molecules' direction and speed abruptly change. The random molecular motions are responsible for gaseous transport processes. In liquids, on the other

---

[11] Barring phenomena unique to liquids such as cavitation, free surface flows, and so forth.

[12] Note that we distinguish between a fluid and a flow being compressible or incompressible. For example, the *flow* of the highly compressible air can be either compressible or incompressible.

hand, the molecules are closely packed though not fixed in one position. In essence, the liquid molecules are always in a *collision* state. Applying a shear force must create a velocity gradient so that the molecules move relative to one another ad infinitum as long as the stress is applied. For liquids, momentum transport due to the random molecular motion is negligible compared with that due to the intermolecular forces. The straining between liquid molecules causes some to separate from their original neighbors, bringing them into the force field of new molecules. Across the plane of the shear stress, the sum of all intermolecular forces must, on the average, balance the imposed shear. Liquids at rest transmit only normal force, but when a velocity gradient occurs, the net intermolecular force will have a tangential component.

The incompressible Navier–Stokes equations describe liquid flows under most circumstances. Liquids, however, do not have as well-advanced a molecular-based theory as that for dilute gases. The concept of mean free path is not very useful for liquids, and the conditions under which a liquid flow fails to be in quasi-equilibrium state are not well defined. There is no Knudsen number for liquid flows to guide us through the maze. We do not know, from first principles, the conditions under which the no-slip boundary condition becomes inaccurate or the point at which the (stress)–(rate of strain) relation or the (heat flux)–(temperature gradient) relation fails to be linear. Certain empirical observations indicate that those simple relations that we take for granted occasionally fail to model liquid flows accurately. For example, it has been shown in rheological studies (Loose and Hess 1989) that non-Newtonian behavior commences when the strain rate approximately exceeds twice the molecular frequency-scale

$$\dot{\gamma} = \frac{\partial u}{\partial y} \geq 2\tau^{-1} \tag{13.59}$$

where the molecular time scale τ is given by

$$\tau = \left[\frac{m\sigma^2}{\epsilon} \right]^{\frac{1}{2}} \tag{13.60}$$

where m is the molecular mass, and σ and ϵ are, respectively, the characteristic length and energy scales for the molecules. For ordinary liquids such as water, this time scale is extremely small, and the threshold shear rate for the onset of non-Newtonian behavior is therefore extraordinarily high. For high-molecular-weight polymers, on the other hand, m and σ are both many orders of magnitude higher than their respective values for water, and the linear stress-strain relation breaks down at realistic values of the shear rate.

The moving contact line, when a liquid spreads on a solid substrate, is an example for which slip flow must be allowed to avoid singular or unrealistic behavior in the Navier–Stokes solutions (Dussan and Davis 1974; Dussan 1976, 1979; Thompson and Robbins 1989). Other examples for which slip flow must be admitted include corner flows (Moffatt 1964; Koplik and Banavar 1995) and extrusion of polymer melts from capillary tubes (Pearson and Petrie 1968; Richardson 1973; Den 1990).

Existing experimental results of liquid flow in microdevices are contradictory. This is not surprising given the difficulty of such experiments and the lack of a guiding rational theory. Pfahler et al. (1990, 1991), Pfahler (1992), and Bau (1994) summarize the relevant literature. For small-length-scale flows, a phenomenological approach for analyzing the data is to define an *apparent* viscosity μ_a calculated so that if it were used in the traditional no-slip Navier–Stokes equations instead of the fluid viscosity μ, the results would be in agreement with experimental observations. Israelachvili (1986) and Gee et al. (1990) found that $\mu_a = \mu$ for thin-film flows as long as the film thickness exceeds 10 molecular layers (≈ 5 nm). For thinner films, μ_a depends on the number of molecular layers and can be as much as 10^5 times larger than μ. Chan and Horn's (1985) results are somewhat different; the apparent viscosity deviates from the fluid viscosity for films thinner than 50 nm.

In polar-liquid flows through capillaries, Migun and Prokhorenko (1987) reported that μ_a increases for tubes smaller than 1 μm in diameter. In contrast, Debye and Cleland (1959) reported μ_a smaller than μ for paraffin flow in porous glass with average pore size several times larger than the molecular length scale. Experimenting with microchannels ranging in depths from 0.5 to 50 μm, Pfahler et al. (1991) found that μ_a is consistently smaller than μ for both liquid (isopropyl alcohol, silicone oil) and gas (nitrogen, helium) flows in microchannels. For liquids, the apparent viscosity decreases with decreasing channel depth. Other researchers using small capillaries report that μ_a is about the same as μ (Anderson and Quinn 1972; Tuckermann and Pease 1981, 1982; Tuckermann 1984; Guvenc 1985; Nakagawa, Shoji, and Esashi 1990).

The preceding contradictory results point to the need for replacing phenomenological models by first-principles ones. The lack of a molecular-based theory of liquids—despite extensive research by the rheology and polymer communities—leaves molecular dynamics simulations as the nearest weapon to the first-principles arsenal. Molecular dynamics simulations offer a unique approach to checking the validity of the traditional continuum assumptions. However, as was pointed out in Section 13.2.6, such simulations are limited to exceedingly minute flow extent.

Thompson and Troian (1997) provided molecular dynamics simulations to quantify the slip-flow boundary condition dependence on shear rate. Recall the linear Navier boundary condition introduced in Section 13.2.5,

$$\Delta u \bigg|_w = u_{\text{fluid}} - u_{\text{wall}} = L_s \frac{\partial u}{\partial y}\bigg|_w \tag{13.61}$$

where L_s is the constant slip length, and $\frac{\partial u}{\partial y}|_w$ is the strain rate computed at the wall. The goal of Thompson and Troian's simulations was to determine the degree of slip at a solid–liquid interface as the interfacial parameters and the shear rate change. In their simulations, a simple liquid underwent planar shear in a Couette cell as shown in Figure 13.9. The typical cell measured $12.51 \times 7.22 \times h$ in units of molecular length scale σ, where the channel depth h varied in the range of 16.71σ–24.57σ, and the corresponding number of molecules simulated ranged from 1152 to 1728. The liquid was treated as an isothermal ensemble of spherical molecules. A shifted Lennard–Jones 6–12 potential was used to model intermolecular interactions with energy and length scales ϵ and σ and cut-off distance $r_c = 2.2\sigma$:

$$V(r) = 4\epsilon \left[\left(\frac{r}{\sigma}\right)^{-12} - \left(\frac{r}{\sigma}\right)^{-6} - \left(\frac{r_c}{\sigma}\right)^{-12} + \left(\frac{r_c}{\sigma}\right)^{-6} \right] \tag{13.62}$$

The truncated potential was set to zero for $r > r_c$.

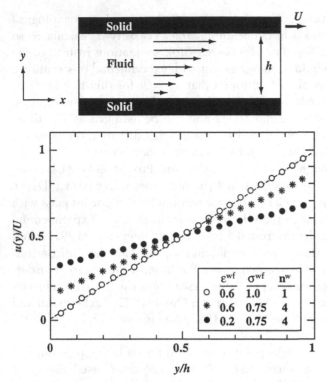

Figure 13.9: Velocity profiles in a Couette flow geometry at different interfacial parameters. All three profiles are for $U = \sigma\tau^{-1}$, and $h = 24.57\sigma$. The dashed line is the no-slip Couette-flow solution [after Thompson and Troian 1997].

The fluid–solid interaction was also modeled with a truncated Lennard–Jones potential with energy and length scales ϵ^{wf} and σ^{wf}, and cut-off distance r_c. The equilibrium state of the fluid was a well-defined liquid phase characterized by number density $n = 0.81\sigma^{-3}$ and temperature $T = 1.1\epsilon/k$, where k is the Boltzmann constant.

The steady-state velocity profiles resulting from Thompson and Troian's (1997) MD simulations are depicted in Figure 13.9 for different values of the interfacial parameters ϵ^{wf}, σ^{wf}, and n^{w}. These parameters, shown in units of the corresponding fluid parameters ϵ, σ, and n, characterize, respectively, the strength of the liquid–solid coupling, the thermal roughness of the interface, and the commensurability of wall and liquid densities. The macroscopic velocity profiles recover the expected flow behavior from continuum hydrodynamics with boundary conditions involving varying degrees of slip. Note that when slip exists, the shear rate $\dot{\gamma}$ no longer equals U/h. The degree of slip increases (i.e., the amount of momentum transfer at the wall–fluid interface decreases) as the relative wall density n^{w} increases or the strength of the wall–fluid coupling σ^{wf} decreases; in other words, when the relative surface energy corrugation of the wall decreases. Conversely, the corrugation is maximized when the wall and fluid densities are commensurate and the strength of the wall–fluid coupling is large. In this case, the liquid *feels* the corrugations in the surface energy of the solid owing to the atomic close-packing. Consequently, there is efficient momentum transfer and the no-slip condition applies, or in extreme cases, a "stick" boundary condition takes hold.

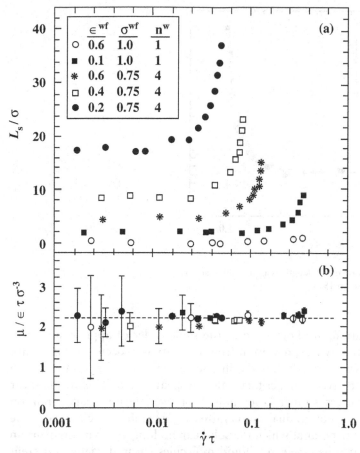

Figure 13.10: Variation of slip length and viscosity as functions of shear rate [after Thompson and Troian 1997].

Variations of the slip length L_s and viscosity μ as functions of shear rate $\dot{\gamma}$ are shown in parts (a) and (b) of Figure 13.10, for five different sets of interfacial parameters. For Couette flow, the slip length is computed from its definition, $L_s = \Delta u|_w/\dot{\gamma} = (U/\dot{\gamma} - h)/2$. The slip length, viscosity, and shear rate are normalized in the figure using the respective molecular scales for length σ, viscosity $\epsilon\tau\sigma^{-3}$, and inverse time τ^{-1}. The viscosity of the fluid is constant over the entire range of shear rates (Figure 13.10b), indicating Newtonian behavior. As indicated earlier, non-Newtonian behavior is expected for $\dot{\gamma} \geq 2\tau^{-1}$, which is well above the shear rates used in Thompson and Troian's simulations.

At low shear rates, the slip-length behavior is consistent with the Navier model (i.e., is independent of the shear rate). Its limiting value L_s^0 ranges from 0 to $\approx 17\sigma$ for the range of interfacial parameters chosen (Figure 13.10a). In general, the amount of slip increases with decreasing surface energy corrugation. Most interestingly, at high shear rates the Navier condition breaks down as the slip length increases rapidly with $\dot{\gamma}$. The critical shear-rate value for the slip length to diverge, $\dot{\gamma}_c$, decreases as the surface energy corrugation decreases. Surprisingly, the boundary condition is nonlinear even though the liquid is still Newtonian. In dilute gases, as discussed in Section 13.2.6, the linear slip condition and the Navier–Stokes equations, with their

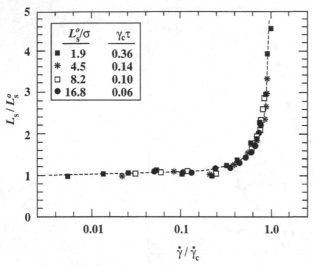

Figure 13.11: Universal relation of slip length as a function of shear rate [after Thompson and Troian 1997].

linear stress–strain relation, are both valid to the same order of approximation in Knudsen number. In other words, deviation from linearity is expected to take place at the same value of $Kn = 0.1$. In liquids, in contrast, the slip length appears to become nonlinear and to diverge at a critical value of shear rate well below the shear rate at which the linear stress–strain relation fails. Moreover, the boundary condition deviation from linearity is not gradual but is rather catastrophic. The critical value of shear rate $\dot{\gamma}_c$ signals the point at which the solid can no longer impart momentum to the liquid. This means that the same liquid molecules sheared against different substrates will experience varying amounts of slip and vice versa.

On the basis of the preceding results, Thompson and Troian (1997) suggested a universal boundary condition at a solid–liquid interface. Scaling the slip length L_s by its asymptotic limiting value L_s^o and the shear rate $\dot{\gamma}$ by its critical value $\dot{\gamma}$ collapses the data in the single curve shown in Figure 13.11. The data points are well described by the relation

$$L_s = L_s^o \left[1 - \frac{\dot{\gamma}}{\dot{\gamma}_c} \right]^{-\frac{1}{2}} \tag{13.63}$$

The nonlinear behavior close to a critical shear rate suggests that the boundary condition can significantly affect flow behavior at macroscopic distances from the wall. Experiments with polymers confirm this observation (Atwood and Schwalter 1989). The rapid change in the slip length suggests that for flows in the vicinity of $\dot{\gamma}_c$, small changes in surface properties can lead to large fluctuations in the apparent boundary condition. Thompson and Troian (1997) concluded that the Navier slip condition is but the low-shear-rate limit of a more generalized universal relationship that is nonlinear and divergent. Their relation provides a mechanism for relieving the stress singularity in spreading contact lines and corner flows, for it naturally allows for varying degrees of slip on approach to regions of higher rate of strain.

To place the preceding results in physical terms, consider water[13] at a temperature of $T = 288$ K. The energy scale in the Lennard–Jones potential is then $\epsilon = 3.62 \times 10^{-21}$ J. For water, $m = 2.99 \times 10^{-26}$ kg, $\sigma = 2.89 \times 10^{-10}$ m, and at standard temperature $n = 3.35 \times 10^{28}$ molecules/m^3. The molecular time scale can thus be computed, $\tau = [m\sigma^2/\epsilon]^{1/2} = 8.31 \times 10^{-13}$ s. For the third case depicted in Figure 13.11 (the open squares), $\dot{\gamma}_c \tau = 0.1$, and the critical shear rate at which the slip condition diverges is thus $\dot{\gamma}_c = 1.2 \times 10^{11}$ s^{-1}. Such an enormous rate of strain[14] may be found in extremely small devices having extremely high speeds. On the other hand, the conditions to achieve a measurable slip of 17σ (the solid circles in Figure 13.10) are not difficult to encounter in microdevices: density of the solid four times that of liquid and energy scale for wall–fluid interaction that is one-fifth of the energy scale for liquid.

The limiting value of slip length is independent of the shear rate and can be computed for water as $L_s^\circ = 17\sigma = 4.91 \times 10^{-9}$ m. Consider a water microbearing having a shaft diameter of 100 μm, rotation rate of 20,000 rpm, and a minimum gap of $h = 1$ μm. In this case, $U = 0.1$ m/s, and the no-slip shear rate is $U/h = 10^5$ s^{-1}. When slip occurs at the limiting value just computed, the shear rate and the wall-slip velocity are computed as follows:

$$\dot{\gamma} = \frac{U}{h + 2L_s^\circ} = 9.90 \times 10^4 \text{ s}^{-1} \tag{13.64}$$

$$\Delta u|_w = \dot{\gamma} L_s = 4.87 \times 10^{-4} \text{ m/s} \tag{13.65}$$

As a result of the Navier slip, the shear rate is reduced by 1% from its no-slip value, and the slip velocity at the wall is about 0.5% of U, which is small but not insignificant.

13.2.8 Surface Phenomena

As mentioned in Section 13.2.1, the surface-to-volume ratio for a machine with a characteristic length of 1 m is 1 m^{-1}, whereas that for a MEMS device having a size of 1 μm is 10^6 m^{-1}. The millionfold increase in surface area relative to the mass of the minute device substantially affects the transport of mass, momentum, and energy through the surface. Obviously, surface effects dominate in small devices. The surface boundary conditions in MEMS flows have already been discussed in Sections 13.2.5 and 13.2.7. In microdevices, it has been shown that it is possible to have measurable slip velocity and temperature jump at a solid–fluid interface. In this subsection, we illustrate other ramifications of the large surface-to-volume ratio unique to MEMS and provide a molecular viewpoint for surface forces.

In microdevices, both radiative and convective heat loss or gain are enhanced by the huge surface-to-volume ratio. Consider a device having a characteristic length L_s. Use of the lumped capacitance method to compute the rate of convective heat transfer, for example, is justified if the Biot number ($\equiv hL_s/\kappa_s$, where h is the convective heat

---

[13] Water molecules are complex ones forming directional, short-range covalent bonds and thus require a more complex potential than the Lennard–Jones to describe the intermolecular interactions. For the purpose of the qualitative example described here, however, we use the computational results of Thompson and Troian (1997), who employed the L–J potential.

[14] Note however that $\dot{\gamma}_c$ for high-molecular-weight polymers would be many orders of magnitude smaller than the value developed here for water.

transfer coefficient of the fluid and κ_s is the thermal conductivity of the solid) is less than 0.1. Small L_s implies small Biot number and a nearly uniform temperature within the solid. Within this approximation, the rate at which heat is lost to the surrounding fluid is given by

$$\rho_s L_s^3 c_s \frac{dT_s}{dt} = -h L_s^2 (T_s - T_\infty) \tag{13.66}$$

where ρ_s and c_s are, respectively, the density and specific heat of the solid, T_s is its (uniform) temperature, and T_∞ is the ambient fluid temperature. Solution of Eq. (13.66) is trivial, and the temperature of a hot surface drops exponentially with time from an initial temperature T_i,

$$\frac{T_s(t) - T_\infty}{T_i - T_\infty} = \exp\left[-\frac{t}{\mathcal{T}}\right] \tag{13.67}$$

where the time constant $\mathcal{T}$ is given by

$$\mathcal{T} = \frac{\rho_s L_s^3 c_s}{h L_s^2} \tag{13.68}$$

For small devices, the time it takes the solid to cool down is proportionally small. Clearly, the millionfold increase in surface-to-volume ratio implies a proportional increase in the rate at which heat escapes. Identical scaling arguments can be made regarding mass transfer.

Another effect of the diminished scale is the increased importance of surface forces and the waning importance of body forces. On the basis of biological studies, Went (1968) concluded that the demarcation length scale is around 1 mm. Below that, surface forces dominate over gravitational forces. A 10-mm piece of paper will fall down when gently placed on a smooth, vertical wall, whereas a 0.1-mm piece will stick. Try it! *Stiction* is a major problem in MEMS applications. Certain structures such as long, thin polysilicon beams and large, thin comb drives have a propensity to stick to their substrates and thus fail to perform as designed (Mastrangelo and Hsu 1992; Tang, Nguyen, and Howe 1989).

Conventional dry friction between two solids in relative motion is proportional to the normal force, which is usually a component of the moving device weight. The friction is independent of the contact-surface area because the van der Waals cohesive forces are negligible relative to the weight of the macroscopic device. In MEMS applications, the cohesive intermolecular forces between two surfaces are significant, and the stiction is independent of the device mass but is proportional to its surface area. The first micromotor did not move—despite large electric current through it—until the contact area between the 100-micron rotor and the substrate was reduced significantly by placing dimples on the rotor's surface (Fan, Tai, and Muller 1988, 1989; Tai and Muller 1989).

One last example of surface effects that to my knowledge has not been investigated for microflows is the adsorbed layer in gaseous wall-bounded flows. It is well known (see, for example, Brunauer 1944; Lighthill 1963b) that when a gas flows in a duct, the gas molecules are attracted to the solid surface by the van der Waals and other forces of cohesion. The potential energy of the gas molecules drops upon reaching the surface. The adsorbed layer partakes of the thermal vibrations of the solid, and the gas molecules can only escape when their energy exceeds the potential energy minimum. In equilibrium, at least part of the solid would be covered by a monomolecular layer of adsorbed gas molecules. Molecular species with significant partial pressure—relative to their vapor pressure—may locally form layers two or more molecules thick. Consider, for example, the flow of a mixture of dry air and water vapor at standard temperature and pressure. The energy of adsorption of water is much larger than that for nitrogen and oxygen, making it more difficult for water molecules to escape the potential energy trap. It follows that the lifetime of water molecules in the adsorbed layer significantly exceeds that for the air molecules (by 60,000 folds, in fact) and, as a result, the thin surface layer would be mostly water. For example, if the proportion of water vapor in the ambient air is 1:1000 (i.e., very low humidity level), the ratio of water to air in the adsorbed layer would be 60:1. Microscopic roughness of the solid surface causes partial condensation of the water along portions having sufficiently strong concave curvature. Thus, surfaces exposed to nondry air flows are mainly liquid water surfaces. In most applications, this thin adsorbed layer has little effect on the flow dynamics despite the density and viscosity of liquid water being far greater than those for air. In MEMS applications, however, the layer thickness may not be an insignificant portion of the characteristic flow dimension, and the water layer may have a measurable effect on the gas flow. A hybrid approach of molecular dynamics and continuum flow simulations or MD–Monte Carlo simulations may be used to investigate this issue.

It should be noted that quite recently Majumdar and Mezic (1998, 1999) have studied the stability and rupture into droplets of thin liquid films on solid surfaces. They pointed out that the free energy of a liquid film consists of a surface tension component as well as highly nonlinear volumetric intermolecular forces resulting from van der Waals, electrostatic, hydration, and elastic strain interactions. For water films on hydrophilic surfaces such as silica and mica, Majumdar and Mezic (1998) estimated the equilibrium film thickness to be about 0.5 nm (2 monolayers) for a wide range of ambient-air relative humidities. The equilibrium thickness grows very sharply, however, as the relative humidity approaches 100%.

Majumdar and Mezic's (1998, 1999) results open many questions. What are the stability characteristics of their water film in the presence of air flow above it? Would this water film affect the accommodation coefficient for microduct air flow? In a modern Winchester-type hard disk, the drive mechanism has a read–write head that floats 50 nm above the surface of the spinning platter. The head and platter together with the air layer in between form a slider bearing. Would the computer performance be affected adversely by the high relative humidity on a particular day when the adsorbed water film is no longer "thin"? If a microduct hauls liquid water, would the water film adsorbed by the solid walls influence the effective viscosity of the water flow? Electrostatic forces can extend to almost 1 micron (the Debye length), and that length is known to be highly pH-dependent. Would the water flow be influenced by the surface and liquid chemistry? Would this explain the contradictory experimental results of liquid flows in microducts discussed in Section 13.2.7?

The few examples above illustrate the importance of surface effects in small devices. From the continuum viewpoint, forces at a solid–fluid interface are the limit of pressure and viscous forces acting on a parallel elementary area displaced into the fluid when the displacement distance is allowed to tend to zero. From the molecular point of view, all macroscopic surface forces are ultimately traced to intermolecular forces, which subject is extensively covered in the book by Israelachvilli (1991) and references therein. Here we provide a very brief introduction to the molecular viewpoint. The four forces in nature are (1) the strong and (2) weak forces describing the interactions between neutrons, protons, electrons, and so forth; (3) the electromagnetic forces between atoms and molecules; and (4) gravitational forces between masses. The range of action of the first two forces is around 10^{-5} nm, and hence neither concerns us overly in MEMS applications. The electromagnetic forces are effective over a much larger though still small distance on the order of the interatomic separations (0.1–0.2 nm). Effects over longer range—several orders of magnitude longer—can and do rise from the short-range intermolecular forces. For example, the rise of the liquid column in capillaries and the action of detergent molecules in removing oily dirt from fabric are the result of intermolecular interactions. Gravitational forces decay with the distance to second power, whereas intermolecular forces decay much quicker, typically with the seventh power. Cohesive forces are therefore negligible once the distance between molecules exceeds few molecular diameters, whereas massive bodies like stars and planets are still strongly interacting, via gravity, over astronomical distances.

Electromagnetic forces are the source of all intermolecular interactions and the cohesive forces holding atoms and molecules together in solids and liquids. They can be classified into (1) purely electrostatic forces arising from the Coulomb force between charges, interactions between charges, permanent dipoles, quadrupoles, and so forth; (2) polarization forces arising from the dipole moments induced in atoms and molecules by the electric field of nearby charges and permanent dipoles; and (3) quantum mechanical forces that give rise to covalent or chemical bonding and to repulsive steric or exchange interactions that balance the attractive forces at very short distances. The Hellman–Feynman theorem of quantum mechanics states that once the spatial distribution of the electron clouds has been determined by solving the appropriate Schrödinger equation, intermolecular forces may be calculated on the basis of classical electrostatics, in effect reducing all intermolecular forces to Coulombic forces. Note, however, that intermolecular forces exist even when the molecules are totally neutral. Solutions of the Schrödinger equation for general atoms and molecules are not easy of course, and alternative modeling is sought to represent intermolecular forces. The van der Waals attractive forces are usually represented with a potential that varies as the inverse-sixth power of distance, whereas the repulsive forces are represented with either a power or an exponential potential.

A commonly used potential between two molecules is the generalized Lennard–Jones (L–J 6–12) pair potential given by

$$V_{ij}(r) = 4\epsilon \left[c_{ij} \left(\frac{r}{\sigma} \right)^{-12} - d_{ij} \left(\frac{r}{\sigma} \right)^{-6} \right] \tag{13.69}$$

where V_{ij} is the potential energy between two particles i and j, r is the distance between the two molecules, ϵ and σ are respectively characteristic energy and length scales, and c_{ij} and d_{ij} are parameters to be chosen for the particular fluid and solid combinations under consideration. The first term on the right-hand side is the strong repulsive force that is felt when two molecules are at extremely close range comparable to the molecular length scale. That short-range repulsion prevents overlap of the molecules in physical space. The second term is the weaker van der Waals attractive force that commences when the molecules are sufficiently close (several times σ). That negative part of the potential represents the attractive polarization interaction of neutral, spherically symmetric particles. The power of 6 associated with this term is derivable from quantum mechanics considerations, whereas the power of the repulsive part of the potential is found empirically. The Lennard–Jones potential is zero at very large distances, has a weak negative peak at r slightly larger than σ, is zero at $r = \sigma$, and is infinite as $r \to 0$.

The force field resulting from this potential is given by

$$F_{ij}(r) = -\frac{\partial V_{ij}}{\partial r} = \frac{48\epsilon}{\sigma}\left[c_{ij}\left(\frac{r}{\sigma}\right)^{-13} - \frac{d_{ij}}{2}\left(\frac{r}{\sigma}\right)^{-7}\right] \qquad (13.70)$$

A typical L–J 6–12 potential and force field are shown in Figure 13.12, for $c = d = 1$. The minimum potential $V_{\min} = -\epsilon$ corresponds to the equilibrium position (zero force) and occurs at $r = 1.12\sigma$. The attractive van der Waals contribution to the minimum potential is -2ϵ, whereas the repulsive energy contribution is $+\epsilon$. Thus, the inverse 12th-power repulsive force term decreases the strength of the binding energy at equilibrium by 50%.

The L–J potential is commonly used in molecular dynamics simulations to model intermolecular interactions between dense gas or liquid molecules and between fluid and solid molecules. As mentioned in Section 13.2.7, such potential is not accurate for complex substances such as water whose molecules form directional covalent bonds. As a result, MD simulations for water are much more involved.

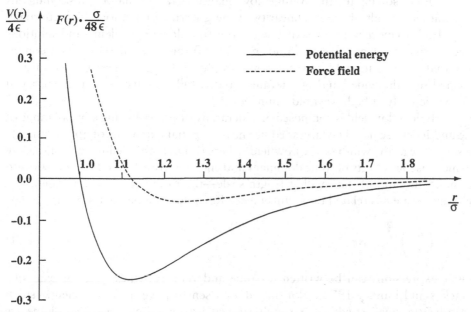

Figure 13.12: Typical Lennard–Jones 6–12 potential and the intermolecular force field resulting from it. Only a small portion of the potential function is shown for clarity.

13.3 Typical Fluid Applications

13.3.1 Introductory Remarks

The physics of fluid flows in microdevices was covered in Section 13.2. In this section, we provide several examples of useful applications of MEMS devices in fluid mechanics. The list is by no means exhaustive but includes the use of MEMS-based sensors and actuators for flow diagnosis and control, a recently developed viscous micropump–microturbine, and analysis of a journal microbearing. The articles by Gad-el-Hak (1999) and Löfdahl and Gad-el-Hak (1999) offer more detail on some of the topics covered in this section. From the perspective of the present monograph, using microsensors and microactuators for flow control is a very important topic and will be deferred to the following chapter.

13.3.2 Turbulence Measurements

Microelectromechanical systems offer great opportunities for better flow diagnosis and control—particularly for turbulent flows. The batch processing fabrication of microdevices makes it possible to produce large numbers of identical transducers within extremely tight tolerance. Microsensors and microactuators are small and inexpensive, combine electronic and mechanical parts, have low energy consumption, and can be distributed over a wide area. In this subsection we discuss the advantages of using MEMS-based sensors for turbulence measurements, and in the following subsection the issue of flow control will be addressed.

Turbulence remains largely an enigma—analytically unapproachable yet practically very important. For a turbulent flow, the dependent variables are random functions of space and time, and no straightforward method exists for analytically obtaining stochastic solutions to the governing nonlinear, partial differential equations. The statistical approach to solving the Navier–Stokes equations sets a more modest aim of solving for the average flow quantities rather than the instantaneous ones. But as a result of the nonlinearity of the governing equations, this approach always leads to more unknowns than equations (the closure problem), and solutions based on first principles are again not possible. Turbulence, therefore, is a conundrum that appears to yield its secrets only to physical and numerical experiments, provided that the wide band of relevant scales is fully resolved—a far-from-trivial task, particularly at high Reynolds numbers.

A turbulent flowfield is composed of a hierarchy of eddies having a broad range of time and length scales. The largest eddies have a spatial extension of approximately the same size as the width of the flowfield, whereas the smallest eddies are of the size at which viscous effects become dominant and energy is transferred from kinetic into internal. The ratio of the smallest length scale—the Kolmogorov microscale η—to the largest scale ℓ is related to the turbulence Reynolds number as follows:

$$\frac{\eta}{\ell} \approx \left(\frac{u\ell}{\nu}\right)^{-\frac{3}{4}} = Re^{-\frac{3}{4}} \tag{13.71}$$

Similar expressions can be written for time and velocity scales (see, for example, Tennekes and Lumley 1972). Not only does a sensor have to be sufficiently small to resolve the smallest eddies, but multisensors distributed over a large volume are needed to detect any flow structures at the largest scale. Clearly the problem worsens as the Reynolds number increases.

In wall-bounded flows, the shear-layer thickness provides a measure of the largest eddies in the flow. The smallest scale is the viscous wall unit. Viscous forces dominate over inertia in the near-wall region. The characteristic scales there are obtained from the magnitude of the mean vorticity in the region and its viscous diffusion away from the wall. Thus, the viscous time scale, t_ν, is given by the inverse of the mean wall vorticity

$$t_\nu = \left[\left.\frac{\partial \overline{U}}{\partial y}\right|_{\mathrm{w}}\right]^{-1} \tag{13.72}$$

where $\overline{U}$ is the mean streamwise velocity. The viscous length scale, ℓ_ν, is determined by the characteristic distance by which the (spanwise) vorticity is diffused from the wall and is thus given by

$$\ell_\nu = \sqrt{\nu t_\nu} = \sqrt{\frac{\nu}{\left.\frac{\partial \overline{U}}{\partial y}\right|_{\mathrm{w}}}} \tag{13.73}$$

where ν is the kinematic viscosity. The wall velocity scale (so-called friction velocity u_τ) follows directly from the time and length scales

$$u_\tau = \frac{\ell_\nu}{t_\nu} = \sqrt{\nu \left.\frac{\partial \overline{U}}{\partial y}\right|_{\mathrm{w}}} = \sqrt{\frac{\tau_{\mathrm{w}}}{\rho}} \tag{13.74}$$

where τ_{w} is the mean shear stress at the wall, and ρ is the fluid density. A wall unit implies scaling with the viscous scales, and the usual $()^+$ notation is used; for example, $y^+ = y/\ell_\nu = y u_\tau/\nu$. In the wall region, the characteristic length for the large eddies is y itself, whereas the Kolmogorov scale is related to the distance from the wall y as follows:

$$\eta^+ \equiv \frac{\eta u_\tau}{\nu} \approx (\kappa y^+)^{\frac{1}{4}} \tag{13.75}$$

where κ is the von Kármán constant (≈ 0.41). As y^+ changes in the range of 1–5 (the extent of the viscous sublayer), η changes from 0.8 to 1.2 wall units.

It is clear from the above that the spatial and temporal resolutions for any probe to be used to resolve high-Reynolds-number turbulent flows are extremely tight. For example, both the Kolmogorov scale and the viscous length scale change from few microns at the typical field Reynolds number—based on the momentum thickness—of 10^6 to a couple of hundred microns at the typical laboratory Reynolds number of 10^3, and MEMS sensors for pressure, velocity, temperature and shear stress are at least one order of magnitude smaller than conventional sensors (Ho and Tai 1996; Löfdahl, Kälvesten, and Stemme 1996). The small size of MEMS sensors improves both the spatial and temporal resolutions of the measurements, typically by a few microns and few microseconds, respectively. For example, a microhot-wire (called hot point) has very small thermal inertia, and the diaphragm of a micropressure-transducer has correspondingly fast dynamic response. Moreover, the microsensors' extreme miniaturization and low energy consumption make them ideal for monitoring the flow state without appreciably affecting it. Lastly, literally hundreds of microsensors can be fabricated on the same silicon chip at a reasonable cost, making them well suited for distributed measurements. The UCLA–Caltech team (see, for example, Ho and Tai 1996, 1998 and references therein) has been very effective in developing many MEMS-based sensors and actuators for turbulence diagnosis and control.

13.3.3 Micropumps

There have been several studies of microfabricated pumps. Some of them use nonmechanical effects. The Knudsen pump mentioned in Section 13.2.5 uses the thermal-creep effect to move rarefied gases from one chamber to another. Ion-drag is used in electrohydrodynamic pumps (Bart et al. 1990; Richter et al. 1991; Fuhr et al. 1992); these rely on the electrical properties of the fluid and are thus not suitable for many applications. Valveless pumping by ultrasound has also been proposed (Moroney, White, and Howe 1991) but produces very little pressure difference.

Mechanical pumps based on conventional centrifugal or axial turbomachinery will not work at micromachine scales at which the Reynolds numbers are typically small (on the order of 1 or less). Centrifugal forces are negligible and, furthermore, the Kutta condition through which lift is normally generated is invalid when inertial forces are vanishingly small. In general there are three ways in which mechanical micropumps can work:

- Positive-displacement pumps. These are mechanical pumps with a membrane or diaphragm actuated in a reciprocating mode and with unidirectional inlet and outlet valves. They work on the same physical principle as their larger cousins. Micropumps with piezoelectric actuators have been fabricated (Van Lintel, Van de Pol, and Bouwstra 1988; Esashi, Shoji, and Nakano 1989; Smits 1990). Other actuators, such as thermopneumatic, electrostatic, electromagnetic, or bimetallic, can be used (Pister, Fearing, and Howe 1990; Döring et al. 1992; Gabriel et al. 1992). These exceedingly minute positive-displacement pumps require even smaller valves, seals, and mechanisms—a not-too-trivial micromanufacturing challenge. In addition there are long-term problems associated with wear or clogging and consequent leaking around valves. The pumping capacity of these pumps is also limited by the small displacement and frequency involved. Gear pumps are a different kind of positive-displacement device.
- Continuous, parallel-axis rotary pumps. A screw-type, three-dimensional device for low Reynolds numbers was proposed by Taylor (1972) for propulsion purposes and shown in his seminal film. The pump has an axis of rotation parallel to the flow direction, implying that the powering motor must be submerged in the flow, the flow be turned through an angle, or that complicated gearing would be needed.
- Continuous, transverse-axis rotary pumps. This is the class of machines that was recently developed by Sen, Wajerski, and Gad-el-Hak (1996). They have shown that a rotating body, asymmetrically placed within a duct, will produce a net flow due to viscous action. The axis of rotation can be perpendicular to the flow direction, and the cylinder can thus be easily powered from outside a duct. A related viscous-flow pump was designed by Odell and Kovasznay (1971) for a water channel with density stratification. However, their design operates at a much higher Reynolds number and is too complicated for microfabrication.

As evidenced from the third item above, it is possible to generate axial fluid motion in open channels through the rotation of a cylinder in a viscous fluid medium. Odell and Kovasznay (1971) studied a pump based on this principle at high Reynolds numbers. Sen et al. (1996) carried out an exper-

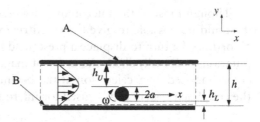

Figure 13.13: Schematic of micropump developed by Sen et al. (1996).

imental study of a different version of such a pump. The novel viscous pump, shown schematically in Figure 13.13, consists simply of a transverse-axis cylindrical rotor eccentrically placed in a channel so that the differential viscous resistance between the small and large gaps causes a net flow along the duct. The Reynolds numbers involved in Sen et al.'s work were low ($0.01 < Re \equiv 2\omega a^2/\nu < 10$, where ω is the radian velocity of the rotor, and a is its radius) and typical of microscale devices but achieved using a macroscale rotor and a very viscous fluid. The bulk velocities obtained were as high as 10% of the surface speed of the rotating cylinder. Sen et al. (1996) have also tried cylinders with square and rectangular cross sections, but the circular cylinder delivered the best pumping performance.

A finite-element solution for low-Reynolds-number, uniform flow past a rotating cylinder near an impermeable plane boundary has already been obtained by Liang and Liou (1995). However detailed two-dimensional Navier–Stokes simulations of the pump described above have been carried out by Sharatchandra, Sen, and Gad-el-Hak (1997), who extended the operating range of Re beyond 100. The effects of varying the channel height H and the rotor eccentricity ε have been studied. It was demonstrated that an optimum plate spacing exists and that the induced flow increases monotonically with eccentricity, the maximum flow rate being achieved with the rotor in contact with a channel wall. Both the experimental results of Sen et al. (1996) and the two-dimensional numerical simulations of Sharatchandra et al. (1997) have verified that, at $Re < 10$, the pump characteristics are linear and therefore kinematically reversible. Sharatchandra et al. (1997, 1998a) also investigated the effects of slip flow on the pump performance as well as the thermal aspects of the viscous device. Wall slip does reduce the traction at the rotor surface and thus lowers the performance of the pump somewhat. However, the slip effects appear to be significant only for Knudsen numbers greater than 0.1, which is encouraging from the point of view of microscale applications.

In an actual implementation of the micropump, several practical obstacles need to be considered. Among these are the larger stiction and seal design associated with rotational motion of microscale devices. Both the rotor and the channel have a finite, in fact rather small, width. DeCourtye, Sen, and Gad-el Hak (1998) numerically investigated the viscous micropump performance as the width of the channel W became exceedingly small. The bulk flow generated by the pump decreased as a result of the additional resistance to the flow caused by the side walls. However, effective pumping was still observed with extremely narrow channels. Finally, Sharatchandra et al. (1998b) used a genetic algorithm to determine the optimum wall shape to maximize the micropump performance. Their genetic algorithm uncovered shapes that were nonintuitive but yielded vastly superior pump performance.

Though most of the micropump discussion above is of flow in the steady state, it should be possible to give the eccentric cylinder a finite number of turns or even a portion of a turn to displace a prescribed minute volume of fluid. Numerical computations will easily show the order of magnitude of the volume discharged and the errors induced by acceleration at the beginning of the rotation and deceleration at the end. Such system can be used for microdosage delivery in medical applications.

13.3.4 Microturbines

DeCourtye et al. (1998) have described the possible utilization of the inverse micropump device (Section 13.3.3) as a turbine. The most interesting application of such a microturbine would be as a microsensor for measuring exceedingly small flow rates on the order of nanoliters per second (i.e., microflow metering for medical and other applications).

The viscous pump described in Section 13.3.3 operates best at low Reynolds numbers and should therefore be kinematically reversible in the creeping-flow regime. A microturbine based on the same principle should, therefore, lead to a net torque in the presence of a prescribed bulk velocity. The results of three-dimensional numerical simulations of the envisioned microturbine are summarized in this subsection. The Reynolds number for the turbine problem is defined in terms of the bulk velocity, because the rotor surface speed is unknown in this case,

$$Re = \frac{\overline{U}(2a)}{v} \tag{13.76}$$

where $\overline{U}$ is the prescribed bulk velocity in the channel, a is the rotor radius, and v is the kinematic viscosity of the fluid.

Figure 13.14 shows the dimensionless rotor speed as a function of the bulk velocity for two dimensionless channel widths $W = \infty$ and $W = 0.6$. In these simulations, the dimensionless channel depth is $H = 2.5$, and the rotor eccentricity is $\varepsilon/\varepsilon_{\max} = 0.9$. The relation is linear, as was the case for the pump problem. The slope of the lines is 0.37 for the two-dimensional turbine and 0.33 for the narrow channel with $W = 0.6$. This means that the induced rotor speed is, respectively, 0.37 and 0.33 of the bulk

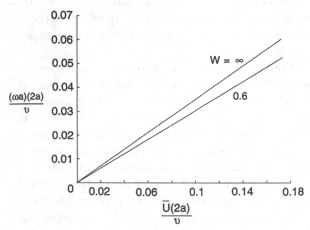

Figure 13.14: Turbine rotation as a function of the bulk velocity in the channel [after DeCourtye et al. 1998].

velocity in the channel.[15] For the pump, the corresponding numbers were 11.11 for the two-dimensional case and 100 for the three-dimensional case. Although it appears that the side walls have bigger influence on the pump performance, it should be noted that in the turbine case a vastly higher pressure drop is required in the three-dimensional duct to yield the same bulk velocity as that in the two-dimensional duct (dimensionless pressure drop of $\Delta p^* \equiv \Delta p 4 a^2 / \rho v^2 = -29$ versus $\Delta p^* = -1.5$).

The turbine characteristics are defined by the relation between the shaft speed and the applied load. A turbine load results in a moment on the shaft, which at steady state balances the torque due to viscous stresses. At a fixed bulk velocity, the rotor speed is determined for different loads on the turbine. Again, the turbine characteristics are linear in the Stokes (creeping) flow regime, but the side walls have weaker, though still adverse, effect on the device performance as compared with the pump case. For a given bulk velocity, the rotor speed drops linearly as the external load on the turbine increases. At large enough loads, the rotor will not spin, and maximum rotation is achieved when the turbine is subjected to zero load.

At present it is difficult to measure flow rates on the order of 10^{-12} m^3/s (1 nl/s). One possible way is to collect the effluent directly over time. This is useful for calibration but is not practical for on-line flow measurement. Another is to use heat transfer from a wire or film to determine the local flow rate, as in a thermal anemometer. Heat transfer from slowly moving fluids is mainly by conduction, and thus temperature gradients can be large. This is undesirable for biological and other fluids easily damaged by heat. The viscous mechanism that has been proposed and verified for pumping may be turned around and used for measuring. As demonstrated in this subsection, a freely rotating cylinder eccentrically placed in a duct will rotate at a rate proportional to the flow rate owing to a turbine effect. In fact, other geometries such as a freely rotating sphere in a cylindrical tube should also behave similarly. The calibration constant, which depends on system parameters such as geometry and bearing friction, should be determined computationally to ascertain the practical viability of such a microflow meter. Geometries that are simplest to fabricate should be explored and studied in detail.

13.3.5 Microbearings

Many of the micromachines use rotating shafts and other moving parts that carry a load and need fluid bearings for support, and most of them operate with air or water as the lubricating fluid. The fluid mechanics of these bearings are very different compared with that of their larger cousins. Their study falls in the area of microfluid mechanics, an emerging discipline that has been greatly stimulated by its applications to micromachines and is the subject of this chapter.

Macroscale journal bearings develop their load-bearing capacity from large pressure differences that are a consequence of the presence of a viscous fluid, an eccentricity between the shaft and its housing, a large surface speed of the shaft, and a small clearance to diameter ratio. Several closed-form solutions of the no-slip flow in a macrobearing have been developed. Wannier (1950) used modified Cartesian coordinates to find an exact solution to the biharmonic equation governing

---

[15] The rotor speed can never, of course, exceed the fluid velocity even if there is no load on the turbine. Without load, the integral of the viscous shear stress over the entire surface area of the rotor is exactly zero, and the turbine achieves its highest, albeit finite, rpm.

two-dimensional journal bearings in the no-slip, creeping flow regime. Kamal (1966) and Ashino and Yoshida (1975) worked in bipolar coordinates; they assumed a general form for the stream function with several constants that were determined using the boundary conditions. Though all these methods work if there is no slip, they cannot be readily adapted to slip flow. The basic reason is that the flow pattern changes if there is slip at the walls, and the assumed form of the solution is no longer valid.

Microbearings are different in the following aspects: (1) being so small, it is difficult to manufacture them with a clearance that is much smaller than the diameter of the shaft; (2) because of the small shaft size, their surface speed, at normal rotational speeds, is also small;[16] and (3) air bearings in particular may be small enough for noncontinuum effects to become important. For these reasons the hydrodynamics of lubrication are very different at microscales. The lubrication approximation that is normally used is no longer directly applicable, and other effects come into play. From an analytical point of view there are three consequences of the preceding characteristics: fluid inertia is negligible, slip flow may be important for air and other gases, and relative shaft clearance need not be small.

In a recent study, Maureau et al. (1997) analyzed microbearings represented as an eccentric cylinder rotating in a stationary housing. The flow Reynolds number was assumed small, the clearance between shaft and housing was not small relative to the overall bearing dimensions, and there was slip at the walls due to nonequilibrium effects. The two-dimensional governing equations were written in terms of the streamfunction in bipolar coordinates. Following the method of Jeffery (1920), Maureau et al. (1997) succeeded in obtaining an exact infinite-series solution of the Navier–Stokes equations for the specified geometry and flow conditions. In contrast to macrobearings and owing to the large clearance, flow in a microbearing is characterized by the possibility of a recirculation zone that strongly affects the velocity and pressure fields. For high values of the eccentricity and low slip factors, the flow develops a recirculation region, as shown in the streamlines plot in Figure 13.15.

From the infinite-series solution the frictional torque and the load-bearing capacity can be determined. The results show that both are similarly affected by the eccentricity and the slip factor: they increase with the former and decrease with the latter. For a given load, there is a corresponding eccentricity that generates a force sufficient to separate shaft from housing (i.e., sufficient to prevent solid-to-solid contact). As the load changes, the rotational center of the shaft shifts a distance necessary for the forces to balance. It is interesting to note that for a weight that is vertically downwards, the equilibrium displacement of the center of the shaft is in the horizontal direction. This can lead to complicated rotor dynamics governed by mechanical inertia, viscous damping, and pressure forces. A study of these dynamics may be of interest. Real microbearings have finite shaft lengths, and end walls and other three-dimensional effects influence the bearing characteristics. Numerical simulations of the three-dimensional problem can readily be carried out and may also be of interest to the designers of microbearings. Other potential research includes determination of a criterion for onset of cavitation in liquid bearings. From the results of these studies, information related to load, rotational speed, and geometry can be generated that would be useful for the designer.

---

[16] The microturbomachines being developed presently at MIT operate at shaft rotational speeds on the order of 1 million rpm and are therefore operating at different flow regimes from those considered here.

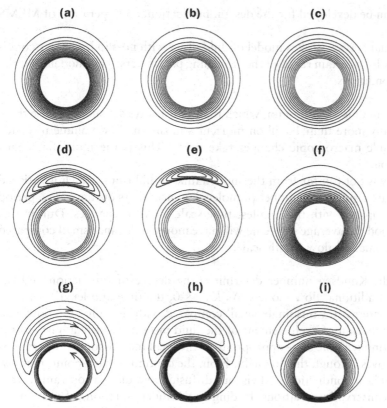

Figure 13.15: Effect of slip factor and eccentricity on the microbearing streamlines. From top to bottom, eccentricity changes as $\varepsilon = 0.2, 0.5, 0.8$. From left to right, slip factor changes as $S \equiv (\frac{2-\sigma_v}{\sigma_v})Kn = 0, 0.1, 0.5$ [after Maureau et al. 1997].

Finally, Piekos et al. (1997) have used full Navier–Stokes computations to study the stability of ultra-high-speed gas microbearings. They concluded that it is possible—despite significant design constraints—to attain stability for specific bearings to be used with the MIT microturbomachines (Epstein and Senturia 1997; Epstein et al. 1997), which incidentally operate at much higher Reynolds numbers (and rpm) than the micropumps, microturbines, and microbearings considered thus far in this and the previous two subsections. According to Piekos et al. (1997), high-speed bearings are more robust than low-speed ones owing to their reduced running eccentricities and the large loads required to maintain them.

13.4 Parting Remarks

The 40-year-old vision of Richard Feynman of building minute machines is now a reality. Microelectromechanical systems have witnessed explosive growth during the last decade and are finding increased applications in a variety of industrial and medical fields. The physics of fluid flows in microdevices and some representative applications have been explored in this chapter. Although we now know considerably more than we did just few years ago, much physics remains to be explored so that rational tools can be developed for the design, fabrication, and operation of MEMS devices.

The traditional Navier–Stokes model of fluid flows with no-slip boundary conditions works only for a certain range of the governing parameters. This model basically demands two conditions:

1. That the fluid is a continuum, which is almost always satisfied because there are usually more than 1 million molecules in the smallest volume in which appreciable macroscopic changes take place. This is the molecular chaos restriction.
2. The flow is not too far from thermodynamic equilibrium, which is satisfied if there are a sufficient number of molecular encounters during a time period small compared with the smallest time scale for flow changes. During this time period the average molecule will have moved a distance small compared with the smallest-flow length scale.

For gases, the Knudsen number determines the degree of rarefaction and the applicability of traditional flow models. As $Kn \to 0$, the time and length scales of molecular encounters are vanishingly small compared with those for the flow, and the velocity distribution of each element of the fluid instantaneously adjusts to the equilibrium thermodynamic state appropriate to the local macroscopic properties as this molecule moves through the flowfield. From the continuum viewpoint, the flow is isentropic and heat conduction and viscous diffusion and dissipation vanish from the continuum conservation relations, leading to the Euler equations of motion. At small but finite Kn, the Navier–Stokes equations describe near-equilibrium continuum flows.

Slip flow must be taken into account for $Kn > 0.001$. The slip boundary condition is at first linear in Knudsen number then nonlinear effects take over beyond a Knudsen number of 0.1. At the same transition regime (i.e., $0.1 < Kn < 10$), the linear (stress)–(rate of strain) and (heat flux)–(temperature gradient) relations—needed to close the Navier–Stokes equations—also break down, and alternative continuum equations (e.g., Burnett) or molecular-based models must be invoked. In the transition regime, provided that the dilute gas and molecular chaos assumptions hold, solutions to the difficult Boltzmann equation are sought, but physical simulations such as Monte Carlo methods are more readily executed in this range of Knudsen number. In the free-molecule flow regime (i.e., $Kn > 10$), the nonlinear collision integral is negligible and the Boltzmann equation is drastically simplified. Analytical solutions are possible in this case for simple geometries, and numerical integration of the Boltzmann equation is straightforward for arbitrary geometries provided that the surface-reflection characteristics are accurately modeled.

Gaseous flows are often compressible in microdevices even at low Mach numbers. Viscous effects can cause sufficient pressure drop and density changes for the flow to be treated as compressible. In a long, constant-area microduct, all Knudsen number regimes may be encountered, and the degree of rarefaction increases along the tube. The pressure drop is nonlinear, and the Mach number increases downstream, limited only by choked-flow condition.

Similar deviation and breakdown of the traditional Navier–Stokes equations occur for liquids as well, but there the situation is more murky. Existing experiments are contradictory. There is no kinetic theory of liquids, and first-principles prediction methods are scarce. Molecular dynamics simulations can be used, but they are limited to extremely small flow extents. Nevertheless, measurable slip is predicted from MD simulations at realistic shear rates in microdevices.

Microelectromechanical systems are finding increased applications in the diagnosis and control of turbulent flows. The use of microsensors and microactuators promises a quantum leap in the performance of reactive-flow control systems and is now in the realm of the possible for future practical devices (Chapter 14). Simple, viscous-based micropumps can be utilized for microdosage delivery, and microturbines can be used for measuring flow rates in the nanoliter per second range. Both of these can be of value in several medical applications. Much nontraditional physics is still to be learned, and many exciting applications of microdevices are yet to be discovered. The future is bright for this emerging field of science and technology. Richard Feynman was right about the possibility of building mite-size machines but was somewhat cautious in forecasting that such machines, although they "would be fun to make," might or might not be useful.

CHAPTER FOURTEEN

Frontiers of Flow Control

There is always an easy solution to every human problem—neat, plausible and wrong.

(Henry Louis Mencken, 1880–1956)

As for the future, your task is not to foresee, but to enable it.

(Antoine-Marie-Roger de Saint-Exupéry, 1900–1944, in Citadelle (The Wisdom of the Sands))

PROLOGUE

In contrast to the early chapters, in the present chapter I will emphasize the frontiers of the field of flow control, pondering mostly the control of turbulent flows. I will review the important advances in the field that have taken place during the past few years and are anticipated to dominate progress in the future, essentially covering the fifth era outlined in Section 1.2. By comparison with laminar flow control or separation prevention, the control of turbulent flow remains a very challenging problem. Flow instabilities quickly magnify near critical flow regimes, and therefore delaying transition or separation is a relatively easier task. In contrast, classical control strategies are often ineffective for fully turbulent flows. Newer ideas for turbulent flow control to achieve skin-friction drag reduction, for example, focus on the direct onslaught on coherent structures. Spurred by the recent developments in chaos control, microfabrication, and soft computing tools, reactive control of turbulent flows through which sensors detect oncoming coherent structures and actuators attempt to modulate these quasi-periodic events favorably is now in the realm of the possible for future practical devices. In this last chapter, I will provide estimates for the number, size, frequency, and energy consumption of the sensor and actuator arrays needed to control the turbulent boundary layer on a full-scale aircraft or submarine.

14.1 Introduction

14.1.1 *The Taming of the Shrew*

When one considers the extreme complexity of the turbulence problem in general and the unattainability of first-principles analytical solutions in particular, it is not surprising that controlling a turbulent flow remains a challenging task mired in empiricism and unfulfilled promises and aspirations. Brute force suppression, or *taming*, of turbulence via active, energy-consuming control strategies is always possible, but the penalty for doing so often exceeds any potential benefits. The artifice is to achieve a desired effect with minimum energy expenditure. This is of course easier said than done. Indeed, suppressing turbulence is as arduous as the taming of the shrew. The former task will be emphasized throughout this chapter, but for now we reflect on a short verse from the latter.

> From William Shakespeare's The Taming of the Shrew:
> Curtis (Petruchio's servant, in charge of his country house): Is she so hot a shrew as she's reported?
> Grumio (Petruchio's personal lackey): She was, good Curtis, before this frost. But thou know'st winter tames man, woman, and beast; for it hath tamed my old master, and my new mistress, and myself, fellow Curtis.

14.1.2 Control of Turbulence

Numerous methods of flow control have already been successfully implemented in practical engineering devices. Delaying laminar-to-turbulence transition to reasonable Reynolds numbers and preventing separation can readily be accomplished using a myriad of passive and predetermined active control strategies. Such classical techniques have been reviewed in the preceding chapters of this book and by, among others, Bushnell (1983, 1994); Wilkinson et al. (1988); Bushnell and McGinley (1989); Gad-el-Hak (1989); Bushnell and Hefner (1990); Fiedler and Fernholz (1990); Gad-el-Hak and Bushnell (1991a,b); Barnwell and Hussaini (1992); Viswanath (1995); and Joslin, Erlebacher, and Hussaini (1996). Yet, very few of the classical strategies are effective in controlling free-shear or wall-bounded turbulent flows. Serious limitations exist for some familiar control techniques when applied to certain turbulent flow situations. For example, in attempting to reduce the skin-friction drag of a body having a turbulent boundary layer by using global suction, the *penalty* associated with the control device often exceeds the *savings* derived from its use. What is needed is a way to reduce this penalty to achieve a more efficient control.

Flow control is most effective when applied near the transition or separation points—in other words, near the critical flow regimes where flow instabilities magnify quickly. Therefore, delaying or advancing laminar-to-turbulence transition and preventing or provoking separation are relatively easier tasks to accomplish. To reduce the skin-friction drag in a nonseparating turbulent boundary layer, where the mean flow is quite stable, is a more challenging problem. Yet, even a modest reduction in the fluid resistance to the motion of, for example, the worldwide commercial airplane fleet is translated into fuel savings estimated to be in the billions of dollars. Newer ideas for turbulent flow control focus on the direct onslaught on coherent structures. Spurred by the recent developments in chaos control, microfabrication, and soft computing tools, reactive control of turbulent flows is now in the realm of the possible for future practical devices.

The primary objective of this chapter is to advance possible scenarios by which viable control strategies of turbulent flows can be realized. As will be argued in the following presentation, future systems for control of turbulent flows in general and turbulent boundary layers in particular could greatly benefit from the merging of the science of chaos control, the technology of microfabrication, and the newest computational tools collectively termed soft computing. Control of chaotic, nonlinear dynamical systems has been demonstrated theoretically as well as experimentally—even for multi-degree-of-freedom systems. Microfabrication is an emerging technology that has the potential for producing inexpensive, programmable sensor or actuator chips that have dimensions of the order of a few microns (see Chapter 13). Soft computing tools include neural networks, fuzzy logic, and genetic algorithms and are now more advanced as well as more widely used as compared with just few years ago. These tools could be very useful in constructing effective adaptive controllers.

Such futuristic systems are envisaged as consisting of a large number of intelligent, interactive, microfabricated wall sensors and actuators arranged in a checkerboard pattern and targeted toward specific organized structures that occur quasi-randomly within a turbulent flow. Sensors detect oncoming coherent structures, and adaptive controllers process the sensors' information and provide control signals to the actuators, which in turn attempt to modulate the quasi-periodic events favorably. Finite numbers of wall sensors perceive only partial information about the entire flowfield above. However, a low-dimensional dynamical model of the near-wall region used in a Kalman filter can make the most of the partial information from the sensors. Conceptually, all of that is not too difficult, but in practice the complexity of such a control system is daunting, and much research and development work still remain.

14.1.3 Outline

In this chapter, I will review the important developments in the field of flow control that took place during the past few years and suggest avenues for future research. The emphasis will be on reactive flow control for future vehicles and other industrial devices. This chapter is organized into ten sections. A particular example of a classical control system, suction, is described in the following section. This will serve as a prelude to introducing the selective suction concept. In Section 14.3, the different hierarchy of coherent structures that dominate a turbulent boundary layer and that constitute the primary target for direct onslaught are briefly recalled. The characteristic lengths and sensor requirements of turbulent flows are discussed in Section 14.4. Reactive flow control and the selective suction concept are described in Section 14.5. The number, size, frequency, and energy consumption of the sensor and actuator units required to tame the turbulence on a full-scale air or water vehicle are estimated in that same section. Section 14.6 introduces the topic of magnetohydrodynamics and suggests a reactive flow control scheme using electromagnetic body forces. Sections 14.7, 14.8, and 14.9 consider, in turn, the emerging areas of chaos control, microfabrication, and soft computing—particularly as they relate to reactive control strategies. Finally, brief concluding remarks are given in Section 14.10.

14.2 Suction

To set the stage for introducing the concept of targeted or selected control in Section 14.5, I will first discuss in this section global control as applied to wall-bounded flows. A viscous fluid that is initially irrotational will acquire vorticity when an obstacle is passed through the fluid. This vorticity controls the nature and structure of the boundary layer flow in the vicinity of the obstacle. For an incompressible, wall-bounded flow, the flux of spanwise or streamwise vorticity at the wall, and hence whether the surface is a sink or a source of vorticity, is affected by the wall motion (e.g., in the case of a compliant coating), transpiration (suction or injection), streamwise or spanwise pressure gradient, wall curvature, normal viscosity gradient near the wall (caused by, for example, heating or cooling of the wall or introduction of a shear-thinning or shear thickening additive into the boundary layer), and body forces such as electromagnetic ones in a conducting fluid (Section 14.6). These alterations separately or collectively control the shape of the instantaneous as well as the mean velocity profiles, which in turn determines the skin friction at the wall, the boundary layer ability to resist transition and separation, and the intensity of turbulence and its structure.

For illustration purposes, I will focus on global wall suction as a generic control tool. The arguments presented here and in subsequent sections are equally valid for other global control techniques such as geometry modification (body shaping), surface heating or cooling, electromagnetic control, and so forth. Transpiration provides a good example of a single control technique that is used to achieve a variety of goals. Suction leads to a fuller velocity profile (vorticity flux away from the wall) and can, therefore, be employed to delay laminar-to-turbulence transition, postpone separation, achieve an asymptotic turbulent boundary layer (i.e., one having constant momentum thickness), or relaminarize an already turbulent flow. Unfortunately, global suction cannot be used to reduce the skin-friction drag in a turbulent boundary layer. The amount of suction required to inhibit boundary-layer growth is too large to effect a net drag reduction. This is a good illustration of a situation in which the penalty associated with a control device may exceed the savings derived from its use.

Small amounts of fluid withdrawn from the near-wall region of a boundary layer change the curvature of the velocity profile at the wall and can dramatically alter the stability characteristics of the flow. Concurrently, suction inhibits the growth of the boundary layer, and thus the critical Reynolds number based on thickness may never be reached. Although laminar flow can be maintained to extremely high Reynolds numbers provided that enough fluid is sucked away, the goal is to accomplish transition delay with the minimum suction flow rate. This will reduce not only the power necessary to drive the suction pump but also the momentum loss due to the additional freestream fluid entrained into the boundary layer as a result of withdrawing fluid from the wall. That momentum loss is, of course, manifested as an increase in the skin-friction drag.

The case of uniform suction from a flat plate at zero incidence is an exact solution of the Navier–Stokes equation. The asymptotic velocity profile in the viscous region is exponential and has a negative curvature at the wall. The displacement thickness has the constant value $\delta^* = \nu/|v_w|$, where ν is the kinematic viscosity and $|v_w|$ is the absolute value of the normal velocity at the wall. In this case, the familiar von Kármán integral equation reads: $C_f = 2C_q$. Bussmann and Münz (1942) computed the critical Reynolds number for the asymptotic suction profile to be $Re_{\delta^*} \equiv U_\infty \delta^*/\nu = 70{,}000$. From the value of δ^* given above, the flow is stable to all small disturbances if $C_q \equiv |v_w|/U_\infty > 1.4 \times 10^{-5}$. The amplification rate of unstable disturbances for the asymptotic profile is an order of magnitude less than that for the Blasius boundary layer (Pretsch 1942). This treatment ignores the development distance from the leading edge needed to reach the asymptotic state. When this is included into the computation, a higher $C_q = 1.18 \times 10^{-4}$ is required to ensure stability (Iglisch 1944; Ulrich 1944).

In a turbulent wall-bounded flow, the results of Eléna (1975, 1984) and Antonia et al. (1988) indicate that suction causes an appreciable stabilization of the low-speed streaks in the near-wall region. The maximum turbulence level at $y^+ \approx 13$ drops from 15 to 12% as C_q varies from 0 to 0.003. More dramatically, the tangential Reynolds stress near the wall drops by a factor of 2 for the same variation of C_q. The dissipation length scale near the wall increases by 40% and the integral length scale by 25% with the suction.

The suction rate necessary for establishing an asymptotic turbulent boundary layer independent of streamwise coordinate $(d\delta_\theta/dx = 0)$ is much lower than the rate required for relaminarization $(C_q \approx 0.01)$ but still not low enough to yield net drag reduction. For Reynolds number based on distance from leading edge $Re_x = \mathcal{O}[10^6]$, Favre et al. (1966), Rotta (1970), and Verollet et al. (1972), among others, reported an asymptotic suction coefficient of $C_q \approx 0.003$. For a zero-pressure-gradient boundary layer on a flat plate, the corresponding skin-friction coefficient is $C_f = 2C_q = 0.006$, indicating higher skin friction than if no suction were applied. To achieve a net skin-friction reduction with suction, the process must be further optimized. One way to accomplish that is to target the suction toward particular organized structures within the boundary layer and not to use it globally as in classical control schemes. This point will be revisited in Section 14.5, but the coherent structures to be targeted and the length scales to be expected are first detailed in Sections 14.3 and 14.4.

14.3   Coherent Structures

The discussion in Section 14.2 indicates that achieving a particular control goal is always possible. The challenge is reaching that goal with a penalty that can be tolerated. Suction, for example, would lead to a net drag reduction if only we could reduce the suction coefficient necessary for establishing an asymptotic turbulent boundary layer to below one-half of the unperturbed skin-friction coefficient. A more efficient way of using suction, or any other global control method, is to target particular coherent structures within the turbulent boundary layer. Before discussing this *selective control* idea, I will briefly describe in this and the following section the different hierarchy of organized structures in a wall-bounded flow and the expected scales of motion. Coherent structures were more fully discussed in Chapter 4.

The classical view that turbulence is essentially a stochastic phenomenon having a randomly fluctuating velocity field superimposed on a well-defined mean has been changed in the last few decades by the realization that the transport properties of all turbulent shear flows are dominated by quasi-periodic, large-scale vortex motions (Laufer 1975; Cantwell 1981; Fiedler 1988; Robinson 1991). Despite the extensive research work in this area, no generally accepted definition of what is meant by coherent motion has emerged. In physics, coherence stands for well-defined phase relationship. For the present purpose we adopt the rather restrictive definition given by Hussain (1986): "A coherent structure is a connected turbulent fluid mass with instantaneously phase-correlated vorticity over its spatial extent." In other words, underlying the random, three-dimensional vorticity that characterizes turbulence, there is a component of large-scale vorticity that is instantaneously coherent over the spatial extent of an organized structure. The apparent randomness of the flowfield is, for the most part, due to the random size and strength of the different types of organized structures comprising that field.

In a wall-bounded flow, a multiplicity of coherent structures have been identified mostly through flow visualization experiments, although some important early discoveries have been made using correlation measurements (e.g., Townsend 1961, 1970; Bakewell and Lumley 1967). Although the literature on this topic is vast, no research-community-wide consensus has been reached—particularly on the issues of the origin of, and interaction between, the different structures, regeneration mechanisms, and Reynolds number effects. What follow are somewhat biased remarks addressing those issues gathered mostly through low-Reynolds number experiments. The interested reader is referred to the book edited by Panton (1997a) and the large number of review articles available (e.g., Kovasznay 1970; Laufer 1975; Willmarth 1975a,b; Saffman 1978; Cantwell 1981; Fiedler 1986, 1988; Blackwelder 1988, 1998; Robinson 1991; Delville et al. 1998). The article by Robinson (1991) in particular summarizes many of the different, sometimes contradictory, conceptual models offered thus far by different research groups. Those models are aimed ultimately at explaining how the turbulence maintains itself and range from the speculative to the rigorous, but not one of them, unfortunately, is self-contained and complete. Furthermore, the structure research dwells largely on the kinematics of organized motion, and little attention is given to the dynamics of the regeneration process.

In a boundary layer, the turbulence production process is dominated by three kinds of quasi-periodic—or, depending on one's viewpoint, quasi-random—eddies: the large outer structures, the intermediate Falco eddies, and the near-wall events. The large, three-dimensional structures scale with the boundary layer thickness, δ, and extend across the entire layer (Kovasznay et al. 1970; Blackwelder and Kovasznay 1972). These eddies control the dynamics of the boundary layer in the outer region, such as entrainment, turbulence production, and so forth. They appear randomly in space and time and seem to be, at least for moderate Reynolds numbers, the residue of the transitional Emmons spots (Zilberman et al. 1977; Gad-el-Hak et al. 1981; Riley and Gad-el-Hak 1985). The Falco eddies are also highly coherent and three-dimensional. Falco (1974, 1977) named them typical eddies because they appear in wakes, jets, Emmons spots, grid-generated turbulence, and boundary layers in zero, favorable, and adverse pressure gradients. They have an intermediate scale of about $100\nu/u_\tau$ (100 wall units; u_τ is the friction velocity and ν/u_τ is the viscous length scale). The Falco eddies appear to be an important link between the large structures and the near-wall events.

The third kind of eddies exists in the near-wall region ($0 < y < 100\nu/u_\tau$) where the Reynolds stress is produced in a very intermittent fashion. Half of the total production of turbulence kinetic energy ($-\overline{uv}\partial\overline{U}/\partial y$) takes place near the wall in the first 5% of the boundary layer at typical laboratory Reynolds numbers (smaller fraction at higher Reynolds numbers), and the dominant sequence of eddy motions there is collectively termed the bursting phenomenon. This dynamically significant process was reviewed by Willmarth (1975a); Blackwelder (1978); Robinson (1991); and more recently Panton (1997a). Qualitatively, the process, according to at least one school of thought, begins with elongated, counterrotating, streamwise vortices having diameters of approximately $40\nu/u_\tau$. The vortices exist in a strong shear and induce low- and high-speed regions between them, as was sketched in Figure 4.10 of Chapter 4. The vortices and the accompanying eddy structures occur randomly in space and time. However, their appearance is regular enough that an average spanwise wavelength of approximately 80 to $100\nu/u_\tau$ has been identified by Kline et al. (1967) and others. Kline et al. also observed that the low-speed regions grow downstream and develop inflectional $U(y)$ profiles. At approximately the same time, the interface between the low- and high-speed fluid begins to oscillate—apparently signaling the onset of a secondary instability. The low-speed region lifts up away from the wall as the oscillation amplitude increases, and then the flow rapidly breaks down into a completely chaotic motion. Because this latter process occurs on a very short time scale, Kline et al. called it a *burst*. Corino and Brodkey (1969) showed that the low-speed regions are quite narrow (i.e., $z = 20\nu/u_\tau$) and may also have significant shear in the spanwise direction. Virtually all of the net production of turbulence kinetic energy in the near-wall region occurs during these bursts.

Considerably more has been learned about the bursting process during the last two decades. For example, Falco (1980, 1983, 1991) has shown that when a typical eddy, which may be formed in part by ejected wall-layer fluid, moves over the wall, it induces a high uv sweep (positive u and negative v). The wall region is continuously bombarded by *pockets* of high-speed fluid originating in the logarithmic and possibly the outer layers of the flow. These pockets tend to promote, enhance, or both, the inflectional velocity profiles by increasing the instantaneous shear, leading to a more rapidly growing instability. Blackwelder and Haritonidis (1983) have shown convincingly that the frequency of occurrence of these events scales with the viscous parameters consistent with the usual boundary layer scaling arguments. An excellent review of the dynamics of turbulent boundary layers has been provided by Sreenivasan (1989). More information about coherent structures in high-Reynolds number boundary layers is given by Gad-el-Hak and Bandyopadhyay (1994). The book edited by Panton (1997a) emphasizes the self-sustaining mechanisms of wall turbulence.

14.4 Scales

In this section, we develop the equations needed to compute the characteristic lengths and sensor requirements of turbulent flows in general and wall-bounded flows in particular. Turbulence is a high-Reynolds-number phenomenon characterized by the existence of numerous length and time scales. The spatial extension of the length scales is bounded from above by the dimensions of the flowfield and from below by the diffusive and dissipative action of the molecular viscosity. If we limit our interest to shear flows, which are basically characterized by two large length scales, one in the streamwise direction (the convective or longitudinal length scale) and the other perpendicular to the flow direction (the diffusive or lateral length scale), we obtain a more well-defined problem. Moreover, at sufficiently high Reynolds numbers, the boundary layer approximation applies, and it is assumed that there is a wide separation between the lateral and the longitudinal length scales. This leads to some attractive simplifications in the equations of motion such as that the elliptical Navier–Stokes equations are transferred to the parabolic boundary-layer equations (e.g., Hinze 1975). Thus, in this approximation, the lateral scale is approximately equal to the extension of the flow perpendicular to the flow direction (the boundary layer thickness), and the largest eddies typically have this spatial extension. These eddies are most energetic and play a crucial role in both the transport of momentum and contaminants. A constant energy supply is needed to maintain the turbulence, and this energy is extracted from the mean flow into the largest, most energetic eddies. The lateral length scale is also the relevant scale for analyzing this energy transfer. However, there is an energy destruction in the flow due to the action of the viscous forces (the dissipation), and for the analysis of this process other smaller length scales are needed.

As the eddy size decreases, viscosity becomes a more significant parameter because one property of viscosity is its effectiveness in smoothing out velocity gradients. The viscous and the nonlinear terms in the momentum equation counteract each other in the generation of small-scale fluctuations. Although the inertial terms try to produce smaller and smaller eddies, the viscous terms check this process and prevent the generation of infinitely small scales by dissipating the small-scale energy into heat. In the early 1940s, the universal equilibrium theory was developed by Kolmogorov (1941a,b). One cornerstone of this theory is that the small-scale motions are statistically independent of the relatively slower large-scale turbulence. An implication of this is that the turbulence at the small scales depends only on two parameters, namely, the rate at which energy is supplied by the large-scale motion and the kinematic viscosity. In addition, it is assumed in the equilibrium theory that the rate of energy supply to the turbulence should be equal to the rate of dissipation. Hence, in the analysis of turbulence at small scales, the dissipation rate per unit mass ε is a relevant parameter together with the kinematic viscosity v. Kolmogorov (1941a) used simple dimensional arguments to derive a length, a time, and a velocity scale relevant for the small-scale motion given respectively by

$$\eta = \left(\frac{v^3}{\varepsilon}\right)^{\frac{1}{4}} \tag{14.1}$$

$$\tau = \left(\frac{v}{\varepsilon}\right)^{\frac{1}{2}} \tag{14.2}$$

$$\upsilon = (v\varepsilon)^{\frac{1}{4}} \tag{14.3}$$

These scales are accordingly called the Kolmogorov microscales, or sometimes the inner scales of the flow. Because they are obtained through a physical argument, these scales are the smallest scales that can exist in a turbulent flow and they are relevant for both free-shear and wall-bounded flows.

In boundary layers, the shear-layer thickness provides a measure of the largest eddies in the flow. The smallest scale in wall-bounded flows is the viscous wall unit, which will be shown below to be of the same order as the Kolmogorov length scale. Viscous forces dominate over inertia in the near-wall region, and the characteristic scales there are obtained from the magnitude of the mean vorticity in the region and its viscous diffusion away from the wall. Thus, the viscous time-scale, t_v, is given by the inverse of the mean wall vorticity

$$t_v = \left[\frac{\partial \overline{U}}{\partial y}\bigg|_w\right]^{-1} \tag{14.4}$$

where $\overline{U}$ is the mean streamwise velocity. The viscous length-scale, ℓ_v, is determined by the characteristic distance by which the (spanwise) vorticity is diffused from the wall and is thus given by

$$\ell_v = \sqrt{v t_v} = \sqrt{\frac{v}{\left.\frac{\partial \overline{U}}{\partial y}\right|_{\mathrm{w}}}} \tag{14.5}$$

where v is the kinematic viscosity. The wall velocity scale (so-called friction velocity, u_τ) follows directly from the time and length scales

$$u_\tau = \frac{\ell_v}{t_v} = \sqrt{v \left.\frac{\partial \overline{U}}{\partial y}\right|_{\mathrm{w}}} = \sqrt{\frac{\tau_{\mathrm{w}}}{\rho}} \tag{14.6}$$

where τ_{w} is the mean shear stress at the wall, and ρ is the fluid density. A wall unit implies scaling with the viscous scales, and the usual $()^+$ notation is used; for example, $y^+ = y/\ell_v = y u_\tau / v$. In the wall region, the characteristic length for the large eddies is y itself, whereas the Kolmogorov scale is related to the distance from the wall y as follows:

$$\eta^+ \equiv \frac{\eta u_\tau}{v} \approx (\kappa y^+)^{\frac{1}{4}} \tag{14.7}$$

where κ is the von Kármán constant (≈ 0.41). As y^+ changes in the range of 1–5 (the extent of the viscous sublayer), η changes from 0.8 to 1.2 wall units.

We now have access to scales for the largest and smallest eddies of a turbulent flow. To continue our analysis of the cascade energy process, it is necessary to find a connection between these diverse scales. One way of obtaining such a relation is to use the fact that at equilibrium the amount of energy dissipating at high wave numbers must equal the amount of energy drained from the mean flow into the energetic large-scale eddies at low wave numbers. In the inertial region of the turbulence kinetic energy spectrum, the flow is almost independent of viscosity, and because the same amount of energy dissipated at the high wave numbers must pass this "inviscid" region, an inviscid relation for the total dissipation may be obtained by the following argument. The amount of kinetic energy per unit mass of an eddy with a wave number in the inertial sublayer is proportional to the square of a characteristic velocity for such an eddy, u^2. The rate of transfer of energy is assumed to be proportional to the reciprocal of one eddy turnover time, u/ℓ, where ℓ is a characteristic length of the inertial sublayer. Hence, the rate of energy that is supplied to the small-scale eddies via this particular wave number is of order u^3/ℓ, and this amount of energy must be equal to the energy dissipated at the highest wavenumber, expressed as

$$\varepsilon \approx \frac{u^3}{\ell} \tag{14.8}$$

Note that this is an inviscid estimate of the dissipation because it is based on large-scale dynamics and does not either involve or contain viscosity. More comprehensive discussion of this issue can be found in Taylor (1935) and Tennekes and Lumley (1972). From an experimental perspective, this is a very important expression because it offers one way of estimating the Kolmogorov microscales from quantities measured in a much lower wave number range.

Because the Kolmogorov length and time scales are the smallest scales occurring in turbulent motion, a central question will be how small these scales can be without violating the continuum hypothesis. By looking at the governing equations, it can be concluded that high dissipation rates are usually associated with large velocities, and this situation is more likely to occur in gases than in liquids, and thus it would be sufficient to show that for gas flows the smallest turbulence scales are normally much larger than the molecular scales of motion. The relevant molecular length scale is the mean free path, $\mathcal{L}$, and the ratio between this length and the Kolmogorov length-scale, η, is the microstructure Knudsen number and can be expressed as (see Corrsin 1959):

$$Kn = \frac{\mathcal{L}}{\eta} \approx \frac{Ma^{\frac{1}{4}}}{Re} \tag{14.9}$$

where the turbulence Reynolds number Re and the turbulence Mach number Ma are used as independent variables. It is obvious that a turbulent flow will interfere with the molecular motion only at high Mach number and low Reynolds number, and this is a very unusual situation occurring only in certain gaseous nebulae.[1] Thus, under normal condition the turbulence Knudsen number falls in the group of continuum flows. However, measurements using extremely thin hot wires, small MEMS sensors, or flows within narrow MEMS channels can generate values in the slip-flow regime and even beyond, and this implies that, for instance, the no-slip condition may be questioned, as was discussed in Section 13.2.5.

14.4.1 Sensor Requirements

It is the ultimate goal of all measurements in turbulent flows to resolve the largest and smallest eddies that occur in the flow. At the lower wave numbers, the largest and most energetic eddies occur, and normally there are no problems associated with resolving these eddies. Basically, this is a question of having access to computers with sufficiently large memory for storing the amount of data that it may be necessary to acquire from a large number of distributed probes, each collecting data for a time period long enough to reduce the statistical error to a prescribed level. However, at the other end of the spectrum, both the spatial and the temporal resolutions are crucial, and this puts severe limitations on the sensors to be used. It is possible to obtain a relation between the small and large scales of the flow by substituting the inviscid estimate of the total dissipation rate, Eq. (14.8), into the expressions for the Kolmogorov microscales, Eqs. (14.1)–(14.3). Thus,

$$\frac{\eta}{\ell} \approx \left(\frac{u\ell}{\nu}\right)^{-\frac{3}{4}} = Re^{-\frac{3}{4}} \tag{14.10}$$

$$\frac{\tau u}{\ell} \approx \left(\frac{u\ell}{\nu}\right)^{-\frac{1}{2}} = Re^{-\frac{1}{2}} \tag{14.11}$$

$$\frac{\upsilon}{u} \approx \left(\frac{u\ell}{\nu}\right)^{-\frac{1}{4}} = Re^{-\frac{1}{4}} \tag{14.12}$$

[1] Note that in microduct flows and the like, the Re is usually too small for turbulence to even exist. Thus the issue of turbulence Knudsen number is mute in those circumstances even if rarefaction effects become strong.

where Re is the Reynolds number based on the speed of the energy containing eddies u and their characteristic length ℓ. Because turbulence is a high-Reynolds-number phenomenon, these relations show that the small length, time, and velocity scales are much less than those of the larger eddies and that the separation in scales widens considerably as the Reynolds number increases. Moreover, this also implies that the assumptions made on the statistical independence and the dynamical equilibrium state of the small structures will be most relevant at high Reynolds numbers. Another interesting conclusion that can be drawn from the preceding relations is that if two turbulent flowfields have the same spatial extension (i.e., same large-scale) but different Reynolds numbers, there would be an obvious difference in the small-scale structure in the two flows. The low-Reynolds-number flow would have a relatively coarse small-scale structure, whereas the high-Re flow would have much finer small eddies.

To resolve the smallest eddies spatially, sensors that are of about the same size as the Kolmogorov length scale for the particular flow under consideration are needed. This implies that, as the Reynolds number increases, smaller sensors are required. For instance, in the self-preserving region of a plane-cylinder wake at a modest Reynolds number, on the basis of the cylinder diameter of 1840, the value of η varies in the range of 0.5–0.8 mm (Aronson and Löfdahl 1994). For this case, conventional hot wires can be used for turbulence measurements. However, an increase in the Reynolds number by a factor of 10 will yield Kolmogorov scales in the micrometer range and call for either extremely small conventional hot wires or MEMS-based sensors. Another example illustrating the Reynolds number effect on the requirement of small sensors is a simple two-dimensional, flat-plate boundary layer. At a momentum thickness Reynolds number of $Re_\theta = 4000$, the Kolmogorov length scale is typically of the order of 50 μm, and to resolve these scales it is necessary to have access to sensors that have a characteristic active measuring length of the same spatial extension.

Severe errors will be introduced in the measurements by using sensors that are too large, for such a sensor will integrate the fluctuations due to the small eddies over its spatial extension, and the energy content of these eddies will be interpreted by the sensor as an average 'cooling.' When measuring fluctuating quantities, this implies that these eddies are counted as part of the mean flow, and their energy is 'lost.' The result will be a lower value of the turbulence parameter, and this will wrongly be interpreted as a measured attenuation of the turbulence (see for example Ligrani and Bradshaw 1987). However, because turbulence measurements deal with statistical values of fluctuating quantities, it may be possible to loosen the spatial constraint of having a sensor of the same size as η to allow a sensor dimensions that are slightly larger than the Kolmogorov scale—say on the order of η.

For boundary layers, the wall unit has been used to estimate the smallest necessary size of a sensor for accurately resolving the smallest eddies. For instance Keith, Hurdis, and Abraham (1992) stated that 10 wall units or less is a relevant sensor dimension for resolving small-scale pressure fluctuations. Measurements of fluctuating velocity gradients, essential for estimating the total dissipation rate in turbulent flows, are another challenging task. Gad-el-Hak and Bandyopadhyay (1994) argued that turbulence measurements with probe lengths greater than the viscous sublayer thickness (about 5 wall units) are unreliable—particularly near the surface. Many studies have been conducted on the spacing between sensors necessary to optimize the formed velocity gradients (see Aronson, Johansson, and Löfdahl 1997 and references therein). A general conclusion from both experiments and direct numerical simulations is that a sensor spacing of 3–5 Kolmogorov lengths is recommended. When designing arrays for correlation measurements or for targeted control, the spacing between the coherent structures will be the determining factor. For example, when targeting the low-speed streaks in a turbulent boundary layer, several sensors must be situated along a lateral distance of 100 wall units, the average spanwise spacing between streaks. All this requires quite small sensors, and many attempts have been made to meet these conditions with conventional sensor designs. However, although conventional sensors like hot wires have been fabricated in the micrometer-size range (for their diameter but not their length), they are usually handmade, difficult to handle, and are too fragile, and here the MEMS technology has really opened a door for new applications.

It is clear from the preceding discussion that the spatial and temporal resolutions for any probe to be used to resolve high-Reynolds-number turbulent flows are extremely tight. For example, the Kolmogorov scale and the viscous length scale change from a few microns at the typical field Reynolds number—based on the momentum thickness—of 10^6, to a couple of hundred microns at the typical laboratory Reynolds number of 10^3. Microelectromechanical systems sensors for pressure, velocity, temperature and shear stress are at least one order of magnitude smaller than conventional sensors (Ho and Tai 1996, 1998; Löfdahl et al. 1996; Löfdahl and Gad-el-Hak 1999). Their small size improves both the spatial and temporal resolutions of the measurements, typically a few microns and a few microseconds, respectively. For example, a micro-hot-wire (called hot point) has very small thermal inertia, and the diaphragm of a micropressure transducer has correspondingly fast dynamic response. Moreover, the microsensors' extreme miniaturization and low energy consumption make them ideal for monitoring the flow state without appreciably affecting it. Lastly, literally hundreds of microsensors can be fabricated on the same silicon chip at a reasonable cost, making them well suited for distributed measurements and control. The UCLA–Caltech team (see, for example, Ho and Tai 1996, 1998 and references therein) has been very effective in developing many MEMS-based sensors and actuators for turbulence diagnosis and control.

14.5 Reactive Control

14.5.1 Introductory Remarks

Targeted control implies sensing and reacting to a particular quasi-periodic structure in the boundary layer. The wall seems to be the logical place for such reactive control because of the relative ease of placing something in there, the sensitivity of the flow in general to surface perturbations, and the proximity and therefore accessibility to the dynamically all important near-wall coherent events. According to Wilkinson (1990), there are very few actual experiments that use embedded wall sensors to initiate a surface actuator response (Alshamani, Livesey, and Edwards 1982; Wilkinson and Balasubramanian 1985; Nosenchuck and Lynch 1985; Breuer, Haritonidis, and Landahl 1989). This 10-year-old assessment is fast changing, however, with the introduction of microfabrication technology that has the potential for producing small, inexpensive, programmable sensor and actuator chips. Witness the more recent reactive control attempts by Kwong and Dowling (1993); Reynolds (1993); Jacobs et al. (1993); Jacobson and Reynolds (1993a,b; 1994, 1995, 1998); Fan, Hofmann, and Herbert (1993); James, Jacobs, and Glezer (1994); and Keefe (1996). Fan et al. and Jacobson and Reynolds even considered the use of self-learning neural networks for increased computational speeds and efficiency. Recent reviews of reactive flow control include those by Gad-el-Hak (1994, 1996b); Lumley (1996b); McMichael (1996); Mehregany, DeAnna, and Reshotko (1996); and Ho and Tai (1996).

Numerous methods of flow control have already been successfully implemented in practical engineering devices. Yet, limitations exist for some familiar control techniques when applied to specific situations. For example, in attempting to reduce the drag or enhance the lift of a body having a turbulent boundary layer using global suction, global heating or cooling, or global application of electromagnetic body forces, the actuator's energy expenditure often exceeds the saving derived from the predetermined active control strategy. What is needed is a way to reduce this penalty to achieve a more efficient control. Reactive control geared specifically toward manipulating the coherent structures in turbulent shear flows, though considerably more complicated than passive control or even predetermined active control, has the potential to do just that. As will be argued in this and the following sections, future systems for control of turbulent flows in general and turbulent boundary layers in particular could greatly benefit from the merging of the science of chaos control, the technology of microfabrication, and the newest computational tools collectively termed soft computing. Such systems are envisaged as consisting of a large number of intelligent, communicative wall sensors and actuators arranged in a checkerboard pattern and targeted toward controlling certain quasi-periodic, dynamically significant coherent structures present in the near-wall region.

14.5.2   Targeted Control

As discussed in Chapter 10, successful techniques to reduce the skin friction in a turbulent flow, such as polymers, particles, or riblets, appear to act indirectly through local interaction with discrete turbulent structures, particularly small-scale eddies, within the flow. Common characteristics of all these methods are increased losses in the near-wall region, thickening of the buffer layer, and lowered production of Reynolds shear stress (Bandyopadhyay 1986b). Methods that act directly on the mean flow, such as suction or lowering of near-wall viscosity, also lead to inhibition of Reynolds stress. However, skin friction is increased when any of these velocity-profile modifiers is applied globally.

Could these seemingly inefficient techniques (e.g., global suction) be used more sparingly and be optimized to reduce their associated penalty? It appears that the more successful drag-reducing methods (e.g., polymers) act selectively on particular scales of motion and are thought to be associated with stabilization of the secondary instabilities. It is also clear that energy is wasted when suction or heating or cooling is used to suppress the turbulence throughout the boundary layer when the main interest is to affect a near-wall phenomenon. One ponders, what would become of wall turbulence if specific coherent structures were to be targeted by the operator through a reactive control scheme for modification? The myriad of organized structures present in all shear flows are instantaneously identifiable, quasi-periodic motions (Cantwell

1981; Robinson 1991). Bursting events in wall-bounded flows, for example, are both intermittent and random in space as well as time. The random aspects of these events reduce the effectiveness of a predetermined active control strategy. If such structures are nonintrusively detected and altered, on the other hand, net performance gain might be achieved. It seems clear, however, that temporal phasing as well as spatial selectivity would be required to achieve proper control targeted toward random events.

A nonreactive version of the preceding idea is the *selective suction technique*, which combines suction to achieve an asymptotic turbulent boundary layer and longitudinal riblets to fix the location of low-speed streaks. Although far from indicating net drag reduction, the available results are encouraging, and further optimization is needed. When implemented via an array of reactive control loops, the selective suction method is potentially capable of skin-friction reduction that approaches 60%.

The genesis of the selective suction concept can be found in the articles by Gad-el-Hak and Blackwelder (1987c, 1989) and the patent by Blackwelder and Gad-el-Hak (1990). These researchers have suggested that one possible means of optimizing the suction rate is to be able to identify where a low-speed streak is located and to apply a small amount of suction under it. Assuming that the production of turbulence kinetic energy is due to the instability of an inflectional $U(y)$ velocity profile, one needs to remove only enough fluid so that the inflectional nature of the profile is alleviated. An alternative technique that could reduce the Reynolds stress is to inject fluid selectively under the high-speed regions. The immediate effect of normal injection would be to decrease the viscous shear at the wall, resulting in less drag. In addition, the velocity profiles in the spanwise direction, $U(z)$, would have a smaller shear, $\partial U/\partial z$, because the suction or injection would create a more uniform flow. Because Swearingen and Blackwelder (1984) and Blackwelder and Swearingen (1990) have found that inflectional $U(z)$ profiles occur as often as inflection points are observed in $U(y)$ profiles, suction under the low-speed streaks, injection under the high-speed regions, or both, would decrease this shear and hence the resulting instability. The combination of selective suction and injection is sketched in Figure 14.1. In Figure 14.1a, the vortices are idealized by a periodic distribution in the spanwise direction. The instantaneous velocity profiles without transpiration at constant y and z locations are shown by the dashed lines in Figures 14.1b and 14.1c, respectively. Clearly, the $U(y_0, z)$ profile is inflectional, having two inflection points per wavelength. At z_1 and z_3, an inflectional $U(y)$ profile is also evident. The same profiles with suction at z_1 and z_3 and injection at z_2 are shown by the solid lines. In all cases, the shear associated with the inflection points would have been reduced. Because the inflectional profiles are all inviscidly unstable with growth rates proportional to the shear, the resulting instabilities would be weakened by the suction or injection process.

The feasibility of the selective suction as a drag-reducing concept has been demonstrated by Gad-el-Hak and Blackwelder (1989) and is indicated in Figure 14.2. Low-speed streaks were artificially generated in a laminar boundary layer using three spanwise suction holes as per the method proposed by Gad-el-Hak and Hussain (1986), and a hot-film probe was used to record the near-wall signature of the streaks. An open, feedforward control loop with a phase lag was used to activate a predetermined suction from a longitudinal slot located in between the spanwise holes and the downstream hot-film probe. An equivalent suction coefficient of $C_q = 0.0006$

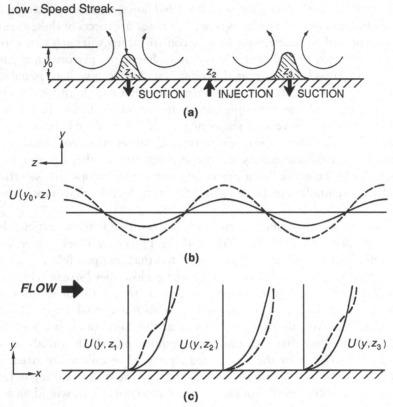

Figure 14.1: Effects of suction or injection on velocity profiles. Broken lines: reference profiles. Solid lines: profiles with transpiration applied. (a) Streamwise vortices in the y–z plane, suction or injection applied at z_1, z_2, and z_3. (b) Resulting spanwise velocity distribution at $y = y_0$. (c) Velocity profiles normal to the surface.

was sufficient to eliminate the artificial events and prevent bursting. This rate is five times smaller than the asymptotic suction coefficient for a corresponding turbulent boundary layer. If this result is sustained in a naturally developing turbulent boundary layer, a skin-friction reduction of close to 60% would be attained. Gad-el-Hak and Blackwelder (1989) proposed to combine suction with nonplanar surface modifications. Minute longitudinal roughness elements, if properly spaced in the spanwise direction, greatly reduce the spatial randomness of the low-speed streaks (Johansen and Smith 1986). By withdrawing the streaks forming near the peaks of the roughness elements, less suction should be required to achieve an asymptotic boundary layer. Experiments by Wilkinson and Lazos (1987) and Wilkinson (1988) combine suction or blowing with thin-element riblets. Although no net drag reduction has yet been attained in these experiments, their results indicate some advantage of combining suction with riblets, as proposed by Gad-el-Hak and Blackwelder (1987c, 1989).

The recent numerical experiments of Choi, Moin, and Kim (1994) also validated the concept of targeting suction or injection to specific near-wall events in a turbulent channel flow. On the basis of complete interior flow information and using the rather simple, heuristic control law proposed earlier by Gad-el-Hak and Blackwelder (1987c), Choi et al.'s (1994) direct numerical simulations indicated a 20% net drag

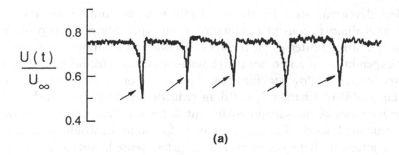

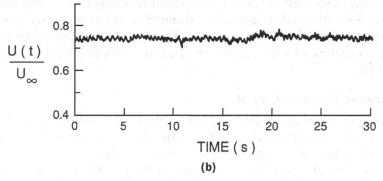

Figure 14.2: Effects of suction from a streamwise slot on five artificially induced burst-like events in a laminar boundary layer (from Gad-el-Hak and Blackwelder 1989). (a) $C_q = 0.0$. (b) $C_q = 0.0006$.

reduction accompanied by significant suppression of the near-wall structures and the Reynolds stress throughout the entire wall-bounded flow. When only wall information was used, a drag reduction of 6% was observed, which is a rather disappointing result if one considers that sensing and actuation took place at *every* grid point along the computational wall. In a practical implementation of this technique, even fewer wall sensors would perhaps be available measuring only a small subset of the accessible information and thus requiring even more sophisticated control algorithms to achieve the same degree of success. Low-dimensional models of the near-wall flow (Section 14.7) and soft computing tools (Section 14.9) can help in constructing more effective control algorithms.

Time sequences of the numerical flowfield of Choi et al. (1994) indicated the presence of two distinct drag-reducing mechanisms when selective suction or injection is used:

1. First, deterring the sweep motion, without modifying the primary streamwise vortices above the wall, and consequently moving the high-shear regions from the surface to the interior of the channel, thus directly reducing the skin friction.
2. Changing the evolution of the wall vorticity layer by stabilizing and preventing lifting of the near-wall spanwise vorticity, thus suppressing a potential source of new streamwise vortices above the surface and interrupting a very important regeneration mechanism of turbulence.

Three modern developments have relevance to the issue at hand. Firstly, the recently demonstrated ability to convert a chaotic system to a periodic one may provide optimal nonlinear control strategies for further reduction in the amount of suction (or the energy expenditure of any other active wall-modulation technique) needed to attain a given degree of flow stabilization. This is important because, as can be seen from Eq. (2.44) of Chapter 2, net drag reduction achieved in a turbulent boundary layer increases as the suction coefficient decreases. Secondly, to remove the randomly occurring low-speed streaks selectively, for example, would ultimately require reactive control. In that case, an event is targeted, sensed, and subsequently modulated. Microfabrication technology provides opportunities for practical implementation of the required large array of inexpensive, programmable sensor–actuator chips. Thirdly, newly introduced soft computing tools include neural networks, fuzzy logic, and genetic algorithms and are now more advanced as well as more widely used as compared to just few years ago. These tools could be very useful in constructing effective adaptive controllers. All three novel developments will be discussed in turn in Sections 14.7–14.9.

14.5.3 Reactive Feedback Control

As was schematically depicted in Figure 3.3 of Chapter 3, a control device can be passive, requiring no auxiliary power, or active, requiring energy expenditure. Active control is further divided into predetermined or reactive. Predetermined control includes the application of steady or unsteady energy input without regard to the particular state of the flow. The control loop in this case is open, as was shown in Figure 3.4a of Chapter 3, and no sensors are required. Because no sensed information is being fed forward, this open control loop is not a feedforward one. Reactive control is a special class of active control in which the control input is continuously adjusted based on measurements of some kind. The control loop in this case can either be an open, feedforward one (Figure 3.4b) or a closed, feedback loop (Figure 3.4c).

The distinction between feedforward and feedback is particularly important when dealing with the control of flow structures that convect over stationary sensors and actuators. In feedforward control, the measured variable and the controlled variable differ. For example, the pressure or velocity can be sensed at an upstream location, and the resulting signal is used together with an appropriate control law to trigger an actuator that in turn influences the velocity at a downstream position. Feedback control, on the other hand, necessitates that the controlled variable be measured, fed back, and compared with a reference input.

Moin and Bewley (1994) categorized reactive feedback control strategies by examining the extent to which they are based on the governing flow equations. Four categories are discerned: adaptive, physical–model-based, dynamical–systems-based, and optimal control (Figure 3.3 of Chapter 3). Note that, except for adaptive control, the other three categories of reactive feedback control can also be used in the feedforward mode or the combined feedforward–feedback mode. Also, in a convective environment such as that for a boundary layer, a controller will perhaps combine feedforward and feedback information and may include elements from each of the four classifications. Each of the four categories is briefly described below.

Adaptive schemes attempt to develop models and controllers via some learning algorithm without regard to the details of the flow physics. System identification is performed independently of the flow dynamics or the Navier–Stokes equations that govern these dynamics. An adaptive controller tries to optimize a specified performance index by providing a control signal to an actuator. To update its parameters, the controller thus requires feedback information relating to the effects of its control. The most recent innovation in adaptive flow-control schemes involves the use of neural networks that relate the sensor outputs to the actuator inputs through functions with variable coefficients and nonlinear, sigmoid saturation functions. The coefficients are updated using the so-called back-propagation algorithm, and complex control laws can be represented with a sufficient number of terms. Hand tuning is required, however, to achieve good convergence properties. The nonlinear adaptive technique has been used with different degrees of success by Fan et al. (1993) and Jacobson and Reynolds (1993b, 1995, 1998) to control, respectively, the transition process and the bursting events in turbulent boundary layers. We will return to this subject in Section 14.9.1.

Heuristic physical arguments can instead be used to establish effective control laws. That approach obviously will work only in situations in which the dominant physics are well understood. An example of this strategy is the active cancellation scheme used by Gad-el-Hak and Blackwelder (1989) in a physical experiment and by Choi et al. (1994) in a numerical experiment to reduce the drag by mitigating the effect of near-wall vortices. As mentioned earlier, the idea is to oppose the near-wall motion of the fluid, caused by the streamwise vortices, with an opposing wall control, thus lifting the high-shear region away from the surface and interrupting the turbulence regeneration mechanism.

Nonlinear dynamical systems theory allows turbulence to be decomposed into a small number of representative modes whose dynamics are examined to determine the best control law. The task is to stabilize the attractors of a low-dimensional approximation of a turbulent chaotic system. The best-known strategy is the Ott–Grebogi–Yorke (OGY) method, which, when applied to simpler, small-number-of-degrees-of-freedom systems, achieves stabilization with minute expenditure of energy. This and other chaos control strategies, especially as applied to the more complex turbulent flows, will be revisited in Section 14.7.2.

Finally, optimal control theory applied directly to the Navier–Stokes equations can, in principle, be used to minimize a cost function in the space of the control. This strategy provides perhaps the most rigorous theoretical framework for flow control. As compared with other reactive control strategies, optimal control applied to the full Navier–Stokes equations is also the most computer-time intensive. In this method, feedback control laws are derived systematically for the most efficient distribution of control effort to achieve a desired goal. Abergel and Temam (1990) developed such an optimal control theory for suppressing turbulence in a numerically simulated, two-dimensional Navier–Stokes flow, but their method requires impractical full flow-field information. Choi, Temam, Moin, and Kim (1993) developed a more practical, wall-information-only, suboptimal control strategy that they applied to the one-dimensional stochastic Burgers equation. Later application of the suboptimal control theory to a numerically simulated turbulent channel flow has been reported by Moin and Bewley (1994) and Bewley et al. (1997, 1998). The recent book edited by Sritharan (1998) provides eight articles that focus on the mathematical aspects of optimal control of viscous flows.

14.5.4 Required Characteristics

The randomness of the bursting events necessitates temporal phasing as well as spatial selectivity to effect selective control. Practical applications of methods targeted at controlling a particular turbulent structure to achieve a prescribed goal would therefore require implementing a large number of surface sensors and actuators together with appropriate control algorithms. This strategy for controlling wall-bounded turbulent flows has been advocated by, among others and in chronological order, Gad-el-Hak and Blackwelder (1987c, 1989); Lumley (1991, 1996); Choi, Moin, and Kim (1992); Reynolds (1993); Jacobson and Reynolds (1993b, 1995); Moin and Bewley (1994); Gad-el-Hak (1994, 1996b, 1998b); McMichael (1996); Mehregany et al. (1996); Blackwelder (1998); Delville et al. (1998); and Perrier (1998).

It is instructive to estimate some representative characteristics of the required array of sensors and actuators. Consider a typical commercial aircraft cruising at a speed of $U_\infty = 300$ m/s and at an altitude of 10 km. The density and kinematic viscosity of air and the unit Reynolds number in this case are, respectively, $\rho = 0.4$ kg/m^3, $\nu = 3 \times 10^{-5}$ m^2/s, and $Re = 10^7$/m. Assume further that the portion of fuselage to be controlled has turbulent boundary layer characteristics that are identical to those for a zero-pressure-gradient flat plate at a distance of 1 m from the leading edge. In this case, the skin-friction coefficient[2] and the friction velocity are, respectively, $C_f = 0.003$ and $u_\tau = 11.62$ m/s. At this location, one viscous wall unit is only $\nu/u_\tau = 2.6$ μm. In order for the surface array of sensors and actuators to be hydraulically smooth, it should not protrude beyond the viscous sublayer, or $5\nu/u_\tau = 13$ μm.

Wall-speed streaks are the most visible, reliable, and detectable indicators of the preburst turbulence production process. The detection criterion is simply low velocity near the wall, and the actuator response should be to accelerate (or to remove) the low-speed region before it breaks down. Local wall motion, tangential injection, suction, heating or electromagnetic body force, all triggered on sensed wall-pressure or wall-shear stress, could be used to cause local acceleration of near-wall fluid.

The recent numerical experiments of Berkooz, Fisher, and Psiaki (1993) indicate that effective control of bursting pairs of rolls may be achieved by using the equivalent of two wall-mounted shear sensors. If the goal is to stabilize or to eliminate *all* low-speed streaks in the boundary layer, a reasonable estimate for the spanwise and streamwise distances between individual elements of a checkerboard array is, respectively, 100 and 1000 wall units,[3] or 260 μm and 2600 μm, for our particular example. A reasonable size for each element is probably one-tenth of the spanwise separation, or 26 μm. A (1 m × 1 m) portion of the surface would have to be covered with about $n = 1.5$ million elements. This is a colossal number, but the density of

---

[2] Note that the skin friction decreases as the distance from the leading increases. It is also strongly affected by such things as the externally imposed pressure gradient. Therefore, the estimates provided here are for illustration only.

[3] These are equal to, respectively, the average spanwise wavelength between two adjacent streaks and the average streamwise extent for a typical low-speed region. One can argue that these estimates are too conservative: once a region is *relaminarized*, it would perhaps stay as such for quite a while as the flow convects downstream. The next row of sensors and actuators may therefore be relegated to a downstream location well beyond 1000 wall units. Relatively simple physical or numerical experiments could settle this issue.

sensors and actuators could be considerably reduced if we moderated our goal of targeting every single bursting event (and also if less conservative assumptions were used).

It is well known that not every low-speed streak leads to a burst. On the average, a particular sensor would detect an incipient bursting event every wall-unit interval of $P^+ = P\, u_\tau^2 / \nu = 250$, or $P = 56\ \mu s$. The corresponding dimensionless and dimensional frequencies are $f^+ = 0.004$ and $f = 18$ kHz, respectively. At different distances from the leading edge and in the presence of nonzero-pressure gradient, the sensors–actuators array would have different characteristics, but the corresponding numbers would still be in the same ballpark, as estimated here.

As a second example, consider an underwater vehicle moving at a speed of $U_\infty = 10$ m/s. Despite the relatively low speed, the unit Reynolds number is still the same as estimated above for the air case, $Re = 10^7/\mathrm{m}$, owing to the much lower kinematic viscosity of water. At 1 m from the leading edge of an imaginary flat plate towed in water at the same speed, the friction velocity is only $u_\tau = 0.39$ m/s, but the wall unit is still the same as in the aircraft example, $\nu/u_\tau = 2.6\ \mu m$. The density of the required sensors–actuators array is the same as computed for the aircraft example, $n = 1.5 \times 10^6$ elements/m². The anticipated average frequency of sensing a bursting event is, however, much lower at $f = 600$ Hz.

Similar calculations have been recently made by Gad-el-Hak (1993, 1994, 1998b); Reynolds (1993); and Wadsworth et al. (1993). Their results agree closely with the estimates made here for typical field requirements. In either the airplane or the submarine case, the actuator's response need not be too large. As will be shown in Section 14.8, wall displacement on the order of 10 wall units (26 μm in both examples), suction coefficient of about 0.0006, or surface cooling or heating on the order of 40 °C/2 °C (in the first and second example, respectively) should be sufficient to stabilize the turbulent flow.

As computed in the two examples above, both the required size for a sensor–actuator element and the average frequency at which an element would be activated are within the presently known capabilities of microfabrication technology. The number of elements needed per unit area is, however, alarmingly large. The unit cost of manufacturing a programmable sensor–actuator element would have to come down dramatically, perhaps matching the unit cost of a conventional transistor,[4] before the idea advocated here would become practical.

A consideration in addition to the size, amplitude, and frequency response is the energy consumed by each sensor–actuator element. Total energy consumption by the entire control system obviously has to be low enough to achieve net savings. Consider the following calculations for the aircraft example: One meter from the leading edge, the skin-friction drag to be reduced is approximately 54 N/m². Engine power needed to overcome this retarding force per unit area is 16 kW/m², or $10^4\ \mu$W/sensor. If a 60% drag-reduction is achieved,[5] this energy consumption is reduced to 4320 μW/sensor. This number will increase by the amount of energy consumption of a sensor–actuator unit, but one hopes not back to the uncontrolled

---

[4] The transistor was invented in 1947. In the mid-1960s, a single transistor sold for around $70. In 1997, Intel's Pentium II processor (microchip) contained 7.5×10^6 transistors and cost around $500, which is less than $0.00007 per transistor!

[5] A not-too-farfetched goal according to the selective suction results discussed earlier.

levels. The voltage across a sensor is typically in the range of $V = 0.1$–1 V, and its resistance in the range of $R = 0.1$–1 MΩ. This means a power consumption by a typical sensor in the range of $\mathcal{P} = V^2/R = 0.1$–$10$ μW, which is well below the anticipated power savings due to reduced drag. For a single actuator in the form of a spring-loaded diaphragm with a spring constant of $k = 100$ N/m oscillating up and down at the bursting frequency of $f = 18$ kHz with an amplitude of $y = 26$ μm, the power consumption is $\mathcal{P} = (1/2)\, ky^2 f = 600\, \mu$W/actuator. If suction is used instead, $C_q = 0.0006$, and if a pressure difference of $\Delta p = 10^4$ N/m^2 is assumed across the suction holes or slots, the corresponding power consumption for a single actuator is $\mathcal{P} = C_q\, U_\infty\, \Delta p/n = 1200\, \mu$W/actuator. It is clear, then, that when the power penalty for the sensor and actuator is added to the lower-level drag, a net savings is still achievable. The corresponding actuator power penalties for the submarine example are even smaller ($\mathcal{P} = 20$ μW/actuator for the wall motion actuator, and $\mathcal{P} = 40$ μW/actuator for the suction actuator), and larger savings are therefore possible.

14.6 Magnetohydrodynamic Control

14.6.1 Introductory Remarks

Magnetohydrodynamics (MHD) is the science underlying the interaction of an electrically conducting fluid with a magnetic field. Several decades ago several researchers from around the world were vigorously attempting to exploit the electric current and thus electric power that results when a fluid conductor moves in the presence of a magnetic field. Thus, the fluid internal energy (or enthalpy) is directly converted into electrical energy, eliminating the traditional intermediate mechanical step. The usual turbine and generator in conventional power plants are therefore combined in a single unit—in essence an electromagnetic turbine with no moving parts. The inverse device is an electromagnetic pump in which, in the presence of a magnetic field, a conducting fluid is caused to move when electric current passes through it. An excellent primer for the topic of MHD can be found in the book by Shercliff (1965).

Examples of conducting fluids (in descending order of electrical conductivities) include liquid metals, plasmas, molten glass, and seawater. The electrical conductivity of mercury is $\sigma \approx 10^6$ mhos/m, and that for seawater is $\sigma \approx 4$ mhos/m. Possible useful applications include MHD power generation, propulsion, and liquid metals used as coolants for magnetically confined fusion reactors. Though fictitious, the Soviet submarine in Tom Clancy's best-seller *The Hunt for Red October* was endowed with an extremely quiet MHD propulsion system. With the collapse of Communism, the interest in this field has shifted somewhat to more peaceful uses such as flow control or, in other words, the ability to manipulate a flowfield to achieve a beneficial goal such as drag reduction, lift enhancement, and mixing augmentation. Here the so-called Lorentz body forces that result when a conducting fluid moves in the presence of a magnetic field are exploited to effect desired changes in the flowfield (e.g., to suppress turbulence).

Of particular interest to this chapter is the development of efficient reactive control strategies that employ the Lorentz forces to enhance the performance of sea vessels.

Methods to delay transition, prevent separation, and reduce skin-friction drag in turbulent boundary layers are sought. Control strategies targeted toward certain coherent structures in a turbulent flow are particularly sought. We start here by developing the von Kármán momentum integral equation and the instantaneous equations of motion at the wall in the presence of Lorentz force. This should prove useful for making quick estimates of system behavior under different operating conditions. Results are usually not as accurate as solving the differential conservation equations themselves but are more accurate than dimensional analysis. This is followed by an outline of a suggested reactive control strategy that exploits electromagnetic body forces.

14.6.2 The von Kármán Equation

The Lorentz force (body force per unit volume) is given by the cross product of the current density vector $\mathbf{j}$ and the magnetic flux density vector $\mathbf{B}$

$$\mathbf{F} = \mathbf{j} \times \mathbf{B} \tag{14.13}$$

where the current is the sum of the applied and induced contributions

$$\mathbf{j} = \sigma(\mathbf{E} + \mathbf{u} \times \mathbf{B}) \tag{14.14}$$

where σ is the electrical conductivity of the fluid, $\mathbf{E}$ is the applied electric field vector, and $\mathbf{u}$ is the fluid velocity vector. Thus, the Lorentz force is in general given by

$$\mathbf{F} = \sigma(\mathbf{E} \times \mathbf{B}) + \sigma(\mathbf{u} \times \mathbf{B}) \times \mathbf{B} \tag{14.15}$$

Owing to the rather low electrical conductivity of seawater, the induced Lorentz force is typically very small in such applications, and the only way to effect a significant body force is to cross the magnetic field with an applied electric field. For such an application, therefore, the term electromagnetic forcing (EMHD) is commonly used in place of MHD.

Assume a two-dimensional boundary-layer flow of a conducting fluid. Further assume that the magnetic and electric fields are generated using alternating electrodes and magnets parallel to the streamwise direction (see Figure 14.3). Basically, both fields are in the plane (y, z). The resulting Lorentz force has components in the streamwise (x), normal (y), and spanwise (z) directions given by, respectively,

$$F_x = \sigma(E_y B_z - E_z B_y) - \sigma u\left(B_z^2 + B_y^2\right) \tag{14.16}$$

$$F_y = 0 - \sigma v B_z^2 \tag{14.17}$$

$$F_z = 0 - \sigma v B_z B_y \tag{14.18}$$

where u and v are, respectively, the streamwise and normal velocity components. Each of the first terms on the the right-hand sides of Eqs. (14.16)–(14.18) is due to the applied electric field. The second term is the induced Lorentz force and is negligible for low-conductivity fluids such as seawater. Therefore, for such an application and particular arrangement of electrodes or magnets, the electromagnetic body force is predominately in the streamwise direction, $F_x \approx \sigma(E_y B_z - E_z B_y)$.

We now write the continuity and streamwise momentum equations. For an incompressible flow and with the gravitational body force neglected

$$\frac{\partial u}{\partial x} + \frac{\partial v}{\partial y} = 0 \tag{14.19}$$

$$\rho \frac{\partial u}{\partial t} + \rho u \frac{\partial u}{\partial x} + \rho v \frac{\partial u}{\partial y} = -\frac{\partial p}{\partial x} + \frac{\partial \tau}{\partial y} + \sigma(E_y B_z - E_z B_y) - \sigma u (B_z^2 + B_y^2) \tag{14.20}$$

where τ is the viscous shear stress. The pressure term can be written in terms of the velocity outside the boundary layer U_o, which in general is a function of x and t. Note that the total drag should be the sum of the usual viscous and form drags plus (or minus) the Lorentz force. In Eq. (14.20) the induced Lorentz force is always negative (for positive u). The applied force can be positive or negative, depending on the direction of the applied electric field **E** relative to the magnetic field **B**. In any case, when the Lorentz force is positive, it constitutes a thrust (negative drag).

At a given x location, the streamwise momentum equation is readily integrated in y from the wall to the edge of the boundary layer. If we invoke the continuity equation and use the usual definitions for the skin friction coefficient (C_f) and the displacement and momentum thicknesses (δ^* and δ_θ), the resulting integral equation reads

$$C_f \equiv \frac{2\tau_w}{\rho U_o^2} = \frac{2}{U_o^2} \frac{\partial (U_o \delta^*)}{\partial t} + 2\frac{\partial \delta_\theta}{\partial x} - 2\frac{v_w}{U_o} + 2\delta_\theta \left(2 + \frac{\delta^*}{\delta_\theta}\right) \frac{1}{U_o} \frac{\partial U_o}{\partial x}$$

$$+ \frac{2\sigma}{\rho U_o^2} \int_0^\infty [E_y B_z - E_z B_y]\, dy - \frac{2\sigma}{\rho U_o^2} \int_0^\infty u[B_z^2 + B_y^2]\, dy \tag{14.21}$$

where v_w is the injection (or suction) velocity normal to the wall. If σ is not constant, it has to be included in the two integrals on the right-hand side of Eq. (14.21).

It is clear that the following effects contribute to an *increase* in the skin friction: temporal acceleration of freestream, growth of momentum thickness, wall suction, favorable pressure gradient, and positive streamwise Lorentz force. When a Lorentz force points opposite to the flow direction, the skin friction decreases, and the velocity profile tends to become more inflectional. But once again remember to add (or subtract) the drag (or thrust) due to the applied as well as induced Lorentz forces.

Note that Eq. (14.21) is valid for steady or unsteady flows, for Newtonian or non-Newtonian fluids, and for laminar or turbulent flows (mean quantities are used in the latter case). The two integrals in Eq. (14.21) can be computed once the electric and magnetic fields are known and an approximate velocity profile is assumed. For seawater, the last integral in Eq. (14.21) is neglected.

14.6.3   Vorticity Flux at the Wall

Another very useful equation is the differential momentum equation written at the wall ($y = 0$). Here we assume a Newtonian fluid but allow viscosity to vary spatially and the flow to be three-dimensional. For a nonmoving wall, the instantaneous streamwise, normal, and spanwise momentum equations read respectively

$$\rho v_{\mathrm{w}} \frac{\partial u}{\partial y}\bigg|_{y=0} + \frac{\partial p}{\partial x}\bigg|_{y=0} - \frac{\partial \mu}{\partial y}\bigg|_{y=0} \frac{\partial u}{\partial y}\bigg|_{y=0} - \sigma [E_y B_z - E_z B_y]_{y=0} = \mu \frac{\partial^2 u}{\partial y^2}\bigg|_{y=0}$$

(14.22)

$$0 + \frac{\partial p}{\partial y}\bigg|_{y=0} - 0 - F_y\bigg|_{y=0} = \mu \frac{\partial^2 v}{\partial y^2}\bigg|_{y=0}$$

(14.23)

$$\rho v_{\mathrm{w}} \frac{\partial w}{\partial y}\bigg|_{y=0} + \frac{\partial p}{\partial z}\bigg|_{y=0} - \frac{\partial \mu}{\partial y}\bigg|_{y=0} \frac{\partial w}{\partial y}\bigg|_{y=0} - F_z\bigg|_{y=0} = \mu \frac{\partial^2 w}{\partial y^2}\bigg|_{y=0}$$

(14.24)

where $F_y|_{y=0}$ and $F_z|_{y=0}$ are, respectively, the wall values of the Lorentz force in the normal and spanwise directions. In Eq. (14.22), the streamwise Lorentz force at the wall $F_x|_{y=0}$ is due only to the applied electric field (the induced force is zero at the wall because u vanishes there for nonmoving walls). The expression for F_x in Eq. (14.22) is for an array of alternating electrodes and magnets parallel to the streamwise direction,[6] but the streamwise Lorentz force can readily be computed for any other configuration of magnets and electrodes. If the electric conductivity varies spatially, the value of σ at the wall should be used in Eq. (14.22).

The right-hand side of Eq. (14.22) is the (negative of) wall flux of spanwise vorticity, whereas that of Eq. (14.24) is the (negative of) wall flux of streamwise vorticity. Much like transpiration, pressure gradient, or viscosity variations, the wall value of the Lorentz force determines the sign and intensity of wall vorticity flux. Streamwise force contributes to spanwise vorticity flux, and spanwise force contributes to streamwise vorticity flux. Take for example a positive streamwise Lorentz force. This is a negative term on the left-hand side of Eq. (14.22) that makes the curvature of the streamwise velocity profile at the wall more negative (i.e., instantaneously fuller velocity profile). The wall is then a source of spanwise vorticity, and the flow is more resistant to transition and separation. In a turbulent flow, a positive streamwise Lorentz force leads to suppression of normal and tangential Reynolds stresses.

14.6.4 EMHD Tiles for Reactive Control

In this subsection, I will outline a reactive control strategy that exploits the Lorentz forces to modulate the flow of an electrically conducting fluid such as seawater. The idea here is to target the low-speed streaks in the near-wall region of a turbulent boundary layer. The electric field is applied only when and where it is needed, and hence the power consumed by the reactive control system is kept far below that consumed by a predetermined active control system. According to recent numerical and experimental results, brute force application of steady or time-dependent Lorentz force—to reduce drag, for example—does not achieve the break-even point. The reason for this is the high energy expenditure by predetermined control systems (see, for example, the two meeting proceedings edited by Gerbeth 1997 and Meng 1998).

---

[6] Note that for such an array, the applied part of both $F_y|_{y=0}$ and $F_z|_{y=0}$ is identically zero, but the induced part is nonzero if $v_{\mathrm{w}} \neq 0$. In seawater applications, the induced Lorentz force is negligible regardless of any suction, injection, or normal wall motion.

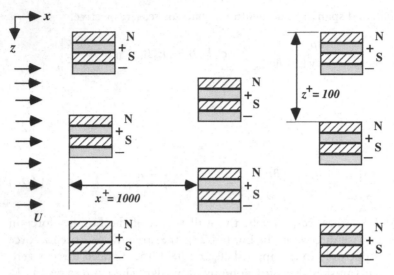

Figure 14.3: EMHD tiles for reactive control of turbulent boundary layers.

The top view in Figure 14.3 depicts the distribution of EMHD tiles to achieve targeted control in a turbulent wall-bounded flow. Each tile consists of two streamwise strips constituting the north and south poles of a permanent magnet and another two strips constituting positive and negative electrodes.[7] The tiles are staggered in a checkerboard configuration and are separated by 100 wall units in the spanwise direction (z) and 1000 wall units in the streamwise direction (x). The length and width of each tile is about 50 wall units.[8]

A shear stress sensor having a spatial resolution of 2–5 wall units is placed just upstream and along the centerline of each tile. Each sensor and corresponding actuator are connected via a closed-feedback control loop. The tile is activated to give a positive or negative streamwise Lorentz force when a low- or high-speed region is detected. The control law used can be based on simple physical arguments or a more complex self-learning neural network. Other possible, albeit more sophisticated, control laws can be based on nonlinear dynamical systems theory or optimal or suboptimal control theory.

For a simpler physical or numerical experiment that does not require sensors and complex closed-loop control, a single low-speed streak is artificially generated in a laminar boundary layer using two suction holes, as depicted in the (x–z) view in Figure 14.4. The two holes are separated in the spanwise direction by 100 wall units, and each is about 0.5 wall unit in diameter. Such a method was successfully used by Gad-el-Hak and Hussain (1986) to generate artificial low-speed streaks as well as simulated bursting events in a laminar environment. In this same experiment, high-speed regions and counterrotating streamwise vortices are also generated, and, if desired, one can readily target any of these simulated events for elimination. The necessary actuation would of course depend on the kind of coherent structure to be targeted.

A single EMHD tile is placed about 50 wall units directly downstream of the suction holes and is activated with a suitable time delay after a suction pulse is

---

[7] The cathode and anode are interchangeable, depending on the desired direction of the Lorentz force.
[8] The length and width of each magnet pole (or electrode) are thus 50 and 12.5 wall units, respectively.

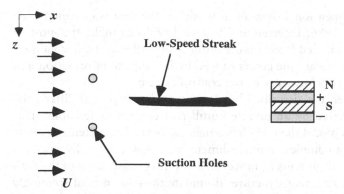

Figure 14.4: Control of a single artificial low-speed streak.

applied—in essence using a simple, predetermined, open-loop control. A single electric pulse[9] triggers a positive Lorentz force to eliminate the artificially generated low-speed streak. The experiment would of course be repeated several times, and the actuator location, phase lag, strength, and duration could be optimized to achieve the desired goal.

The streamwise Lorentz force necessary to eliminate the resulting artificial low-speed streak could be computed from the measurements, and thus the energy expenditure necessary to eliminate all the low-speed streaks in a real turbulent boundary layer could readily be estimated.

14.7 Chaos Control

14.7.1 Nonlinear Dynamical Systems Theory

In the theory of dynamical systems, the so-called butterfly effect denotes sensitive dependence of nonlinear differential equations on initial conditions with phase–space solutions initially very close together separating exponentially. The solution of nonlinear dynamical systems of three or more degrees of freedom may be in the form of a strange attractor whose intrinsic structure contains a well-defined mechanism to produce a chaotic behavior without requiring random forcing. Chaotic behavior is complex, aperiodic and, though deterministic, appears to be random.

A question arises naturally. Just as small disturbances can radically grow within a deterministic system to yield rich, unpredictable behavior, can minute adjustments to a system parameter be used to reverse the process and control (i.e., regularize) the behavior of a chaotic system? Recently, that question was answered in the affirmative theoretically as well as experimentally—at least for system orbits that reside on low-dimensional strange attractors (see the review by Lindner and Ditto 1995). Before describing such strategies for controlling chaotic systems, we first summarize the recent attempts to construct a low-dimensional dynamical systems representation of turbulent boundary layers. Such construction is a necessary first step to be able to use chaos-control strategies for turbulent flows. Additionally, as argued by

---

[9] Note that in this simulated environment the electric pulse would have the same polarity if only low-speed streaks were to be obliterated.

Lumley (1996b), a low-dimensional dynamical model of the near-wall region used in a Kalman filter (Banks 1986; Petersen and Savkin 1999) can make the most of the partial information assembled from a finite number of wall sensors. Such a filter minimizes in a least-squares sense the errors caused by incomplete information and thus globally optimizes the performance of the control system.

Boundary layer turbulence is described by a set of nonlinear partial differential equations and is characterized by an infinite number of degrees of freedom. This makes it rather difficult to model the turbulence using a dynamical systems approximation. The notion that a complex, infinite-dimensional flow can be decomposed into several low-dimensional subunits is, however, a natural consequence of the realization that quasi-periodic coherent structures dominate the dynamics of seemingly random turbulent shear flows. This implies that low-dimensional, localized dynamics can exist in formally infinite-dimensional extended systems such as open turbulent flows. Reducing the flow physics to finite-dimensional dynamical systems enables a study of its behavior through an examination of the fixed points and the topology of their stable and unstable manifolds. From the dynamical systems theory viewpoint, the meandering of low-speed streaks is interpreted as hovering of the flow state near an unstable fixed point in the low-dimensional state space. An intermittent event that produces high wall stress—a burst—is interpreted as a jump along a heteroclinic cycle to a different unstable fixed point that occurs when the state has wandered too far from the first unstable fixed point. Delaying this jump by holding the system near the first fixed point should lead to lower momentum transport in the wall region and, therefore, to lower skin-friction drag. Reactive control means sensing the current local state and through appropriate manipulation keeping the state close to a given unstable fixed point, thereby preventing further production of turbulence. Reducing the bursting frequency by say 50% may lead to a comparable reduction in skin-friction drag. For a jet, relaminarization may lead to a quiet flow and very significant noise reduction.

In one significant attempt the proper orthogonal, or Karhunen–Loève, decomposition method has been used to extract a low-dimensional dynamical system from experimental data of the wall region (Aubry et al. 1988; Aubry 1990). Aubry et al. (1988) expanded the instantaneous velocity field of a turbulent boundary layer using experimentally determined eigenfunctions which are in the form of streamwise rolls. They expanded the Navier–Stokes equations using these optimally chosen, divergence-free, orthogonal functions applied a Galerkin projection and then truncated the infinite-dimensional representation to obtain a 10-dimensional set of ordinary differential equations. These equations represent the dynamical behavior of the rolls and are shown to exhibit a chaotic regime as well as an intermittency due to a burst-like phenomenon. However, Aubry et al.'s (1988) 10-mode dynamical system displays a *regular* intermittency in contrast both to that in actual turbulence as well as to the chaotic intermittency encountered by Pomeau and Manneville (1980) in which event durations are distributed stochastically. Nevertheless, the major conclusion of Aubry et al.'s (1988) study is that the bursts appear to be produced autonomously by the wall region even without turbulence but are triggered by turbulent pressure signals from the outer layer. More recently, Berkooz et al. (1991) generalized the class of wall-layer models developed by Aubry et al. (1988) to permit uncoupled evolution of streamwise and cross-stream disturbances. Berkooz et al.'s results suggest that the intermittent events observed in Aubry et al.'s representation do not arise solely because of the effective closure assumption incorporated but are rather rooted deeper in the dynamical phenomena of the wall region. The book by Holmes et al. (1996) details the Cornell research group's attempts at describing turbulence as a low-dimensional dynamical system.

In addition to the reductionist viewpoint exemplified by the work of Aubry et al. (1988) and Berkooz et al. (1991), attempts have been made to determine directly the dimension of the attractors underlying specific turbulent flows. Again, the central issue here is whether or not turbulent solutions to the infinite-dimensional Navier–Stokes equations can be asymptotically described by a finite number of degrees of freedom. Grappin and Léorat (1991) computed the Lyapunov exponents and the attractor dimensions of two- and three-dimensional periodic turbulent flows without shear. They found that the number of degrees of freedom contained in the large scales establishes an upper bound for the dimension of the attractor. Deane and Sirovich (1991) and Sirovich and Deane (1991) numerically determined the number of dimensions needed to specify chaotic Rayleigh–Bénard convection over a moderate range of Rayleigh numbers, Ra. They suggested that the *intrinsic* attractor dimension is $\mathcal{O}[Ra^{2/3}]$.

The corresponding dimension in wall-bounded flows appears to be dauntingly high. Keefe, Moin, and Kim (1992) determined the dimension of the attractor underlying turbulent Poiseuille flows with spatially periodic boundary conditions. Using a coarse-grained numerical simulation, they computed a lower bound on the Lyapunov dimension of the attractor to be approximately 352 at a pressure-gradient Reynolds number of 3200. Keefe et al. (1992) argued that the attractor dimension in fully-resolved turbulence is unlikely to be much larger than 780. This suggests that periodic turbulent shear flows are deterministic chaos and that a strange attractor does underlie solutions to the Navier–Stokes equations. Temporal unpredictability in the turbulent Poiseuille flow is thus due to the exponential spreading property of such attractors. Although finite, the computed dimension invalidates the notion that the global turbulence can be attributed to the interaction of a *few* degrees of freedom. Moreover, in a physical channel or boundary layer, the flow is not periodic and is open. The attractor dimension in such a case is not known but is believed to be even higher than the estimate provided by Keefe et al. for the periodic *(quasi-closed)* flow.

In contrast to closed, absolutely unstable flows such as Taylor–Couette systems in which the number of degrees of freedom can be small, local measurements in open, convectively unstable flows, such as boundary layers, do not express the global dynamics, and the attractor dimension in that case may inevitably be too large to be determined experimentally. According to the estimate provided by Keefe et al. (1992), the colossal data required (about 10^D, where D is the attractor dimension) for measuring the dimension simply exceed current computer capabilities. Turbulence near transition or near a wall is an exception to that bleak picture. In those special cases, a relatively small number of modes are excited, and the resulting *simple* turbulence can therefore be described by a dynamical system of a reasonable number of degrees of freedom.

14.7.2 Chaos Control

There is another question of greater relevance here. Given a dynamical system in the chaotic regime, is it possible to stabilize its behavior through some kind of active control? Although other alternatives have been devised (e.g., Fowler 1989;

Hübler and Lüscher 1989; Huberman 1990; Huberman and Lumer 1990), the recent method proposed by workers at the University of Maryland (Ott, Grebogi, and Yorke 1990a,b; Shinbrot et al. 1990, 1992a,b, 1998; Shinbrot, Ditto, Grebogi, Ott, Spano, and Yorke 1992; Romeiras et al. 1992) promises to be a significant breakthrough. Comprehensive reviews and bibliographies of the emerging field of chaos control can be found in the articles by Shinbrot et al. (1993); Shinbrot (1993, 1995, 1998); and Lindner and Ditto (1995).

Ott et al. (1990a) demonstrated, through numerical experiments with the Henon map, that it is possible to stabilize a chaotic motion about any prechosen unstable orbit through the use of relatively small perturbations. The procedure consists of applying minute time-dependent perturbations to one of the system parameters to control the chaotic system around one of its many unstable periodic orbits. In this context, targeting refers to the process whereby an arbitrary initial condition on a chaotic attractor is steered toward a prescribed point (target) on this attractor. The goal is to reach the target as quickly as possible using a sequence of small perturbations (Kostelich et al. 1993a).

The success of the Ott–Grebogi–Yorke (OGY) strategy for controlling chaos hinges on the fact that beneath the apparent unpredictability of a chaotic system lies an intricate but highly ordered structure. Left to its own recourse, such a system continually shifts from one periodic pattern to another, creating the appearance of randomness. An appropriately controlled system, on the other hand, is locked into one particular type of repeating motion. With such reactive control the dynamical system becomes one with a stable behavior. The OGY method can be simply illustrated by the schematic in Figure 14.5. The state of the system is represented as the intersection of a stable manifold and an unstable one. The control is applied intermittently whenever the system departs from the stable manifold by a prescribed tolerance; otherwise, the control is shut off. The control attempts to put the system back onto the stable manifold so that the state converges toward the desired trajectory. Unmodeled dynamics cause noise in the system and a tendency for the state to wander off in the unstable direction. The intermittent control prevents that, and the desired trajectory is achieved. This efficient control is not unlike trying to balance a ball in the center of a horse saddle (Moin and Bewley 1994). There is one stable direction (front or back) and one unstable direction (left or right). The restless horse is the unmodeled dynamics intermittently causing the ball to move in the wrong direction. The OGY control needs only to be applied, in the most direct manner possible, whenever the ball wanders off in the left or right direction.

The OGY method has been successfully applied in a relatively simple experiment by Ditto, Rauseo, and Spano (1990) and Ditto and Pecora (1993) at the Naval Surface Warfare Center in which reverse chaos was obtained in a parametrically driven, gravitationally buckled, amorphous magnetoelastic ribbon. Garfinkel et al. (1992) applied the

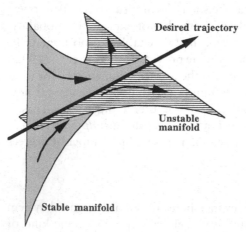

Figure 14.5: The OGY method for controlling chaos.

same control strategy to stabilize drug-induced cardiac arrhythmias in sections of a rabbit ventricle. Other extensions, improvements, and applications of the OGY strategy include higher-dimensional targeting (Auerbach et al. 1992; Kostelich et al. 1993b); controlling chaotic scattering in Hamiltonian (i.e., nondissipative, area conservative) systems (Lai, Deng, and Grebogi 1993; Lai, Tél, and Grebogi 1993); synchronizing identical chaotic systems that govern communication, neural, or biological processes (Lai and Grebogi 1993); using chaos to transmit information (Hayes, Grebogi, and Ott 1994; Hayes, Grebogi, Ott, and Mark 1994); controlling transient chaos (Lai et al. 1994); and taming spatiotemporal chaos using a sparse array of controllers (Chen, Wolf, and Chang 1993; Qin, Wolf, and Chang 1994; Auerbach 1994).

In a more complex system, such as a turbulent boundary layer, there exist numerous interdependent modes and many stable as well as unstable manifolds (directions). The flow can then be modeled as coherent structures plus a parameterized turbulent background. The proper orthogonal decomposition (POD) is used to model the coherent part because POD guarantees the minimum number of degrees of freedom for a given model accuracy. Factors that make turbulence control a challenging task are the potentially quite large perturbations caused by the unmodeled dynamics of the flow, the nonstationary nature of the desired dynamics, and the complexity of the saddle shape describing the dynamics of the different modes. Nevertheless, the OGY-control strategy has several advantages that are of special interest in the control of turbulence: (1) the mathematical model for the dynamical system need not be known, (2) only *small* changes in the control parameter are required, and (3) noise can be tolerated (with appropriate penalty).

Recently, Keefe (1993a,b) made a useful comparison between two nonlinear control strategies as applied to fluid problems: Ott–Grebogi–Yorke's feedback method described above and the model-based control strategy originated by Hübler (see, for example, Hübler and Lüscher 1989; Lüscher and Hübler 1989), the H-method. Both novel control methods are essentially generalizations of the classical perturbation cancellation technique: apply a prescribed forcing to subtract the undesired dynamics and impose the desired one. The OGY strategy exploits the sensitivity of chaotic systems to stabilize existing periodic orbits and steady states. Some feedback is needed to steer the trajectories toward the chosen fixed point, but the required control signal is minuscule. In contrast, Hübler's scheme does not explicitly make use of the system sensitivity. It produces general control response (periodic or aperiodic) and needs little or no feedback, but its control inputs are generally large. The OGY strategy exploits the nonlinearity of a dynamical system; indeed, the presence of a strange attractor and the extreme sensitivity of the dynamical system to initial conditions are essential to the success of the method. In contrast, the H-method works equally for both linear and nonlinear systems.

Keefe (1993a) first numerically examined the two schemes as applied to fully developed and transitional solutions of the Ginzburg–Landau equation, an evolution equation that governs the initially weakly nonlinear stages of transition in several flows and that possesses both transitional and fully chaotic solutions. The Ginzburg–Landau equation has solutions that display either absolute or convective instabilities and is thus a reasonable model for both closed and open flows. Keefe's main conclusion is that control of nonlinear systems is best obtained by making the maximum use possible of the underlying natural dynamics. If the goal dynamics is an unstable nonlinear solution of the equation and the flow is nearby at the instant control is applied, both methods perform reliably and at low-energy cost in reaching and maintaining this goal. Predictably, the performance of both control strategies degrades owing to noise and the spatially discrete nature of realistic forcing. Subsequently, Keefe (1993b) extended the numerical experiment in an attempt to reduce the drag in a channel flow with spatially periodic boundary conditions. The OGY method reduces the skin friction to 60–80% of the uncontrolled value at a mass-flux Reynolds number of 4408. The H-method fails to achieve any drag reduction when starting from a fully turbulent initial condition but shows potential for suppressing or retarding laminar-to-turbulence transition. Keefe (1993a) suggested that the H-strategy might be more appropriate for boundary layer control, whereas the OGY-method might best be used for channel flows.

It is also relevant to note here the work of Bau and his colleagues at the University of Pennsylvania (Singer, Wang, and Bau 1991; Wang, Singer, and Bau 1992) who devised a feedback control to stabilize (*relaminarize*) the naturally occurring chaotic oscillations of a toroidal thermal convection loop heated from below and cooled from above. On the basis of a simple mathematical model for the thermosyphon, Bau and his colleagues constructed a reactive control system that was used to alter the flow characteristics inside the convection loop significantly. Their linear control strategy, perhaps a special version of the OGY chaos control method, consists simply of sensing the deviation of fluid temperatures from desired values at a number of locations inside the thermosyphon loop and then altering the wall heating either to suppress or to enhance such deviations. Wang et al. (1992) also suggested extending their theoretical and experimental method to more complex situations such as those involving Bénard convection (Tang and Bau 1993a,b). Hu and Bau (1994) used a similar feedback control strategy to demonstrate that the critical Reynolds number for the loss of stability of planar Poiseuille flow can be significantly increased or decreased.

Other attempts to use low-dimensional dynamical systems representation for flow control include the work of Berkooz et al. (1993); Corke, Glauser, and Berkooz (1994); and Coller, Holmes, and Lumley (1994a,b). Berkooz et al. (1993) applied techniques of modern control theory to estimate the phase-space location of dynamical models of the wall-layer coherent structures and used these estimates to control the model dynamics. Because discrete wall sensors provide incomplete knowledge of phase-space location, Berkooz et al. maintained that a nonlinear observer, which incorporates past information and the equations of motion into the estimation procedure, is required. Using an extended Kalman filter, they achieved effective control of a bursting pair of rolls with the equivalent of two wall-mounted shear sensors. Corke et al. (1994) used a low-dimensional dynamical system based on the proper orthogonal decomposition to guide control experiments for an axisymmetric jet. By sensing the downstream velocity and actuating an array of miniature speakers located at the lip of the jet, their feedback control succeeded in converting the near-field instabilities from spatial-convective to temporal-global. Coller et al. (1994a,b) developed a feedback control strategy for strongly nonlinear dynamical systems, such as turbulent flows, subject to small random perturbations that kick the system intermittently from one saddle point to another along heteroclinic cycles. In essence, their approach is to use local, weakly nonlinear feedback control to keep a solution near a saddle point as long as possible but then to let the natural, global nonlinear dynamics run their course when *bursting* (in a low-dimensional model) does occur. Though conceptually related to the OGY strategy, Coller et al.'s method does not actually stabilize the state but merely holds the system near the desired point longer than it would otherwise stay.

Shinbrot and Ottino (1993a,b) offered yet another strategy presumably most suited to controlling coherent structures in area-preserving turbulent flows. Their geometric method exploits the premise that the dynamical mechanisms that produce the organized structures can be remarkably simple. By repeated stretching and folding of "horseshoes" that are present in chaotic systems, Shinbrot and Ottino have demonstrated numerically as well as experimentally the ability to create, destroy, and manipulate coherent structures in chaotic fluid systems. The key idea to create such structures is to place folds of horseshoes near low-order periodic points intentionally. In a dissipative dynamical system, volumes contract in state space, and the colocation of a fold with a periodic point leads to an isolated region that contracts asymptotically to a point. Provided that the folding is done properly, it counteracts stretching. Shinbrot and Ottino (1993a) applied the technique to three prototypical problems: a one-dimensional chaotic map, a two-dimensional one, and a chaotically advected fluid. Shinbrot (1995, 1998) and Shinbrot et al. (1998) provided recent reviews of the stretching and folding as well as other chaos control strategies.

14.8 Microfabrication

This topic was covered in more detail in the previous chapter. Manufacturing processes that can create extremely small machines have been developed in recent years (Angell et al. 1983; Gabriel et al. 1988, 1992; O'Connor 1992; Gravesen et al. 1993; Bryzek et al. 1994; Gabriel 1995; Hogan 1996; Ho and Tai 1996, 1998; Tien 1997; Busch-Vishniac 1998; Amato 1998). In this emerging microfabrication technology, under intensive development only since 1990, electronic and mechanical components are combined on a single silicon chip using photolithographic micromachining techniques. Motors, electrostatic actuators, pneumatic actuators, valves, gears, and tweezers of typical size $\mathcal{O}\,[10\ \mu\mathrm{m}]$ have been fabricated. These have been used as sensors for pressure, temperature, velocity, mass flow, or sound, and as actuators for linear and angular motions. Current usage for microelectromechanical systems (MEMS) includes accelerometers for airbags and guidance systems, pressure sensors for engine air intake and blood analysis, rate gyroscopes for antilock brakes, microrelays and microswitches for semiconductor automatic test equipment, and microgrippers for surgical procedures (O'Connor 1992; Hogan 1996; Paula 1996; Ouellette 1996; Ashley 1996; Robinson, Helvajian, and Jansen 1996a,b). There is considerable work under way to include other applications, one example being the microsteam engine described by Lipkin (1993). A second example is the 3 cm × 1.5 cm digital light processor that contains 0.5–2 million individually addressable micromirrors, each measuring 16 μm×μm. Texas Instruments, Inc., is currently producing such a device with a resolution of 2000 × 1000 pixels, for high-definition televisions and other display equipments. The company maintains that when mass produced, such device costs on the order of $100 (i.e., less than $0.0001 per actuator).

The new *Journal of Microelectromechanical Systems* and *Journal of Micromechanics and Microengineering* are dedicated to this technology, and the older *Sensors and Actuators* is increasingly allotting more of its pages to MEMS. Entire sessions in scientific meetings have been increasingly assigned to MEMS applications in fluid mechanics (see, for example, the presentations by McMichael, Tai, Mehregany, Mastrangelo, and Yun, all made at the AIAA Third Shear Flow Control Conference,

Orlando, Florida, 6–9 July 1993, and the volumes edited by Bandyopadhyay, Breuer, and Blechinger 1994 and Breuer, Bandyopadhyay, and Gad-el-Hak 1996). Recent reviews of the use, or potential use, of MEMS in flow control include those by Gad-el-Hak (1994, 1996b); Lumley (1996); McMichael (1996); Mehregany et al. (1996); and Ho and Tai (1996, 1998).

Microelectromechanical systems would be ideal for the reactive flow-control concept advocated in this chapter. Methods of flow control targeted toward specific coherent structures involve nonintrusive detection and subsequent modulation of events that occur randomly in space and time. To achieve proper targeted control of these quasi-periodic vortical events, temporal phasing as well as spatial selectivity are required. Practical implementation of such an idea necessitates the use of a large number of intelligent, communicative wall sensors and actuators arranged in a checkerboard pattern. Section 14.5 provides estimates for the number, characteristics, and energy consumption of such elements required to modulate the turbulent boundary layer that develops along a typical commercial aircraft or nuclear submarine. An upper-bound number to achieve total turbulence suppression is about one million sensors or actuators per square meter of the surface, although as argued earlier the actual number needed to achieve effective control could perhaps be 1–2 orders of magnitude below that.

The sensors would be expected to measure the amplitude, location, and phase or frequency of the signals impressed upon the wall by incipient bursting events. Instantaneous wall-pressure or wall-shear stress can be sensed, for example. The normal or in-plane motion of a minute membrane is proportional to the respective point force of primary interest. For measuring wall pressure, microphone-like devices respond to the motion of a vibrating surface membrane or an internal elastomer. Several types are available, including variable capacitance (condenser or electret), ultrasonic, optical (e.g., optical-fiber and diode-laser), and piezoelectric devices (see, for example, Löfdahl et al. 1993, 1994; Löfdahl and Gad-el-Hak 1999). A potentially useful technique for our purposes has been tried at MIT (Warkentin et al. 1987; Young et al. 1988; Haritonidis et al. 1990a,b). An array of extremely small (0.2 mm in diameter) laser-powered microphones (termed picophones) was machined in silicon using integrated-circuit fabrication techniques and used for field measurement of the instantaneous surface pressure in a turbulent boundary layer. The wall-shear stress, though smaller and therefore more difficult to measure than pressure, provides a more reliable signature of the near-wall events.

Actuators are expected to produce a desired change in the targeted coherent structures. The local acceleration action needed to stabilize an incipient bursting event can be in the form of adaptive wall, transpiration, wall-heat transfer, or electromagnetic body force. Traveling surface waves can be used to modify a locally convecting pressure gradient such that the wall motion follows that of the coherent event causing the pressure change. Surface motion in the form of a Gaussian hill with height $y^+ = \mathcal{O}[10]$ should be sufficient to suppress typical incipient bursts (Lumley 1991b; Carlson and Lumley 1996). Such time-dependent alteration in wall geometry can be generated by driving a flexible skin using an array of piezoelectric devices (which dilate or contract depending on the polarity of current passing through them), electromagnetic actuators, magnetoelastic ribbons (made of nonlinear materials that change their stiffness in the presence of varying magnetic fields), or Terfenol-d rods (a novel metal composite developed at Grumman Corporation that changes its length when subjected to a magnetic field). Other exotic materials that can be used for actuation should also be noted. For example, electrorheological fluids (Halsey and Martin 1993) instantly solidify when exposed to an electric field and may thus be useful for the present application. Recently constructed microactuators specifically designed for flow control include those by Wiltse and Glezer (1993); James et al. (1994); Jacobson and Reynolds (1995); Vargo and Muntz (1996); and Keefe (1996).

Suction or injection at many discrete points can be achieved simply by connecting a large number of minute streamwise slots arranged in a checkerboard pattern to a low-pressure or high-pressure reservoir located underneath the working surface. The transpiration through each individual slot is turned on and off using a corresponding number of independently controlled microvalves. Alternatively, positive-displacement or rotary micropumps (see, for example, Sen et al. 1996; Sharatchandra et al. 1997) can be used for blowing or sucking fluid through small holes or slits. On the basis of the results of Gad-el-Hak and Blackwelder (1989), equivalent suction coefficients of about 0.0006 should be sufficient to stabilize the near-wall region. If it is assumed that the skin-friction coefficient in the uncontrolled boundary layer is $C_f = 0.003$ and further assumed that the suction used is sufficient to establish an asymptotic boundary layer ($d\delta_\theta/dx = 0$, where δ_θ is the momentum thickness), the skin friction in the reactively controlled case is then $C_f = 0 + 2\,C_q = 0.0012$, or 40% of the original value. The net benefit will, of course, be reduced by the energy expenditure of the suction pump (or micropumps) as well as the array of microsensors and microvalves.

Finally, if the bursting events are to be eliminated by lowering the near-wall viscosity, direct electric-resistance heating can be used in liquid flows, and thermoelectric devices based on the Peltier effect can be used for cooling in the case of gaseous boundary layers. The absolute viscosity of water at 20 °C decreases by approximately 2% for each 1 °C rise in temperature, whereas for room-temperature air, μ decreases by about 0.2% for each 1 °C drop in temperature. The streamwise momentum equation written at the wall can be used to show that a suction coefficient of 0.0006 has approximately the same effect on the wall curvature of the instantaneous velocity profile as a surface heating of 2 °C in water or a surface cooling of 40 °C in air (Liepmann and Nosenchuck 1982; Liepmann et al. 1982).

Sensors and actuators of the types discussed in this section can be combined on individual electronic chips using microfabrication technology. The chips can be interconnected in a communications network that is controlled by a massively parallel computer or a self-learning neural network (Section 14.9), perhaps each sensor or actuator unit communicating only with its immediate neighbors. In other words, it may not be necessary for one sensor or actuator to exchange signals with another far-away unit. Factors to be considered in an eventual field application of chips produced using microfabrication processes include sensitivity of sensors, sufficiency and frequency response of actuators' action, fabrication of large arrays at affordable prices, survivability in the hostile field environment, and energy required to power the sensors or actuators. As argued by Gad-el-Hak (1994, 1996b) and in Chapter 13, sensor and actuator chips currently produced are small enough for typical field application, and they can be programmed to provide a sufficiently large or fast action in response to a certain sensor output (see also Jacobson and Reynolds 1995). Present prototypes are, however, still quite expensive as well as delicate. But so was the transistor when first introduced! It is hoped that the unit price of future sensor and actuator elements will follow the same dramatic trends witnessed in case of the simple transistor and even the much more complex integrated circuit. The price anticipated by Texas Instruments for an array of one million mirrors hints that the technology is well in its way to mass produce phenomenally inexpensive microsensors and microactuators. Additionally, current automotive applications are a rigorous proving ground for MEMS: under-the-hood sensors can already withstand harsh conditions such as intense heat, shock, continual vibration, corrosive gases, and electromagnetic fields.

14.9 Soft Computing

The term soft computing was coined by Lotfi Zadeh of the University of California, Berkeley, to describe several ingenious modes of computations that exploit tolerance for imprecision and uncertainty in complex systems to achieve tractability, robustness, and low cost (Yager and Zadeh 1992; Bouchon-Meunier, Yager, and Zadeh 1995a,b; Jang, Sun, and Mizutani 1997). The principle of complexity provides the impetus for soft computing: as the complexity of a system increases, the ability to predict its response diminishes until a threshold is reached beyond which precision and relevance become almost mutually exclusive (Noor and Jorgensen 1996). In other words, precision and certainty carry a cost. By employing modes of reasoning— probabilistic reasoning—that are approximate rather than exact, soft computing can help in searching for globally optimal design or achieving effectual control while taking into account system uncertainties and risks.

Soft computing refers to a domain of computational intelligence that loosely lies in between purely numerical (hard) computing and purely symbolic computations. Alternatively, one can think about symbolic computations as a form of artificial intelligence lying in between biological intelligence and computational intelligence (soft computing). The schematic in Figure 14.6 illustrates the general idea. Artificial intelligence relies on symbolic information-processing techniques and uses logic as representation and inference mechanisms. It attempts to approach the high level of human cognition. In contrast, soft computing is based on modeling low-level cognitive processes and strongly emphasizes modeling of uncertainty as well as learning. Computational intelligence mimics the ability of the human brain to employ modes of reasoning that are approximate. Soft computing provides a machinery for the numeric representation of the types of constructs developed in the symbolic artificial intelligence. The boundaries between these paradigms are of course *fuzzy*.

The principal constituents of soft computing are neurocomputing, fuzzy logic, and genetic algorithms, as depicted in Figure 14.6. These elements, together with probabilistic reasoning, can be combined in hybrid arrangements, resulting in better systems in terms of parallelism, fault tolerance, adaptivity, and uncertainty management. To my knowledge, only neurocomputing has been employed for fluid-flow control, but the other tools of soft computing may be just as useful for constructing powerful controllers and have in fact been used as such in other fields such as large-scale subway controllers and video cameras. A brief description of these three constituents follows.

Neurocomputing is inspired by the neurons of the human brain and how they work. Neural networks are information-processing devices that can learn by adapting synaptic weights to changes in the surrounding environment; can handle imprecise,

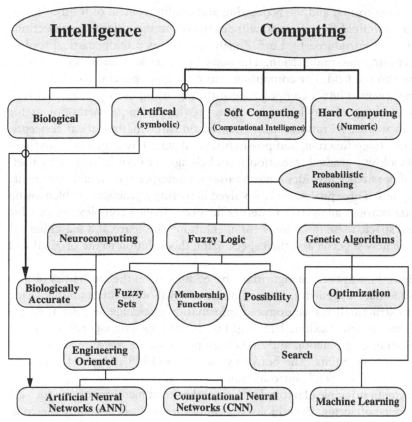

Figure 14.6: Tools for soft computing.

fuzzy, noisy, and probabilistic information; and can generalize from known tasks (examples) to unknown ones. Actual engineering-oriented hardware are termed artificial neural networks (ANN), whereas algorithms are called computational neural networks (CNN). The nonlinear, highly parallel networks can perform any of the following tasks: classification, pattern matching, optimization, control, and noise removal. As modeling and optimization tools, neural networks are particularly useful when good analytic models are either unknown or extremely complex.

An artificial neural network consists of a large number of highly interconnected processing elements—essentially equations known as "transfer functions"—that are analogous to neurons and are tied together with weighted connections that are analogous to synapses. A processing unit takes weighted signals from other units, possibly combines them, and gives a numeric result. The behavior of neural networks—how they map input data—is influenced primarily by the transfer functions of the processing elements, how the transfer functions are interconnected, and the weights of those interconnections. Learning typically occurs by example—through exposure to a set of input–output data for which the training algorithm adjusts the connection weights (synapses). These connection weights store the knowledge necessary to solve specific problems. As an example, it is now possible to use neural networks to sense (*smell*) odors in many different applications (Ouellette 1999). The electronic noses (*e-noses*) are on the verge of finding commercial applications in medical diagnostics, environmental monitoring, and the processing and quality control of foods. Neural networks, as used in fluid-flow control, will be covered in the following subsection.

Fuzzy logic was introduced by Lotfi Zadeh in 1965 as a mathematical tool to deal with uncertainty and imprecision. The book by Yager and Zadeh (1992) is an excellent primer to the field. For computing and reasoning, general concepts (such as size) are implemented into a computer algorithm by using mostly words (such as small, medium, or large). Fuzzy logic, therefore, provides a unique methodology for computing with words. Its ratiocination is based on three mathematical concepts: fuzzy sets, membership function, and possibility. As dictated by a membership function, fuzzy sets allow a gradual transition from belonging to not belonging to a set. The concept of possibility provides a mechanism for interpreting factual statements involving fuzzy sets. Three processes are involved in solving a practical problem using fuzzy logic: fuzzification, analysis, and defuzzification. Given a complex, unsolvable problem in real space, these three steps entail enlarging the space and searching for a solution in the new superset and then specializing this solution to the original real constraints.

Genetic algorithms are search algorithms based loosely on the mechanics of natural selection and natural genetics. They combine survival of the fittest among string structures with structured yet randomized information exchange and are used for search, optimization, and machine learning. For control, genetic algorithms aim at achieving minimum cost function and maximum performance measure while satisfying the problem constraints. The books by Goldberg (1989); Davis (1991); and Holland (1992) provide a gentle introduction to the field.

In the Darwinian principle of natural selection, the fittest members of a species are favored to produce offspring. Even biologists cannot help but be awed by the complexity of life observed to evolve in the relatively short time suggested by the fossil records. A living being is an amalgam of characteristics determined by the (typically tens of thousands) genes in its chromosomes. Each gene may have several forms or alternatives called alleles that produce differences in the set of characteristics associated with that gene. The chromosomes are therefore the organic devices through which the structure of a creature is encoded, and this living being is created partly through the process of decoding those chromosomes. Genes transmits hereditary characters and form specific parts of a self-perpetuated deoxyribonucleic acid (DNA) in a cell nucleus. Natural selection is the link between the chromosomes and the performance of their decoded structures. Simply put, the process of natural selection causes those chromosomes that encode successful structures to reproduce more often than those that do not.

In an attempt to solve difficult problems, John H. Holland of the University of Michigan introduced in the early 1970s the man-made version of the procedure of natural evolution. The candidate solutions to a problem are ranked by the genetic algorithm according to how well they satisfy a certain criterion, and the fittest members are the most favored to combine among themselves to form the next generation of the members of the *species*. Fitter members presumably produce even fitter offspring and therefore better solutions to the problem at hand. Solutions are represented by binary strings, and each trial solution is coded as a vector called a chromosome. The elements of a chromosome are described as genes, and their varying values at specific positions are called alleles. Good solutions are selected for reproduction based on a fitness function using genetic recombination operators such as crossover and mutation. The main advantage of genetic algorithms is their global parallelism in which the search efforts to many regions of the search area are simultaneously allocated.

Genetic algorithms have been used for the control of different dynamical systems such as the optimization of robot trajectories. But to my knowledge and at the time of writing this chapter (mid-1999), the control of fluid flow is yet to benefit from this powerful soft computing tool. In particular, when a finite number of sensors are used to gather information about the state of the flow, a genetic algorithm, perhaps combined with a neural network, can adapt and learn to use current information to eliminate the uncertainty created by insufficient a priori information.

14.9.1 Neural Networks for Flow Control

Biologically inspired neural networks are finding increased applications in many fields of science and technology. Modeling of complex dynamical systems, adaptive noise canceling in telephones and modems, bomb sniffers, mortgage-risk evaluators, sonar classifiers, and word recognizers are but a few of the existing usages of neural nets. The book by Nelson and Illingworth (1991) provides a lucid introduction to the field, and the review article by Antsaklis (1993) focuses on the use of neural nets for the control of complex dynamical systems. For flow-control applications, neural networks offer the possibility of adaptive controllers that are simpler and potentially less sensitive to parameter variations as compared with conventional controllers. Moreover, if a colossal number of sensors and actuators is to be used, the massively parallel computational power of neural nets will surely be needed for real-time control.

The basic elements of a neural network are schematically shown in Figure 14.7. Several inputs are connected to the nodes (neurons or processing elements) that form the input layer. There are one or more hidden layers followed by an output layer. Note that the number of connections is higher than the total number of nodes. Both numbers are chosen based on the particular application and can be arbitrarily large for complex tasks. Simply put, the multitask—albeit simple—job of each processing element is to evaluate each of the input signals to that particular element, calculate the weighted sum of the combined inputs, compare that total to some threshold level,

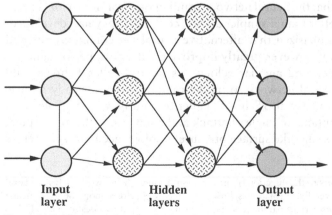

Input layer   **Hidden layers**   **Output layer**

Figure 14.7: Elements of a neural network.

and finally determine what the output should be. The various weights are the adaptive coefficients, which vary dynamically as the network *learns* to perform its assigned task; some inputs are more important than others. The threshold, or transfer, function is generally nonlinear; the most common one is the continuous sigmoid, or S-shaped, curve, which approaches a minimum and maximum value at the asymptotes. If the sum of the weighted inputs is larger than the threshold value, the neuron generates a signal; otherwise, no signal is *fired*. Neural networks can operate in feedforward or feedback mode.[10] Complex systems, for which dynamical equations may not be known or may be too difficult to solve, can be modeled using neural nets.

For flow control, neural networks provide convenient, fast, nonlinear adaptive algorithms to relate sensor outputs to actuator inputs via variable-coefficient functions and nonlinear, sigmoid saturation functions. With no prior knowledge of the pertinent dynamics, a self-learning neural network develops a model for those dynamics through observations of the applied control and sensed measurements. The network is by nature nonlinear and can therefore better handle nonlinear dynamical systems, which is a difficult task when classical (linear or weakly nonlinear) control strategies are attempted. The feedforward type of neural network acts as a nonlinear filter forming an output from a set of input data. The output can then be compared with some desired output, and the difference (error) is typically used in a back-propagation algorithm that updates the network parameters.

The number of researchers using neural networks to control fluid flows is growing rapidly. Herein, we provide only a small sample. Using a pretrained neural network, Fan et al. (1993) conducted a conceptual reactive-flow-control experiment to delay laminar-to-turbulence transition. Numerical simulations of their flow-control system demonstrate almost complete cancellation of single and multiple artificial wave disturbances. Their controller also successfully attenuated a natural disturbance signal with developing wave packets from an actual wind-tunnel experiment.

Jacobson and Reynolds (1993b, 1995, 1998) used neural networks to minimize the boundary velocity gradient of three model flows: the one-dimensional stochastic Burgers equation, a two-dimensional computational model of the near-wall region of a turbulent boundary layer, and a real-time turbulent flow with a spanwise array of wall actuators together with upstream and downstream wall-sensors. For all three problems, the neural network successfully learned about the flow and developed into proficient controllers. For the laboratory experiments, however, Jacobson and Reynolds (1995) reported that the neural network training time was much longer and the performance was no better than a simpler ad hoc controller they had developed. Jacobson and Reynolds emphasized that alternative neural net configurations and convergence algorithms may, however, greatly improve the network performance.

Using the angle of attack and angular velocity as inputs, Faller, Schreck, and Luttges (1994) trained a neural network to model the measured unsteady surface pressure over a pitching airfoil (see also Schreck, Faller, and Luttges 1995). Following training and using the instantaneous angle of attack and pitch rate as the only inputs, their network was able to predict the surface pressure topology accurately as well as

---

[10] Note that this terminology refers to the direction of information through the network. When a neural net is used as a controller, the overall control loop is, however, a feedback, closed loop: the self-learning network dynamically updates its various parameters by comparing its output to a desired output, thus requiring feedback information relating to the effect of its control.

the time-dependent aerodynamic forces and moments. The model was then used to develop a neural network controller for wing-motion actuator signals, which in turn provided direct control of the lift-to-drag ratio across a wide range of time-dependent motion histories.

As a final example, Kawthar-Ali and Acharya (1996) developed a neural network controller for use in suppressing the dynamic-stall vortex that periodically develops in the leading edge of a pitching airfoil. On the basis of the current state of the unsteady pressure field, their control system specified the optimum amount of leading-edge suction to achieve complete vortex suppression.

14.10   Parting Remarks

In this chapter, I have emphasized the frontiers of the field of flow control, reviewing the important advances that took place during the past few years and providing a blueprint for future progress. In two words, the future of flow control is in taming turbulence by targeting its coherent structures: *reactive control*. Recent developments in chaos control, microfabrication, and soft computing tools are making it more feasible to perform reactive control of turbulent flows to achieve drag reduction, lift enhancement, mixing augmentation, and noise suppression. Field applications, however, have to await further progress in those three modern areas. Other less complex control schemes, passive as well as active, are more market-ready and are also witnessing a resurgence of interest.

The outlook for reactive control is quite optimistic. Soft computing tools and nonlinear dynamical systems theory are developing at a fast pace. Microelectro-mechanical systems technology is improving even faster. The ability of Texas Instruments to produce an array of one million individually addressable mirrors for around 0.01 cent per actuator is a foreteller of the spectacular advances anticipated in the near future. Existing automotive applications of MEMS have already proven the ability of such devices to withstand the harsh environment under the hood. For the first time, targeted control of turbulent flows is in the realm of the possible for future practical devices. What is needed is a focused, well-funded research and development program to make it all come together for field application of reactive flow-control systems.

In parting, it may be worth recalling that a mere 10% reduction in the total drag of an aircraft translates into a saving of one billion dollars in annual fuel cost (at 1999 prices) for the commercial fleet of aircraft in the United States alone. Contrast this benefit to the annual worldwide expenditure of perhaps a few million dollars for all basic research in the broad field of flow control. Taming turbulence, though arduous, will pay for itself in gold. Reactive control, as difficult as it seems, is neither impossible nor a pie in the sky. Besides, lofty goals require strenuous efforts. Easy solutions to difficult problems are likely to be wrong, as implied by Henry Mencken's words of wisdom quoted at the beginning of this last chapter.

A parting verse from William Shakespeare's *The Taming of the Shrew*:
Hortensio (a gentleman of Padua): Now go thy ways, thou hast tam'd a curst shrew.
Lucentio (a gentleman of Pisa): 'Tis a wonder, by your leave, she will be tam'd so.

Epilogue

> To be conscious that we are perceiving or thinking is to be conscious of
> our own existence.
>
> *(Aristotle, 384–322 B.C.)*
>
> Minds are like parachutes. They only function when they are open.
>
> *(Sir James Dewar, 1842–1923)*

The field of flow control is broad, practically very important, and rich in scientific and technological challenges. Though the field is as old as man himself, its potential for improving our lives will keep it going strong for yet another millennium. In this book, I have made a modest attempt to place the field in a unifying framework and to categorize the different control strategies properly. At a minimum, I hope to have provided a useful navigation tool through the colossal literature in the field of flow control and its intricately related subfields such as control theory, chaos, microelectromechanical systems, soft computing, and so forth. Review articles were favored for citation because they themselves usually contain many useful citations.

As is clear from all the previous chapters, there is no lack of flow-control methods to achieve a particular goal for free-shear as well as wall-bounded flows across the entire range of Mach and Reynolds numbers. Ranging from simple to complex, from inexpensive to expensive, from passive to active to reactive, and from market-ready to futuristic, a great variety of control devices are at the disposal of the fluids engineer. Flow control is most effective when applied near the transition or separation points; in other words, near the critical-flow regimes where flow instabilities magnify quickly. Therefore, delaying or advancing laminar-to-turbulence transition and preventing or provoking separation can readily be accomplished. To reduce the skin-friction drag in a nonseparating turbulent boundary layer, where the mean flow is quite stable, is a more challenging problem. In all cases, because flow-control goals are often adversely interrelated, tough design compromises are always in the forefront. When designing a flow-control device to achieve a particular goal such as skin-friction reduction, the engineer's foremost task is to ensure a minimum and most benign trade-off.

Market-ready techniques include passive and predetermined active control. Shaping, suction, heating or cooling, Lorentz body force, and compliant coatings can be used to delay transition by an order of magnitude in Reynolds number, and, except for the latter technique (at least at this time), can also be used to prevent boundary layer separation. The use of polymers, microbubbles, riblets, and LEBUs can lead to skin-friction reduction in turbulent boundary layers. Numerous other techniques are available to reduce form drag, induced drag, and wave drag. Remaining issues for field application of market-ready techniques include cost, maintenance, and reliability. Potential further improvements in classical flow-control techniques will perhaps involve combining more than one technique to aim at achieving a favorable effect that is greater than the sum. Examples include combining suction or polymer injection with riblets for increased effectiveness and savings. Owing to its obvious difficulties, synergism has not been extensively studied in the past but deserves future consideration.

Classical control techniques, though spectacularly successful in the past, have been extended to near their physical limits. Conventional strategies are often ineffective for turbulent flows. Substantial gains are potentially possible, however, when reactive flow-control methods are used to target specific coherent structures for modulation. Particularly for turbulent flows, reactive control requires large numbers of sensors and actuators and will not become practical until the technology for manufacturing inexpensive, robust microsensors and microactuators becomes available. Autonomous control algorithms and associated computers to handle the required colossal data in real time must also be developed. Further research is needed in dynamical systems theory—particularly chaos control, microelectromechanical systems, alternative neural network configurations, convergence algorithms, and distributed control. The difficulties are daunting, but the potential payoffs are enormous.

Did we benefit from flow control thus far? The simple act of streamlining represents one of the oldest and most spectacular achievements of flow control. Streamlined spears were perhaps the difference between life and death to our Stone Age ancestors about 400,000 years ago (see Chapter 1). Pressure drag due to separation has virtually been eliminated from modern airplanes and submarines, making it possible to fly or sail as well as saving countless resources. Automobiles to date have significantly lower aerodynamic drag than they did a generation ago. Sweeping back the wings of all large airliners made it possible for them to fly just below the speed of sound without encountering the tremendous wave-drag penalty associated with shock-wave formation, and thus also saving countless resources. Dilute polymer solutions can reduce the skin-friction drag by up to 80% in wall-bounded flows of most liquids of practical interest. In fact, polymers are in actual use in several pipe flows, including the crude oil trans-Alaskan pipeline and the Iraq–Turkey pipeline. Jet airplanes would not have been possible without the significant noise reduction made possible by clever design. High-lift devices are common on all sort of aircraft, making it possible for jumbo jets to take off and land at safe, moderate speeds and to use relatively short runways. Mostly laminar wings are very close to being viable for commercial aircraft, promising further savings in fuel consumption. In my view, the most significant recent advance in the field of flow control is its gradual—perhaps even not so gradual—shift from being an art to being a science. More control devices are now designed using first principles. This trend will undoubtedly accelerate as computer power continues to increase exponentially and, with the advent of molecular electronics (see Section 1.3), seemingly indefinitely.

As for the future? The French pilot, warrior, and writer Antoine de Saint-Exupéry wrote in *Citadelle (The Wisdom of the Sands)*: "As for the future, your task is not to foresee, but to enable it." Despite certain practical difficulties mostly related to the low price of fuel and the high price and volume of polymers, there is little fundamental reason not to be able to benefit from polymer injections to reduce dramatically the skin-friction drag of the turbulent boundary layers developing around underwater vehicles and the submerged portion of ship hulls. A stated goal of NASA is to double the lift-to-drag ratio for the United States commercial fleet of aircraft in the next two decades. This lofty goal will obviously best be achieved by simultaneously lowering the drag and increasing the lift. In the future, reactive control will target certain coherent structures to tame turbulence in practical vehicles and industrial devices. Drag will be reduced to near the lowest level allowed by the laws of nature, and thus valuable resources will be saved and a better environment will be preserved for future generations. My personal wish list includes the utilization of advanced flow-control strategies to make practical such seeming pies in the sky as the commercial submarine, the reusable spaceplane, and the personal skycar—all briefly mentioned in Section 1.3. In fact, prototypes of the latter two dreams were scheduled for testing during 1999 (Ashley 1998; May 1999).

Now my friends it is time to end, at least for now, the adventure that we have started. But where do we go from here? I leave you with the words of the mathematician Charles Lutwidge Dodgson (1832–1898), who created the two memorable stories *Alice's Adventures in Wonderland* and *Through the Looking-Glass* merely to amuse the young daughter—Alice—of an acquaintance, and whose nom de plume was Lewis Carroll.

> "Cheshire Puss," she began, rather timidly, as she did not at all know whether it would like the name: however, it only grinned a little wider. "Come, it's pleased so far," thought Alice, and she went on, "Would you tell me, please, which way I ought to walk from here?".
>
> "That depends a good deal on where you want to get to," said the Cat.
>
> "I don't much care where— —" said Alice.
>
> "Then it doesn't matter which way you walk," said the Cat.
>
> "— —so long as I get somewhere," Alice added as an explanation.
>
> "Oh, you're sure to do that," said the Cat, "if you only walk long enough."
>
> (From Lewis Carroll's *Alice's Adventures in Wonderland,* 1865)

Bibliography

Abbott, I.H., and von Doenhoff, A.E. (1959) *Theory of Wing Sections*, Dover, New York.

Abergel, F., and Temam, R. (1990) "On Some Control Problems in Fluid Mechanics," *Theor. Comput. Fluid Dyn.* **1**, pp. 303–325.

Acarlar, M.S., and Smith, C.R. (1987a) "A Study of Hairpin Vortices in a Laminar Boundary Layer. Part 1. Hairpin Vortices Generated by a Hemisphere Protuberance," *J. Fluid Mech.* **175**, pp. 1–41.

Acarlar, M.S., and Smith, C.R. (1987b) "A Study of Hairpin Vortices in a Laminar Boundary Layer. Part 2. Hairpin Vortices Generated by Fluid Injection," *J. Fluid Mech.* **175**, pp. 43–83.

Acharya, N., Sen, M., and Chang, H.-C. (1992) "Heat Transfer Enhancement in Coiled Tubes by Chaotic Mixing," *Int. J. Heat Mass Transfer* **35**, pp. 2475–2489.

Achia, B.U., and Thompson, D.W. (1977) "Structure of the Turbulent Boundary Layer in Drag-Reducing Pipe Flow," *J. Fluid Mech.* **81**, pp. 439–464.

Ackeret, J., Ras, M., and Pfenninger, W. (1941) "Verhinderung des Turbulentwerdens einer Grenzschicht durch Absaugung," *Naturwissenschaften* **29**, pp. 622–623.

Adkins, R.C. (1975) "A Short Diffuser with Low Pressure Loss," *J. Fluids Eng.* **97**, pp. 297–302.

Adkins, R.C. (1977) "Diffusers and Their Performance Improvement by Means of Boundary Layer Control," in *AGARD Special Course on Concepts for Drag Reduction*, AGARD-R-654, paper no. 6, Rhode-Saint-Génèse, Belgium.

Adkins, R.C., Mathaus, D.S., and Yost, J.O. (1980) "The Hybrid Diffuser," ASME Paper No. 80-GT-136, New York.

Adleman, L.M. (1994) "Molecular Computation of Solutions to Combinatorial Problems," *Science* **266**, no. 5187, 11 November, pp. 1021–1024.

Afzal, N. (1976) "Millikan's Argument at Moderately Large Reynolds Number," *Phys. Fluids* **1**, pp. 600–602.

Afzal, N., and Bush, W.B. (1985) "A Three-Layer Asymptotic Analysis of Turbulent Channel Flow," in *Proc. Indian Acad. Sci. (Math. Sci.)* **94**, pp. 135–148.

Agarwal, R., Yun, K., and Balakrishnan, R. (1999) "Beyond Navier Stokes: Burnett Equations for Flow Simulations in Continuum-Transition Regime," AIAA Paper No. 99-3580, Reston, Virginia.

Ahuja, K.K., and Burrin, R.H. (1984) "Control of Flow Separation by Sound," AIAA Paper No. 84-2298, New York.

Ahuja, K.K., Hayden, W., and Entrekin, A. (1993) "Mixing Enhancement and Noise Reduction Through Taps Plus Ejectors," AIAA Paper No. 93-4347, Washington, D.C.

Ahuja, K.K., Whipkey, R.R., and Jones, G.S. (1983) "Control of Turbulent Boundary Layer Flows by Sound," AIAA Paper No. 83-0726, New York.

Akhavan, R., Jung, W.J., and Mangiavacchi, N. (1992) "Turbulence Control of Wall Bounded Flows by Spanwise Oscillations," *Appl. Scientific Res.* **51**, pp. 299–303.

Alder, B.J., and Wainwright, T.E. (1957) "Studies in Molecular Dynamics," *J. Chem. Phys.* **27**, pp. 1208–1209.

Alder, B.J., and Wainwright, T.E. (1958) "Molecular Dynamics by Electronic Computers," in *Transport Processes in Statistical Mechanics*, ed. I. Prigogine, pp. 97–131, Interscience, New York.

Alder, B.J., and Wainwright, T.E. (1970) "Decay of the Velocity Auto-Correlation Function," *Phys. Rev. A* **1**, pp. 18–21.

Aleyev, Yu.G. (1977) *Nekton*, Dr. W. Junk b.v. Publishers, The Hague, The Netherlands.

Alfredsson, P.H., and Johansson, A.V. (1984) "On the Detection of Turbulence-Generating Events," *J. Fluid Mech.* **139**, pp. 325–345.

Allan, J.F., and Cutland, R.S. (1953) "Wake Studies of Plane Surfaces," *Trans. North East Coast Institute of Engineers and Shipbuilders* **69**, pp. 245–266.

Allen, M.P., and Tildesley, D.J. (1987) *Computer Simulation of Liquids*, Clarendon Press, Oxford, Great Britain.

Alrefai, M., and Acharya, M. (1995) "Controlled Leading-Edge Suction for the Management of Unsteady Separation over Pitching Airfoils," AIAA Paper No. 95-2188, Washington, D.C.

Alshamani, K.M.M., Livesey, J.L., and Edwards, F.J. (1982) "Excitation of the Wall Region by Sound in Fully Developed Channel Flow," *AIAA J.* **20**, pp. 334–339.

Alvarez-Calderon, A. (1964) "Rotating Cylinder Flaps of V/STOL Aircraft," *Aircraft Eng.* **36**, pp. 304–309.

Alving, A.E., Smits, A.J., and Watmuff, J.H. (1990) "Turbulent Boundary Layer Relaxation from Convex Curvature," *J. Fluid Mech.* **211**, pp. 529–556.

Amato, I. (1998) "Formenting a Revolution, in Miniature," *Science* **282**, no. 5388, 16 October, pp. 402–405.

Anders, J.B. (1990a) "Boundary Layer Manipulators at High Reynolds Numbers," in *Structure of Turbulence and Drag Reduction*, ed. A. Gyr, pp. 475–482, Springer–Verlag, Berlin.

Anders, J.B. (1990b) "Outer-Layer Manipulators for Turbulent Drag Reduction," in *Viscous Drag Reduction in Boundary Layers*, eds. D.M. Bushnell and J.N. Hefner, pp. 263–284, AIAA, Washington, D.C.

Anders, J.B., and Watson, R.D. (1985) "Airfoil Large-Eddy Breakup Devices for Turbulent Drag Reduction," AIAA Paper No. 85-0520, New York.

Anders, J.B., Hefner, J.N., and Bushnell, D.M. (1984) "Performance of Large-Eddy Breakup Devices at Post-Transitional Reynolds Numbers," AIAA Paper No. 84-0345, New York.

Anders, J.B., Walsh, M.J., and Bushnell, D.M. (1988) "The Fix for Tough Spots," *Aerospace America* **26**, January, pp. 24–27.

Anderson, J.L., and Quinn, J.A. (1972) "Ionic Mobility in Microcapillaries," *J. Chem. Phys.* **27**, pp. 1208–1209.

Andreopoulos, J., Durst, F., Zaric, Z., and Jovanovic, J. (1984) "Influence of Reynolds Number on Characteristics of Turbulent Wall Boundary Layers," *Exp. Fluids* **2**, pp. 7–16.

Angell, J.B., Terry, S.C., and Barth, P.W. (1983) "Silicon Micromechanical Devices," *Faraday Transactions I* **68**, pp. 744–748.

Antonia, R.A., Bisset, D.K., and Browne, L.W.B. (1990) "Effect of Reynolds Number on the Topology of the Organized Motion in a Turbulent Boundary Layer," *J. Fluid Mech.* **213**, pp. 267–286.

Antonia, R.A., Fulachier, L., Krishnamoorthy, L.V., Benabid, T., and Anselmet, F. (1988) "Influence of Wall Suction on the Organized Motion in a Turbulent Boundary Layer," *J. Fluid Mech.* **190**, pp. 217–240.

Antonia, R.A., Rajagopalan, S., Subramanian, C.S., and Chambers, A.J. (1982) "Reynolds-Number Dependence of the Structure of a Turbulent Boundary Layer," *J. Fluid Mech.* **121**, pp. 123–140.

Antonia, R.A., Teitel, M., Kim, J., and Browne, L.W.B. (1992) "Low-Reynolds-Number Effects in a Fully Developed Turbulent Channel Flow," *J. Fluid Mech.* **236**, pp. 579–605.

Antsaklis, P.J. (1993) "Control Theory Approach," in *Mathematical Approaches to Neural Networks*, ed. J.G. Taylor, pp. 1–23, Elsevier, Amsterdam.

Anyiwo, J.C., and Bushnell, D.M. (1982) "Turbulence Amplification in Shock Wave–Boundary Layer Interaction," *AIAA J.* **20**, pp. 893–899.

Aref, H. (1984) "Stirring by Chaotic Advection," *J. Fluid Mech.* **143**, pp. 1–21.

Aref, H., and El Naschie, M.S., editors (1995) *Chaos Applied to Fluid Mixing*, Pergamon Press, Oxford, Great Britain.

Arkilic, E.B. (1997) "Measurement of the Mass Flow and Tangential Momentum Accommodation Coefficient in Silicon Micromachined Channels," Ph.D. Thesis, Massachusetts Institute of Technology, Cambridge, Massachusetts.

Arkilic, E.B., Schmidt, M.A., and Breuer, K.S. (1995) "Slip Flow in Microchannels," in *Rarefied Gas Dynamics 19*, eds. J. Harvey and G. Lord, Oxford University Press, Oxford, Great Britain.

Arkilic, E.B., Schmidt, M.A., and Breuer, K.S. (1997a) "Gaseous Slip Flow in Long Microchannels," *J. MEMS* **6**, pp. 167–178.

Arkilic, E.B., Schmidt, M.A., and Breuer, K.S. (1997b) "TMAC Measurement in Silicon Micromachined Channels," in *Rarefied Gas Dynamics 20*, ed. C. Shen, 6 pages, Beijing University Press, Beijing, China.

Aroesty, J., and Berger, S.A. (1975) "Controlling the Separation of Laminar Boundary Layers in Water: Heating and Suction," RAND Corporation Report No. R-1789-ARPA, Santa Monica, California. (Also available from U.S. NTIS; Document Number AD-AO20026.)

Aronson, D., and Löfdahl, L. (1994) "The Plane Wake of a Cylinder: An Estimate of the Pressure Strain Rate Tensor," *Phys. Fluids* **6**, pp. 2716–2721.

Aronson, D., Johansson, A.V., and Löfdahl, L. (1997) "A Shear-Free Turbulent Boundary Layer— Experiments and Modeling," *J. Fluid Mech.* **338**, pp. 363–385.

Ashino, I., and Yoshida, K. (1975) "Slow Motion between Eccentric Rotating Cylinders," *Bulletin JSME* **18**, no. 117, pp. 280–285.

Ashkenas, H.I., and Riddell, F.R. (1955) "Investigation of the Turbulent Boundary Layer on a Yawed Flat Plate," NACA Technical Note No. TN-3383, Washington, D.C.

Ashley, S. (1996) "Getting a Microgrip in the Operating Room," *Mech. Eng.* **118**, September, pp. 91–93.

Ashley, S. (1998) "Bringing Launch Costs Down to Earth," *Mech. Eng.* **120**, no. 10, October, pp. 62–68.

Aslanov, P.V., Maksyutenko, S.N., Povkh, I.L., Simonenko, A.P., and Stupin, A.B. (1980) "Turbulent Flows of Solutions of Surface-Active Substances," *Izvestiya Akademii Nauk SSSR, Mekhanika Zhidkosti i Gaza*, no. 1, pp. 36–43.

Atassi, H.M., editor (1993) *Unsteady Aerodynamics, Aeroacoustics, and Aeroelasticity of Turbomachines and Propellers*, Springer–Verlag, New York.

Atassi, H.M., and Gebert, G.A. (1987) "Modification of Turbulent Boundary Layer Structure by Large-Eddy Breakup Devices," in *Proc. Int. Conf. on Turbulent Drag Reduction by Passive Means*, vol. 2, pp. 432–456, Royal Aeronautical Society, London, Great Britain.

Atta, R., and Rockwell, D. (1990) "Leading-Edge Vortices Due to Low Reynolds Number Flow Past a Pitching Delta Wing," *AIAA J.* **28**, pp. 995–1004.

Atwood, B.T., and Schowalter, W.R. (1989) "Measurements of Slip at the Wall During Flow of High-Density Polyethylene through a Rectangular Conduit," *Rheologica Acta* **28**, pp. 134–146.

Au, D., and Weihs, D. (1980) "At High Speeds Dolphins Save Energy by Leaping," *Nature* **284**, no. 5756, 10 April, pp. 548–550.

Aubry, N. (1990) "Use of Experimental Data for an Efficient Description of Turbulent Flows," *Appl. Mech. Rev.* **43**, pp. S240–S245.

Aubry, N., Holmes, P., Lumley, J.L., and Stone, E. (1988) "The Dynamics of Coherent Structures in the Wall Region of a Turbulent Boundary Layer," *J. Fluid Mech.* **192**, pp. 115–173.

Auerbach, D. (1994) "Controlling Extended Systems of Chaotic Elements," *Phys. Rev. Lett.* **72**, pp. 1184–1187.

Auerbach, D., Grebogi, C., Ott, E., and Yorke, J.A. (1992) "Controlling Chaos in High Dimensional Systems," *Phys. Rev. Lett.* **69**, pp. 3479–3482.

Ayers, R.F., and Wilde, M.R. (1956) "An Experimental Investigation of the Aerodynamic Characteristics of a Low Aspect Ratio Swept Wing with Blowing in a Spanwise Direction from the Tips," College of Aeronautics, Note No. 57, Cranfield, Great Britain.

Bahi, L., Ross, J.M., and Nagamatsu, H.T. (1983) "Passive Shock Wave/Boundary Layer Control for Transonic Airfoil Drag Reduction," AIAA Paper No. 83-0137, New York.

Baker, G.L., and Gollub, J.P. (1990) *Chaotic Dynamics—An Introduction*, Cambridge University Press, Cambridge, Great Britain.

Bakewell, H.P., and Lumley, J.L. (1967) "Viscous Sublayer and Adjacent Wall Region in Turbulent Pipe Flow," *Phys. Fluids* **10**, pp. 1880–1889.

Ball, K.O. (1971) "Flap Span Effects on Boundary Layer Separation," *AIAA J.* **9**, pp. 2080–2081.

Bandyopadhyay, P.R. (1980) "Large Structure with a Characteristic Upstream Interface in Turbulent Boundary Layers," *Phys. Fluids* **23**, pp. 2326–2327.

Bandyopadhyay, P.R. (1982) "Period Between Bursting in Turbulent Boundary Layers," *Phys. Fluids* **25**, pp. 1751–1754.

Bandyopadhyay, P.R. (1983) "Turbulent Spot-Like Features of a Boundary Layer," *Annals New York Acad. Sci.* **404**, pp. 393–395.

Bandyopadhyay, P.R. (1986a) "Drag Reducing Outer-Layer Devices in Rough Wall Turbulent Boundary Layers," *Exp. Fluids* 4, pp. 247–256.

Bandyopadhyay, P.R. (1986b) "Review—Mean Flow in Turbulent Boundary Layers Disturbed to Alter Skin Friction," *J. Fluids Eng.* 108, pp. 127–140.

Bandyopadhyay, P.R. (1987) "Rough-Wall Turbulent Boundary Layers in the Transition Regime," *J. Fluid Mech.* 180, pp. 231–266.

Bandyopadhyay, P.R. (1988) "Resonant Flow in Small Cavities Submerged in a Boundary Layer," *Proc. R. Soc. London A* 420, pp. 219–245.

Bandyopadhyay, P.R. (1989) "Effect of Abrupt Pressure Gradients on the Structure of Turbulent Boundary Layers," *Proc. Tenth Australasian Fluid Mechanics Conf.*, eds. A.E. Perry et al., vol. I, pp. 1.1–1.4, University of Melbourne, Melbourne, Australia.

Bandyopadhyay, P.R. (1990) "Convex Curvature Concept of Viscous Drag Reducion," in *Viscous Drag Reduction in Boundary Layers*, eds. D.M. Bushnell and J.N. Hefner, pp. 285–324, AIAA, Washington, D.C.

Bandyopadhyay, P.R. (1991) "Comments on Reynolds Number Effects in Wall-Bounded Shear Layers," AIAA Paper No. 91-0231, New York.

Bandyopadhyay, P.R. (1992) "Reynolds Number Dependence of the Freestream Turbulence Effects on Turbulent Boundary Layers," *AIAA J.* 30, pp. 1910–1912.

Bandyopadhyay, P.R., and Ahmed, A. (1993) "Turbulent Boundary Layers Subjected to Multiple Curvatures and Pressure Gradients," *J. Fluid Mech.* 246, pp. 503–527.

Bandyopadhyay, P.R., and Balasubramanian, R. (1993) "A Vortex Model for Calculating Wall Pressure Fluctuations in Turbulent Boundary Layers," in *ASME Symposium on Flow Noise Modeling, Measurement and Control*, eds. T.M. Farabee, W.L. Keith, and R.M. Lueptow, NCA-Vol. 15/FED-Vol. 168, pp. 13–24, New York.

Bandyopadhyay, P.R., and Balasubramanian, R. (1995) "Vortex Reynolds Number in Turbulent Boundary Layers," *Theor. Comput. Fluid Dynamics* 7, pp. 101–117.

Bandyopadhyay, P.R., and Balasubramanian, R. (1996) "Structural Modeling of the Wall Effects of Lorentz Force," *J. Fluids Eng.* 118, pp. 412–414.

Bandyopadhyay, P.R., and Hussain, A.K.M.F. (1984) "The Coupling Between Scales in Shear Flows," *Phys. Fluids* 27, pp. 2221–2228.

Bandyopadhyay, P.R., and Watson, R.D. (1988) "Structure of Rough-Wall Turbulent Boundary Layers," *Phys. Fluids* 31, pp. 1877–1883.

Bandyopadhyay, P.R., Breuer, K.S., and Blechinger, C.J., editors (1994) *Application of Microfabrication to Fluid Mechanics*, FED Vol. 197, New York.

Banks, S.P. (1986) *Control Systems Engineering*, Prentice–Hall International, Englewood Cliffs, New Jersey.

Barenblatt, G.I. (1979) *Similarity, Self-Similarity, and Intermediate Hypothesis*, Plenum Press, New York.

Barenblatt, G.I. (1993) "Scaling Laws for Fully Developed Turbulent Shear Flows. Part 1. Basic Hypothesis and Analysis," *J. Fluid Mech.* 248, pp. 513–520.

Barenblatt, G.I., and Prostokishin, V.M. (1993) "Scaling Laws for Fully Developed Turbulent Shear Flows. Part 2. Processing of Experimental Data," *J. Fluid Mech.* 248, pp. 521–529.

Barenblatt, G.I., Chorin A.J., and Prostokishin, V.M. (1997) "Scaling Laws for Fully Developed Turbulent Flows in Pipes," *Appl. Mech. Rev.* 50, pp. 413–429.

Barger, J.E., and von Winkle, W.A. (1961) "Evaluation of a Boundary Layer Stabilization Coating," *J. Acoustical Soc. of America* 33, p. 836.

Barker, R.A. (1986) "The Aerodynamic Effects of a Serrated Strip Near the Leading Edge of an Airfoil," M.S. Thesis, Royal Air Force College, Report No. ETN-87-99480, Cranwell, Great Britain.

Barker, S.J., and Gile, D. (1981) "Experiments on Heat-Stabilized Laminar Boundary Layers in Water," *J. Fluid Mech.* 104, pp. 139–158.

Barnwell, R.W., and Hussaini, M.Y., editors (1992) *Natural Laminar Flow and Laminar Flow Control*, Springer–Verlag, New York.

Barnwell, R.W., Bushnell, D.M., Nagamatsu, H.T., Bahi, L., and Ross, J. (1985) "Passive Drag Control of Airfoils at Transonic Speeds," U.S. Patent No. 4,522,360.

Bar-Sever, A. (1989) "Separation Control on an Airfoil by Periodic Forcing," *AIAA J.* 27, pp. 820–821.

Bart, S.F., Tavrow, L.S., Mehregany, M., and Lang, J.H. (1990) "Microfabricated Electrohydro dynamic Pumps," *Sensors & Actuators A* **21–23**, pp. 193–197.

Bastedo, W.G., Jr., and Mueller, T.J. (1985) "Performance of Finite Wings at Low Reynolds Numbers," in *Proc. Conf. on Low Reynolds Number Airfoil Aerodynamics*, ed. T.J. Mueller, pp. 195–205, University of Notre Dame, Notre Dame, Indiana.

Bastedo, W.G., Jr., and Mueller, T.J. (1986) "Spanwise Variation of Laminar Separation Bubbles on Wings at Low Reynolds Numbers," *J. Aircraft* **23**, pp. 687–694.

Batchelor, G.K. (1967) *An Introduction to Fluid Dynamics*, Cambridge University Press, Cambridge, Great Britain.

Batchelor, G.K., and Green, J.T. (1972) "The Determination of the Bulk Stress in a Suspension of Spherical Particles to Order c^2," *J. Fluid Mech.* **56**, pp. 401–427.

Bau, H.H. (1994) "Transport Processes Associated with Micro-Devices," *Thermal Sci. Eng.* **2**, pp. 172–178.

Bauer, S.X.S., and Hernandez, G. (1988) "Reduction of Cross-Flow Shock-Induced Separation with a Porous Cavity at Supersonic Speeds," AIAA Paper No. 88-2567, New York.

Baullinger, N., and Page, V. (1989) "High Altitude Long Endurance (HALE) RPV," AIAA Paper No. 89-2014, New York.

Bechert, D.W., Hoppe, G., and Reif, W.-E. (1985) "On the Drag Reduction of the Shark Skin,"AIAA Paper No. 85-0546, New York.

Becker, S., Durst, F., and Lienhart, H. (1999) "Laser Doppler Anemometer for In-Flight Velocity Measurements on Airplane Wings," *AIAA J.* **37**, pp. 680–687.

Benjamin, T.B. (1960a) "Effects of a Flexible Boundary on Hydrodynamic Stability," *J. Fluid Mech.* **9**, pp. 513–532.

Benjamin, T.B. (1960b) "Fluid Flow with Flexible Boundaries, in *Proc. Eleventh Int. Cong. of Appl. Mech.*, ed. H. Görtler, pp. 109–128, Springer–Verlag, Berlin.

Benjamin, T.B. (1963) "The Threefold Classification of Unstable Disturbances in Flexible Surfaces Bounding Inviscid Flows," *J. Fluid Mech.* **16**, pp. 436–450.

Benjamin, T.B. (1966) "Fluid Flow with Flexible Boundaries," in *Proc. Eleventh Int. Cong. of Appl. Mech.*, ed. H. Görtler, pp. 109–128, Springer–Verlag, Berlin.

Bergles, A.E. (1978) "Enhancement of Heat Transfer," in *Proc. Sixth Int. Heat Trans. Conf.*, vol. 6, pp. 89–108, Hemisphere, Washington, D.C.

Bergles, A.E. (1981) "Principles of Heat Transfer Augmentation," in *Heat Exchangers, Thermal-Hydraulic Fundamentals and Design*, eds. S. Kakaç, A.E. Bergles, and F. Mayinger, pp. 819–842, Hemisphere, New York.

Bergles, A.E. (1985) "Techniques to Augment Heat Transfer," in *Handbook of Heat Transfer Applications*, second edition, eds. W.M. Rohsenow, J.P. Hartnett, and E.N. Ganić, pp. 3.1–3.80, McGraw-Hill, New York.

Bergles, A.E., and Morton, L.H. (1965) "Survey and Evaluation of Techniques to Augment Convective Heat Transfer," Report No. EPL 5382-34, Department of Mechanical Engineering, Massachusetts Institute of Technology, Cambridge, Massachusetts.

Bergles, A.E., and Webb, R.L. (1985) "A Guide to the Literature on Convective Heat Transfer Augmentation," in *Twenty-Third National Heat Transfer Conf.: Advances in Enhanced Heat Transfer*, Denver, Colorado.

Berkooz, G., Fisher, M., and Psiaki, M. (1993) "Estimation and Control of Models of the Turbulent Wall Layer," *Bull. Am. Phys. Soc.* **38**, p. 2197.

Berkooz, G., Holmes, P., and Lumley, J.L. (1991) "Intermittent Dynamics in Simple Models of the Turbulent Boundary Layer," *J. Fluid Mech.* **230**, pp. 75–95.

Berman, N.S. (1978) "Drag Reduction by Polymers," *Ann. Rev. Fluid Mech.* **10**, pp. 47–64.

Berman, N.S., and George, W.K. (1974) "Time Scale and Molecular Weight Distribution Contributions to Dilute Polymer Solution Fluid Mechanics," in *Proc. Heat Trans. Fluid Mech. Inst.*, eds. L.R. Davis and R.E. Wilson, pp. 348–364, Stanford University Press, Stanford, California.

Bernard, J.J., and Siestrunck, R. (1959) "Échanges de Chaleur dans les Écoulements Présentant des Décollements," in *Proc. First Int. Cong. Aero. Sci.*, eds. Th. von Kármán et al., pp. 314–332, *Adv. Aero. Sci.*, vol. 1, Pergamon Press, London, Great Britain.

Bertelrud, A., Truong, T.V., and Avellan, F. (1982) "Drag Reduction in Turbulent Boundary Layers Using Ribbons," AIAA Paper No. 82-1370, New York.

Beskok, A. (1994) "Simulation of Heat and Momentum Transfer in Complex Micro-Geometries," M.Sc. Thesis, Princeton University, Princeton, New Jersey.

Beskok, A. (1996) "Simulations and Models of Gas Flows in Microgeometries," Ph.D. Thesis, Princeton University, Princeton, New Jersey.

Beskok, A., and Karniadakis, G.E. (1994) "Simulation of Heat and Momentum Transfer in Complex Micro-Geometries," *J. Thermophys. & Heat Transfer* **8**, pp. 355–370.

Beskok, A., and Karniadakis, G.E. (1999) "A Model for Flows in Channels, Pipes and Ducts at Micro and Nano Scales," *Microscale Thermophys. Eng.* **3**, pp. 43–77.

Beskok, A., Karniadakis, G.E., and Trimmer, W. (1996) "Rarefaction and Compressibility Effects in Gas Microflows," *J. Fluids Eng.* **118**, pp. 448–456.

Betchov, K. (1960) "Simplified Analysis of Boundary-Layer Oscillations," *J. Ship Res.* **4**, pp. 37–54.

Bewley, T.R., Moin, P., and Temam, R. (1997) "Optimal and Robust Approaches for Linear and Nonlinear Regulation Problems in Fluid Mechanics," AIAA Paper No. 97-1872, Reston, Virginia.

Bewley, T.R., Temam, R., and Ziane, M. (1998) "A General Framework for Robust Control in Fluid Mechanics," Center for Turbulence Research No. CTR Manuscript 169, Stanford University, Stanford, California.

Bhatnagar, P.L., Gross, E.P., and Krook, M. (1954) "A Model for Collision Processes in Gases. I. Small Amplitude Processes in Charged and Neutral One-Component Systems," *Phys. Rev.* **94**, pp. 511–524.

Bhatt, W.V. (1971) "Flight Test Measurement of Exterior Turbulent Boundary Layer Pressure Fluctuations on Boeing Model 737 Airplane," *J. Sound Vibr.* **14**, pp. 439–457.

Bhattacharjee, S., Scheelke, B., and Troutt, T.R. (1985) "Modification of Vortex Interactions in a Reattaching Separated Flow," AIAA Paper No. 85-0555, New York.

Bird, G.A. (1963) "Approach to Translational Equilibrium in a Rigid Sphere Gas," *Phys. Fluids* **6**, pp. 1518–1519.

Bird, G.A. (1965) "The Velocity Distribution Function within a Shock Wave," *J. Fluid Mech.* **30**, pp. 479–487.

Bird, G.A. (1976) *Molecular Gas Dynamics*, Clarendon Press, Oxford, Great Britain.

Bird, G.A. (1978) "Monte Carlo Simulation of Gas Flows," *Ann. Rev. Fluid Mech.* **10**, pp. 11–31.

Bird, G.A. (1994) *Molecular Gas Dynamics and the Direct Simulation of Gas Flows*, Clarendon Press, Oxford, Great Britain.

Biringen, S. (1984) "Active Control of Transition by Periodic Suction-Blowing," *Phys. Fluids* **27**, pp. 1345–1347.

Biringen, S., and Maestrello, L. (1984) "Development of Spot-Like Turbulence in Plane Channel Flow," *Phys. Fluids* **27**, pp. 318–321.

Black, T.J. (1966) "Some Practical Applications of a New Theory of Wall Turbulence," in *Proc. 1966 Heat Transfer & Fluid Mechanics Institute*, eds. M.A. Saad and J.A. Miller, pp. 366–386, Stanford University Press, Stanford, California.

Black, T.J. (1968) "An Analytical Study of the Measured Wall Pressure Field under Supersonic Turbulent Boundary Layers," NASA Contractor Report No. CR-888, Washington, D.C.

Blackwelder, R.F. (1978) "The Bursting Process in Turbulent Boundary Layers," in *Workshop on Coherent Structure of Turbulent Boundary Layers*, eds. C.R. Smith and D.E. Abbott, pp. 211–227, Lehigh University, Bethlehem, Pennsylvania.

Blackwelder, R.F. (1988) "Coherent Structures Associated with Turbulent Transport," in *Transport Phenomena in Turbulent Flows*, eds. M. Hirata and N. Kasagi, pp. 69–88, Hemisphere, New York.

Blackwelder, R.F. (1989) "Some Ideas on the Control of Near-Wall Eddies," AIAA Paper No. 89-1009, New York.

Blackwelder R.F. (1998) "Some Notes on Drag Reduction in the Near-Wall Region," in *Flow Control: Fundamentals and Practices*, eds. M. Gad-el-Hak, A. Pollard, and J.-P. Bonnet, pp. 155–198, Springer–Verlag, Berlin.

Blackwelder, R.F., and Eckelmann, H. (1978) "The Spanwise Structure of the Bursting Phenomenon," in *Structure and Mechanisms of Turbulence I*, ed. H. Fiedler, pp. 190–204, Springer–Verlag, Berlin.

Blackwelder, R.F., and Eckelmann, H. (1979) "Streamwise Vortices Associated with the Bursting Phenomenon," *J. Fluid Mech.* **94**, pp. 577–594.

Blackwelder R.F., and Gad-el-Hak M. (1990) "Method and Apparatus for Reducing Turbulent Skin Friction," United States Patent No. 4,932,612.

Blackwelder, R.F., Gad-el-Hak, M., and Srnsky, R.A. (1987) "Method and Apparatus for Controlling Bound Vortices in the Vicinity of Lifting Surfaces," U.S. Patent No. 4,697,769.

Blackwelder, R.F., and Haritonidis, J.H. (1983) "Scaling of the Bursting Frequency in Turbulent Boundary Layers," *J. Fluid Mech.* **132**, pp. 87–103.

Blackwelder, R.F., and Kaplan, R.E. (1976) "On the Wall Structure of the Turbulent Boundary Layer," *J. Fluid Mech.* **76**, pp. 89–112.

Blackwelder, R.F., and Kovasznay, L.S.G. (1972) "Time-Scales and Correlations in a Turbulent Boundary Layer," *Phys. Fluids* **15**, pp. 1545–1554.

Blackwelder, R.F., and Swearingen, J.D. (1990) "The Role of Inflectional Velocity Profiles in Wall Bounded Flows," in *Near-Wall Turbulence: 1988 Zoran Zarić Memorial Conference*, eds. S.J. Kline and N.H. Afgan, pp. 268–288, Hemisphere, New York.

Blair, M.F. (1983) "Influence of Free-Stream Turbulence on Turbulent Boundary Layer Heat Transfer and Mean Profile Development, Part II—Analysis of Results," *J. Heat Transfer* **105**, pp. 41–47.

Blair, M.F., and Werle, M.J. (1980) "The Influence of Free-Stream Turbulence on the Zero Pressure Gradient Fully Turbulent Boundary Layer," United Technologies Research Center Report No. R80-914388-12, Hartford, Connecticut.

Blake, W.K. (1986) *Mechanics of Flow-Induced Sound and Vibration*, Academic Press, New York.

Blake, W.K., and Gershfeld, J.L. (1989) "The Aeroacoustics of Trailing Edges," in *Frontiers in Experimental Fluid Mechanics*, ed. M. Gad-el-Hak, pp. 457–532, Springer-Verlag, New York.

Blick, E.F., and Walters, R.R. (1968) "Turbulent Boundary-Layer Characteristics of Compliant Surfaces," *J. Aircraft* **5**, pp. 11–16.

Bogdevich, V.G., and Malyuga, A.G. (1976) "The Distribution of Skin Friction in a Turbulent Boundary Layer of Water Beyond the Location of Gas Injection," in *Studies on the Boundary Layer Control* (in Russian), eds. S.S. Kutateladze and G.S. Migirenko, p. 62, Institute of Thermophysics, Novosibirsk, U.S.S.R.

Bogdevich, V.G., Evseev, A.R., Malyuga, A.G., and Migirenko, G.S. (1977) "Gas-Saturation Effect on Near-Wall Turbulence Characteristics," in *Second Int. Conf. on Drag Reduction*, Paper No. D2, BHRA Fluid Engineering, Cranfield, Great Britain.

Bolton, J.S., editor (1988) *Noise Control Design: Methods and Practice (Noise-Con 88 Proceedings)*, Noise Control Foundation, Poughkeepsie, New York.

Bouchon-Meunier, B., Yager, R.R., and Zadeh, L.A., editors (1995a) *Fuzzy Logic and Soft Computing*, World Scientific, Singapore.

Bouchon-Meunier, B., Yager, R.R., and Zadeh, L.A., editors (1995b) *Advances in Intelligent Computing—IPMU'94, Lecture Notes in Computer Science*, vol. 945, Springer-Verlag, Berlin.

Bradley, R.G., and Wray, W.O. (1974) "A Conceptual Study of Leading-Edge-Vortex Enhancement by Blowing," *J. Aircraft* **11**, pp. 33–38.

Bradshaw, P. (1967) "'Inactive' Motion and Pressure Fluctuations in Turbulent Boundary Layers," *J. Fluid Mech.* **30**, pp. 241–258.

Bradshaw, P. (1969) "A Note on Reverse Transition," *J. Fluid Mech.* **35**, pp. 387–390.

Bradshaw, P. (1994) "Turbulence: The Chief Outstanding Difficulty of Our Subject," *Exp. Fluids* **16**, pp. 203–216.

Bradshaw, P., and Huang, G.P. (1995) "The Law of the Wall in Turbulent Flow," *Proc. R. Soc. London A* **451**, pp. 165–188.

Bradshaw, P., Ferriss, D.H., and Atwell, N.P. (1967) "Calculations of Boundary-Layer Development Using the Turbulent Energy Equation," *J. Fluid Mech.* **28**, pp. 593–616.

Bradshaw, P., Ferriss, D.H., and Johnson, R.F. (1964) "Turbulence in the Noise-Producing Region of a Circular Jet," *J. Fluid Mech.* **19**, pp. 591–624.

Bragg, M.B., and Gregorek, G.M. (1987) "Experimental Study of Airfoil Performance with Vortex Generators," *J. Aircraft* **24**, pp. 305–309.

Braslow, A.L., Burrows, D.L., Tetervin, N., and Visconti, F. (1951) "Experimental and Theoretical Studies of Area Suction for the Control of Laminar Boundary Layer," NACA Report No. 1025, Washington, D.C.

Bremhorst, K., and Walker, T.B. (1974) "Spectral Measurements of Turbulent Momentum Transfer in Fully Developed Pipe Flow," *J. Fluid Mech.* **61**, pp. 173–186.

Brendel, M., and Mueller, T.J. (1988) "Boundary Layer Measurements on an Airfoil at a Low Reynolds Number in an Oscillating Freestream," *AIAA J.* **26**, pp. 257–263.

Breuer, K.S., Bandyopadhyay, P.R., and Gad-el-Hak, M., editors (1996) *Application of Silicon Microfabrication to Fluid Mechanics*, DSC-Volume 59, ASME, New York.

Breuer, K.S., Haritonidis, J.H., and Landahl, M.T. (1989) "The Control of Transient Disturbances in a Flat Plate Boundary Layer through Active Wall Motion," *Phys. Fluids A* **1**, pp. 574–582.

Briedenthal, R.E., Jr., and Russell, D.A. (1988) "Aerodynamics of Vortex Generators," NASA CR-182511, Washington, D.C.

Brookes, M. (1997) "On a Wing and a Vortex," *New Scientist* **156**, 11 October, pp. 24–27.

Brown, A.C., Nawrocki, H.F., and Paley, P.N. (1968) "Subsonic Diffusers Designed Integrally with Vortex Generators," *J. Aircraft* **5**, pp. 221–229.

Brown, G.L., and Roshko, A. (1971) "The Effect of Density Difference on the Turbulent Mixing Layer," in *Turbulent Shear Flows*, pp. 23.1–23.12, AGARD-CP-93, Rhode-Saint-Génèse, Belgium.

Brown, G.L., and Roshko, A. (1974) "On Density Effects and Large Strucure in Turbulent Mixing Layers," *J. Fluid Mech.* **64**, pp. 775–816.

Brown, G.L., and Thomas, A.S.W. (1977) "Large Structure in a Turbulent Boundary Layer," *Phys. Fluids* **20**, no. 10, part II, pp. S243–S252.

Brunauer, S. (1944) *Physical Adsorption of Gases and Vapours*, Oxford University Press, Oxford, Great Britain.

Bryzek, J., Peterson, K., and McCulley, W. (1994) "Micromachines on the March," *IEEE Spectrum* **31**, May, pp. 20–31.

Buckingham, A.C., Hall, M.S., and Chun, R.C. (1985) "Numerical Simulations of Compliant Material Response to Turbulent Flow," *AIAA J.* **23**, pp. 1046–1052.

Burd, J.E. (1981) "Flow Control for a High Energy Laser Turret Using Trapped Vortices Stabilized by Suction," M.Sc. Thesis, Naval Postgraduate School, Monterey, CA. (Also available from U.S. NTIS; Document Number AD-A115263.)

Burns, T.F. (1981) "Experimental Studies of Eppler 61 and Pfenninger 048 Airfoils at Low Reynolds Numbers," M.Sc. Thesis, University of Notre Dame, Notre Dame, Indiana.

Burton, T.E. (1974) "The Connection Between Intermittent Turbulent Activity Near the Wall of a Turbulent Boundary Layer with Pressure Fluctuations at the Wall," Massachusetts Institute of Technology Report No. 70208-10, Cambridge, Massachusetts.

Busch-Vishniac, I.J. (1998) "Trends in Electromechanical Transduction," *Physics Today* **51**, July, pp. 28–34.

Bush, V. (1945) *Science—The Endless Frontier*, Government Printing Office, Washington, D.C.

Bushnell, D.M. (1983) "Turbulent Drag Reduction for External Flows," AIAA Paper No. 83-0227, New York.

Bushnell, D.M. (1989) "Applications and Suggested Directions of Transition Research," in *Fourth Symp. on Numerical and Physical Aspects of Aerodynamic Flows*, Long Beach, CA, 16–19 January.

Bushnell, D.M. (1990) "Supersonic Aircraft Drag Reduction," AIAA Paper No. 90-1596, New York.

Bushnell, D.M. (1994) "Viscous Drag Reduction in Aeronautics," in *Proc. Nineteenth Cong. of the Int. Council of the Aeronaut. Sci.*, vol. 1, pp. XXXIII–LVI, paper no. ICAS-94-0.1, AIAA, Washington, D.C.

Bushnell, D.M. (1998) "Frontiers of the 'Responsibly Imaginable' in Aeronautics," Dryden Lecture, AIAA Paper No. 98-0001, Reston, Virginia.

Bushnell, D.M., and Donaldson, C.D. (1990) "Control of Submersible Vortex Flows," NASA Tech. Memo. 102693, Washington, D.C.

Bushnell, D.M., and Greene, G.C. (1991) "High-Reynolds-Number Test Requirements in Low-Speed Aerodynamics," in *High Reynolds Number Flows Using Liquid and Gaseous Helium*, ed. R.J. Donnelly, pp. 79–85, Springer–Verlag, New York.

Bushnell, D.M., and Hefner, J.N., editors (1990) *Viscous Drag Reduction in Boundary Layers*, AIAA, Washington, D.C.

Bushnell, D.M., and Malik, M.R. (1988) "Compressibility Influences on Boundary-Layer Transition," in *Symp. on Physics of Compressible Turbulent Mixing*, 25–27 October, Princeton, New Jersey.

Bushnell, D.M., and McGinley, C.B. (1989) "Turbulence Control in Wall Flows," *Ann. Rev. Fluid Mech.* **21**, pp. 1–20.

Bushnell, D.M., and Moore, K.J. (1991) "Drag Reduction in Nature," *Ann. Rev. Fluid Mech.* **23**, pp. 65–79.

Bushnell, D.M., and Trimpi, R.L. (1986) "Optimum Supersonic Wind Tunnel," AIAA Paper No. 86-0773, New York.

Bushnell, D.M., Cary, A.M., and Holley, B.B. (1975) "Mixing Length in Low Reynolds Number Compressible Turbulent Boundary Layers," *AIAA J.* **13**, pp. 1119–1121.

Bushnell, D.M., Hefner, J.N., and Ash, R.L. (1977) "Effect of Compliant Wall Motion on Turbulent Boundary Layers," *Phys. Fluids* **20**, no. 10, part II, pp. S31–S48.

Bushnell, D.M., Malik, M.R., and Harvey, W.D. (1989) "Transition Prediction in External Flows via Linear Stability Theory," in *IUTAM Symp. Transsonicum III*, eds. J. Zierep and H. Oertel, pp. 225–242, Springer–Verlag, Berlin.

Bushnell, D.M., Yip, L.P., Yao, C.-S., Lin, J.C., Lawing, P.L., Batina, J.T., Hardin, J.C., Horvath, T.J., Fenbert, J.W., and Domack, C.S. (1993) "Reynolds Number Influences in Aeronautics," *NASA Technical Memorandum* No. TM-107730, Washington, D.C.

Bussmann, K., and Münz, H. (1942) "Die Stabilität der laminaren Reibungsschicht mit Absaugung," *Jahrb. Dtsch. Luftfahrtforschung* **1**, pp. 36–39.

Butler, K.M., and Farrell, B.F. (1993) "Optimal Perturbations and Streak Spacing in Wall-Bounded Shear Flow," *Phys. Fluids A* **5**, pp. 774–777.

Butter, D.J. (1984) "Recent Progress on Development and Understanding of High Lift Systems," in *AGARD Conference on Improvement of Aerodynamic Performance Through Boundary Layer Control and High Lift Systems*, AGARD-CP-365, pp. 1.1–1.26, Brussels, Belgium.

Calarese, W., Crisler, W.P., and Gustafson, G.L. (1985) "Afterbody Drag Reduction by Vortex Generators," AIAA Paper No. 85-0354, New York.

Cantwell, B.J. (1981) "Organized Motion in Turbulent Flow," *Ann. Rev. Fluid Mech.* **13**, pp. 457–515.

Carlson, H.A., and Lumley, J.L. (1996) "Flow over an Obstacle Emerging from the Wall of a Channel," *AIAA J.* **34**, pp. 924–931.

Carmichael, B.H. (1974) "Application of Sailplane and Low-Drag Underwater Vehicle Technology to the Long-Endurance Drone Problem," AIAA Paper No. 74-1036, New York.

Carmichael, B.H. (1981) "Low Reynolds Number Airfoil Survey," NASA CR-165803, Washington, D.C.

Carpenter, P.W. (1988) "The Optimization of Compliant Surfaces for Transition Delay, in *Turbulence Management and Relaminarisation*, eds. H.W. Liepmann and R. Narasimha, pp. 305–313, Springer–Verlag, Berlin.

Carpenter, P.W. (1990) "Status of Transition Delay Using Compliant Walls," in *Viscous Drag Reduction in Boundary Layers*, eds. D.M. Bushnell and J.N. Hefner, pp. 79–113, AIAA, Washington, D.C.

Carpenter, P.W. (1993) "Optimization of Multiple-Panel Compliant Walls for Delay of Laminar-Turbulent Transition," *AIAA J.* **31**, pp. 1187–1188.

Carpenter, P.W. (1997) "The Right Sort of Roughness," *Nature* **388**, 21 August, pp. 713–714.

Carpenter, P.W. (1998) "Current Status of the Use of Wall Compliance for Laminar-Flow Control," *Exp. Thermal & Fluid Sci.* **16**, pp. 133–140.

Carpenter, P.W., and Garrad, A.D. (1985) "The Hydrodynamic Stability of Flow over Kramer-Type Compliant Surfaces. Part 1. Tollmien–Schlichting Instabilities," *J. Fluid Mech.* **155**, pp. 465–510.

Carpenter, P.W., and Garrad, A.D. (1986) "The Hydrodynamic Stability of Flow over Kramer-Type Compliant Surfaces. Part 2. Flow-Induced Surface Instabilities," *J. Fluid Mech.* **170**, pp. 199–232.

Cary, A.M., Jr., Weinstein, L.M., and Bushnell, D.M. (1980) "Drag Reduction Characteristics of Small Amplitude Rigid Surface Waves," in *Viscous Flow Drag Reduction*, ed. G.R. Hough, pp. 144–167, AIAA, New York.

Castro, I.P. (1984) "Effects of Free Stream Turbulence on Low Reynolds Number Boundary Layers," *J. Fluids Eng.* **106**, pp. 298–306.

Cebeci, T. (1989) "Calculation of Flow over Iced Airfoils," *AIAA J.* **27**, pp. 853–861.

Cebeci, T., and Chang, K.C. (1978) "Calculation of Incompressible Rough-Wall Boundary-Layer Flows," *AIAA J.* **16**, pp. 730–735.

Cebeci, T., and Egan, D.A. (1989) "Prediction of Transition due to Isolated Roughness," *AIAA J.* **27**, pp. 870–875.

Cercignani, C. (1988) *The Boltzmann Equation and Its Applications*, Springer–Verlag, Berlin.

Chambers, J.R. (1986) "High-Angle-Of-Attack Aerodynamics: Lessons Learned," AIAA Paper No. 86-1774-CP, New York.

Chan, D.Y.C., and Horn, R.G. (1985) "Drainage of Thin Liquid Films," *J. Chem. Phys.* 83, pp. 5311–5324.

Chang, H.-C., and Sen, M. (1995) "Application of Chaotic Advection to Heat Transfer," in *Chaos Applied to Fluid Mixing*, eds. H. Aref and M.S. El Naschie, pp. 211–231, Pergamon Press, Oxford, Great Britain.

Chang, P.K. (1970) *Separation of Flow*, Pergamon Press, Oxford, Great Britain.

Chang, P.K. (1976) *Control of Flow Separation*, Hemisphere, Washington, D.C.

Chapman, S., and Cowling, T.G. (1970) *The Mathematical Theory of Non-Uniform Gases*, third edition, Cambridge University Press, Cambridge, Great Britain.

Cheeseman, I.C., and Seed, A.R. (1967) "The Application of Circulation Control by Blowing to Helicopter Rotors," *J. R. Aeronaut. Soc.* 71, pp. 451–467.

Chen, C.-C., Wolf, E.E., and Chang, H.-C. (1993) "Low-Dimensional Spatiotemporal Thermal Dynamics on Nonuniform Catalytic Surfaces," *J. Phys. Chemistry* 97, pp. 1055–1064.

Chen, C.P., Goland, Y., and Reshotko (1979) "Generation Rate of Turbulent Patches in the Laminar Boundary Layer of a Submersible," in *Viscous Flow Drag Reduction*, ed. G.R. Hough, pp. 73–89, AIAA, New York.

Chen, D., and Shi Ying, Z. (1989) "Control of Separation in Diffusers Using Forced Unsteadiness," AIAA Paper No. 89-1015, New York.

Cheng, H.K. (1993) "Perspectives on Hypersonic Viscous Flow Research," *Ann. Rev. Fluid Mech.* 25, pp. 455–484.

Cheng, H.K., and Emmanuel, G. (1995) "Perspectives on Hypersonic Nonequilibrium Flow," *AIAA J.* 33, pp. 385–400.

Chiwanga, S.C., and Ramaprian, B.R. (1993) "The Effect of Convex Wall Curvature on the Large-Scale Structure of the Turbulent Boundary Layer," *Exp. Thermal & Fluid Sci.* 6, pp. 168–176.

Choi, H., Moin, P., and Kim, J. (1992) "Turbulent Drag Reduction: Studies of Feedback Control and Flow Over Riblets," Department of Mechanical Engineering Report No. TF-55, Stanford University, Stanford, California.

Choi, H., Moin, P., and Kim, J. (1993) "Direct Numerical Simulations of Turbulent Flow over Riblets," *J. Fluid Mech.* 255, pp. 503–539.

Choi, H., Moin, P., and Kim, J. (1994) "Active Turbulence Control for Drag Reduction in Wall-Bounded Flows," *J. Fluid Mech.* 262, pp. 75–110.

Choi, H., Temam, R., Moin, P., and Kim, J. (1993) "Feedback Control for Unsteady Flow and its Application to the Stochastic Burgers Equation," *J. Fluid Mech.* 253, pp. 509–543.

Choi, K.-S. (1989) "Near-Wall Structure of a Turbulent Boundary Layer with Riplets," *J. Fluid Mech.* 208, pp. 417–458.

Choi, K.-S., editor (1991) *Recent Developments in Turbulence Management*, Kluwer, Dordrecht, The Netherlands.

Choi, K.-S., and Clayton, B.R. (1998) "The Mechanism of Turbulent Drag Reduction with Wall Oscillation," in *Proc. Int. Symp. on Sea Water Drag Reduction*, ed. J.C.S. Meng, pp. 229–235, July 22–23, Naval Undersea Warfare Center, Newport, Rhode Island.

Choi, K.-S., and Graham, M. (1998) "Drag Reduction of Turbulent Pipe Flows by Circular-Wall Oscillation," *Phys. Fluids* 10, pp. 7–9.

Choi, K.-S., DeBisschop, J.-R., and Graham, M. (1998) "Turbulent Boundary-Layer Control by Means of Spanwise-Wall Oscillation," *AIAA J.* 36, pp. 1157–1163.

Choi, K.-S., Yang, X., Clayton, B.R., Glover, E.J., Atlar, M., Semenov, B.N., and Kulik, V.M. (1997) "Turbulent Drag Reduction Using Compliant Surfaces," *Proc. R. Soc. London A* 453, 2229–2240.

Chow, C.Y., Chen, C.L., and Huang, M.K. (1985) "Trapping of Free Vortex by Airfoils with Surface Suction," AIAA Paper No. 85-0446, New York.

Chu, H.H., and Blick, E.F. (1969) "Compliant Surface Drag as a Function of Speed," *J. Spacecraft & Rockets* 6, pp. 763–764.

Ciccotti, G., and Hoover, W.G., editors (1986) *Molecular Dynamics Simulation of Statistical Mechanics Systems*, North Holland, Amsterdam, The Netherlands.

Cichy, D.R., Harris, J.W., and MacKay, J.K. (1972) "Flight Tests of a Rotating Cylinder Flap on a North American Rockwell YOV-10A Aircraft," NASA CR-2135, Washington, D.C.

Clauser, F.H. (1954) "Turbulent Boundary Layers in Adverse Pressure Gradient," *J. Aeronaut. Sci.* **21**, pp. 91–108.

Clauser, F.H. (1956) "The Turbulent Boundary Layer," *Adv. Appl. Mech.* **4**, pp. 1–51.

Coles, D. (1956) "The Law of the Wake in the Turbulent Boundary Layer," *J. Fluid Mech.* **1**, pp. 191–226.

Coles, D. (1978) "A Model for Flow in the Viscous Sublayer," in *Workshop on Coherent Structure of Turbulent Boundary Layers*, eds. C.R. Smith and D.E. Abbott, pp. 462–475, Lehigh University, Bethlehem, Pennsylvania.

Coles, D.E. (1962) "The Turbulent Boundary Layer in a Compressible Fluid," Rand Corporation Report No. R-403-PR, Santa Monica, California.

Coles, D.E. (1969) "The Young Person's Guide to the Data," in *Proc. 1968 AFOSR-IFP-Stanford Conference on Computation of Turbulent Boundary Layers*, eds. D.E. Coles and E.A. Hirst, vol. 2, pp. 1–45, Thermosciences Division, Stanford University, Stanford, California.

Colin, P.E., and Williams, J. (1971) *Assessment of Lift Augmentation Devices*, AGARD-LS-43-71, AGARD, NATO, Rhode-Saint-Génèse, Belgium.

Coller, B.D., Holmes, P., and Lumley, J.L. (1994a) "Control of Bursting in Boundary Layer Models," *Appl. Mech. Rev.* **47**, no. 6, part 2, pp. S139–S143.

Coller, B.D., Holmes, P., and Lumley, J.L. (1994b) "Control of Noisy Heteroclinic Cycles," *Physica D* **72**, pp. 135–160.

Collier, C.P., Wong, E.W., Belohradský, M., Raymo, F.M., Stoddart, J.F., Kuekes, P.J., Williams, R.S., and Heath, J.R. (1999) "Electronically Configurable Molecular-Based Logic Gates," *Science* **285**, no. 5426, 16 July, pp. 391–394.

Collins, F.G. (1979) "Boundary Layer Control on Wings Using Sound and Leading Edge Serrations," AIAA Paper No. 79-1875, New York.

Collins, F.G. (1981) "Boundary Layer Control on Wings Using Sound and Leading-Edge Serrations," *AIAA J.* **19**, pp. 129–130.

Collins, F.G., and Zelenevitz, J. (1975) "Influence of Sound upon Separated Flow over Wings," *AIAA J.* **13**, pp. 408–410.

Compton, D.A., and Johnston, J.P. (1991) "Streamwise Vortex Development by Pitched and Skewed Jets in a Turbulent Boundary Layer," AIAA Paper No. 91-0038, New York.

Comte-Bellot, G. (1963) "Contribution à l'Étude de la Turbulence de Conduite," Doctoral Thesis, University of Grenoble, France.

Comte-Bellot, G. (1965) "Ecoulement Turbulent entre Deux Parois Paralleles," Publications Scienifques et Techniques du Ministere de l'Air No. 419, Paris, France.

Cook, W.L., Mickey, D.M., and Quigley, H.G. (1974) "Aerodynamics of Jet Flap and Rotating Cylinder Flap STOL Concepts," AGARD Fluid Dynamics Panel on V/STOL Aerodynamics, paper no. 10, Delft, The Netherlands.

Cooke, J.C., and Brebner, G.G. (1961) "The Nature of Separation and Its Prevention by Geometric Design in a Wholly Subsonic Flow," in *Boundary Layer and Flow Control*, ed. G.V. Lachmann, vol. 1, pp. 144–185, Pergamon Press, Oxford, Great Britain.

Cooper, A.J., and Carpenter, P.W. (1995) "The Effects of Wall Compliance on Instability in Rotating Disc Flow," AIAA Paper No. 95-2257, Washington, D.C.

Cooper, A.J., and Carpenter, P.W. (1997a) "The Stability of Rotating-Disc Boundary-Layer Flow over a Compliant Wall. Part 1. Type I and II Instabilities," *J. Fluid Mech.* **350**, pp. 231–259.

Cooper, A.J., and Carpenter, P.W. (1997b) "The Stability of Rotating-Disc Boundary-Layer Flow over a Compliant Wall. Part 2. Absolute Instability," *J. Fluid Mech.* **350**, pp. 261–270.

Cooper, A.J., and Carpenter, P.W. (1997c) "The Effect of Wall Compliance on Inflexion Point Instability in Boundary Layers," *Phys. Fluids* **9**, pp. 468–470.

Corino, E.R., and Brodkey, R.S. (1969) "A Visual Investigation of the Wall Region in Turbulent Flow," *J. Fluid Mech.* **37**, pp. 1–30.

Corke, T.C., and Cavalieri, D.A. (1997) "Controlled Experiments on Instabilities and Transition to Turbulence in Supersonic Boundary Layers," AIAA Paper No. 97-1817, Washington, D.C.

Corke, T.C., Cavalieri, D.A., and Matlis, E. (1999) "Boundary Layer Instability on a Sharp Cone at Mach 3.5 with Controlled Input," *AIAA J.*, to appear.

Corke, T.C., Glauser, M.N., and Berkooz, G. (1994) "Utilizing Low-Dimensional Dynamical Systems Models to Guide Control Experiments," *Appl. Mech. Rev.* **47**, no. 6, part 2, pp. S132–S138.

Corke, T.C., Guezennec, Y., and Nagib, H.M. (1980) "Modification in Drag of Turbulent Boundary Layers Resulting from Manipulation of Large-Scale Structures," in *Viscous Flow Drag Reduction*, ed. G.R. Hough, pp. 128–143, AIAA, New York.

Corke, T.C., and Mangano, R.A. (1989) "Resonant Growth of Three-Dimensional Modes in Transitioning Blasius Boundary Layers," *J. Fluid Mech.* **209**, pp. 93–150.

Corke, T.C., Nagib, H.M., and Guezennec, Y. (1981) "A New View on Origin, Role and Manipulation of Large Scales in Turbulent Boundary Layers," NASA Contractor Report No. 165861, Washington, D.C.

Corrsin, S. (1957) "Some Current Problems in Turbulent Shear Flow," in *Symp. on Naval Hydrodynamics*, ed. F.S. Sherman, pp. 373–400, National Academy of Sciences/National Research Council Publication No. 515, Washington, D.C.

Corrsin, S. (1959) "Outline of Some Topics in Homogenous Turbulent Flow," *J. Geophys. Research* **64**, pp. 2134–2150.

Corrsin, S. (1974) "Limitations of Gradient Transport Models in Random Walks and Turbulence," in *Turbulent Diffusion in Environmental Pollution*, eds. F.N. Frenkiel and R.E. Munn, pp. 25–60, Academic Press, New York.

Crabtree, L.F. (1954) "The Formation of Regions of Separated Flow on Wing Surfaces, Part I: Low-Speed Tests on a Two-Dimensional Unswept Wing with a 10% Thick RAE—101 Section," Royal Aircraft Establishment Report Aero. 2528, Farnborough, Great Britain.

Crabtree, L.F. (1957) "Effects of Leading-Edge Separation on Thin Wings in Two-Dimensional Incompressible Flow," *J. Aeronaut. Sci.* **24**, pp. 597–604.

Crighton, D.G. (1984) "Long-Range Acoustic Scattering by Surface Inhomogeneities Beneath a Turbulent Boundary Layer," *J. Vibration, Acoustics, Stress & Reliability in Design* **106**, pp. 376–382.

Currie, I.G. (1993) *Fundamental Mechanics of Fluids*, second edition, McGraw–Hill, New York.

Cutler, A., and Bradshaw, P. (1986) "The Interaction Between a Strong Longitudinal Vortex and a Turbulent Boundary Layer," AIAA Paper No. 86-1071, New York.

Cutler, A., and Bradshaw, P. (1989) "Vortex/Boundary-Layer Interactions," AIAA Paper No. 89-0083, New York.

Daniel, A.P., Gaster, M., and Willis, G.J.K. (1987) "Boundary Layer Stability on Compliant Surfaces," Final Report No. 35020, British Maritime Technology Ltd., Teddington, Great Britain.

Darby, R.A. (1954) "An Aircraft Manufacturer Looks at Boundary Layer Control," Report No. RR-1, Fairchild Aircraft Division, Fairchild Engine and Airplane Corp., Hagerstown, Maryland.

Davidson, C.J. (1985) "The Experimental Investigation of the Effects of Roughness upon Aerofoil Characteristics at Low Reynolds Numbers," M.Sc. Thesis, Cranfield Institute of Technology, Cranfield, Great Britain.

Davies, C., and Carpenter, P.W. (1997a) "Numerical Simulation of the Evolution of Tollmien–Schlichting Waves over Finite Compliant Panels," *J. Fluid Mech.* **335**, pp. 361–392.

Davies, C., and Carpenter, P.W. (1997b) "Instabilities in a Plane Channel Flow Between Compliant Walls," *J. Fluid Mech.* **352**, pp. 205–243.

Davis, L., editor (1991) *Handbook of Genetic Algorithms*, Van Nostrand Reinhold, New York.

Deane, A.E., and Sirovich, L. (1991) "A Computational Study of Rayleigh-Bénard Convection. Part 1. Rayleigh-Number Scaling," *J. Fluid Mech.* **222**, pp. 231–250.

Debye, P., and Cleland, R.L. (1959) "Flow of Liquid Hydrocarbons in Porous Vycor," *J. Appl. Phys.* **30**, pp. 843–849.

Decken, J.V. (1971) "Aerodynamics of Pneumatic High-Lift Devices," in *AGARD Conference on Assessment of Lift Augmentation Devices*, eds. P.E. Colin and J. Williams, AGARD-LS-43-71, paper no. 2, Rhode-Saint-Génèse, Belgium.

DeCourtye, D., Sen, M., and Gad-el-Hak, M. (1998) "Analysis of Viscous Micropumps and Microturbines," *Int. J. Computat. Fluid Dyn.* **10**, pp. 13–25.

De Gennes, P.-G. (1990) *Introduction to Polymer Dynamics*, Cambridge University Press, Cambridge, Great Britain.

Delery, J.M. (1985) "Shock Wave/Turbulent Boundary Layer Interaction and Its Control," *Prog. Aero. Sci.* **22**, pp. 209–280.

Delville, J., Cordier, L., and Bonnet, J.-P. (1998) "Large-Scale-Structure Identification and Control in Turbulent Shear Flows," in *Flow Control: Fundamentals and Practices*, eds. M. Gad-el-Hak, A. Pollard, and J.-P. Bonnet, pp. 199–273, Springer–Verlag, Berlin.

DeMeis, R. (1986) "Sounding a Happy Note for Lift," *Aerospace America* **24**, August, pp. 10–11.

Den, L.M. (1990) "Issues in Viscoelastic Fluid Mechanics," *Ann. Rev. Fluid Mech.* **22**, pp. 13–34.

Dennell, R. (1997) "The World's Oldest Spears," *Nature* **385**, 27 February, pp. 767–768.

Despard, R.A., and Miller, J.A. (1971) "Separation in Oscillating Laminar Boundary-Layer Flows," *J. Fluid. Mech.* **47**, pp. 21–31.

Didden, N., and Ho, C.-M. (1985) "Unsteady Separation in a Boundary Layer Produced by an Impinging Jet," *J. Fluid Mech.* **160**, pp. 235–256.

Dillner, B., May, F., and McMasters, J.H. (1984) "Aerodynamic Issues in the Design of High-Lift Systems for Transport Aircraft," in *AGARD Conference on Improvement of Aerodynamic Performance Through Boundary Layer Control and High Lift Systems*, AGARD-CP-365, paper no. 9, Brussels, Belgium.

Dimotakis, P.E. (1991) "Turbulent Free Shear Layer Mixing and Combustion," in *High-Speed Flight Propulsion Systems*, eds. S.N.B. Murthy and E.T. Curran, pp. 265–340, AIAA, Washington, D.C.

Dimotakis, P.E. (1993) "Some Issues on Turbulent Mixing and Turbulence," Guggenheim Aeronautical Laboratory Report No. FM93-1a, California Institute of Technology, Pasadena, California.

Dinkelacker, A. (1966) "Preliminary Experiments on the Influence of Flexible Walls on Boundary Layer Turbulence," *J. Sound Vibr.* **4**, pp. 187–214.

DiPrima, R.C., and Swinney, H.L. (1985) "Instabilities and Transition in Flow Between Concentric Rotating Cylinders, in *Hydrodynamic Instabilities and the Transition to Turbulence*, eds. H.L. Swinney and J.P. Gollub, second edition, pp. 139–180, Springer–Verlag, Berlin.

Ditto, W.L., and Pecora, L.M. (1993) "Mastering Chaos," *Scientific American* **269**, August, pp. 78–84.

Ditto, W.L., Rauseo, S.N., and Spano, M.L. (1990) "Experimental Control of Chaos," *Phys. Rev. Lett.* **65**, pp. 3211–3214.

Dixon, A.E., Lucey, A.D., and Carpenter, P.W. (1994) "The Optimization of Viscoelastic Compliant Walls for Transition Delay," *AIAA J.* **32**, pp. 256–267.

Domaradzki, J.A., and Metcalfe, R.W. (1987) "Stabilization of Laminar Boundary Layers by Compliant Membranes," *Phys. Fluids* **30**, pp. 695–705.

Donnelly, R.J. (1991) *High Reynolds Number Flows Using Liquid and Gaseous Helium*, Springer–Verlag, New York.

Donohue, G.L., Tiederman, W.G., and Reischman, M.M. (1972) "Flow Visualization of the Near-Wall Region in a Drag-Reducing Channel Flow," *J. Fluid Mech.* **56**, pp. 559–575.

Donovan, J.F., and Selig, M.S. (1989) "Low Reynolds Number Airfoil Design and Wind Tunnel Testing at Princeton University," in *Low Reynolds Number Aerodynamics*, ed. T.J. Mueller, pp. 39–57, Springer–Verlag, Berlin.

Döring, C., Grauer, T., Marek, J., Mettner, M.S., Trah, H.-P., and Willmann, M. (1992) "Micromachined Thermoelectrically Driven Cantilever Structures for Fluid Jet Deflection," in *Proc. IEEE Micro Electro Mechanical Systems '92*, pp. 12–18, 4–7 February, Travemünde, Germany.

Dougherty, N.S., and Fisher, D.F. (1980) "Boundary Layer Transition on a 10-Degree Cone," AIAA Paper No. 80-0154, New York.

Dowling, A.P. (1983) "Flow-Acoustic Interaction Near a Flexible Wall," *J. Fluid Mech.* **128**, pp. 181–198.

Dowling, A.P. (1985) "The Effect of Large-Eddy Breakup Devices on Oncoming Vorticity," *J. Fluid Mech.* **160**, pp. 447–463.

Dowling, A.P. (1986) "Mean Flow Effects on the Low-Wavenumber Pressure Spectrum on a Flexible Surface," *J. Fluids Eng.* **108**, pp. 104–108.

Dowling, A.P., and Ffowcs Williams, J.E. (1983) *Sound and Sources of Sound*, Ellis Horwood, Chichester, Great Britain.

Drazin, P., and Reid, W. (1981) *Hydrodynamic Stability*, Cambridge University Press, Cambridge, Great Britain.

Dreyden, H.L. (1948) "Recent Advances in the Mechanics of Boundary Layer Flow," in *Advances in Applied Mechanics*, eds. R. von Mises and Th. von Kármán, vol. 1, pp. 1–40, Academic Press, Boston, Massachusetts.

Driver, D.M. (1989) "Experimental Study of a Three-Dimensional Shear-Driven Turbulent Boundary Layer with Streamwise Adverse Pressure Gradient," Ph.D. Thesis, Stanford University, Stanford, CA.

Duncan, J.H. (1986) "The Response of an Incompressible, Viscoelastic Coating to Pressure Fluctuations in a Turbulent Boundary Layer," *J. Fluid Mech.* **171**, pp. 339–363.

Duncan, J.H. (1987) "A Comparison of Wave Propagation on the Surfaces of Simple Membrane Walls and Elastic Coatings Bounded by a Fluid Flow," *J. Sound Vibr.* **119**, pp. 565–573.

Duncan, J.H. (1988) "The Dynamics of Waves at the Interface Between a Two-Layer Viscoelastic Coating and a Fluid Flow," *J. Fluids Structures* **2**, pp. 35–51.

Duncan, J.H., and Sirkis, J.S. (1992) "The Generation of Wave Patterns on Isotropic Coatings by Pressure Fluctuations in a Turbulent Boundary Layer," *J. Sound Vibr.* **157**, pp. 243–264.

Duncan, J.H., and Hsu, C.C. (1984) "The Response of a Two-Layer Viscoelastic Coating to Pressure Disturbances from a Turbulent Boundary Layer," AIAA Paper No. 84-0535, New York.

Duncan, J.H., Waxman, A.M., and Tulin, M.P. (1985) "The Dynamics of Waves at the Interface Between a Viscoelastic Coating and a Fluid Flow," *J. Fluid Mech.* **158**, pp. 177–197.

Durbin, P.A., and McKinzie, D.J. (1987) "Corona Anemometry for Qualitative Measurement of Reversing Surface Flow with Application to Separation Control by External Excitation," in *Proc. Forum on Unsteady Flow Separation*, ed. K.N. Ghia, pp. 15–18, ASME, New York.

Dussan, E.B. (1976) "The Moving Contact Line: the Slip Boundary Condition," *J. Fluid Mech.* **77**, pp. 665–684.

Dussan, E.B. (1979) "On the Spreading of Liquids on Solid Surfaces: Static and Dynamic Contact Lines," *Annu. Rev. Fluid Mech.* **11**, pp. 371–400.

Dussan, E.B., and Davis, S.H. (1974) "On the Motion of Fluid–Fluid Interface Along a Solid Surface," *J. Fluid Mech.* **65**, pp. 71–95.

Dutton, R.A. (1955) "Experimental Studies of the Turbulent Boundary Layer on a Flat Plate With and Without Distributed Suction," Ph.D. Dissertation, Cambridge University, Cambridge, Great Britain.

Dutton, R.A. (1960) "The Effects of Distributed Suction on the Development of Turbulent Boundary Layers," ARC R&M No. 3155, Aeronautical Research Council, London, Great Britain.

Eckelmann, H. (1974) "The Structure of the Viscous Sublayer and the Adjacent Wall Region in a Turbulent Channel Flow," *J. Fluid Mech.* **65**, pp. 439–459.

Eléna, M. (1975) "Etude des Champs Dynamiques et Thermiques d'un Ecoulement Turbulent en Conduit avec Aspiration à la Paroi," Thèse de Doctorat des Sciences, Université d'Aix-Marseille, Marseille, France.

Eléna, M. (1984) "Suction Effects on Turbulence Statistics in a Heated Pipe Flow," *Phys. Fluids* **27**, pp. 861–866.

Eléna, M., and Dumas, R. (1978) "Champs Dynamiques et Thermiques d'un Ecoulement Turbulent en Conduit avec Aspiration à la Paroi," in *Sixth Int. Heat Trans. Conf.*, vol. 5, pp. 239–244, Hemisphere, Washington, D.C.

Ellington, C.P., van den Berg, C., Willmott, A.P., and Thomas, A.L.R. (1996) "Leading-Edge Vortices in Insect Flight," *Nature* **384**, 19/26 December, pp. 626–629.

Ely, W.L., and Berrier, F.C. (1975) "Performance of Steady and Intermittent Blowing Jet Flaps and Spanwise Upper Surface Slots," Air Force Flight Dynamics Laboratory Report No. AFFDL-TR-75-128, Wright-Patterson Air Force Base, Ohio.

Emmons, H.W. (1951) "The Laminar-Turbulent Transition in a Boundary Layer. Part I," *J. Aero. Sci.* **18**, pp. 490–498.

Eppler, R., and Somers, D.M. (1985) "Airfoil Design for Reynolds Numbers Between 50,000 and 500,000," in *Proc. Conf. on Low Reynolds Number Airfoil Aerodynamics*, ed. T.J. Mueller, pp. 1–14, University of Notre Dame, Notre Dame, Indiana.

Epstein, A.H., and Senturia, S.D. (1997) "Macro Power from Micro Machinery," *Science* **276**, 23 May, p. 1211.

Epstein, A.H., Senturia, S.D., Al-Midani, O., Anathasuresh, G., Ayon, A., Breuer, K., Chen, K.-S., Ehrich, F.F., Esteve, E., Frechette, L., Gauba, G., Ghodssi, R., Groshenry, C., Jacobson, S.A., Kerrebrock, J.L., Lang, J.H., Lin, C.-C., London, A., Lopata, J., Mehra, A., Mur Miranda, J.O., Nagle, S., Orr, D.J., Piekos, E., Schmidt, M.A., Shirley, G., Spearing, S.M., Tan, C.S., Tzeng, Y.-S., and Waitz, I.A. (1997) "Micro-Heat Engines, Gas Turbines, and Rocket Engines—The MIT Microengine Project," AIAA Paper No. 97-1773, AIAA, Reston, Virginia.

Ericsson, L.E. (1967) "Comment on Unsteady Airfoil Stall," *J. Aircraft* **4**, pp. 478–480.

Ericsson, L.E. (1971) "Unsteady Airfoil Stall and Stall Flutter," NASA CR-111906, Washington, D.C.

Ericsson, L.E. (1988) "Moving Wall Effects in Unsteady Flow," *J. Aircraft* 25, pp. 977–990.

Ericsson, L.E., and Reding, J.P. (1971) "Unsteady Airfoil Stall, Review and Extension," *J. Aircraft* 8, pp. 609–616.

Ericsson, L.E., and Reding, J.P. (1984) "Unsteady Flow Concepts for Dynamic Stall Analysis," *J. Aircraft* 21, pp. 601–606.

Ericsson, L.E., and Reding, J.P. (1986) "Fluid Dynamics of Unsteady Separated Flow. Part I. Bodies of Revolution," *Prog. Aero. Sci.* 23, pp. 1–84.

Ericsson, L.E., and Reding, J.P. (1987) "Fluid Dynamics of Unsteady Separated Flow. Part II. Lifting Surfaces," *Prog. Aero. Sci.* 24, pp. 249–356.

Eriksson, L.J., Allie, M.C., Bremigan, C.D., and Gilbert, J.A. (1988) "Active Noise Control and Specifications for Fan Noise Problems," in *Noise Control Design: Methods and Practice (Noise-Con 88 Proceedings)*, ed. J.S. Bolton, pp. 273–278, Noise Control Foundation, Poughkeepsie, New York.

Erm, L.P., Smits, A.J., and Joubert, P.N. (1987) "Low Reynolds Number Turbulent Boundary Layers on a Smooth Flat Surface in a Zero Pressure Gradient," in *Fifth Symposium on Turbulent Shear Flows*, eds. F. Durst et al., pp. 2.13–2.18, Springer–Verlag, Berlin.

Esashi, M., Shoji, S., and Nakano, A. (1989) "Normally Closed Microvalve Fabricated on a Silicon Wafer," *Sensors & Actuators* 20, pp. 163–169.

Falco, R.E. (1974) "Some Comments on Turbulent Boundary Layer Structure Inferred from the Movements of a Passive Contaminant," AIAA Paper No. 74-99, New York.

Falco, R.E. (1977) "Coherent Motions in the Outer Region of Turbulent Boundary Layers," *Phys. Fluids* 20, no. 10, part II, pp. S124–S132.

Falco, R.E. (1980) "The Production of Turbulence Near a Wall," AIAA Paper No. 80-1356, New York.

Falco, R.E. (1983) "New Results, a Review and Synthesis of the Mechanism of Turbulence Production in Boundary Layers and its Modification," AIAA Paper No. 83-0377, New York.

Falco, R.E. (1991) "A Coherent Structure Model of the Turbulent Boundary Layer and Its Ability to Predict Reynolds Number Dependence," *Phil. Trans. R. Soc. London A* 336, pp. 103–129.

Faller, W.E., Schreck, S.J., and Luttges, M.W. (1994) "Real-Time Prediction and Control of Three-Dimensional Unsteady Separated Flow Fields Using Neural Networks," AIAA Paper No. 94-0532, Washington, D.C.

Fan, L.-S., Tai, Y.-C., and Muller, R.S. (1988) "Integrated Movable Micromechanical Structures for Sensors and Actuators," in *IEEE Transactions on Electronic Devices*, vol. 35, pp. 724–730.

Fan, L.-S., Tai, Y.-C., and Muller, R.S. (1989) "IC-Processed Electrostatic Micromotors," *Sensors & Actuators* 20, pp. 41–47.

Fan, X., Hofmann, L., and Herbert, T. (1993) "Active Flow Control with Neural Networks," AIAA Paper No. 93-3273, Washington, D.C.

Farabee, T.M., and Casarella, M.J. (1991) "Spectral Features of Wall Pressure Fluctuations Beneath Turbulent Boundary Layers," *Phys. Fluids A* 3, pp. 2410–2420.

Farge M. (1992) "Wavelet Transforms and Their Applications to Turbulence," *Ann. Rev. Fluid Mech.* 24, pp. 395–457.

Favre, A., Dumas, R., Verollet, E., and Coantic, M. (1966) "Couche Limite Turbulente sur Paroi Poreuse avec Aspiration," *J. Mécanique* 5, pp. 3–28.

Feir, J.B. (1965) "The Effects of an Arrangement of Vortex Generators Installed to Eliminate Wind Tunnel Diffuser Separation," UTIAS Technical Note No. 87, Institute for Aerospace Studies, University of Toronto, Toronto, Canada.

Ferziger, J.H., and Perić, M. (1996) *Computational Methods for Fluid Dynamics*, Springer–Verlag, New York.

Feynman, R.P. (1961) "There's Plenty of Room at the Bottom," in *Miniaturization*, ed. H.D. Gilbert, pp. 282–296, Reinhold Publishing, New York.

Ffowcs Williams, J.E. (1965) "Sound Radiation from Turbulent Boundary Layers Formed on Compliant Surfaces," *J. Fluid Mech.* 22, pp. 347–358.

Ffowcs Williams, J.E. (1977) "Aeroacoustics," *Ann. Rev. Fluid Mech.* 9, pp. 447–468.

Ffowcs Williams, J.E. (1984) "Anti-Sound," *Proc. R. Soc. London A* 395, pp. 63–88.

Fiedler, H.E. (1986) "Coherent Structures," in *Advances in Turbulence*, eds. G. Comte-Bellot and J. Mathieu, pp. 320–336, Springer–Verlag, Berlin.

Fiedler, H.E. (1988) "Coherent Structures in Turbulent Flows," *Prog. Aero. Sci.* 25, pp. 231–269.

Fiedler H.E. (1998) "Control of Free Turbulent Shear Flows," in *Flow Control: Fundamentals and Practices*, eds. M. Gad-el-Hak, A. Pollard, and J.-P. Bonnet, pp. 335–429, Springer–Verlag, Berlin.

Fiedler, H.E., and Fernholz, H.-H. (1990) "On Management and Control of Turbulent Shear Flows," *Prog. Aero. Sci.* **27**, pp. 305–387.

Fiedler, H.E., and Head, M.R. (1966) "Intermittency Measurements in the Turbulent Boundary Layer," *J. Fluid Mech.* **25**, pp. 719–735.

Fiedler, H.E., Glezer, A., and Wygnanski, I. (1988) "Control of Plane Mixing Layer: Some Novel Experiments," in *Current Trends in Turbulence Research*, eds. H. Branover, M. Mond, and Y. Unger, pp. 30–64, AIAA, Washington, D.C.

Fischer, M.C., and Ash, R.L. (1974) "A General Review of Concepts for Reducing Skin Friction, Including Recommendations for Future Studies," NASA Technical Memorandum No. X-2894, Washington, D.C.

Fisher, D.H., and Blick, E.F. (1966) "Turbulent Damping by Flabby Skins," *J. Aircraft* **3**, pp. 163–164.

Fitzgerald, E.J., and Mueller, T.J. (1990) "Measurements in a Separation Bubble on an Airfoil Using Laser Velocimetry," *AIAA J.* **28**, pp. 584–592.

Flatt, J. (1961) "The History of Boundary Layer Control Research in the United States of America," in *Boundary Layer and Flow Control*, ed. G.V. Lachmann, vol. 1, pp. 122–143, Pergamon Press, New York.

Flettner, A. (1924) "Die Anwendung der Erkenntnisse der Aerodynamik zum Windantrieb von Schiffen," *Jb. Schiffbautech. Ges.* **25**, pp. 222–251.

Fontaine, A.A., Deutsch, S., Brungart, T.A., Petrie, H.L., and Fenstermacker, M. (1999) "Drag Reduction by Coupled Systems: Microbubble Injection with Homogenous Polymer and Surfactant Solutions," *Exp. Fluids* **26**, pp. 397–403.

Foreman, J.E.K. (1989) *Sound Analysis and Noise Control*, Van Nostrand Reinhold, New York.

Fortuna, G., and Hanratty, T.J. (1971) "The Influence of Drag-Reducing Polymers on Turbulence in the Viscous Sublayer," *J. Fluid Mech.* **53**, pp. 575–586.

Fowler, T.B. (1989) "Application of Stochastic Control Techniques to Chaotic Nonlinear Systems," *IEEE Trans. Autom. Control* **34**, pp. 201–205.

Francis, M.S., Keesee, J.E., Lang, J.D., Sparks, G.W., and Sisson, G.E. (1979) "Aerodynamic Characteristics of an Unsteady Separated Flow," *AIAA J.* **17**, pp. 1332–1339.

Fraser, L.A., and Carpenter, P.W. (1985) "A Numerical Investigation of Hydroelastic and Hydrodynamic Instabilities in Laminar Flows over Compliant Surfaces Comprising One or Two Layers of Viscoelastic Material," in *Numerical Methods in Laminar and Turbulent Flow*, eds. C. Taylor et al., pp. 1171–1181, Pineridge Press, Swansea, Great Britain.

Frick, C.W., and McCullough, C.B. (1942) "Tests of a Heated Low Drag Airfoil," NACA ARR, December, Washington, D.C.

Frisch, U. (1995) *Turbulence: The Legacy of A.N. Kolmogorov*, Cambridge University Press, Cambridge, Great Britain.

Frisch, U., and Orszag, S.A. (1990) "Turbulence: Challenges for Theory and Experiment," *Phys. Today* **43**, no. 1, January, pp. 24–32.

Fuhr, G., Hagedorn, R., Müller, T., Benecke, W., and Wagner, B. (1992) "Microfabricated Electrohydrodynamic (EHD) Pumps for Liquids of Higher Conductivity," *J. MEMS* **1**, pp. 141–145.

Fujimasa, I. (1996) *Micromachines: A New Era in Mechanical Engineering*, Oxford University Press, Oxford, Great Britain.

Fung, K.-Y., editor (1994) *Symp. on Aerodynamics & Aeracoustics*, World Scientific, Singapore.

Gabriel, K.J. (1995) "Engineering Microscopic Machines," *Scientific American* **260**, September, pp. 150–153.

Gabriel, K.J., Jarvis, J., and Trimmer, W., editors (1988) *Small Machines, Large Opportunities: A Report on the Emerging Field of Microdynamics*, National Science Foundation, published by AT&T Bell Laboratories, Murray Hill, New Jersey.

Gabriel, K.J., Tabata, O., Shimaoka, K., Sugiyama, S., and Fujita, H. (1992) "Surface-Normal Electrostatic/Pneumatic Actuator," in *Proc. IEEE Micro Electro Mechanical Systems '92*, pp. 128–131, 4–7 February, Travemünde, Germany.

Gadd, G.E. (1960) "Boundary Layer Separation in the Presence of Heat Transfer," AGARD Report No. 280, AGARD, NATO, Rhode-Saint-Génèse, Belgium.

Gadd, G.E., Cope, W.F., and Attridge, J.L. (1958) "Heat-Transfer and Skin-Friction Measurements at a Mach Number of 2.44 for a Turbulent Boundary Layer on a Flat Surface and in Regions of Separated Flow," ARC R&M 3148, Aeronautical Research Council, London, Great Britain.

Gad-el-Hak, M. (1986a) "Boundary Layer Interactions with Compliant Coatings: An Overview," *Appl. Mech. Rev.* **39**, pp. 511–524.

Gad-el-Hak, M. (1986b) "The Response of Elastic and Viscoelastic Surfaces to a Turbulent Boundary Layer," *J. Appl. Mech.* **53**, pp. 206–212.

Gad-el-Hak, M. (1986c) "The Use of the Dye-Layer Technique for Unsteady Flow Visualization," *J. Fluids Eng.* **108**, pp. 34–38.

Gad-el-Hak, M. (1987a) "Compliant Coatings Research: A Guide to the Experimentalist," *J. Fluids Structures* **1**, pp. 55–70.

Gad-el-Hak, M. (1987b) "Unsteady Separation on Lifting Surfaces," *Appl. Mech. Rev.* **40**, pp. 441–453.

Gad-el-Hak, M. (1988a) "Review of Flow Visualization Techniques for Unsteady Flows," in *Flow Visualization IV*, ed. C. Véret, pp. 1–12, Hemisphere, Washington, D.C.

Gad-el-Hak, M. (1988b) "Visualization Techniques for Unsteady Flows: An Overview," *J. Fluids Eng.* **110**, pp. 231–243.

Gad-el-Hak, M. (1989) "Flow Control," *Appl. Mech. Rev.* **42**, pp. 261–293.

Gad-el-Hak, M. (1990) "Control of Low-Speed Airfoil Aerodynamics," *AIAA J.* **28**, pp. 1537–1552.

Gad-el-Hak, M. (1992) "Splendor of Fluids in Motion," *Prog. Aero. Sci.* **29**, pp. 81–123.

Gad-el-Hak, M. (1993) "Innovative Control of Turbulent Flows," AIAA Paper No. 93-3268, Washington, D.C.

Gad-el-Hak, M. (1994) "Interactive Control of Turbulent Boundary Layers: A Futuristic Overview," *AIAA J.* **32**, pp. 1753–1765.

Gad-el-Hak, M. (1995) "Questions in Fluid Mechanics: Stokes' Hypothesis for a Newtonian, Isotropic Fluid," *J. Fluids Eng.* **117**, pp. 3–5.

Gad-el-Hak, M. (1996a) "Compliant Coatings: A Decade of Progress," *Appl. Mech. Rev.* **49**, no. 10, part 2, pp. S1–S11.

Gad-el-Hak, M. (1996b) "Modern Developments in Flow Control," *Appl. Mech. Rev.* **49**, pp. 365–379.

Gad-el-Hak, M. (1997) "The Last Conundrum," *Appl. Mech. Rev.* **50**, no. 12, part 1, pp. 1–2.

Gad-el-Hak, M. (1998a) "Fluid Mechanics from the Beginning to the Third Millennium," *Int. J. Engng. Ed.* **14**, pp. 177–185.

Gad-el-Hak, M. (1998b) "Frontiers of Flow Control," in *Flow Control: Fundamentals and Practices*, eds. M. Gad-el-Hak, A. Pollard, and J.-P. Bonnet, pp. 109–153, Springer–Verlag, Berlin.

Gad-el-Hak, M. (1999) "The Fluid Mechanics of Microdevices—The Freeman Scholar Lecture," *J. Fluids Eng.* **121**, pp. 1–29.

Gad-el-Hak, M., and Bandyopadhyay, P.R. (1994) "Reynolds Number Effects in Wall-Bounded Flows," *Appl. Mech. Rev.* **47**, pp. 307–365.

Gad-el-Hak, M., and Blackwelder, R.F. (1985) "The Discrete Vortices from a Delta Wing," *AIAA J.* **23**, pp. 961–962.

Gad-el-Hak, M., and Blackwelder, R.F. (1987a) "Simulation of Large-Eddy Structures in a Turbulent Boundary Layer," *AIAA J.* **25**, pp. 1207–1215.

Gad-el-Hak, M., and Blackwelder, R.F. (1987b) "Control of the Discrete Vortices from a Delta Wing," *AIAA J.* **25**, pp. 1042–1049.

Gad-el-Hak, M., and Blackwelder, R.F. (1987c) "A Drag Reduction Method for Turbulent Boundary Layers," AIAA Paper No. 87-0358, New York.

Gad-el-Hak, M., and Blackwelder, R.F. (1989) "Selective Suction for Controlling Bursting Events in a Boundary Layer," *AIAA J.* **27**, pp. 308–314.

Gad-el-Hak, M., Blackwelder, R.F., and Riley, J.J. (1981) "On the Growth of Turbulent Regions in Laminar Boundary Layers," *J. Fluid Mech.* **110**, pp. 73–95.

Gad-el-Hak, M., Blackwelder, R.F., and Riley, J.J. (1984) "On the Interaction of Compliant Coatings with Boundary Layer Flows," *J. Fluid Mech.* **140**, pp. 257–280.

Gad-el-Hak, M., Blackwelder, R.F., and Riley, J.J. (1985) "Visualization Techniques for Studying Transitional and Turbulent Flows," in *Flow Visualization III*, ed. W.J. Yang, pp. 568–575, Hemisphere, Washington, D.C.

Gad-el-Hak, M., and Bushnell, D.M. (1991a) "Status and Outlook of Flow Separation Control," AIAA Paper No. 91-0037, New York.

Gad-el-Hak, M., and Bushnell, D.M. (1991b) "Separation Control: Review," *J. Fluids Eng.* **113**, pp. 5–30.

Gad-el-Hak, M., Davis, S.H., McMurray, J.T., and Orszag, S.A. (1984) "On the Stability of the Decelerating Boundary Layer," *J. Fluid Mech.* **138**, pp. 297–323.

Gad-el-Hak, M., and Ho, C.-M. (1985) "The Pitching Delta Wing," *AIAA J.* **23**, pp. 1660–1665.

Gad-el-Hak, M., and Ho, C.-M. (1986a) "Unsteady Vortical Flow Around Three-Dimensional Lifting Surfaces," *AIAA J.* **24**, pp. 713–721.

Gad-el-Hak, M., and Ho, C.-M. (1986b) "Unsteady Flow Around an Ogive-Cylinder," *J. Aircraft* **23**, pp. 520–528.

Gad-el-Hak, M., and Hussain, A.K.M.F. (1986) "Coherent Structures in a Turbulent Boundary Layer. Part 1. Generation of 'Artificial' Bursts," *Phys. Fluids* **29**, pp. 2124–2139.

Gad-el-Hak, M., Pollard, A., and Bonnet, J.-P. (1998) *Flow Control: Fundamentals and Practices*, Springer–Verlag, Berlin.

Gad-el-Hak, M., and Sen, M. (1996) "Fluid Mechanics in the Next Century," *Appl. Mech. Rev.* **49**, no. 3, pp. III–IV.

Gadetskii, V.M., Serebriiskii, I.A.M., and Fomin, V.M. (1972) "Investigation of the Influence of Vortex Generators on Turbulent Boundary Layer Separation," *Uchenye Zapiski TSAGI* **3**, pp. 22–28.

Gallagher, J.A., and Thomas, A.S.W. (1984) "Turbulent Boundary Layer Characteristics over Streamwise Grooves," AIAA Paper No. 84-2185, New York.

Galperin, B., and Orszag, S.A., editors (1993) *Large Eddy Simulations of Complex Engineering and Geophysical Flows*, Cambridge University Press, Cambridge, Great Britain.

Gampert, B., Homann, K., and Rieke, H.B. (1980) "The Drag Reduction in Laminar and Turbulent Boundary Layers by Prepared Surfaces with Reduced Momentum Transfer," *Israel J. Technology* **18**, pp. 287–292.

Garcia, E.J., and Sniegowski, J.J. (1993) "The Design and Modelling of a Comb-Drive-Based Microengine for Mechanism Drive Applications," in *Proc. Seventh Int. Conf. on Solid-State Sensors and Actuators (Transducers '93)*, pp. 763–766, Yokohama, Japan, 7–10 June.

Garcia, E.J., and Sniegowski, J.J. (1995) "Surface Micromachined Microengine," *Sensors & Actuators A* **48**, pp. 203–214.

Garfinkel, A., Spano, M.L., Ditto, W.L., and Weiss, J.N. (1992) "Controlling Cardiac Chaos," *Science* **257**, pp. 1230–1235.

Garrad, A.D., and Carpenter, P.W. (1982) "A Theoretical Investigation of Flow-Induced Instabilities in Compliant Coatings," *J. Sound Vibr.* **85**, pp. 483–500.

Gartling, D.K. (1970) "Tests of Vortex Generators to Prevent Separation of Supersonic Flow in a Compression Corner," Applied Research Laboratory Report No. ARL-TR-70-44, University of Texas, Austin, TX. (Also available from U.S. NTIS; Document Number AD-734154.)

Gaster, M. (1988) "Is the Dolphin a Red Herring?," in *Turbulence Management and Relaminarisation*, eds. H.W. Liepmann and R. Narasimha, pp. 285–304, Springer–Verlag, Berlin.

Gates, B., Myhrvold, N., and Rinearson, P. (1995) *The Road Ahead*, Viking, New York.

Gebert, G.A. (1988) "Turbulent Boundary Layer Modification by Streamlined Devices," Ph.D. Thesis, University of Notre Dame, Notre Dame, Indiana.

Gedney, C.J. (1983) "The Cancellation of a Sound-Excited Tollmien–Schlichting Wave with Plate Vibration," *Phys. Fluids* **26**, pp. 1158–1160.

Gee, M.L., McGuiggan, P.M., Israelachvili, J.N., and Homola, A.M. (1990) "Liquid to Solidlike Transitions of Molecularly Thin Films Under Shear," *J. Chem. Phys.* **93**, pp. 1895–1906.

George, W.K., and Castillo, L. (1993) "Boundary Layers with Pressure Gradient: Another Look at the Equilibrium Theory," in *Near-Wall Turbulent Flows*, eds. R.M.C. So, C.G. Speziale, and B.E. Launder, pp. 901–910, Elsevier, Amsterdam.

George, W.K., Castillo L. (1997) "Zero-Pressure-Gradient Turbulent Boundary Layer," *Appl. Mech. Rev.* **50**, no. 12, part 1, pp. 689–729.

George, W.K., Castillo, L., and Knecht, P. (1992) "The Zero-Pressure Gradient Turbulent Boundary Layer Revisited," in *Thirteenth Symposium on Turbulence*, ed. X.B. Reed, 21–23 April, University of Missouri, Rolla, Missouri.

George, W.K., Castillo, L., and Knecht, P. (1993) "The Zero-Pressure Gradient Turbulent Boundary Layer," in *William C. Reynolds Anniversary Symposium on Turbulence*, 22–23 March, Asilomar, California.

George, W.K., Castillo, L., and Knecht, P. (1994) "The Zero-Pressure Gradient Turbulent Boundary Layer," unpublished manuscript submitted to *Phys. Fluids.*

Gerbeth, G., editor (1997) in *Proc. Int. Workshop on Electromagnetic Boundary Layer Control (EBLC) for Saltwater Flows*, 7–8 July, Forschungszentrum Rossendorf, Dresden, Germany.

Germano, M., Piomelli, U., Moin, P., and Cabot, W.H. (1991) "A Dynamic Subgrid-Scale Eddy Viscosity Model," *Phys. Fluids A* **3**, pp. 1760–1765.

Ghosh, S., Chang, H.-C., and Sen, M. (1992) "Heat-Transfer Enhancement due to Slender Recirculation and Chaotic Transport Between Counter-Rotating Eccentric Cylinders," *J. Fluid Mech.* **238**, pp. 119–154.

Gillis, J.C. (1980) "Turbulent Boundary Layer on a Convex, Curved Surface," Ph.D. Dissertation, Stanford University, Stanford, California.

Glanz, J. (1995) "Computer Scientists Rethink Their Discipline's Foundations," *Science* **269**, no. 5229, 8 September, pp. 1363–1364.

Gleick, J. (1987) *Chaos—Making a New Science*, Viking, New York.

Goldberg, D.E. (1989) *Genetic Algorithms in Search, Optimization, and Machine Learning*, Addison–Wesley, Reading, Massachusetts.

Gol'd Fel'd, M.A., and Zatoloka, V.V. (1979) "On the Improvement of Separating Properties of a Turbulent Boundary Layer as a Result of the Effect of a Shock Wave," *Izvestiya Sibirskogo Otdeleniya, Akademii Nank* **3**, pp. 40–47.

Goldstein, M.E. (1984) "Generation of Instability Waves in Flows Separating from Smooth Surfaces," *J. Fluid Mech.* **145**, pp. 71–94.

Goldstein, M.E., and Hultgren, L.S. (1989) "Boundary-Layer Receptivity to Long-Wave Free-Stream Disturbances," *Ann. Rev. Fluid Mech.* **21**, pp. 137–166.

Goldstein, S. (1948) "On Laminar Boundary Layer Flow Near a Position of Separation," *Q. J. Mech. Appl. Math.* **1**, pp. 43–69.

Goodman, W.L. (1985) "Emmons Spot Forcing for Turbulent Drag Reduction," *AIAA J.* **23**, pp. 155–157.

Görtler, H. (1955) "Dreidimensionales zur Stabilitätstheorie laminarer Grenzschichten," *ZAMM* **35**, pp. 362–363.

Grant, H.L. (1958) "The Large Eddies of Turbulent Motion," *J. Fluid Mech.* **4**, pp. 149–190.

Granville, P.S. (1977) "Drag and Turbulent Boundary Layer of Flat Plates at Low Reynolds Numbers," *J. Ship Research* **21**, pp. 30–39.

Granville, P.S. (1979) "Drag of Underwater Bodies," in *Hydroballistics Design Handbook*, vol. 1, pp. 309–341, SEAHAC TR 79-1, Naval Sea Systems Command, Washington, D.C.

Grappin, R., and Léorat, J. (1991) "Lyapunov Exponents and the Dimension of Periodic Incompressible Navier-Stokes Flows: Numerical Measurements," *J. Fluid Mech.* **222**, pp. 61–94.

Grass, A.J. (1971) "Structural Features of Turbulent Flow over Smooth and Rough Boundaries," *J. Fluid Mech.* **50**, pp. 233–255.

Gratzer, L.B. (1971) "Analysis of Transport Applications for High Lift Schemes," in *AGARD Course on Assessment of Lift Augmentation Devices*, eds. P.E. Colin and J. Williams, AGARD-LS-43-71, paper no. 7, Rhode-Saint-Génèse, Belgium.

Gravesen, P., Branebjerg, J., and Jensen, O.S. (1993) "Microfluidics—A Review," *J. Micromech. Microeng.* **3**, pp. 168–182.

Gray, J. (1936) "Studies in Animal Locomotion. VI. The Propulsive Powers of the Dolphin," *J. Exp. Biol.* **13**, pp. 192–199.

Gregory, N., Stuart, J.T., and Walker, W.S. (1955) "On the Stability of Three-Dimensional Boundary Layers with Applications to the Flow due to a Rotating Disk," *Phil. Trans. R. Soc. London A* **248**, pp. 155–199.

Grigson, C.W.B. (1992) "An Accurate Smooth Friction Line for Use in Performance Prediction," in *Proc. Roy. Inst. Nav. Arch.* **W5**, pp. 1–9.

Gupta, A.K., and Kaplan, R.E. (1972) "Statistical Characteristics of Reynolds Stress in a Turbulent Boundary Layer," *Phys. Fluids* **15**, pp. 981–985.

Gutmark, E.J., and Grinstein, F.F. (1999) "Flow Control with Noncircular Jets," *Ann. Rev. Fluid Mech.* **31**, pp. 239–272.

Gutmark, E.J., and Ho, C.-M. (1986) "Visualization of a Forced Elliptical Jet," *AIAA J.* **24**, pp. 684–685.

Gutmark, E.J., Schadow, K.C., and Yu, K.H. (1995) "Mixing Enhancement in Supersonic Free Shear Flows," *Ann. Rev. Fluid Mech.* **27**, pp. 375–417.

Guvenc, M.G. (1985) "V-Groove Capillary for Low Flow Control and Measurement," in *Micromachining and Micropackaging of Transducers*, eds. C.D. Fung, P.W. Cheung, W.H. Ko, and D.G. Fleming, pp. 215–223, Elsevier, Amsterdam, the Netherlands.

Gyr, A., and Bewersdorff, H.-W. (1995) *Drag Reduction of Turbulent Flows by Additives*, Kluwer, Dordrecht, the Netherlands.

Haight, C.H., Reed, T.D., and Morland, B.T. (1974) "Design Studies of Transonic and STOL Airfoils with Active Diffusion Control," Advanced Technology Center Report No. ATC-B-94300/4CR-24, Dallas, Texas. (Also available from U.S. NTIS; Document Number AD-A011928/9.)

Haile, E., Delabroy, O., Durox, D., Lacas, F., and Candel, S. (1998) "Combustion Enhancement by Active Control," in *Flow Control: Fundamentals and Practices*, eds. M. Gad-el-Hak, A. Pollard, and J.-P. Bonnet, pp. 467–499, Springer–Verlag, Berlin.

Haile, J.M. (1993) *Molecular Dynamics Simulation: Elementary Methods*, Wiley, New York.

Halsey, T.C., and Martin, J.E. (1993) "Electrorheological Fluids," *Scientific American* **269**, October, pp. 58–64.

Hama, F.R. (1947) "The Turbulent Boundary Layer Along a Flat Plate, Parts I and II," *Reports of the Institute of Science and Technology* **1**, pp. 13–16 and 49–50, University of Tokyo, Tokyo, Japan.

Hama, F.R. (1954) "Boundary-Layer Characteristics for Smooth and Rough Surfaces," *Trans. Soc. Nav. Archt. Marine Engrs.* **62**, pp. 333–358.

Hancock, P.E. (1980) "The Effect of Free-Stream Turbulence on Turbulent Boundary Layers," Ph.D. Dissertation, Imperial College, London, Great Britain.

Hancock, P.E., and Bradshaw, P. (1983) "The Effect of Free-Stream Turbulence on Turbulent Boundary Layers," *J. Fluids Eng.* **105**, pp. 284–289.

Hanratty, T.J. (1990) "A Conceptual Model of the Viscous Wall Region," in *Near-Wall Turbulence: 1988 Zoran Zarić Memorial Conference*, eds. S.J. Kline and N.H. Afgan, pp. 81–103, Hemisphere, New York.

Hansen, R.J., and Hunston, D.L. (1974) "An Experimental Study of Turbulent Flows over Compliant Surfaces," *J. Sound Vibr.* **34**, pp. 297–308.

Harder, K.J., and Tiederman, W.G. (1991) "Drag Reduction and Turbulent Structure in Two-Dimensional Channel Flows," *Phil. Trans. R. Soc. London A* **336**, pp. 19–34.

Haritonidis, J.H., Senturia, S.D., Warkentin, D.J., and Mehregany, M. (1990a) "Optical Micropressure Transducer," United States Patent number 4,926,696.

Haritonidis, J.H., Senturia, S.D., Warkentin, D.J., and Mehregany, M. (1990b) "Pressure Transducer Apparatus," United States Patent number 4,942,767.

Harley, J.C., Huang, Y., Bau, H.H., and Zemel, J.N. (1995) "Gas Flow in Micro-Channels," *J. Fluid Mech.* **284**, pp. 257–274.

Harris, C.D., and Bartlett, D.W. (1972) "Wind-Tunnel Investigation of Effects of Underwing Leading Edge Vortex Generators on a Supercritical-Wing Research Airplane Configuration," NASA TMX-2471, Washington, D.C.

Harvey, J.K., and Perry, F.J. (1971) "Flowfield Produced by Trailing Vortices in the Vicinity of the Ground," *AIAA J.* **9**, pp. 1659–1660.

Harvey, W.D. (1986) "Low-Reynolds Number Aerodynamics Research at NASA Langley Research Center," in *Proc. Int. Conf. on Aerodynamics at Low Reynolds Numbers*, vol. II, pp. 19.1–19.49, Royal Aeronautical Society, London, Great Britain.

Harvey, W.D., Bushnell, D.M., and Beckwith, I.E. (1969) "On the Fluctuating Properties of Turbulent Boundary Layers for Mach Numbers up to 9.0," NASA TND-5496, Washington, D.C.

Hayakawa, L., and Squire, L.C. (1982) "The Effect of the Upstream Boundary Layer State on the Shock Interaction at a Compression Corner," *J. Fluid Mech.* **122**, pp. 369–394.

Hayes, S., Grebogi, C., and Ott, E. (1994) "Communicating with Chaos," *Phys. Rev. Lett.* **70**, pp. 3031–3040.

Hayes, S., Grebogi, C., Ott, E., and Mark, A. (1994) "Experimental Control of Chaos for Communication," *Phys. Rev. Lett.* **73**, pp. 1781–1784.

Hazen, D.C. (1984) "Aeronautics Technology Possibilities for 2000: Report of a Workshop." Aeronautics and Space Engineering Board, Commission on Engineering and Technical Systems, U.S. National Research Council, Washington, D.C.

Head, M.R., and Bandyopadhyay, P.R. (1978) "Combined Flow Visualization and Hot-Wire Measurements," in *Coherent Structure of Turbulent Boundary Layers*, eds. C.R. Smith and D.E. Abbott, pp. 98–129, Lehigh University, Bethlehm, Pennsylvania.

Head, M.R., and Bandyopadhyay, P.R. (1981) "New Aspects of Turbulent Boundary-Layer Structure," *J. Fluid Mech.* **107**, pp. 297–338.

Heckl, M. (1988) "The Use of Mathematical Methods in Noise Control Design," in *Noise Control Design: Methods and Practice (Noise-Con 88)*, ed. J.S. Bolton, pp. 27–38, Noise Control Foundation, Poughkeepsie, New York.

Hefner, J.N. (1988) "Dragging Down Fuel Costs," *Aerospace America* **26**, January, pp. 14–16.

Hefner, J.N., and Sabo, F.E., editors (1987) "Research in Natural Laminar Flow and Laminar Flow Control, Part 1," NASA CP-2487, Washington, D.C.

Hefner, J.N., Anders, J.B., and Bushnell, D.M. (1983) "Alteration of Outer Flow Structures for Turbulent Drag Reduction," AIAA Paper No. 83-0293, New York.

Hefner, J.N., Weinstein, L.M., and Bushnell, D.M. (1980) "Large-Eddy Breakup Scheme for Turbulent Viscous Drag Reduction," in *Viscous Flow Drag Reduction*, ed. G.R. Hough, pp. 110–127, AIAA, New York.

Hendricks, E.W., and Ladd, D.M. (1983) "Effect of Surface Roughness on the Delayed Transition on 9:1 Heated Ellipsoid," *AIAA J.* **21**, pp. 1406–1409.

Hendricks, E.W., Lawler, J.V., Horne, M.P., Handler, R.A., and Swearingen, J.D. (1989) "Experiments in Drag-Reducing Polymer Flows," in *Advances in Fluid Mechanics Measurements*, ed. M. Gad-el-Hak, pp. 535–568, Springer–Verlag, New York.

Henry, J.R., Wood, C.C., and Wilbur, S.W. (1956) "Summary of Subsonic Diffuser Data," NACA RML-56F05, Washington, D.C.

Hess, D.E. (1990) "An Experimental Investigation of a Compliant Surface Beneath a Turbulent Boundary Layer," Ph.D. Dissertation, The Johns Hopkins University, Baltimore, Maryland.

Hess, D.E., Peattie, R.A., and Schwarz, W.H. (1993) "A Nonintrusive Method for the Measurement of Flow-Induced Surface Displacement of a Compliant Surface," *Exp. Fluids* **14**, pp. 78–84.

Hinch, F.J. (1977) "Mechanical Models of Dilute Polymer Solutions in Strong Flows," *Phys. Fluids* **20**, no. 10, part II, pp. S22–S30.

Hinze, J.O. (1975) *Turbulence*, second edition, McGraw–Hill, New York.

Hirschfelder, J.O., Curtiss, C.F., and Bird, R.B. (1954) *Molecular Theory of Gases and Liquids*, Wiley, New York.

Ho, C.-M. (1986) "An Alternative Look at the Unsteady Separation Phenomenon," in *Recent Advances in Aerodynamics*, eds. A. Krothappalli and C.A. Smith, pp.165–178, Springer–Verlag, New York.

Ho, C.-M., and Gutmark, E.J. (1987) "Vortex Induction and Mass Entrainment in a Small Aspect-Ratio Elliptic Jet," *J. Fluid Mech.* **179**, pp. 385–405.

Ho, C.-M., and Huang, L.-S. (1982) "Subharmonics and Vortex Merging in Mixing Layers," *J. Fluid Mech.* **119**, pp. 443–473.

Ho, C.-M., and Huerre, P. (1984) "Perturbed Free Shear Layers," *Ann. Rev. Fluid Mech.* **16**, pp. 365–424.

Ho, C.-M., and Tai, Y.-C. (1996) "Review: MEMS and Its Applications for Flow Control," *J. Fluids Eng.* **118**, pp. 437–447.

Ho, C.-M., and Tai, Y.-C. (1998) "Micro-Elctro-Mechanical Systems (MEMS) and Fluid Flows," *Ann. Rev. Fluid Mech.* **30**, pp. 579–612.

Ho, C.-M., Tung, S., Lee, G.B., Tai, Y.-C., Jiang, F., and Tsao, T. (1997) "MEMS—A Technology for Advancements in Aerospace Engineering," AIAA Paper No. 97-0545, Reston, Virgina.

Hoffmann, J.A. (1981) "Effects of Free Stream Turbulence on Diffuser Performance," *J. Fluids Eng.* **103**, pp. 385–390.

Hoffmann, J.A., Kassir, S.M., and Larwood, S.M. (1988) "The Influence of Free Stream Turbulence on Turbulent Boundary Layers with Mild Adverse Pressure Gradients," NASA CR-184677, Washington, D.C.

Hogan, H. (1996) "Invasion of the Micromachines," *New Scientist* **29**, June, pp. 28–33.

Holland, J.H. (1992) *Adaptation in Natural and Artificial Systems*, MIT Press, Cambridge, Massachusetts.

Holmes, B.J. (1988) "NLF Technology is Ready to Go," *Aerospace America* **26**, January, pp. 16–20.

Holmes, P., Lumley, J.L., and Berkooz, G. (1996) *Turbulence, Coherent Structures, Dynamical Systems and Symmetry*, Cambridge University Press, Cambridge, Great Britain.

Holstein, H. (1940) "Messungen zur Laminarhaltung der Grenzschicht an einem Flügel," *Lilienthal-Bericht* S10, pp. 17–27.

Hooshmand, A., Youngs, R., Wallace, J.M., and Balint, J.-L. (1983) "An Experimental Study of Changes in the Structure of a Turbulent Boundary Layer Due to Surface Geometry Changes," AIAA Paper No. 83-0230, New York.

Hopkins, E.J., Keener, E.R., and Polek, T.E. (1972) "Hypersonic Turbulent Skin Friction and Boundary Layer Profiles on Nonadiabatic Flat Plates," *AIAA J.* **10**, pp. 40–48.

Horstmann, K.-H., and Quast, A. (1981) "Widerstandsverminderung durch Blasturbulatoren," DFVLR Report No. FB-81-33, Braunschweig, West Germany.

Hough, G.R., editor (1980) *Viscous Flow Drag Reduction*, AIAA, New York.

Howard, F.G., and Goodman, W.L. (1985) "Axisymmetric Bluff-Body Drag Reduction Through Geometrical Modification," *J. Aircraft* **22**, pp. 516–522.

Howard, F.G., and Goodman, W.L. (1987) "Drag Reduction on a Bluff Body at Yaw Angles to 30 Degrees," *J. Spacecraft & Rockets* **24**, pp. 179–181.

Howard, F.G., Hefner, J.N., and Srokowski, A.J. (1975) "Multiple Slot Skin Friction Reduction," *J. Aircraft* **12**, pp. 753–754.

Howard, F.G., Quass, B.F., Weinstein, L.M., and Bushnell, D.M. (1981) "Longitudinal Afterbody Grooves and Shoulder Radiusing for Low Speed Bluff Body Drag Reduction," ASME Paper No. 81-WA/FE-5, New York.

Howarth, L. (1938) "On the Solution of the Laminar Boundary Layer Equations," *Proc. R. Soc. London A* **164**, pp. 547–579.

Hoyt, J.W. (1972) "Turbulent Flow of Drag-Reducing Suspensions," Report No. TP 299, Naval Undersea Center, San Diego, California.

Hoyt, J.W. (1979) "Polymer Drag Reduction—A Literature Review," in *Second Int. Conf. on Drag Reduction*, Paper No. A1, BHRA Fluid Engineering, Cranfield, Great Britain.

Hsiao, F.-B., Liu, C.-F., and Shyu, J.-Y. (1990) "Control of Wall-Separated Flow by Internal Acoustic Excitation," *AIAA J.* **28**, pp. 1440–1446.

Hu, H.H., and Bau, H.H. (1994) "Feedback Control to Delay or Advance Linear Loss of Stability in Planar Poiseuille Flow," *Proc. Roy. Soc. London A* **447**, pp. 299–312.

Huang, L.S., Maestrello, L., and Bryant, T.D. (1987) "Separation Control Over an Airfoil at High Angles of Attack by Sound Emanating from the Surface," AIAA Paper No. 87-1261, New York.

Huang, P.G., Bradshaw, P., and Coakley, T.J. (1993) "Skin Friction and Velocity Profile Family for Compressible Turbulent Boundary Layers," *AIAA J.* **31**, pp. 1600–1604.

Huberman, B. (1990) "The Control of Chaos," in *Proc. Workshop on Applications of Chaos*, 4–7 December, San Francisco, California.

Huberman, B.A., and Lumer, E. (1990) "Dynamics of Adaptive Systems," *IEEE Trans. Circuits Syst.* **37**, pp. 547–550.

Hübler, A., and Lüscher, E. (1989) "Resonant Stimulation and Control of Nonlinear Oscillators," *Naturwissenschaften* **76**, pp. 67–69.

Huerre, P., and Monkewitz, P.A. (1990) "Local and Global Instabilities in Spatially Developing Flows, *Ann. Rev. Fluid Mech.* **22**, pp. 473–537.

Huffman, G.D., and Bradshaw, P. (1972) "A Note on von Kármán's Constant in Low Reynolds Number Turbulent Flows," *J. Fluid Mech.* **53**, pp. 45–60.

Hunt, J.C.R., Carruthers, D.J., and Fung, J.C.H. (1991) "Rapid Distortion Theory as a Means of Exploring the Structure of Turbulence," in *New Perspectives in Turbulence*, ed. L. Sirovich, pp. 55–103, Springer–Verlag, Berlin.

Hunter, P.A., and Johnson, H.I. (1954) "A Flight Investigation of the Practical Problems Associated with Porous Leading-Edge Suction," NACA TN-3062, Washington, D.C.

Hurley, D.G. (1961) "The Use of Boundary Layer Control to Establish Free Streamline Flows," in *Boundary Layer and Flow Control*, ed. G.V. Lachmann, vol. 1, pp. 295–341, Pergamon Press, Oxford, Great Britain.

Husain, H.S., and Hussain, A.K.M.F. (1983) "Controlled Excitation of Elliptic Jets," *Phys. Fluids* **26**, pp. 2763–2766.

Husain, H.S., and Hussain, F. (1991) "Elliptic Jets. Part 2. Dynamics of Coherent Structures: Pairing," *J. Fluids Mech.* **233**, pp. 439–482.

Husain, H.S., and Hussain, F. (1993) "Elliptic Jets. Part 3. Dynamics of Preferred Mode Coherent Structures," *J. Fluids Mech.* **248**, pp. 315–361.

Hussain, A.K.M.F. (1983) "Coherent Structures—Reality and Myth," *Phys. Fluids* **26**, pp. 2816–2850.

Hussain, A.K.M.F. (1986) "Coherent Structures and Turbulence," *J. Fluid Mech.* **173**, pp. 303–356.

Hussain, A.K.M.F., and Hasan, M.A.Z. (1985) "Turbulence Suppression in Free Turbulent Shear Flows Under Controlled Excitation. Part 2. Jet-Noise Reduction," *J. Fluid Mech.* **150**, pp. 159–168.

Hussain, A.K.M.F., and Reynolds, W.C. (1975) "Measurements in Fully Developed Turbulent Channel Flow," *J. Fluids Eng.* **97**, pp. 568–580.

Hussain, F., and Husain, H.S. (1989) "Elliptic Jets. Part 1. Characteristics of Unexcited and Excited Jets," *J. Fluids Mech.* **208**, pp. 257–320.

Hussaini, M.Y., editor (1997) *Collected Papers of Sir James Lighthill*, vols. I–IV, Oxford University Press, Oxford, Great Britain.

Huyer, S.A., Robinson, M.C., and Luttges, M.W. (1990) "Unsteady Aerodynamic Loading Produced by a Sinusoidally Oscillating Delta Wing," AIAA Paper No. 90-1536, New York.

Iglisch, R. (1944) "Exakte Berechnung der laminaren Reibungsschicht an der längsangeströmten ebenen Platte mit homogener Absaugung," *Schr. Dtsh. Akad., Luftfahrtforschung* **8B**, pp. 1–51.

Illingworth, C.R. (1954) "The Effect of Heat Transfer on the Separation of a Compressible Laminar Boundary Layer," *Quart. J. Mech. Appl. Math.* **7**, pp. 8–34.

Inger, G.R., and Siebersma, T. (1988) "Computational Simulation of Vortex Generator Effects on Transonic Shock/Boundary Layer Interaction," AIAA Paper No. 88-2590, New York.

Isomoto, K., and Honami, S. (1989) "The Effect of Inlet Turbulence Intensity on the Reattachment Process over a Backward-Facing Step," *J. Fluids Eng.* **111**, pp. 87–92.

Israelachvili, J.N. (1986) "Measurement of the Viscosity of Liquids in Very Thin Films," *J. Colloid Interface Sci.* **110**, pp. 263–271.

Israelachvili, J.N. (1991) *Intermolecular and Surface Forces*, second edition, Academic Press, New York.

Itoh, N. (1987) "Another Route to the Three-Dimensional Development of Tollmien–Schlichting Waves with Finite Amplitude," *J. Fluid Mech.* **181**, pp. 1–16.

Izakson, A. (1937) "Formula for the Velocity Distribution Near a Wall," *Zh. Eksper. Teor. Fiz.* **7**, pp. 919–924.

Jacobs, J., James, R., Ratliff, C., and Glazer, A. (1993) "Turbulent Jets Induced by Surface Actuators," AIAA Paper No. 93-3243, Washington, D.C.

Jacobson, S.A., and Reynolds, W.C. (1993a) "Active Control of Boundary Layer Wall Shear Stress Using Self-Learning Neural Networks," AIAA Paper No. 93-3272, AIAA, Washington, D.C.

Jacobson, S.A., and Reynolds W.C. (1993b) "Active Boundary Layer Control Using Flush-Mounted Surface Actuators," *Bull. Am. Phys. Soc.* **38**, p. 2197.

Jacobson, S.A., and Reynolds, W.C. (1994) "Active Control of Transition and Drag in Boundary Layers," *Bull. Am. Phy. Soc.* **39**, p. 1894.

Jacobson, S.A., and Reynolds, W.C. (1995) "An Experimental Investigation Towards the Active Control of Turbulent Boundary Layers," Department of Mechanical Engineering Report No. TF-64, Stanford University, Stanford, California.

Jacobson, S.A., and Reynolds, W.C. (1998) "Active Control of Streamwise Vortices and Streaks in Boundary Layers," *J. Fluid Mech.* **360**, pp. 179–211.

Jaffe, N.A., Okamura, T.T., and Smith, A.M.O. (1970) "Determination of Spatial Amplification Factors and Their Application to Predicting Transition," *AIAA J.* **8**, pp. 301–308.

James, R.D., Jacobs, J.W., and Glezer, A. (1994) "Experimental Investigation of a Turbulent Jet Produced by an Oscillating Surface Actuator," *Appl. Mech. Rev.* **47**, no. 6, part 2, pp. S127–S1131.

Jang, J.-S.R., Sun, C.-T., and Mizutani, E. (1997) *Neuro-Fuzzy and Soft Computing*, Prentice–Hall, Upper Saddle River, New Jersey.

Jeffery, E. (1940) "Nothing Left to Invent," *J. Patent Office Society* **22**, pp. 479–481.

Jeffery, G.B. (1920) "Plane Stress and Plane Strain in Bipolar Co-ordinates," *Phil. Trans. R. Soc. Ser. A* **221**, pp. 265–289.

Johansen, J.B., and Smith, C.R. (1986) "The Effects of Cylindrical Surface Modifications on Turbulent Boundary Layers," *AIAA J.* **24**, pp. 1081–1087.

Johansson, A.V., and Alfredsson, P.H. (1982) "On the Structure of Turbulent Channel Flow," *J. Fluid Mech.* **122**, pp. 295–314.

Johansson, A.V., and Alfredsson, P.H. (1983) "Effects of Imperfect Spatial Resolution on Measurements of Wall-Bounded Turbulent Shear Flows," *J. Fluid Mech.* **137**, pp. 409–421.

Johansson, A.V., Her, J.-Y., and Haritonidis, J.H. (1987) "On the Generation of High-Amplitude Wall-Pressure Peaks in Turbulent Boundary Layers and Spots," *J. Fluid Mech.* **175**, pp. 119–142.

Johari, H., and McManus, K.R. (1997) "Visualization of Pulsed Vortex Generator Jets for Active Control of Boundary Layer Separation," AIAA Paper No. 97-2021, Washington, D.C.

Johnson, R.P. (1980) "Review of Compliant Coating Research of M.O. Kramer, RPJ Associates Technical Report No. RPJA-TR-0955-001, Palos Verdes Peninsula, California.

Johnson, W.S., Tennant, J.S., and Stamps, R.E. (1975) "Leading-Edge Rotating Cylinder for Boundary Layer Control on Lifting Surfaces," *J. Hydronaut* **9**, pp. 76–78.

Johnston, J., and Nishi, M. (1989) "Vortex Generator Jets—A Means for Passive and Active Control of Boundary Layer Separation," AIAA Paper No. 89-0564, New York.

Johnston, J., and Nishi, M. (1990) "Vortex Generator Jets—Means for Flow Separation Control," *AIAA J.* **28**, pp. 989–994.

Jones, B.M. (1934) "Stalling," *J. R. Aero. Soc.* **38**, pp. 753–770.

Joslin, R.D., Erlebacher, G., Hussaini, M.Y. (1996) "Active Control of Instabilities in Laminar Boundary Layers—Overview and Concept Validation," *J. Fluids Eng.* **118**, pp. 494–497.

Joslin, R.D. (1998) "Aircraft Laminar Flow Control," *Ann. Rev. Fluid Mech.* **30**, pp. 1–29.

Joslin, R.D., and Morris, P.J. (1992) "Effect of Compliant Walls on Secondary Instabilities in Boundary-Layer Transition," *AIAA J.* **30**, pp. 332–339.

Jung, W.J., Mangiavacchi, N., and Akhavan, R. (1992) "Suppression of Turbulence in Wall Bounded Flows by High-Frequency Spanwise Oscillations," *Phys. Fluids A* **8**, pp. 1605–1607.

Juvet, P., and Reynolds, W.C. (1989) "Entrainment Control in an Acoustically Controlled Shrouded Jet," AIAA Paper No. 89-0969, New York.

Kachanov, Y.S., Koslov, V.V., and Levchenko, Ya.V. (1974) "Experimental Study of the Influence of Cooling on the Stability of Laminar Boundary Layers," *Izvestia Sibirskogo Otdielenia Ak. Nauk SSSR, Seria Technicheskikh Nauk, Novosibirsk*, no. 8-2, pp. 75–79.

Kailasnath, P. (1993) "Reynolds Number Effects and the Momentum Flux in Turbulent Boundary Layers," Ph.D. Thesis, Yale University, New Haven, Connecticut.

Kamal, M.M. (1966) "Separation in the Flow Between Eccentric Rotating Cylinders," *Trans. ASME Ser. D* **88**, pp. 717–724.

Kandil, O.A., and Chuang, H.A. (1988) "Unsteady Vortex-Dominated Flows Around Maneuvering Wings Over a Wide Range of Mach Numbers," AIAA Paper No. 88-0317, New York.

Kandil, O.A., and Chuang, H.A. (1990a) "Unsteady Navier–Stokes Computations Past Oscillating Delta Wing at High Incidence," *AIAA J.* **28**, pp. 1565–1572.

Kandil, O.A., and Chuang, H.A. (1990b) "Computation of Vortex-Dominated Flow for a Delta Wing Undergoing Pitching Oscillation," *AIAA J.* **28**, pp. 1589–1595.

Kannberg, L.D. (1988) "The Urgency Will Return," *Mechanical Engineering* **110**, p. 33.

Kaplan, R.E. (1964) "The Stability of Laminar Incompressible Boundary Layers in the Presence of Compliant Boundaries," Sc.D. Thesis, Massachusetts Institute of Technology, Cambridge, Massachusetts.

Karim, M.A., and Acharya, M. (1994) "Control of the Dynamic-Stall Vortex Over a Pitching Airfoil by Leading-Edge Suction," *AIAA J.* **32**, pp. 1647–1655.

Karlsson, R.I., and Johansson, T.G. (1986) "LDV Measurements of Higher Order Moments of Velocity Fluctuations in a Turbulent Boundary Layer," in *Proc. Third Int. Symp. on Applications of Laser Anemometry to Fluid Mechanics*, eds. D.F.G. Durão et al., paper no. 12.1, July 7–9, Lisbon, Portugal.

Karniadakis, G. Em (1999) "Simulating Turbulence in Complex Geometries," *Fluid Dyn. Res.* **24**, pp. 343–362.

Karniadakis, G. Em, and Orszag, S.A. (1993) "Nodes, Modes and Flow Codes," *Physics Today* **46**, no. 3, March, pp. 34–42.

Kastrinakis, E.G., and Eckelmann, H. (1983) "Measurement of Streamwise Vorticity Fluctuations in a Turbulent Channel Flow," *J. Fluid Mech.* **137**, pp. 165–186.

Katz, Y., Nishri, B., and Wygnanski, I. (1989a) "The Delay of Turbulent Boundary Layer Separation by Oscillatory Active Control," AIAA Paper No. 89-0975, New York.

Katz, Y., Nishri, B., and Wygnanski, I. (1989b) "The Delay of Turbulent Boundary Layer Separation by Oscillatory Active Control," *Phys. Fluids* **1**, pp. 179–181.

Kawthar-Ali, M.H., and Acharya, M. (1996) "Artificial Neural Networks for Suppression of the Dynamic-Stall Vortex Over Pitching Airfoils," AIAA Paper No. 96-0540, Washington, D.C.

Kays, W.M., and Crawford, M.E. (1993) *Convective Heat and Mass Transfer*, third edition, McGraw–Hill, New York.

Keefe, L.R. (1993a) "Two Nonlinear Control Schemes Contrasted in a Hydrodynamic Model," *Phys. Fluids A* **5**, pp. 931–947.

Keefe, L.R. (1993b) "Drag Reduction in Channel Flow Using Nonlinear Control," AIAA Paper No. 93-3279, Washington, D.C.

Keefe, L.R. (1996) "A MEMS-Based Normal Vorticity Actuator for Near-Wall Modification of Turbulent Shear Flows," in *Proc. Workshop on Flow Control: Fundamentals and Practices*, eds. J.-P. Bonnet, M. Gad-el-Hak, and A. Pollard, pp. 1–21, 1–5 July, Institut d'Etudes Scientifiques des Cargèse, Corsica, France.

Keefe, L.R., Moin, P., and Kim, J. (1992) "The Dimension of Attractors Underlying Periodic Turbulent Poiseuille Flow," *J. Fluid Mech.* **242**, pp. 1–29.

Keith, W.L., Hurdis, D.A., and Abraham, B.M. (1992) "A Comparison of Turbulent Boundary Layer Wall-Pressure Spectra," *J. Fluids Eng.* **114**, pp. 338–347.

Kendall, J.M. (1970) "The Turbulent Boundary Layer over a Wall with Progressive Surface Waves," *J. Fluid Mech.* **41**, pp. 259–281.

Kennard, E.H. (1938) *Kinetic Theory of Gases*, McGraw–Hill, New York.

Kentfield, J.A.C. (1985a) "Drag Reduction of Controlled Separated Flows," AIAA Paper No. 85-1800, New York.

Kentfield, J.A.C. (1985b) "Short, Multi-Step, Afterbody Fairings," *J. Aircraft* **21**, pp. 351–352.

Kidd, J.A., Wikoff, D., and Cottrell, C.J. (1990) "Drag Reduction by Controlling Flow Separation Using Stepped Afterbodies," *J. Aircraft* **27**, pp. 564–566.

Kim, H.T., Kline, S.J., and Reynolds, W.C. (1971) "The Production of Turbulence Near a Smooth Wall in a Turbulent Boundary Layer," *J. Fluid Mech.* **50**, pp. 133–160.

Kim, J., and Moin, P. (1986) "The Structure of the Vorticity Field in Turbulent Channel Flow. Part 2. Study of Ensemble-Averaged Fields," *J. Fluid Mech.* **162**, pp. 339–363.

Kim, J., Moin, P., and Moser, R.D. (1987) "Turbulence Statistics in Fully-Developed Channel Flow at Low Reynolds Number," *J. Fluid Mech.* **177**, pp. 133–166.

Kim, J.H., and Stringer, J., editors (1992) *Applied Chaos*, Wiley, New York.

Kind, R.J. (1967) "A Proposed Method of Circulation Control," Ph.D. Dissertation, Cambridge University, Cambridge, Great Britain.

Kirsch, J.W. (1974) "Drag Reduction of Trucks with S^3 Air Vane," in *Proc. Second Symp. on Aerodynamics of Sports and Competition Automobiles*, pp. 251–263, AIAA, New York.

Klebanoff, P.S. (1954) "Characteristics of Turbulence in a Boundary Layer with Zero Pressure Gradient," NACA Report No. R-1247, Washington, D.C.

Klebanoff, P.S., and Diehl, Z.W. (1952) "Some Features of Artificially Thickened Fully Developed Turbulent Boundary Layers with Zero Pressure Gradient," NACA Report No. 1110, Washington, D.C.

Klebanoff, P.S., Schubauer, G.B., and Tidstrom, K.D. (1955) "Measurements of the Effect of Two-Dimensional and Three-Dimensional Roughness Elements on Boundary-Layer Transition," *J. Aero. Sci.* **22**, pp. 803–804.

Klebanoff, P.S., Tidstrom, K.D., and Sargent, L.M. (1962) "The Three-Dimensional Nature of Boundary Layer Instability," *J. Fluid Mech.* **12**, pp. 1–34.

Klewicki, J.C., Murray, J.A., and Falco, R.E. (1994) "Vortical Motion Contributions to Stress Transport in Turbulent Boundary Layers," *Phys. Fluids* **6**, pp. 277–286.

Kline, S.J. (1967) "Observed Structure Features in Turbulent and Transitional Boundary Layers," in *Fluid Mechanics of Internal Flow*, ed. G. Sovran, pp. 27–79, Elsevier, Amsterdam, The Netherlands.

Kline, S.J., and Afgan, N.H. (1990) *Near-Wall Turbulence: 1988 Zoran Zarić Memorial Conference*, Hemisphere, New York.

Kline, S.J., and Runstadler, P.W. (1959) "Some Preliminary Results of Visual Studies of the Flow Model of the Wall Layers of the Turbulent Boundary Layer," *J. Appl. Mech.* **26**, pp. 166–170.

Kline, S.J., Reynolds, W.C., Schraub, F.A., and Runstadler, P.W. (1967) "The Structure of Turbulent Boundary Layers," *J. Fluid Mech.* **30**, pp. 741–773.

Knight, J. (1999a) "Cunning Plumbing," *New Scientist* **161**, no. 2172, 6 February, pp. 32–37.

Knight, J. (1999b) "Dust Mite's Dilemma," *New Scientist* **162**, no. 2180, 29 May, pp. 40–43.

Knudsen, M. (1909) "Die Gesetze der Molekularströmung und der inneren Reibungsströmung der Gase durch Röhren," *Annalen der Physik* **28**, pp. 75–130.

Koga, D.J., Reisenthel, P., and Nagib, H.M. (1984) "Control of Separated Flowfields Using Forced Unsteadiness," Fluids & Heat Transfer Report No. R84-1, Illinois Institute of Technology, Chicago, Illinois.

Kogan, M.N. (1969) *Rarefied Gas Dynamics*, Nauka, Moscow. Translated from Russian, ed. L. Trilling, Plenum, New York.

Kogan, M.N. (1973) "Molecular Gas Dynamics," *Ann. Rev. Fluid Mech.* **5**, pp. 383–404.

Kolmogorov, A.N. (1941a) "The Local Structure of Turbulence in Incompressible Viscous Fluid for Very Large Reynolds Number," *Dokl. Akad. Nauk SSSR* **30**, pp. 301–305. (Reprinted in *Proc. R. Soc. London A* **434**, pp. 9–13, 1991.)

Kolmogorov, A.N. (1941b) "On Degeneration of Isotropic Turbulence in an Incompressible Viscous Liquid," *Dokl. Akad. Nauk SSSR* **31**, pp. 538–540.

Kolmogorov, A.N. (1941c) "Dissipation of Energy in Locally Isotropic Turbulence," *Dokl. Akad. Nauk SSSR* **32**, pp. 16–18. (Reprinted in *Proc. R. Soc. London A* **434**, pp. 15–17, 1991.)

Kolmogorov, A.N. (1962) "A Refinement of Previous Hypothesis Concerning the Local Structure of Turbulence in a Viscous Incompressible Fluid at High Reynolds Number," *J. Fluid Mech.* **13**, pp. 82–85.

Koplik, J., and Banavar, J.R. (1995) "Continuum Deductions from Molecular Hydroynamics," *Ann. Rev. Fluid Mech.* **27**, pp. 257–292.

Korkan, K.D., Cross, E.J., and Cornell, C.C. (1986) "Experimental Aerodynamic Characteristics of an NACA 0012 Airfoil with Simulated Ice," *J. Aircraft* **22**, pp. 130–134.

Kosecoff, M.A., Ko, D.R.S., and Merkle, C.L. (1976) "An Analytical Study of the Effect of Surface Roughness on the Stability of a Heated Water Boundary Layer," Final Report PDT 76-131, Dynamics Technology, Inc., Torrance, California.

Koshland, D.E., Jr. (1995) "The Crystal Ball and the Trumpet Call," *Science* **267**, 17 March, p. 1575.

Koskie, J.E., and Tiederman, W.G. (1991) "Turbulence Structure and Polymer Drag Reduction in Adverse Pressure Gradient Boundary Layers," School of Mechanical Engineering Report No. PME-FM-91-3, Purdue University, West Lafayette, Indiana.

Kostelich, E.J., Grebogi, C., Ott, E., and Yorke, J.A. (1993a) "Targeting from Time Series," *Bul. Am. Phys. Soc.* **38**, p. 2194.

Kostelich, E.J., Grebogi, C., Ott, E., and Yorke, J.A. (1993b) "Higher-Dimensional Targeting," *Phys. Rev. E* **47**, pp. 305–310.

Koval'nogov, S.A., Fomin, V.M., and Shapovalov, G.K. (1987) "Experimental Study of the Possibility of Passive Control of Shock-Boundary Layer Interactions," *Uchemye kZapiski TSAGI* **18**, pp. 112–116.

Kovasznay, L.S.G. (1970) "The Turbulent Boundary Layer," *Ann. Rev. Fluid Mech.* **2**, pp. 95–112.

Kovasznay, L.S.G., Kibens, V., and Blackwelder, R.F. (1970) "Large-Scale Motion in the Intermittent Region of a Turbulent Boundary Layer," *J. Fluid Mech.* **41**, pp. 283–325.

Krall, K.M., and Haight, C.H. (1972) "Wind Tunnel Tests of a Trapped Vortex-High Lift Airfoil," Advanced Technology Center Report No. ATC-B-94300/3TR-10, Dallas, Texas. (Also available from U.S. NTIS; Document Number AD-762 077.)

Kramer, M.O. (1957) "Boundary-Layer Stabilization by Distributed Damping," *J. Aeronaut. Sci.* **24**, pp. 459–460.

Kramer, M.O. (1960a) "Boundary-Layer Stabilization by Distributed Damping," *J. Aero/Space Sci.* **27**, p. 69.

Kramer, M.O. (1960b) "Boundary Layer Stabilization by Distributing Damping," *J. Am. Soc. Nav. Engrs.* **72**, pp. 25–33.

Kramer, M.O. (1960c) "The Dolphin's Secret," *New Scientist* **7**, 5 May, pp. 1118–1120.

Kramer, M.O. (1961) "The Dolphin's Secret," *J. Am. Soc. Nav. Engrs.* **73**, pp. 103–107.

Kramer, M.O. (1962) "Boundary Layer Stabilization by Distributed Damping," *J. Am. Soc. Nav. Engrs.* **74**, pp. 341–348.

Kramer, M.O. (1965) "Hydrodynamics of the Dolphin," in *Advances in Hydroscience*, ed. V.T. Chow, vol. 2, pp. 111–130, Academic Press, New York.

Kramer, M.O. (1969) "Die Widerstandsverminderung schneller Unterwasserkörper Mittels künstlicher Delphinhaut," in *Jahrbuch 1969 der Deutschen Gesellschaft für Luft- und Raumfahrt*, eds. H. Blenk and W. Schulz, pp. 1–9, Cologne, Federal Republic of Germany.

Kravchenko, A.G., Choi, H., and Moin, P. (1993) "On the Relation of Near-Wall Streamwise Vortices to Wall Skin Friction in Turbulent Boundary Layers," *Phys. Fluids A* 5, pp. 3307–3309.

Kreplin, H.-P., and Eckelmann, H. (1979) "Behaviour of the Three Fluctuating Velocity Components in the Wall Region of a Turbulent Channel Flow," *Phys. Fluids* 22, pp. 1233–1239.

Kudva, A.K., and Sesonske, A. (1972) "Structure of Turbulent Velocity and Temperature Fields in Ethylene Glycol Pipe Flow at Low Reynolds Number," *Int. J. Heat Mass Transfer* 15, pp. 127–145.

Kuethe, A.M. (1973) "Boundary Layer Control of Flow Separation and Heat Exchange," U.S. Patent No. 3,741,285.

Kukainis, J. (1969) "Effects of Three-Dimensional Boundary Layer Control Devices on a Quasi-Two-Dimensional Swept Wing at High Subsonic Speeds," Arnold Engineering Development Center Report No. AEDC-TR-69-251, Arnold Air Forc Base, Tennessee.

Kundu, P.K. (1990) *Fluid Mechanics*, Academic Press, New York.

Kutateladze, S.S., and Khabakhpasheva, E.M. (1978) "Structure of Wall Boundary Layer (Forced Flow, Thermal Convection)," Institute of Thermodynamics, Siberian Branch of the USSR Academy of Science, Novosibirsk, USSR.

Kwong, A., and Dowling, A. (1993) "Active Boundary Layer Control in Diffusers," AIAA Paper No. 93-3255, Washington, D.C.

Laadhari, F., Skandaji, L., and Morel, R. (1994) "Turbulence Reduction in a Boundary Layer by a Local Spanwise Oscillating Surface," *Phys. Fluids* 6, pp. 3218–3220.

Lachmann, G.V., editor (1961) *Boundary Layer and Flow Control*, vols. 1 and 2, Pergamon Press, Oxford, Great Britain.

Ladd, D.M., and Hendricks, E.W. (1988) "Active Control of 2-D Instability Waves on an Axisymmetric Body," *Exp. Fluids* 6, pp. 69–70.

Lai, Y.-C., and Grebogi, C. (1993) "Synchronization of Chaotic Trajectories Using Control," *Phys. Rev. E* 47, pp. 2357–2360.

Lai, Y.-C., Deng, M., and Grebogi, C. (1993) "Controlling Hamiltonian Chaos," *Phys. Rev. E* 47, pp. 86–92.

Lai, Y.-C., Grebogi, C., and Tél, T. (1994) "Controlling Transient Chaos in Dynamical Systems," in *Towards the Harnessing of Chaos*, ed. M. Yamaguchi, Elsevier, Amsterdam, The Netherlands.

Lai, Y.-C., Tél, T., and Grebogi, C. (1993) "Stabilizing Chaotic-Scattering Trajectories Using Control," *Phys. Rev. E* 48, pp. 709–717.

Lakshmanan, B., Tiwari, S.N., and Hussaini, M.Y. (1988) "Control of Supersonic Interaction Flow Fields Through Filleting and Sweep," AIAA Paper No. 88-3534, New York.

Lamb, H. (1895) *Hydrodynamics*, second edition, Cambridge University Press, Cambridge, Great Britain.

Landahl, M.T. (1962) "On the Stability of a Laminar Incompressible Boundary Layer Over a Flexible Surface," *J. Fluid Mech.* 13, pp. 609–632.

Landahl, M.T. (1967) "A Wave-Guide Model for Turbulent Shear Flow," *J. Fluid Mech.* 29, pp. 441–459.

Landahl, M.T. (1972) "Wave Mechanics of Breakdown," *J. Fluid Mech.* 56, pp. 775–802.

Landahl, M.T. (1973) "Drag Reduction by Polymer Addition," in *Proc. Thirteenth IUTAM Congress*, eds. E. Becker and G.K. Mikhailov, pp. 177–199, Springer–Verlag, Berlin.

Landahl, M.T. (1977) "Dynamics of Boundary Layer Turbulence and the Mechanism of Drag Reduction," *Phys. Fluids* 20, no. 10, part II, pp. S55–S63.

Landahl, M.T. (1980) "A Note on an Algebraic Instability of Inviscid Parallel Shear Flows," *J. Fluid Mech.* 98, pp. 243–251.

Landahl, M.T. (1990) "On Sublayer Streaks," *J. Fluid Mech.* 212, pp. 593–614.

Landau, L.D., and Lifshitz, E.M. (1987) *Fluid Mechanics*, second edition, Pergamon Press, Oxford, Great Britain.

Landweber, L. (1953) "The Frictional Resistance of Flat Plates at Zero Pressure Gradient, "*Trans. Soc. Nav. Arch. Mar. Eng.* 61, pp. 5–32.

Landweber, L., and Siao, T.T. (1958) "Comparison of Two Analyses of Boundary-Layer Data on a Flat Plate," *J. Ship Research* **1**, pp. 21–33.

Lang, T.G. (1963) "Porpoise, Whales, and Fish: Comparison of Predicted and Observed Speeds," *Nav. Eng. J.* **75**, pp. 437–441.

Lange, R.H. (1954) "Present Status of Information Relative to the Prediction of Shock-Induced Boundary Layer Separation," NACA TN-3065, Washington, D.C.

Lankford, J.L. (1960) "Investigation of the Flow over an Axisymmetric Compression Surface at High Mach Numbers," Report No. 6866, U.S. Naval Ordnance Laboratory, Corona, California.

Lankford, J.L. (1961) "The Effect of Heat Transfer on the Separation of Laminar Flow Over Axisymmetric Compression Surfaces. Preliminary Results at Mach No. 6.78," Report No. 7402, U.S. Naval Ordnance Laboratory, Corona, California.

Lauchle, G.C., and Daniels, M.A. (1987) "Wall-Pressure Fluctuations in Turbulent Pipe Flow," *Phys. Fluids* **30**, pp. 3019–3024.

Lauchle, G.C., and Gurney, G.B. (1984) "Laminar Boundary-Layer Transition on a Heated Underwater Body," *J. Fluid Mech.* **144**, pp. 79–101.

Laufer, J. (1951) "Investigation of Turbulent Flow in a Two-Dimensional Channel," NACA Report No. R-1053, Washington, D.C.

Laufer, J. (1954) "The Structure of Turbulence in Fully Developed Pipe Flow," NACA Report No. R-1174, Washington, D.C.

Laufer, J. (1975) "New Trends in Experimental Turbulence Research," *Ann. Rev. Fluid Mech.* **7**, pp. 307–326.

Launder, B.E., and Spalding, D.B. (1974) "The Numerical Computation of Turbulent Flows," *Comput. Methods Appl. Mech. Eng.* **3**, pp. 269–289.

Lee, C.S., Tavella, D.A., Wood, N.J., and Roberts, L. (1986) "Flow Structure of Lateral Wing-Tip Blowing," AIAA Paper No. 86-1810, New York.

Lee, C.S., Tavella, D.A., Wood, N.J., and Roberts, L. (1989) "Flow Structure and Scaling Laws in Lateral Wing-Tip Blowing," *AIAA J.* **27**, pp. 1002–1007.

Lee, D.G. (1974) "Subsonic Force Characteristics of a Low Aspect Ratio Wing Incorporating a Spinning Cylinder," DTNSRDC Report No. ASED-329, Bethesda, Maryland. (Also available from U.S. NTIS; Document Number AD-AOO11135.)

Lee, M., and Ho, C.-M. (1989) "Vortex Dynamics of Delta Wings," in *Frontiers in Experimental Fluid Mechanics*, ed. M. Gad-el-Hak, pp. 365–428, Springer–Verlag, New York.

Lee, M., and Ho, C.-M. (1990) "Lift Force of Delta Wings," *Appl. Mech. Rev.* **43**, pp. 209–221.

Lee, M., and Reynolds, W.C. (1985) "Bifurcating and Blooming Jets," Thermosciences Division Report No. TR-22, Department of Mechanical Engineering, Stanford University, Stanford, California.

Lee, R.E., Yanta, W.J., and Leonas, A.C. (1969) "Velocity Profile, Skin Friction Balance and Heat Transfer Measurements of the Turbulent Boundary Layer at M 5 and Zero Pressure Gradient," National Ordnance Laboratory Report No. NOL-TR69-106, White Oak, Maryland.

Lee, T., Fisher, M., and Schwarz, W.H. (1993a) "Investigation of the Stable Interaction of a Passive Compliant Surface with a Turbulent Boundary Layer," *J. Fluid Mech.* **257**, pp. 373–401.

Lee, T., Fisher, M., and Schwarz, W.H. (1993b) "The Measurement of Flow-Induced Surface Displacement on a Compliant Surface by Optical Holographic Interferometry," *Exp. Fluids* **14**, pp. 159–168.

Lee, T., Fisher, M., and Schwarz, W.H. (1995) "Investigation of the Effects of a Compliant Surface on Boundary-Layer Stability," *J. Fluid Mech.* **288**, pp. 37–58.

Lee, T., Fisher, M., and Schwarz, W.H. (1997) "An Experimental Study of the Boundary-Layer Nonlinear Instability Over Compliant Walls," unpublished manuscript originally submitted to *J. Fluid Mech.*

Lee, W.K., Vaseleski, R.C., and Metzner, A.B. (1974) "Turbulent Drag Reduction in Polymeric Solutions Containing Suspended Fibers," *AIChE J.* **20**, pp. 128–133.

Lees, L. (1947) "The Stability of the Laminar Boundary Layer in a Compressible Fluid," NACA Report No. 876, Washington, D.C.

Legner, H.H. (1984) "A Simple Model for Gas Bubble Drag Reduction," *Phys. Fluids* **27**, pp. 2788–2790.

Lerner, E.J. (1999) "Next-Genration Lithography," *Industrial Physicist* **5**, no. 3, June, pp. 18–21.

Lewkowicz, A.K. (1982) "An Improved Universal Wake Function for Turbulent Boundary Layers and Some of Its Consequences," Z. Flugwiss. Weltraumforsch. 6, pp. 261–266.

Liandrat, J., Aupoix, B., and Cousteix, T. (1986) "Calculation of Longitudinal Vortices Embedded in a Turbulent Boundary Layer," in Fifth Symposium on Turbulent Shear Flows, eds. F. Durst, B.E. Launder, F.W. Schmidt, and J.H. Whitelaw, pp. 7.17–7.22, Springer–Verlag, New York.

Liang, W.J., and Liou, J.A. (1995) "Flow Around a Rotating Cylinder Near a Plane Boundary," J. Chinese Institute of Engineers 18, pp. 35–50.

Libby, P.A. (1954) "Method for Calculation of Compressible Laminar Boundary Layer with Axial Pressure Gradient and Heat Transfer," NACA TN No. 3157, Washington, D.C.

Liebeck, R.H. (1978) "Design of Subsonic Airfoils for High Lift," J. Aircraft 15, pp. 547–561.

Liebeck, R.H., and Camacho, P.P. (1985) "Airfoil Design at Low Reynolds Number with Constrained Pitching Moment," in Proc. Conf. on Low Reynolds Number Airfoil Aerodynamics, ed. T.J. Mueller, pp. 27–51, University of Notre Dame, Notre Dame, Indiana.

Liepmann, H.W. (1952) "Aspects of the Turbulence Problem. Part II," Z. Angew. Math. Phys. 3, pp. 407–426.

Liepmann, H.W. (1962) "Free Turbulent Flows," Mécanique de la Turbulence, Int. Symp. Nat. Sci. Res. Centre, pp. 211–227, Marseille 1961, CNRS, Paris, France.

Liepmann, H.W. (1979) "The Rise and Fall of Ideas in Turbulence," American Scientist 67, no. 2, pp. 221–228.

Liepmann, H.W. (1997) "Boundary Layer Transition: The Early Days," Appl. Mech. Rev. 50, no. 2, pp. R1–R4.

Liepmann, H.W., and Fila, G.H. (1947) "Investigations of Effects of Surface Temperature and Single Roughness Elements on Boundary Layer Transition," NACA Report No. 890, Washington, D.C.

Liepmann, H.W., Brown, G.L., and Nosenchuck, D.M. (1982) "Control of Laminar Instability Waves Using a New Technique," J. Fluid Mech. 118, pp. 187–200.

Liepmann, H.W., and Nosenchuck, D.M. (1982) "Active Control of Laminar–Turbulent Transition," J. Fluid Mech. 118, pp. 201–204.

Lighthill, M.J. (1952) "On Sound Generated Aerodynamically—I. General Theory," Proc. Roy. Soc. London A 211, pp. 564–587.

Lighthill, M.J. (1954) "On Sound Generated Aerodynamically—II. Turbulence as a Source of Sound," Proc. Roy. Soc. London A 222, pp. 1–32.

Lighthill, M.J. (1962) "The Bekerian Lecture, 1961—Sound Generated Aerodynamically," Proc. Roy. Soc. London A 267, pp. 147–182.

Lighthill, M. J. (1963a) "Introduction. Boundary Layer Theory," in Laminar Boundary Layers, ed. L. Rosenhead, pp. 46–113, Clarendon Press, Oxford, Great Britain.

Lighthill, M.J. (1963b) "Introduction. Real and Ideal Fluids," in Laminar Boundary Layers, ed. L. Rosenhead, pp. 1–45, Clarendon Press, Oxford, Great Britain.

Lighthill, M.J. (1973) "On the Weis–Fogh Mechanism of Lift Generation," J. Fluid Mech. 60, pp. 1–17.

Lighthill, M.J. (1975) "Aerodynamic Aspects of Animal Flight," in Swimming and Flying in Nature, eds. T.Y. Wu, C.J. Brokaw, and C. Brennen, vol. 2, pp. 423–491, Plenum, New York.

Ligrani, P.M., and Bradshaw, P. (1987) "Spatial Resolution and Measurements of Turbulence in the Viscous Sublayer Using Subminiature Hot-Wire Probes," Exp. Fluids 5, pp. 407–417.

Lin, C.C. (1945) "On the Stability of Two-Dimensional Parallel Flows," Parts I, II and III, Q. Appl. Maths. 3, pp. 117–142, 218–234, 277–301.

Lin, J.C., and Ash, R.L. (1986) "Wall Temperature Control of Low-Speed Body Drag," J. Aircraft 23, pp. 93–94.

Lin, J.C., and Howard, F.G. (1989) "Turbulent Flow Separation Control Through Passive Techniques," AIAA Paper No. 89-0976, New York.

Lin, J.C., Howard, F.G., Bushnell, D.M., and Selby, G.V. (1990) "Investigation of Several Passive and Active Methods for Turbulent Flow Separation Control," AIAA Paper No. 90-1598, New York.

Lin, J.C., Howard, F.G., and Selby, G.V. (1990) "Control of Turbulent Separated Flow over a Rearward-Facing Ramp Using Longitudinal Grooves," J. Aircraft 27, pp. 283–285.

Lin, J.C., Robinson, S.K., McGhee, R.J., and Valarezo, W.O. (1992) "Separation Control on High Reynolds Number Multi-Element Airfoils," AIAA Paper No. 92-2636, Washington, D.C.

Lin, J.C., Weinstein, L.M., Watson, R.D., and Balasubramanian, R. (1983) "Turbulent Drag Characteristic of Small Amplitude Rigid Surface Waves," AIAA Paper No. 83-0228, New York.

Lindner, J.F., and Ditto, W.L. (1995) "Removal, Suppression and Control of Chaos by Nonlinear Design," *Appl. Mech. Rev.* **48**, pp. 795–808.

Linke, W. (1942) "Über den Strömungswiderstand einer beheizten ebenen Platte," *Luftfahrtforschung* **19**, pp. 157–160.

Lipkin, R. (1993) "Micro Steam Engine Makes Forceful Debut," *Science News* **144**, September, p. 197.

Lissaman, P.B.S. (1983) "Low-Reynolds-Number Airfoils," *Ann. Rev. Fluid Mech.* **15**, pp. 223–239.

Lissaman, P.B.S., and Harris, G.L. (1969) "Turbulent Skin Friction on Compliant Surfaces," AIAA Paper No. 69-164, New York.

Liu, C.K., Kline, S.J., and Johnston, J.P. (1966) "Experimental Study of Turbulent Boundary Layer on Rough Walls," Department of Mechanical Engineering Report No. MD-15, Stanford University, Stanford, California.

Liu, J., Tai, Y.C., Lee, J., Pong, K.C., Zohar, Y., and Ho, C.M. (1993) "In-Situ Monitoring and Universal Modeling of Sacrificial PSG Etching Using Hydrofluoric Acid," in *Proc. IEEE Micro Electro Mechanical Systems '93*, pp. 71–76, IEEE, New York.

Liu, J., Tai, Y.C., Pong, K., and Ho, C.M. (1995) "MEMS for Pressure Distribution Studies of Gaseous Flows in Microchannels," in *Proc. IEEE Micro Electro Mechanical Systems '95*, pp. 209–215, IEEE, New York.

Liu, J.T.C. (1988) "Contributions to the Understanding of Large-Scale Coherent Structures in Developing Free Turbulent Shear Flows," in *Advances in Applied Mechanics*, eds. J.W. Hutchinson and T.Y. Wu, vol. 26, pp. 183–309, Academic Press, Boston, Massachusetts.

Loeb, L.B. (1961) *The Kinetic Theory of Gases*, third edition, Dover, New York.

Löfdahl, L., and Gad-el-Hak, M. (1999) "MEMS Applications in Turbulence and Flow Control," *Prog. Aero. Sci.* **35**, pp. 101–203.

Löfdahl, L., Glavmo, M., Johansson, B., and Stemme, G. (1993) "A Silicon Transducer for the Determination of Wall-Pressure Fluctuations in Turbulent Boundary Layers," *Appl. Scientific Res.* **51**, pp. 203–207.

Löfdahl, L., Kälvesten, E., and Stemme, G. (1994) "Small Silicon Based Pressure Transducers for Measurements in Turbulent Boundary Layers," *Exp. Fluids* **17**, pp. 24–31.

Löfdahl, L., Kälvesten, E., and Stemme, G. (1996) "Small Silicon Pressure Transducers for Space-Time Correlation Measurements in a Flat Plate Boundary Layer," *J. Fluids Eng.* **118**, pp. 457–463.

Löfdahl, L., Stemme, G., and Johansson, B. (1989) "A Sensor Based on Silicon Technology for Turbulence Measurements," *J. Phys. E. Sci. Instum.* **22**, pp. 391–393.

Löfdahl, L., Stemme, G., and Johansson, B. (1991) "Reynolds Stress Measurements Using Direction Sensitive Double-Chip Silicon Sensors," *Meas. Sci. Technol.* **2**, pp. 369–373.

Löfdahl, L., Stemme, G., and Johansson, B. (1992) "Silicon Based Flow Sensors Used for Mean Velocity and Turbulence Measurements," *Exp. Fluids* **12**, pp. 270–276.

Long, R.R., and Chen, T.-C. (1981) "Experimental Evidence for the Existence of the 'Mesolayer' in Turbulent Systems," *J. Fluid Mech.* **105**, pp. 19–59.

Looney, R.W., and Blick, E.F. (1966) "Skin Friction Coefficients of Compliant Surfaces in Turbulent Flow," *J. Spacecraft & Rockets* **3**, pp. 1562–1564.

Loose, W., and Hess, S. (1989) "Rheology of Dense Fluids via Nonequilibrium Molecular Hydrodynamics: Shear Thinning and Ordering Transition," *Rheologica Acta* **28**, pp. 91–101.

Lowell, R.L., and Reshotko, E. (1974) "Numerical Study of the Stability of a Heated Water Boundary Layer," Case Western University, Report No. FTAS/TR-73-93, Cleveland, Ohio.

Lucey, A.D., and Carpenter, P.W. (1993) "The Hydroelastic Stability of Three-Dimensional Disturbances of a Finite Compliant Panel," *J. Sound Vibr.* **165**, pp. 527–552.

Lucey, A.D., and Carpenter, P.W. (1995) "Boundary Layer Instability over Compliant Walls: Comparison between Theory and Experiment," *Phys. Fluids* **7**, pp. 2355–2363.

Luchik, T.S., and Tiederman, W.G. (1986) "Effect of Spanwise Probe Volume Length on Laser Velocimeter Measurments in Wall Bounded Turbulent Flows," *Exp. Fluids* **3**, pp. 339–341.

Ludwig, G.R. (1964) "An Experimental Investigation of Laminar Separation from a Moving Wall," AIAA Paper No. 64-6, New York.

Lumley, J.L. (1969) "Drag Reduction by Additives," *Ann. Rev. Fluid Mech.* **1**, pp. 367–384.

Lumley, J.L. (1973) "Drag Reduction in Turbulent Flow by Polymer Additives," *J. Polym. Sci.: Macromol. Rev.* 7, pp. 263–290.

Lumley, J.L. (1977) "Drag Reduction in Two Phase and Polymer Flows," *Phys. Fluids* 20, no. 10, part II, pp. S64–S71.

Lumley, J.L. (1978a) "Two-Phase and Non-Newtonian Flows," in *Topics in Applied Physics*, vol. 12, ed. P. Bradshaw, second edition, pp. 289–324, Springer–Verlag, Berlin.

Lumley, J.L. (1978b) "Computational Modeling of Turbulent Flows," *Adv. Appl. Mech.* 18, pp. 123–176.

Lumley J.L. (1981) "Coherent Structures in Turbulence," in *Transition and Turbulence*, ed. R.E. Mcyer, pp. 215–242, Academic Press, New York.

Lumley, J.L. (1983) "Turbulence Modeling," *J. Appl. Mech.* 50, pp. 1097–1103.

Lumley, J.L. (1987) "Turbulence Modeling," in *Proc. Tenth U.S. National Cong. of Applied Mechanics*, ed. J.P. Lamb, pp. 33–39, ASME, New York.

Lumley, J.L. (1991a) "Order and Disorder in Turbulent Flows," in *New Perspectives in Turbulence*, ed. L. Sirovich, pp. 105–122, Springer–Verlag, Berlin.

Lumley, J.L. (1991b) "Control of the Wall Region of a Turbulent Boundary Layer," in *Turbulence: Structure and Control*, ed. J.M. McMichael, pp. 61–62, 1–3 April, Ohio State University, Columbus, Ohio.

Lumley, J.L. (1992) "Some Comments on Turbulence," *Phys. Fluids A* 4, pp. 203–211.

Lumley, J.L. (1996a) "Turbulence and Turbulence Modeling," in *Research Trends in Fluid Dynamics*, eds. J.L. Lumley, A. Acrivos, L.G. Leal, and S. Leibovich, pp. 167–177, American Institute of Physics, Woodbury, New York.

Lumley, J.L. (1996b) "Control of Turbulence," AIAA Paper No. 96-0001, Washington, D.C.

Lumley, J.L., Acrivos, A., Leal, L.G., and Leibovich, S. (1996) *Research Trends in Fluid Dynamics*, American Institute of Physics, Woodbury, New York.

Lumley, J., and Blossey, P. (1998) "Control of Turbulence," *Ann. Rev. Fluid Mech.* 30, pp. 311–327.

Lumley, J.L., and Kubo, I. (1984) "Turbulent Drag Reduction by Polymer Additives: A Survey," Sibley School of Mechanical and Aerospace Engineering Report No. FDA-84-07, Cornell University, Ithaca, New York.

Lüscher, E., and Hübler, A. (1989) "Resonant Stimulation of Complex Systems," *Helv. Phys. Acta* 62, pp. 544–551.

Luttges, M.W. (1989) "Accomplished Insect Fliers," in *Frontiers in Experimental Fluid Mechanics*, ed. M. Gad-el-Hak, pp. 429–456, Springer–Verlag, New York.

Luttges, M.W., Somps, C., Kliss, M., and Robinson, M. (1984) "Unsteady Separated Flows: Generation and Use by Insects," in *Unsteady Separated Flows*, eds. M.S. Francis and M.W. Luttges, pp. 127–1136, U.S. Air Force Academy, Colorado Springs, Colorado.

Mabey, D.G. (1979) "Influence of the Wake Component on Turbulent Skin Friction at Subsonic and Supersonic Speeds," *Aeronaut. Quart.* 30, pp. 590–606.

Mabey, D.G. (1988) "Design Features Which Influence Flow Separations on Aircraft," *Aero. J.* 92, pp. 409–415.

Mabey, D.G., Meier, H.U., and Sawyer, W.G. (1976) "Some Boundary Layer Measurements on a Flat Plate at Mach Numbers from 2.5 to 4.5," Royal Aeronautical Establishment Technical Report No. RAE-74127, London, Great Britain.

MacCurdy, E. (1938) *The Notebooks of Leonardo da Vinci*, vols. I and II, Reynal & Hitchcock, New York.

Macha, J.M., Norton, D.J., and Young, J.C. (1972) "Surface Temperature Effect on Subsonic Stall," AIAA Paper No. 72-960, New York.

Madavan, N.K., Deutsch, S., and Merkle, C.L. (1984) "Reduction of Turbulent Skin Friction by Microbubbles," *Phys. Fluids* 27, pp. 356–363.

Madavan, N.K., Deutsch, S., and Merkle, C.L. (1985) "Measurements of Local Skin Friction in a Microbubble-Modified Turbulent Boundary Layer," *J. Fluid Mech.* 156, pp. 237–256.

Maestrello, L., Badavi, F.F., and Noonan, K.W. (1988) "Control of the Boundary Layer Separation About an Airfoil by Active Surface Heating," AIAA Paper No. 88-3545-CP, New York.

Magarvey, R.H., and McLatchy, C.S. (1964) "The Disintegration of Vortex Rings," *Can. J. Phys.* 42, pp. 684–689.

Magnus, G. (1852) "On the Deflection of a Projectile," Abhandlung der Akademie der Wissenschaften, Berlin, Germany.

Majumdar, A., and Mezic, I. (1998) "Stability Regimes of Thin Liquid Films," *Microscale Thermophys. Eng.* **2**, pp. 203–213.

Majumdar, A., and Mezic, I. (1999) "Instability of Ultra-Thin Water Films and the Mechanism of Droplet Formation on Hydrophilic Surfaces," in *Proc. ASME-JSME Thermal Engineering and Solar Energy Joint Conference*, San Diego, California, 15–19 March. Also to appear in *J. Heat Transfer*.

Malik, M.R., Weinstein, L.M., and Hussaini, M.Y. (1983) "Ion Wind Drag Reduction," AIAA Paper No. 83-0231, New York.

Malkus, W.V.R. (1956) "Outline of a Theory of Turbulent Shear Flow," *J. Fluid Mech.* **1**, pp. 521–539.

Malkus, W.V.R. (1979) "Turbulent Velocity Profiles from Stability Criteria," *J. Fluid Mech.* **90**, pp. 401–414.

Maltby, R.L. (1962) "Flow Visualization in Wind Tunnels Using Indicators," AGARDograph No. 70, NATO Advisory Group for Aerospace Research and Development, Rhode-Saint-Génèse, Belgium.

Mangalam, S.M., Bar-Sever, A., Zaman, K.B.M.Q., and Harvey, W.D. (1986) "Transition and Separation Control on a Low-Reynolds Number Airfoil," in *Proc. Int. Conf. on Aerodynamics at Low Reynolds Numbers*, vol. I, pp. 10.1–10.19, Royal Aeronautical Society, London, Great Britain.

Mansour, N.N., Kim, J., and Moin, P. (1988) "Reynolds-Stress and Dissipation-Rate Budgets in a Turbulent Channel Flow," *J. Fluid Mech.* **194**, pp. 15–44.

Marchman, J.F., Manor, D., and Plentovich, E.B. (1980) "Performance Improvement of Delta Wings at Subsonic Speeds Due to Vortex Flaps," AIAA Paper No. 80-1802, New York.

Markoff, J. (1999) "Tiniest Circuits Hold Prospect of Explosive Computer Speeds," *The New York Times* **CXLVIII**, no. 51,585, 16 July, p. A1 and p. C17.

Maskell, E.C. (1955) "Flow Separation in Three Dimensions," RAE Report Aero. 2565, Royal Aircraft Establishment, Farnborough, Great Britain.

Mastrangelo, C. (1993) "Integration, Partition, and Reliability of Microelectromechanical Systems," invited oral presentation at *AIAA Third Flow Control Conference*, 6–9 July, Orlando, Florida.

Mastrangelo, C., and Hsu, C.H. (1992) "A Simple Experimental Technique for the Measurement of the Work of Adhesion of Microstructures," in *Technical Digest IEEE Solid-State Sensors and Actuators Workshop*, pp. 208–212, IEEE, New York.

Maureau, J., Sharatchandra, M.C., Sen, M., and Gad-el-Hak, M. (1997) "Flow and Load Characteristics of Microbearings with Slip," *J. Micromech. Microeng.* **7**, pp. 55–64.

Maxwell, J.C. (1879) "On Stresses in Rarefied Gases Arising from Inequalities of Temperature," *Phil. Trans. R. Soc. Part 1* **170**, pp. 231–256.

Maxworthy, T. (1979) "Experiments on the Weis–Fogh Mechanism of Lift Generation by Insects in Hovering Flight. Part 1. Dynamics of the 'Fling'," *J. Fluid Mech.* **93**, pp. 47–63.

Maxworthy, T. (1981) "The Fluid Dynamics of Insect Flight," *Ann. Rev. Fluid Mech.* **13**, pp. 329–350.

May, M. (1999) "I'm Just Flying Down to the Supermarket," *New Scientist* **162**, no. 2188, 29 May, pp. 24–27.

McComb, W.D., and Chan, K.T.J. (1979) "Drag Reduction in Fibre Suspensions: Transitional Behavior Due to Fibre Degradation," *Nature* **280**, pp. 45–46.

McComb, W.D., and Chan, K.T.J. (1981) "Drag Reduction in Fibre Suspension," *Nature* **292**, pp. 520–522.

McComb, W.D., and Rabie, L.H. (1979) "Development of Local Turbulent Drag Reduction Due to Nonuniform Polymer Concentration," *Phys. Fluids* **22**, pp. 183–185.

McCormick, M.E., and Bhattacharyya, R. (1973) "Drag Reduction of a Submersible Hull by Electrolysis," *Nav. Eng. J.* **85**, pp. 11–16.

McCroskey, W.J. (1977) "Some Current Research in Unsteady Fluid Dynamics," *J. Fluids Eng.* **99**, pp. 8–39.

McCroskey, W.J. (1982) "Unsteady Airfoils," *Ann. Rev. Fluid Mech.* **14**, pp. 285–311.

McCullough, G.B. (1955) "The Effect of Reynolds Number on the Stalling Characteristics and Pressure Distributions of Four Moderately Thin Airfoil Sections," NACA TN-3524, Washington, D.C.

McCullough, G.B., and Gault, D.E. (1951) "Examples of Three Representative Types of Airfoil-Section Stall at Low Speed," NACA TN No. 2502, Washington, D.C.

McCune, J.E., and Kerrebrock, J.L. (1973) "Noise from Aircraft Turbomachinery," *Ann. Rev. Fluid Mech.* 5, pp. 281–300.

McInville, R.M., Hassan, H.A., and Goodman, W.L. (1985) "Mixing Layer Control for Tangential Slot Injection in Turbulent Flows," AIAA Paper No. 85-0541, New York.

McLachlan, B.G. (1989) "Study of a Circulation Control Airfoil with Leading/Trailing-Edge Blowing, *J. Aircraft* 26, pp. 817–821.

McLean, J.D., and Herring, H.J. (1974) "Use of Multiple Discrete Wall Jets for Delaying Boundary Layer Separation," NASA CR-2389, Washington, D.C.

McManus, K., and Magill, J. (1996) "Separation Control in Incompressible and Compressible Flows Using Pulsed Jets," AIAA Paper No. 96-1948, Washington, D.C.

McManus, K., and Magill, J. (1997) "Airfoil Performance Enhancement Using Pulsed Jet Separation Control," AIAA Paper No. 97-1971, Washington, D.C.

McManus, K., Ducharme, A., Goldey, C., and Magill, J. (1996) "Pulsed Jet Actuators for Suppressing Flow Separation," AIAA Paper No. 96-0442, Washington, D.C.

McManus, K.R., Legner, H.H., and Davis, S.J. (1994) "Pulsed Vortex Generator Jets for Active Control of Flow Separation," AIAA Paper No. 94-2218, Washington, D.C.

McMasters, J.H., and Henderson, M.I. (1980) "Low Speed Single Element Airfoil Synthesis," *Tech. Soaring* 6, pp. 1–21.

McMichael, J.M. (1993) "MEMS and Challenges of Flow Control," invited oral presentation at *AIAA Third Flow Control Conference*, 6–9 July, Orlando, Florida.

McMichael, J.M. (1996) "Progress and Prospects for Active Flow Control Using Microfabricated Electromechanical Systems (MEMS)," AIAA Paper No. 96-0306, Washington, D.C.

McMichael, J.M., Klebanoff, P.S., and Meese, N.E. (1980) "Experimental Investigation of Drag on a Compliant Surface," in *Viscous Flow Drag Reduction*, ed. G.R. Hough, pp. 410–438, AIAA, New York.

McMurray, J.T., Metcalfe, R.W., and Riley, J.J. (1983) "Direct Numerical Simulations of Active Stabilization of Boundary Layer Flows," in *Proc. Eighth Biennial Symp. on Turbulence*, eds. J.L. Zakin and G.K. Patterson, paper no. 36, University of Missouri, Rolla, Missouri.

Mehregany, M. (1993) "Overview of Microelectromechanical Systems," invited oral presentation at *AIAA Third Flow Control Conference*, 6–9 July, Orlando, Florida.

Mehregany, M., DeAnna, R.G., and Reshotko, E. (1996) "Microelectromechanical Systems for Aerodynamics Applications," AIAA Paper No. 96-0421, Washington, D.C.

Mehta, R.D. (1985a) "Aerodynamics of Sports Balls," *Ann. Rev. Fluid Mech.* 17, pp. 151–189.

Mehta, R.D. (1985b) "Effect of a Longitudinal Vortex on a Separated Turbulent Boundary Layer," AIAA Paper No. 85-0530, New York.

Mehta, R.D. (1988) "Vortex/Separated Boundary-Layer Interactions at Transonic Mach Numbers," *AIAA J.* 26, pp. 15–26.

Meng, J.C.S., editor (1998) *Proceedings of the International Symposium on Sea Water Drag Reduction*, 22–23 July, Naval Undersea Warfare Center, Newport, Rhode Island.

Merkle, C.L., and Deutsch, S. (1985) "Drag Reduction by Microbubbles: Current Research Status," AIAA Paper No. 85-0537, New York.

Merkle, C.L., and Deutsch, S. (1989) "Microbubble Drag Reduction," in *Frontiers in Experimental Fluid Mechanics*, ed. M. Gad-el-Hak, pp. 291–336, Springer–Verlag, New York.

Metcalfe, R.W. (1994) "Boundary Layer Control: A Brief Review," in *Computational Fluid Dynamics '94*, eds. J. Periaux and E. Hirschel, pp. 52–60, Wiley, New York.

Metcalfe, R.W., Battistoni, F., Ekeroot, J., and Orszag, S.A. (1991) "Evolution of Boundary Layer Flow over a Compliant Wall During Transition to Turbulence," in *Proc. Boundary Layer Transition and Control Conference*, pp. 36.1–36.14, Royal Aeronautical Society, Cambridge, Great Britain.

Metcalfe, R.W., Rutland, C.J., Duncan, J.H., and Riley, J.J. (1986) "Numerical Simulations of Active Stabilization of Laminar Boundary Layers," *AIAA J.* 24, pp. 1494–1501.

Miau, J.J., Chen, M.H., and Chow, J.H. (1988) "Flow Structures of a Vertically Oscillating Plate Immersed in a Flat-Plate Turbulent Boundary Layer," in *Proc. Eleventh Symp. on Turbulence*, paper no. A28, University of Missouri, Rolla, Missouri.

Mickley, H.S., and Davis, R.S. (1957) "Momentum Transfer for Flow over a Flat Plate with Blowing," NACA Technical Note No. TN-4017, Washington, D.C.

Migay, V.K. (1960a) "Diffuser with Transverse Fins (English translation from Russian)," *Energomashinostroenie*, no. 4, p. 31.

Migay, V.K. (1960b) "On Improving the Effectiveness of Diffuser Flows with Separation (English translation from Russian)," *Mekhanika i Mashinostroyeniye*, no. 4, pp. 171–173.

Migay, V.K. (1961) "Increasing the Efficiency of Diffusers by Fitting Transverse Fins (English translation from Russian)," *Teploenergetika* , no. 4, pp. 41–43.

Migay, V.K. (1962a) "The Efficiency of a Cross-Ribbed Curvilinear Diffuser (English translation from Russian)," *Energomashinostroenie*, no. 1, pp. 45–46.

Migay, V.K. (1962b) "The Aerodynamic Effectiveness of a Discontinuous Surface (English translation from Russian)," *Inzhenerno-Fizicheskiy Zhurnal* 5, pp. 20–24.

Migay, V.K. (1962c) "Investigating Finned Diffusers: Effects of Geometry on Effectiveness of Finned Body Diffusers (English translation from Russian)," *Teploenergetika*, no. 10, pp. 55–59.

Migun, N.P., and Prokhorenko, P.P. (1987) "Measurement of the Viscosity of Polar Liquids in Microcapillaries," *Colloid J. of the USSR* 49, pp. 894–897.

Millikan, C.B. (1939) "A Critical Discussion of Turbulent Flows in Channels and Circular Tubes," in *Proc. Fifth Int. Cong. Appl. Mech.*, eds. J.P. Den Hartog and H. Peters, pp. 386–392, Wiley, New York.

Milling, R.W. (1981) "Tollmien–Schlichting Wave Cancellation," *Phys. Fluids* 24, pp. 979–981.

Milne-Thomson, L.M. (1968) *Theoretical Hydrodynamics*, fifth edition, Macmillan, London, Great Britain.

Modi, V.J., Fernando, M., and Yokomizo, T. (1990) "Drag Reduction of Bluff Bodies Through Moving Surface Boundary Layer Control," AIAA Paper No. 90-0298, New York.

Modi, V.J., Mokhtarian, F., Fernando, M., and Yokomizo, T. (1989) "Moving Surface Boundary Layer Control as Applied to 2-D Airfoils," AIAA Paper No. 89-0296, New York.

Modi, V.J., Sun, J.L.C., Akutsu, T., Lake, P., McMillan, K., Swinton, P.G., and Mullins, D. (1980) "Moving Surface Boundary Layer Control for Aircraft Operations at High Incidence," AIAA Paper No. 80-1621, New York.

Modi, V.J., Sun, J.L.C., Akutsu, T., Lake, P., McMillian, K., Swinton, P.G., and Mullins, D. (1981) "Moving Surface Boundary Layer Control for Aircraft Operation at High Incidence," *J. Aircraft* 18, pp. 963–968.

Moffatt, H.K. (1964) "Viscous and Resistive Eddies Near a Sharp Corner," *J. Fluid Mech.* 18, pp. 1–18.

Moin, P., and Bewley, T. (1994) "Feedback Control of Turbulence," *Appl. Mech. Rev.* 47, no. 6, part 2, pp. S3–S13.

Moin, P., and Kim, J. (1982) "Numerical Investigation of Turbulent Channel Flow," *J. Fluid Mech.* 118, pp. 341–377.

Moin, P., and Kim, J. (1985) "The Structure of the Vorticity Field in Turbulent Channel Flow. Part 1. Analysis of Instantaneous Fields and Statistical Correlations," *J. Fluid Mech.* 155, pp. 441–464.

Moin, P., and Mahesh, K. (1998) "Direct Numerical Simulation: A Tool in Turbulence Research," *Ann. Rev. Fluid Mech.* 30, pp. 539–578.

Mokhtarian, F., and Modi, V.J. (1988) "Fluid Dynamics of Airfoils with Moving Surface Boundary-Layer Control," *J. Aircraft* 25, pp. 163–169.

Mokhtarian, F., Modi, V.J., and Yokomizo, T. (1988a) "Effect of Moving Surfaces on the Airfoil Boundary Layer Control," AIAA Paper No. 88-4337-CP, New York.

Mokhtarian, F., Modi, V.J., and Yokomizo, T. (1988b) "Rotating Air Scoop as Airfoil Boundary-Layer Control," *J. Aircraft* 25, pp. 973–975.

Monin, A.S., and Yaglom, A.M. (1971) *Statistical Fluid Mechanics*, vol. I, MIT Press, Cambridge, Massachusetts.

Moon, F.C. (1992) *Chaotic and Fractal Dynamics—An Introduction for Applied Scientists and Engineers*, Wiley, New York.

Moore, F.K. (1958) "On the Separation of the Unsteady Laminar Boundary Layer," in *Boundary-Layer Research*, ed. H. Görtler, pp. 296–310, Springer–Verlag, Berlin.

Moran, M.J., and Shapiro, H.N. (1995) *Fundamentals of Engineering Thermodynamics*, third edition, Wiley, New York.

Morduchow, M., and Grape, R.G. (1955) "Separation, Stability, and Other Properties of

Compressible Laminar Boundary Layer with Pressure Gradient and Heat Transfer," NACA TN No. 3296, Washington, D.C.

Morkovin, M.V. (1969) "Critical Evaluation of Transition from Laminar to Turbulent Shear Layers with Emphasis on Hypersonically Traveling Bodies," Air Force Flight Dynamics Laboratory Report No. AFFDL-TR-68-149, Wright-Patterson AFB, Ohio.

Morkovin, M.V. (1984) "Bypass Transition to Turbulence and Research Desiderata," in *Transition in Turbines Symp.*, NASA CP-2386, Washington, D.C.

Morkovin, M.V. (1988) "Recent Insights into Instability and Transition to Turbulence in Open-Flow Systems," AIAA Paper No. 88-3675, New York.

Moroney, R.M., White, R.M., and Howe, R.T. (1991) "Ultrasonically Induced Microtransport," in *Proc. IEEE Micro Electro Mechanical Systems '91*, pp. 277–282, Nara, Japan, IEEE, New York.

Morrison, W.R.B. (1969) "Two-Dimensional Frequency-Wavenumber Spectra and Narrow-Band Shear Stress Correlations in Turbulent Pipe Flow," Ph.D. Thesis, University of Queensland, Queensland, Australia.

Morrison, W.R.B., Bullock, K.J., and Kronauer, R.E. (1971) "Experimental Evidence of Waves in the Sublayer," *J. Fluid Mech.* **47**, pp. 639–656.

Mueller, T.J. (1985a) "Low Reynolds Number Vehicles," AGARDograph No. 288, AGARD, NATO, Neuilly-sur-Seine, France.

Mueller, T.J., editor (1985b) *Proceedings of the Conference on Low Reynolds Number Airfoil Aerodynamics*, University of Notre Dame, Notre Dame, Indiana.

Mueller, T.J. (1985c) "The Influence of Laminar Separation and Transition on Low Reynolds Number Airfoil Hysteresis, *J. Aircraft* **22**, pp. 763–770.

Mueller, T.J., editor (1989) *Proceedings of the Conference on Low Reynolds Number Aerodynamics*, University of Notre Dame, Notre Dame, Indiana.

Mueller, T.J., and Burns, T.F. (1982) "Experimental Studies of the Eppler 61 Airfoil at Low Reynolds Numbers," AIAA Paper No. 82-0345, New York.

Muirhead, V.U., and Saltzman, E.G. (1979) "Reduction of Aerodynamic Drag and Fuel Consumption for Tractor-Trailer Vehicles," *J. Energy* **3**, pp. 279–284.

Muller, D.A., Sorsch, T., Moccio, S., Baumann, F.H., Evans-Lutterodt, K., and Timp, G. (1999) "The Electronic Structure at the Atomic Scale of Ultrathin Gate Oxides," *Nature* **399**, no. 6738, 24 June, pp. 758–761.

Mullin, T., Greated, C.A., and Grant, I. (1980) "Pulsating Flow over a Step," *Phys. Fluids* **23**, pp. 669–674.

Muntz, E.P. (1989) "Rarefied Gas Dynamics," *Ann. Rev. Fluid Mech.* **21**, pp. 387–417.

Murlis, J., Tsai, H.M., and Bradshaw, P. (1982) "The Structure of Turbulent Boundary Layers at Low Reynolds Numbers," *J. Fluid Mech.* **122**, pp. 13–56.

Nadolink, R.H., and Haigh, W.W. (1995) "Bibliography on Skin Friction Reduction with Polymers and Other Boundary-Layer Additives," *Appl. Mech. Rev.* **48**, pp. 351–459.

Nagamatsu, H.T., Dyer, R., and Ficarra, R.V. (1985) "Supercritical Airfoil Drag Reduction by Passive Shock Wave/Boundary Layer Control in the Mach Number Range .75 to .9," AIAA Paper No. 85-0207, New York.

Nagamatsu, H.T., Trilling, T.W., and Bossard, J.A. (1987) "Passive Drag Reduction on a Complete NACA 0012 Airfoil at Transonic Mach Numbers," AIAA Paper No. 87-1263, New York.

Nagano, Y., and Tagawa, M. (1990) "A Structural Turbulence Model for Triple Products of Velocity and Scalar," *J. Fluid Mech.* **215**, pp. 639–657.

Nagel, A.L., Alford, W.J., Jr., and Dugan, J.F. (1975) "Future Long-Range Transports—Prospects for Improved Fuel Efficiency," NASA Technical Memorandum No. X-72659, Washington, D.C.

Naguib, A.M. (1992) "Inner- and Outer-Layer Effects on the Dynamics of a Turbulent Boundary Layer," Ph.D. Thesis, Illinois Institute of Technology, Chicago, Illinois.

Naguib, A.M., and Wark, C.E. (1992) "An Investigation of Wall-Layer Dynamics Using a Combined Temporal Filtering and Correlation Techniques," *J. Fluid Mech.* **243**, pp. 541–560.

Nakagawa, S., Shoji, S., and Esashi, M. (1990) "A Micro-Chemical Analyzing System Integrated on Silicon Chip," in *Proc. IEEE: Micro Electro Mechanical Systems*, Napa Valley, California, IEEE 90CH2832-4, IEEE, New York.

Nakayama, W. (1986) "Thermal Management of Electronic Equipment," *App. Mech. Rev.* **39**, pp. 1847–1868.

Narasimha, R., and Kailas, S.V. (1986) "Energy Events in the Atmospheric Boundary Layer," Indian Institute of Science Report No. 86-AS-8, Bangalore, India.

Narasimha, R., and Kailas, S.V. (1987) "Energy Events in the Atmospheric Boundary Layer," in *Perspectives in Turbulent Studies*, eds. H.U. Meier and P. Bradshaw, pp. 188–222, Springer–Verlag, Berlin.

Narasimha, R., and Kailas, S.V. (1990) "Turbulent Bursts in the Atmosphere," *Atmospheric Environment* **24A**, pp. 1635–1645.

Narasimha, R., and Liepmann, H.W. (1988) "Introduction," in *Turbulence Management and Relaminarisation*, eds. H.W. Liepmann and R. Narasimha, pp. xii–xx, Springer–Verlag, Berlin.

Narasimha, R., and Ojha, S.K. (1967) "Effect of Longitudinal Surface Curvature on Boundary Layers," *J. Fluid Mech.* **29**, pp. 187–199.

Narasimha, R., and Sreenivasan, K.R. (1973) "Relaminarization in Highly Accelerated Turbulent Boundary Layers," *J. Fluid Mech.* **61**, pp. 417–447.

Narasimha, R., and Sreenivasan, K.R. (1979) "Relaminarization of Fluid Flows," in *Advances in Applied Mechanics*, vol. 19, ed. C.-S. Yih, pp. 221–309, Academic Press, New York.

Narasimha, R., and Sreenivasan, K.R. (1988) "Flat Plate Drag Reduction by Turbulence Manipulation," *Sādhānā* **12**, pp. 15–30.

Nayfeh, A.H., and Balachandran, B. (1995) *Applied Nonlinear Dynamics—Analytical, Computational, and Experimental Methods*, Wiley, New York.

Nayfeh, A.H., Ragab, S.A., and Al-Maaitah, A. (1986) "Effects of Roughness on the Stability of Boundary Layers," AIAA Paper No. 86-1044, New York.

Nelson, C.F., Koga, D.J., and Eaton, J.K. (1987) "Control of the Unsteady Separated Flow Behind an Oscillating Two-Dimensional Flap," AIAA Paper No. 89-1027, New York.

Nelson, C.F., Koga, D.J., and Eaton, J.K. (1990) "Unsteady, Separated Flow behind an Oscillating, Two-Dimensional Spoiler," *AIAA J.* **28**, pp. 845–852.

Nelson, M.M., and Illingworth, W.T. (1991) *A Practical Guide to Neural Nets*, Addison–Wesley, Reading, Massachusetts.

Neuburger, D., and Wygnanski, I. (1988) "The Use of a Vibrating Ribbon to Delay Separation on Two-Dimensional Airfoils: Some Preliminary Observations," in *Proc. Workshop II on Unsteady Separated Flow*, ed. J.M. Walker, pp. 333–341, Frank J. Seiler Research Laboratory Report No. FJSRL-TR-88-0004, U.S. Air Force Systems Command, Colorado Springs, Colorado.

Newman, G.R. (1974) "An Experimental Study of Coherent Structures in the Turbulent Boundary Layer," Post-Graduate Study Dissertation, Cambridge University, Cambridge, Great Britain.

Nickerson, J.D. (1986) "A Study of Vortex Generators at Low Reynolds Numbers," AIAA Paper No. 86-0155, New York.

Nikuradse, J. (1932) "Gesetzmässigkeit der turbulenten Strömung in glatten Röhren," *Forschg. Arb. Ing.-Wes.*, no. 356, Germany.

Nikuradse, J. (1933) "Strömungsgesetze in rauhen Röhren," *Forschg. Arb. Ing.-Wes.*, no. 361, Germany.

Nisewanger, C.R. (1964) "Flow Noise and Drag Measurements of Vehicle with Compliant Coating," U.S. Naval Ordnance Test Station Report No. 8518, NOTS No. TP-3510, China Lake, California.

Noor, A., and Jorgensen, C.C. (1996) "A Hard Look at Soft Computing," *Aerospace America* **34**, September, pp. 34–39.

Norman, J.R., and Fraser, F.C. (1937) *Giant Fishes, Whales and Dolphins*, Putnam, London, Great Britain.

Nosenchuck, D.M., and Lynch, M.K. (1985) "The Control of Low-Speed Streak Bursting in Turbulent Spots," AIAA Paper No. 85-0535, New York.

Novak, C.J., Cornelius, K.C., and Roads, R.K. (1987) "Experimental Investigations of the Circular Wall Jet on a Circulation Control Airfoil," AIAA Paper No. 87-0155, New York.

O'Connor, L. (1992) "MEMS: Micromechanical Systems," *Mechanical Engineering* **114**, February, pp. 40–47.

Odell, G.M., and Kovasznay, L.S.G. (1971) "A New Type of Water Channel with Density Stratification," *J. Fluid Mech.* **50**, pp. 535–543.

Offen, G.R., and Kline, S.J. (1974) "Combined Dye-Streak and Hydrogen-Bubble Visual Observations of a Turbulent Boundary Layer," *J. Fluid Mech.* **62**, pp. 223–239.

Offen, G.R., and Kline, S.J. (1975) "A Proposed Model of the Bursting Process in Turbulent Boundary Layers," *J. Fluid Mech.* **70**, pp. 209–228.

Ogorodnikov, D.A., Grin, V.T., and Zakharov, N.N. (1972) "Boundary Layer Control of Hypersonic Air Inlets," NASA TTF-13927, Washington, D.C.

Oldaker, D.K., and Tiederman, W.G. (1977) "Spatial Structure of the Viscous Sublayer in Drag-Reducing Channel Flows," *Phys. Fluids* 20, no. 10, part II, pp. S133–144.

Oran, E.S., Oh, C.K., and Cybyk, B.Z. (1998) "Direct Simulation Monte Carlo: Recent Advances and Applications," *Ann. Rev. Fluid Mech.* 30, pp. 403–441.

Orszag, S.A. (1971) "Accurate Solution of the Orr-Sommerfeld Stability Equation," *J. Fluid Mech.* 50, pp. 689–703.

Orszag, S.A. (1979) "Prediction of Compliant Wall Drag Reduction," NASA Contractor Report No. 3071, Washington, D.C.

Ott, E. (1993) *Chaos in Dynamical Systems*, Cambridge University Press, Cambridge, Great Britain.

Ott, E., Grebogi, C., and Yorke, J.A. (1990a) "Controlling Chaos," *Phys. Rev. Lett.* 64, pp. 1196–1199.

Ott, E., Grebogi, C., and Yorke, J.A. (1990b) "Controlling Chaotic Dynamical Systems," in *Chaos: Soviet–American Perspectives on Nonlinear Science*, ed. D.K. Campbell, pp. 153–172, American Institute of Physics, New York.

Ottino, J.M. (1989a) *The Kinematics of Mixing: Stretching, Chaos, and Transport*, Cambridge University Press, Cambridge, Great Britain.

Ottino, J.M. (1989b) "The Mixing of Fluids," *Scientific American* 260, January, pp. 56–67.

Ottino, J.M. (1990) "Mixing, Chaotic Advection, and Turbulence," *Ann. Rev. Fluid Mech.* 22, pp. 207–253.

Ouellette, J. (1996) "MEMS: Mega Promise for Micro Devices," *Mechanical Engineering* 118, October, pp. 64–68.

Ouellette, J. (1999) "Electronic Noses Sniff Out New Markets," *Industrial Physicist* 5, no. 1, pp. 26–29.

Owen, F.K., and Horstman, C.C. (1972) "On the Structure of Hypersonic Turbulent Boundary Layers," *J. Fluid Mech.* 53, pp. 611–636.

Owen, F.K., Horstman, C.C., and Kussoy, M.I. (1975) "Mean and Fluctuating Flow Measurements of a Fully-Developed, Non-Adiabatic, Hypersonic Boundary Layer," *J. Fluid Mech.* 70, pp. 393–413.

Oyler, T.E., and Palmer, W.E. (1972) "Exploratory Investigation of Pulse Blowing for Boundary Layer Control," Report No. NR 72H-12, Columbus Aircraft Division, North American Rockwell Corp, Columbus, Ohio. (Also available from U.S. NTIS; Document Number AD-742 085.)

Paizi, S.T., and Schwarz, W.H. (1974) "An Investigation of the Topography and Motion of the Turbulent Interface," *J. Fluid Mech.* 63, pp. 315–343.

Panton, R.L. (1990a) "Scaling Turbulent Wall layers," *J. Fluid Eng.* 112, pp. 425–432.

Panton, R.L. (1990b) "The Role of Dimensional Analysis in Matched and Composite Asymptotic Expansions," in *Ocean Waves Mechanics, Computational Fluid Dynamics and Mathematical Modelling*, ed. M. Rahman, pp. 363–379, Computational Mechanics Publications, Boston, Massachusetts.

Panton, R.L. (1990c) "Effects of a Contoured Apex on Vortex Breakdown," *J. Aircraft* 27, pp. 285–288.

Panton, R.L. (1990d) "Inner-Outer Structure of the Wall-Pressure Correlation Function," in *Near-Wall Turbulence: 1988 Zoran Zarić Memorial Conference*, eds. S.J. Kline and N.H. Afgan, pp. 381–396, Hemisphere, New York.

Panton, R.L. (1991) "The Effects of Reynolds Number on Turbulent Wall Flows," in *Proc. Eighth Symposium on Turbulent Shear Flows*, paper no. I-15, 9–11 September, Technical University of Munich, Munich, Germany.

Panton, R.L. (1996) *Incompressible Flow*, second edition, Wiley–Interscience, New York.

Panton, R.L., editor (1997a) *Self-Sustaining Mechanisms of Wall Turbulence*, Computational Mechanics Publications, Southampton, Great Britain.

Panton, R.L. (1997b) "A Reynolds Stress Function for Wall Layers," *J. Fluids Eng.* 119, pp. 325–330.

Panton, R.L. (1998) "On the Wall-Pressure Fluctuations under a Three-Dimensional Boundary Layer," *J. Fluids Eng.* 120, pp. 407–410.

Panton, R.L., Goldman, A.L., Lowery, R.L., and Reischman, M.M. (1980) "Low-Frequency Pressure Fluctuations in Axisymmetric Turbulent Boundary Layers," *J. Fluid Mech.* 97, pp. 299–319.

Panton, R.L., and Linebarger, J.H. (1974) "Wall Pressure Spectra Calculations for Equilibrium Boundary Layers," *J. Fluid Mech.* **65**, pp. 261–287.

Panton, R.L., and Robert, G. (1993) "The Wall-Pressure Spectrum under a Turbulent Boundary Layer: Part 2, Theory and Results," in *ASME Fluid Engineering Conference*, eds. M.J. Morris and B.F. Carrol, FED-Vol. 155, pp. 43–48, ASME, New York.

Panton, R.L., and Robert, G. (1994) "The Wavenumber-Phase Velocity Representation for the Turbulent Wall-Pressure Spectrum," *J. Fluids Eng.* **116**, pp. 477–483.

Papell, S.S. (1984) "Vortex Generating Flow Passage Design for Increased Film-Cooling Effectiveness and Surface Coverage," NASA TM-83617, Washington, D.C.

Patel, V.C., and Head, M.R. (1968) "Reversion of Turbulent to Laminar Flow," *J. Fluid Mech.* **34**, pp. 371–392.

Patera, A.T. (1986) "Spectral Element Simulation of Flow in Grooved Channels: Cooling Chips with Tollmien-Schlichting Waves," in *Supercomputers and Fluid Dynamics*, eds. K. Kuwahara, R. Mendez, and S.A. Orszag, pp. 41–51, Springer–Verlag, Berlin.

Patera, A.T., and Mikić, B.B. (1986) "Exploiting Hydrodynamic Instabilities. Resonant Heat Transfer Enhancement," *Int. J. Heat Mass Transfer* **29**, pp. 1127–1138.

Patterson, G.K., Zakin, J.L., and Rodriguez, J.M. (1969) "Drag Reduction. Polymer Solutions, Soap Solutions, and Solid Particle Suspensions in Pipe Flow," *Indus. & Eng. Chem.* **61**, pp. 22–30.

Paula, G. (1996) "MEMS Sensors Branch Out," *Aerospace America* **34**, September, pp. 26–32.

Pearcey, H.H. (1961) "Shock Induced Separation and Its Prevention by Design and Boundary Layer Control," in *Boundary Layer and Flow Control*, ed. G.V. Lachmann, vol. 2, pp. 1166–1344, Pergamon Press, Oxford, Great Britain.

Pearson, J.R.A., and Petrie, C.J.S. (1968) "On Melt Flow Instability of Extruded Polymers," in *Polymer Systems: Deformation and Flow*, eds. R.E. Wetton and R.W. Whorlow, pp. 163–187, Macmillian, London, Great Britain.

Perrier, P. (1998) "Multiscale Active Flow Control," in *Flow Control: Fundamentals and Practices*, eds. M. Gad-el-Hak, A. Pollard, and J.-P. Bonnet, pp. 275–334, Springer–Verlag, Berlin.

Perry, A.E., and Abell, C.J. (1975) "Scaling Laws for Pipe-Flow Turbulence," *J. Fluid Mech.* **67**, pp. 257–271.

Perry, A.E., and Abell, C.J. (1977) "Asymptotic Similarity of Turbulence Structures in Smooth- and Rough-Walled Pipes," *J. Fluid Mech.* **79**, pp. 785–799.

Perry, A.E., and Chong, M.S. (1982) "On the Mechanism of Wall Turbulence," *J. Fluid Mech.* **119**, pp. 173–217.

Perry, A.E., Henbest, S.M., and Chong, M.S. (1986) "A Theoretical and Experimental Study of Wall Turbulence," *J. Fluid Mech.* **165**, pp. 163–199.

Perry, A.E., Li, J.D., Henbest, S., and Marusic, I. (1990) "The Attached Eddy Hypothesis in Wall Turbulence," in *Near-Wall Turbulence: 1988 Zoran Zarić Memorial Conference*, eds. S.J. Kline and N.H. Afgan, pp. 715–735, Hemisphere, New York.

Perry, A.E., Lim, T.T., and Teh, E.W. (1981) "A Visual Study of Turbulent Spots," *J. Fluid Mech.* **104**, pp. 387–405.

Peters, H. (1938) "A Study in Boundary Layers," in *Proc. Fifth Int. Cong. Appl. Mech.*, eds. J.P. Den Hartog and H. Peters, pp. 393–395, Wiley, New York.

Petersen, I.R., and Savkin, A.V. (1999) *Robust Kalman Filtering for Signals and Systems with Large Uncertainties*, Birkhäuser, Boston, Massachusetts.

Petroski, H. (1997) "Development and Research," *American Scientist* **85**, no. 3, pp. 210–213.

Pfahler, J. (1992) "Liquid Transport in Micron and Submicron Size Channels," Ph.D. Thesis, University of Pennsylvania, Philadelphia, Pennsylvania.

Pfahler, J., Harley, J., Bau, H., and Zemel, J.N. (1990) "Liquid Transport in Micron and Submicron Channels," *Sensors & Actuators A* **21–23**, pp. 431–434.

Pfahler, J., Harley, J., Bau, H., and Zemel, J.N. (1991) "Gas and Liquid Flow in Small Channels," in *Symp. on Micromechanical Sensors, Actuators, and Systems*, eds. D. Cho et al., ASME DSC-Vol. 32, pp. 49–60, ASME, New York.

Pfeffer, R., and Rosetti, S.J. (1971) "Experimental Determination of Pressure Drop and Flow Characteristics of Dilute Gas-Solid Suspensions," NASA Contractor Report No. 1894, Washington, D.C.

Pfenninger, W. (1946) "Untersuchungen über Reibungsverminderung an Tragflügeln, insbesondere mit Hilfe von Grenzschichtabsaugung," Reports of the Inst. of Aerodynamics, no. 13, ETH Zürich, Zürich, Switzerland.

Pfenninger, W., and Vemuru, C.S. (1990) "Design of Low Reynolds Number Airfoils," *J. Aircraft* **27**, pp. 204–210.

Phillips, O.M. (1979) *The Last Chance Energy Book*, Johns Hopkins University Press, Baltimore, Maryland.

Phillips, W.R.C. (1987) "The Wall Region of a Turbulent Boundary Layer," *Phys. Fluids* **30**, pp. 2354–2361.

Phillips, W.R.C., and Ratnanather, J.T. (1990) "The Outer Region of a Turbulent Boundary Layer," *Phys. Fluids A* **2**, pp. 427–434.

Piekos, E.S., and Breuer, K.S. (1996) "Numerical Modeling of Micromechanical Devices Using the Direct Simulation Monte Carlo Method," *J. Fluids Eng.* **118**, pp. 464–469.

Piekos, E.S., Orr, D.J., Jacobson, S.A., Ehrich, F.F., and Breuer, K.S. (1997) "Design and Analysis of Microfabricated High Speed Gas Journal Bearings," AIAA Paper No. 97-1966, AIAA, Reston, Virginia.

Pinebrook, W.E., and Dalton, C. (1983) "Drag Minimization on a Body of Revolution Through Evolution," *Computer Methods in Appl. Mech. and Eng.* **39**, pp. 179–197.

Pister, K.S.J., Fearing, R.S., and Howe, R.T. (1990) "A Planar Air Levitated Electrostatic Actuator System," IEEE Paper No. CH2832-4/90/0000-0067, IEEE, New York.

Plesniak, M.W., and Nagib, H.M. (1985) "Net Drag Reduction in Turbulent Boundary Layers Resulting from Optimized Manipulation," AIAA Paper No. 85-0518, New York.

Poisson-Quinton, Ph. (1950) "On the Mechanism and the Application of the Control of the Boundary Layer of Airplanes," in *Int. Colloquium on Mechanics*, vol. II, Poitiers, France.

Pollard, A. (1997) "Passive and Active Control of Near-Wall Turbulence," *Prog. Aero. Sci.* **33**, pp. 689–708.

Pollard A. (1998) "Near-Wall Turbulence Control," *Flow Control: Fundamentals and Practices*, eds. M. Gad-el-Hak, A. Pollard, and J.-P. Bonnet, pp. 431–466, Springer–Verlag, Berlin.

Pomeau, Y., and Manneville, P. (1980) "Intermittent Transition to Turbulence in Dissipative Dynamical Systems," *Commun. Math. Phys.* **74**, pp. 189–197.

Pong, K.-C., Ho, C.-M., Liu, J., and Tai, Y.-C. (1994) "Non-Linear Pressure Distribution in Uniform Microchannels," in *Application of Microfabrication to Fluid Mechanics 1994*, eds. P.R. Bandyopadhyay, K.S. Breuer, and C.J. Belchinger, ASME FED-Vol. 197, pp. 47–52, ASME, New York.

Pope, S.B. (1996) "Reacting Flows and Combustion," in *Research Trends in Fluid Mechanics*, eds. J.L. Lumley, A. Acrivos, L.G. Leal, and S. Leibovich, pp. 229–243, American Institute of Physics, Woodbury, New York.

Porodnov, B.T., Suetin, P.E., Borisov, S.F., and Akinshin, V.D. (1974) "Experimental Investigation of Rarefied Gas Flow in Different Channels," *J. Fluid Mech.* **64**, pp. 417–437.

Povkh, I.L., Bolonov, N.I., and Eidel'man, A.Ye. (1979) "The Average Velocity Profile and the Frictional Loss in Turbulent Flow of an Aqueous Suspension of Clay," *Fluid Mech. Soviet Research* **8**, pp. 118–124.

Prandtl, L. (1904) "Über Flüssigkeitsbewegung bei sehr kleiner Reibung," in *Proc. Third Int. Math. Cong.*, pp. 484–491, Heidelberg, Germany.

Prandtl, L. (1925a) "Bericht über Untersuchungen zur ausgebildeten Turbulenz," *Z. angew. Math. Mech.* **5**, pp. 136–139.

Prandtl, L. (1925b) "Magnuseffeckt und Windkraftschiff," *Naturwissenschaften* **13**, pp. 93–108.

Prandtl, L. (1935) "The Mechanics of Viscous Fluids," in *Aerodynamic Theory*, ed. W.F. Durand, vol. III, pp. 34–208, Springer–Verlag, Berlin.

Praturi, A.K., and Brodkey, R.S. (1978) "A Stereoscopic Visual Study of Coherent Structures in Turbulent Shear Flows," *J. Fluid Mech.* **89**, pp. 251–272.

Preston, J.H. (1958) "The Minimum Reynolds Number for a Turbulent Boundary Layer and the Selection of a Transition Device," *J. Fluid Mech.* **3**, pp. 373–384.

Preston, J.H., and Sweeting, N.E. (1944) "The Velocity Distribution in the Boundary Layer of a Plane Wall at High Reynolds Numbers with Suggestions for Further Experiments," Aeronautical Research Council Report No. ARC-FM-671, London, Great Britain.

Pretsch, J. (1942) "Umschlagbeginn und Absaugung," *Jahrb. Dtsch. Luftfahrtforschung* **1**, pp. 54–71.

Prud'homme, R.K., Chapman, T.W., and Bowen, J.R. (1986) "Laminar Compressible Flow in a Tube," *Appl. Scientific Res.* **43**, pp. 67–74.

Purohit, S.C. (1987) "Effect of Vectored Suction on a Shock-Induced Separation," *AIAA J.* **25**, pp. 759–760.

Purshouse, M. (1976) "On the Damping of Unsteady Flow by Compliant Boundaries," *J. Sound Vibr.* **49**, pp. 423–436.

Purshouse, M. (1977) "Interaction of Flow with Compliant Surfaces," Ph.D. Thesis, Cambridge University, Cambridge, Great Britain.

Purtell, L.P., Klebanoff, P.S., and Buckley, F.T. (1981) "Turbulent Boundary Layer at Low Reynolds Number, *Phys. Fluids* **24**, pp. 802–811.

Puryear, F.W. (1962) "Boundary Layer Control: Drag Reduction by Use of Compliant Coatings," David Taylor Model Basin Report No. 1668, Bethesda, Maryland.

Qin, F., Wolf, E.E., and Chang, H.-C. (1994) "Controlling Spatiotemporal Patterns on a Catalytic Wafer," *Phys. Rev. Lett.* **72**, pp. 1459–1462.

Radin, I. (1974) "Solid-Fluid Drag Reduction," Ph.D. Thesis, University of Missouri, Rolla, Missouri.

Radin, I., Zakin, J.L., and Patterson, G.K. (1975) "Drag Reduction in Solid–Fluid Systems," *AIChE J.* **21**, pp. 358–371.

Ragab, S.A., and Nayfeh, A.H. (1980) "A Comparison of the Second-Order Triple-Deck Theory and Interacting Boundary Layers for Incompressible Flows Past a Hump," AIAA Paper No. 80-0072, New York.

Raghunathan, S. (1985) "Passive Control of Shock-Boundary Layer Interaction," *Prog. Aero. Sci.* **25**, pp. 271–296.

Rajagopalan, S., and Antonia, R.A. (1984) "Conditional Averages Associated with the Fine Structure in a Turbulent Boundary Layer," *Phys. Fluids* **27**, pp. 1966–1973.

Rajagopalan, S., and Antonia, R.A. (1993) "RMS Spanwise Vorticity Measurements in a Turbulent Boundary Layer," *Exp. Fluids* **14**, pp. 142–144.

Rao, D.M. (1979) "Leading-Edge Vortex Flap Experiments on a 74-Deg. Delta Wing," NASA CR-159161, Washington, D.C.

Rao, D.M., and Kariya, T.T. (1988) "Boundary-Layer Submerged Vortex Generators for Separation Control—An Exploratory Study," AIAA Paper No. 88-3546-CP, New York.

Rao, K.N., Narasimha, R., and Badri Narayanan, M.A. (1971) "The 'Bursting' Phenomenon in a Turbulent Boundary Layer," *J. Fluid Mech.* **48**, pp. 339–352.

Ras, M., and Ackeret, J. (1941) "Über Verhinderung der Grenzschicht-Turbulenz durch Absaugung," *Helv. Phys. Acta* **14**, p. 323.

Raspet, A. (1952) "Boundary-Layer Studies on a Sailplane," *Aeronaut. Eng. Rev.* **11**, pp. 52–60.

Raupach, M.R., Antonia, R.A., and Rajagopalan, S. (1991) "Rough-Wall Turbulent Boundary Layers," *Appl. Mech. Rev.* **44**, pp. 1–25.

Rayleigh, Lord (1880) "On the Stability, or Instability, of Certain Fluid Motions," *Proc. London Math. Soc.* **11**, pp. 57–70.

Rayleigh, Lord (1887) "On Waves Propagated along the Plane Surface of an Elastic Solid," *Proc. London Math. Soc.* **XVII**, pp. 4–11.

Reed, H.L., and Nayfeh, A.H. (1986) "Numerical-Perturbation Technique for Stability of Flat-Plate Boundary Layers with Suction," *AIAA J.* **24**, pp. 208–214.

Reed, H.L., and Saric, W.S. (1987) "Stability and Transition of Three-Dimensional Flows," in *Proc. Tenth U.S. Nat. Cong. of Applied Mechanics*, ed. J.P. Lamb, pp. 457–468, ASME, New York.

Reed, H.L., and Saric, W.S. (1989) "Stability of Three-Dimensional Boundary Layers," *Ann. Rev. Fluid Mech.* **21**, pp. 235–284.

Reischman, M.M., and Tiederman, W.G. (1975) "Laser-Doppler Anemometer Measurements in Drag-Reducing Channel Flows," *J. Fluid Mech.* **70**, pp. 369–392.

Reisenthel, P.H., Nagib, H.M., and Koga, D.J. (1985) "Control of Separated Flows Using Forced Unsteadiness," AIAA Paper No. 85-0556, New York.

Reisenthel, P.H., Xiong, Y., and Nagib, H.M. (1991) "The Preferred Mode in an Axisymmetric Jet With and Without Enhanced Feedback," AIAA Paper No. 91-0315, Washington, D.C.

Render, P.M. (1984) "The Experimental and Theoretical Aerodynamics of Aerofoil Sections Suitable for Remotely Piloted Vehicles," Ph.D. Thesis, Cranfield Institute of Technology, Cranfield, Great Britain.

Render, P.M., Stollery, J.L., and Williams, B.R. (1985) "Aerofoils at Low Reynolds Numbers— Prediction and Experiments," in *Numerical and Physical Aspects of Aerodynamic Flows III*, ed. T. Cebeci, pp. 155–167, Springer–Verlag, New York.

Reshotko, E. (1976) "Boundary-Layer Stability and Transition," *Ann. Rev. Fluid Mech.* **8**, pp. 311–349.

Reshotko, E. (1979) "Drag Reduction by Cooling in Hydrogen-Fueled Aircraft," *J. Aircraft* **16**, pp. 584–590.

Reshotko, E. (1985) "Control of Boundary Layer Transition," AIAA Paper No. 85-0562.

Reshotko, E. (1987) "Stability and Transition, How Much Do We Know?," in *Proc. Tenth U.S. National Cong. of Applied Mechanics*, ed. J.P. Lamb, pp. 421–434, ASME, New York.

Reynolds, G.A., and Saric, W.S. (1986) "Experiments on the Stability of the Flat-Plate Boundary Layer with Suction," *AIAA J.* **24**, pp. 202–207.

Reynolds, O. (1883) "An Experimental Investigation of the Circumstances Which Determine Whether the Motion of Water Shall be Direct or Sinuous, and of the Law of Resistence in Parallel Channels," *Phil. Trans. Roy. Soc. London A* **174**, pp. 935–982.

Reynolds, O. (1895) "On the Dynamical Theory of Incompressible Viscous Fluids and the Determination of the Criterion," *Phil. Trans. Roy. Soc. London A* **186**, pp. 123–164.

Reynolds, W.C. (1976) "Computation of Turbulent Flows," *Ann. Rev. Fluid Mech.* **8**, pp. 183–208.

Reynolds, W.C. (1993) "Sensors, Actuators, and Strategies for Turbulent Shear-Flow Control," invited oral presentation at *AIAA Third Flow Control Conference*, 6–9 July, Orlando, Florida.

Reynolds, W.C. (1996) "Control of Turbulent Flows," in *Research Trends in Fluid Mechanics*, eds. J.L. Lumley, A. Acrivos, L.G. Leal, and S. Leibovich, pp. 253–262, American Institute of Physics, Woodbury, New York.

Reynolds, W.C., and Carr, L.W. (1985) "Review of Unsteady, Driven, Separated Flows," AIAA Paper No. 85-0527, New York.

Reynolds, W.C., Eaton, J.K., Johnston, J.P., Hesselink, L., Powell, D.J., Roberts, L., and Kroo, E. (1988) "Flow Control for Unsteady and Separated Flows and Turbulent Mixing," AFOSR-TR-89-0232, Washington, D.C. (Also available from U.S. NTIS; Document Number AD-A205989.)

Rice, E.J., and Raman, G. (1993) "Mixing Noise Reduction for Rectangular Supersonic Jets by Nozzle Shaping and Induced Screech Mixing," AIAA Paper No. 93-4322, Washington, D.C.

Richardson, L.F. (1920) "The Supply of Energy from and to Atmospheric Eddies," *Proc. R. Soc. London A* **97**, pp. 354–373.

Richardson, L.F. (1922) *Weather Prediction by Numerical Process*, Cambridge University Press, Cambridge, Great Britain.

Richardson, S. (1973) "On the No-Slip Boundary Condition," *J. Fluid Mech.* **59**, pp. 707–719.

Richter, A., Plettner, A., Hofmann, K.A., and Sandmaier, H. (1991) "A Micromachined Electro-hydrodynamic (EHD) Pump," *Sensors & Actuators A* **29**, pp. 159–168.

Riley, J.J., and Gad-el-Hak, M. (1985) "The Dynamics of Turbulent Spots," in *Frontiers in Fluid Mechanics*, eds. S.H. Davis and J.L. Lumley, pp. 123–155, Springer–Verlag, Berlin.

Riley, J.J., Gad-el-Hak, M., and Metcalfe, R.W. (1988) "Compliant Coatings," *Ann. Rev. Fluid Mech.* **20**, pp. 393–420.

Ringleb, E.O. (1961) "Separation Control by Trapped Vortices," in *Boundary Layer and Flow Control*, ed. G.V. Lachmann, vol. 1, pp. 265–294, Pergamon Press, Oxford, Great Britain.

Ritter, H., and Messum, L.T. (1964) "Water Tunnel Measurements of Turbulent Skin Friction on Six Different Compliant Surfaces of 1 Ft Length," British Admiralty Research Laboratory Report No. ARL/N4/GHY/9/7, ARL/G/N9, London, Great Britain.

Ritter, H., and Porteous, J.S. (1964) "Water Tunnel Measurements of Skin Friction on a Compliant Coating," British Admiralty Research Laboratory Report No. ARL/N3/G/HY/9/7, London, Great Britain.

Robert, G. (1993) "The Wall-Pressure Spectrum under a Turbulent Boundary Layer: Part 1, Experiments," in *ASME Fluid Engineering Conference*, eds. M.J. Morris and B.F. Carrol, FED-Vol. 155, pp. 37–42, ASME, New York.

Robinson, E.Y., Helvajian, H., and Jansen, S.W. (1996a) "Small and Smaller: The World of MNT," *Aerospace America* **34**, September, pp. 26–32.

Robinson, E.Y., Helvajian, H., and Jansen, S.W. (1996b) "Big Benefits from Tiny Technologies," *Aerospace America* **34**, October, pp. 38–43.

Robinson, M.C., and Luttges, M.W. (1984) "Unsteady Separated Flow: Forced and Common Vorticity About Oscillating Airfoils," in *Unsteady Separated Flows*, eds. M.S. Francis and M.W. Luttges, pp. 117–126, U.S. Air Force Academy, Colorado Springs, Colorado.

Robinson, S.K. (1990) "A Review of Vortex Structures and Associated Coherent Motions in Turbulent Boundary layers," in *Structure of Turbulence and Drag Reduction*, ed. A. Gyr, pp. 23–50, Springer–Verlag, Berlin.

Robinson, S.K. (1991) "Coherent Motions in the Turbulent Boundary Layer," *Ann. Rev. Fluid Mech.* **23**, pp. 601–639.

Robinson, S.K., Kline, S.J., and Spalart, P.R. (1989) "A Review of Quasi-Coherent Structures in a Numerically Simulated Turbulent Boundary Layer," NASA Technical Memorandum No. TM-102191, Washington, D.C.

Rogallo, R.S., and Moin, P. (1984) "Numerical Simualtion of Turbulent Flows," *Ann. Rev. Fluid Mech.* **16**, pp. 99–137.

Romeiras, F.J., Grebogi, C., Ott, E., and Dayawansa, W.P. (1992) "Controlling Chaotic Dynamical Systems," *Physica D* **58**, pp. 165–192.

Roos, F.W. (1996) " 'Microblowing' for High-Angle-of-Attack Vortex Flow Control on a Fighter Aircraft," AIAA Paper No. 96-0543, Washington, D.C.

Roos, F.W., and Kegelman, J.T. (1986) "Control of Coherent Structures in Reattaching Laminar and Turbulent Shear Layers," *AIAA J.* **24**, pp. 1956–1963.

Rosenhead, L. (1963) *Laminar Boundary Layers*, Clarendon Press, Oxford, Great Britain.

Roshko A. (1976) "Structure of Turbulent Shear Flows: A New Look," *AIAA J.* **14**, pp. 1349–1357.

Roshko, A. (1992) "Bluff-Body and Other Shear Flows," in *Proc. NUWC Division Newport Seminar Series on Turbulence and Its Control*, compiled by P.R. Bandyopadhay and J.C.S. Meng, pp. 4.1–4.34, Naval Undersea Warfare Center Technical Memorandum No. NUWC-NPT TM 922089, Newport, Rhode Island.

Rott, N. (1956) "Unsteady Viscous Flow in the Vicinity of a Stagnation Point," *Q. Appl. Math.* **13**, pp. 444–451.

Rotta, J.C. (1951) "Statistische Theorie nichthomogener Turbulenz," *Z. Phys.* **129**, pp. 547–572.

Rotta, J.C. (1962) "Turbulent Boundary Layers in Incompressible Flow," *Prog. Aeronautical Sci.* **2**, pp. 1–219.

Rotta, J.C. (1970) "Control of Turbulent Boundary Layers by Uniform Injection and Suction of Fluid," in *Seventh Cong. of the International Council of the Aeronautical Sciences*, paper no. 70-10, ICAS, Rome, Italy.

Runstadler, P.G., Kline, S.J., and Reynolds, W.C. (1963) "An Experimental Investigation of Flow Structure of the Turbulent Boundary Layer," Department of Mechanical Engineering Report No. MD-8, Stanford University, Stanford, California.

Runyan, L.J., and Steers, L. L. (1980) "Boundary Layer Stability Analysis of a Natural Laminar Flow Glove on the F-111 TACT Airplane," in *Viscous Flow Drag Reduction*, ed. G.R. Hough, pp. 17–32, AIAA, New York.

Ryskin, G. (1987) "Turbulent Drag Reduction by Polymers: A Quantitative Theory," *Phys. Rev. Lett.* **59**, pp. 2059–2062.

Sabadell, L.A. (1988) "Effects of a Drag Reducing Additive on Turbulent Boundary Layer Structure," M.Sc. Thesis, Princeton University, Princeton, New Jersey.

Saddoughi, S.G., and Veeravalli, S.V. (1994) "Local Isotropy in Turbulent Boundary Layers at High Reynolds Number," *J. Fluid Mech.* **268**, pp. 333–372.

Saffman, P.G. (1978) "Problems and Progress in the Theory of Turbulence," in *Structure and Mechanisms of Turbulence II*, ed. H. Fiedler, pp. 273–306, Springer–Verlag, Berlin.

Saffman, P.G., and Baker, G.R. (1979) "Vortex Interactions," *Ann. Rev. Fluid Mech.* **11**, pp. 95–122.

Sajben, M., Chen, C.P., and Kroutil, J.C. (1976) "A New, Passive Boundary Layer Control Device," AIAA Paper No. 76-700, New York.

Saric, W.S., and Reed, H.L. (1986) "Effect of Suction and Weak Mass Injection on Boundary-Layer Transition," *AIAA J.* **24**, pp. 383–389.

Saripalli, K.R., and Simpson, R.L. (1980) "Investigation of Blown Boundary Layers with an Improved Wall Jet System," NASA CR-3340, Washington, D.C.

Sasaki, K., and Kiya, M. (1985) "Effect of Free-Stream Turbulence on Turbulent Properties of a Separation-Reattachment Flow," *Bulletin JSME* **28**, pp. 610–616.

Sasaki, K., and Kiya, M. (1991) "Three-Dimensional Vortex Structure in a Leading-Edge Separation Bubble at Moderate Reynolds Numbers," *J. Fluids Eng.* **113**, pp. 405–410.

Savas, Ö, and Coles, D. (1985) "Coherence Measurements in Synthetic Turbulent Boundary Layers," *J. Fluid Mech.* **160**, pp. 421–446.

Savins, J.G. (1967) "A Stress-Controlled Drag-Reduction Phenomenon," *Rheologica Acta* **6**, pp. 323–330.

Savu, G., and Trifu, O. (1984) "Porous Airfoils in Transonic Flow," *AIAA J.* **22**, pp. 989–991.

Sawyers, D.R., Sen, M., and Chang, H.-C. (1996) "Effect of Chaotic Interfacial Stretching on Bimolecular Chemical Reaction in Helical-Coil Reactors," *Chem. Eng. J.* **64**, pp. 129–139.

Sawyers, D.R., Sen, M., and Chang, H.-C. (1998) "Heat Transfer Enhancement in Three-Dimensional Corrugated Channel Flow," *Int. J. Heat Mass Transfer* **41**, pp. 3559–3573.

Schaaf, S.A., and Chambré, P.L. (1961) *Flow of Rarefied Gases*, Princeton University Press, Princeton, New Jersey.

Schewe, G. (1983) "On the Structure and Resolution of Wall-Pressure Fluctuations Associated with Turbulent Boundary-Layer Flow," *J. Fluid Mech.* **134**, pp. 311–328.

Schildknecht, M., Miller, J.A., and Meir, G.E.A. (1979) "The Influence of Suction on the Structure of Turbulence in Fully Developed Pipe Flow," *J. Fluid Mech.* **90**, pp. 67–107.

Schilz, W. (1965/66) "Experimentelle Untersuchungen zur Akustischen Beeinflussung der Strömungsgrenzschicht in Luft," *Acustica* **16**, pp. 208–223.

Schlichting, H. (1959) "Einige neuere Ergebnisse über Grenzschichtbeeinflussung," in *Proc. First Int. Cong. Aero. Sci.*, eds. Th. von Kármán et al., pp. 563–586, *Adv. Aero. Sci.*, vol. 2, Pergamon Press, London, Great Britain.

Schlichting, H. (1979) *Boundary-Layer Theory*, seventh edition, McGraw–Hill, New York.

Schlichting, H., and Pechau, W. (1959) "Auftriebserhöhung von Tragflügeln durch kontinuierlich verteilte Absaugung," *ZFW* 7, pp. 113–119.

Schlichting, H., and Ulrich, A. (1940) "Zur Berechnung des Umschlages laminar-turbulent," *Jahrb. Dtsch. Luftfahrtforschung* **1**, pp. 8–35.

Schmitz, F.W. (1967) "Aerodynamics of the Model Airplane, Part I, Airfoil Measurements," NASA-TM-X-60976, Washington, D.C.

Schneider, P.E.M. (1980) "Sekundärwirbelbildung bei Ringwirbeln und in Freistrahlen," *Z. Flugwiss.* **4**, pp. 307–318.

Schofield, W.H. (1985) "Turbulent Boundary Layer Development in an Adverse Pressure Gradient After an Interaction with a Normal Shock Wave," *J. Fluid Mech.* **154**, pp. 43–62.

Schoppa, W., and Hussain, F. (1998) "A Large-Scale Control Strategy for Drag Reduction in Turbulent Boundary Layers," *Phys. Fluids* **10**, pp. 1049–1051.

Schreck, S.J., Faller, W.E., and Luttges, M.W. (1995) "Neural Network Prediction of Three-Dimensional Unsteady Separated Flow Fields," *J. Aircraft* **32**, pp. 178–185.

Schubauer, G.B., and Skramstad, H.K. (1947) "Laminar Boundary-Layer Oscillations and Stability of Laminar Flow," *J. Aero. Sci.* **14**, pp. 69–78.

Schubauer, G.B., and Skramstad, H.K. (1948) "Laminar Boundary-Layer Oscillations and Transition on a Flat Plate," NACA Report No. 909, Washington, D.C.

Schubauer, G.B., and Spangenberg, W.G. (1960) "Forced Mixing in Boundary Layers," *J. Fluid Mech.* **8**, pp. 10–32.

Schulz, M. (1999) "The End of the Road for Silicon," *Nature* **399**, no. 6738, 24 June, pp. 729–730.

Scott, M.R., and Watts, H.A. (1977) "Computational Solution of Linear Two-Point Boundary Value Problems via Orthonormalization," *J. Numerical Analysis* **14**, pp. 40–70.

Sears, W.R. (1956) "Some Recent Developments in Airfoil Theory," *J. Aeronaut. Sci.* **23**, pp. 490–499.

Sears, W.R., and Telionis, D.P. (1972a) "Unsteady Boundary-Layer Separation," in *Recent Research on Unsteady Boundary Layers*, ed. E.A. Eichelbrenner, vol. 1, pp. 404–442, Presses de l'Universite Laval, Quebec, Canada.

Sears, W.R., and Telionis, D.P. (1972b) "Two Dimensional Laminar Boundary Layer Separation for Unsteady Flow or Flow Past Moving Walls, Considering Singularity Due to Bifurcating Wake Bubble," in *Recent Research on Unsteady Boundary Layers*, ed. E.A. Eichelbrenner, vol. 1, pp. 443–447, Presses de l'Universite Laval, Quebec, Canada.

Sears, W.R., and Telionis, D.P. (1975) "Boundary-Layer Separation in Unsteady Flow," *J. Appl. Math.* **28**, pp. 215–235.

Seidl, M., and Steinheil, E. (1974) "Measurement of Momentum Accommodation Coefficients on Surfaces Characterized by Auger Spectroscopy, SIMS and LEED," in *Rarefied Gas Dynamics 9*, eds. M. Becker and M. Fiebig, pp. E9.1–E9.2, DFVLR-Press, Porz–Wahn, Germany.

Selby, G.V., and Miandoab, F.H. (1990) "Effect of Surface Grooves on Base Pressure for a Blunt Trailing-Edge Airfoil," *AIAA J.* **28**, pp. 1133–1135.

Sen, M. (1989) "The Influence of Developments in Dynamical Systems Theory on Experimental Fluid Mechanics," in *Frontiers in Experimental Fluid Mechanics*, ed. M. Gad-el-Hak, pp. 1–24, Springer–Verlag, New York.

Sen, M., Wajerski, D., and Gad-el-Hak, M. (1996) "A Novel Pump for MEMS Applications," *J. Fluids Eng.* **118**, pp. 624–627.

Sen, P.K., and Arora, D.S. (1988) "On the Stability of Laminar Boundary-Layer Flow over a Flat Plate with a Compliant Surface," *J. Fluid Mech.* **197**, pp. 201–240.

Service, R.F. (1999) "Organic Molecule Rewires Chip Design," *Science* **285**, no. 5426, 16 July, pp. 313–315.

Shah, D.A., and Antonia, R.A. (1989) "Scaling of the 'Bursting' Period in Turbulent Boundary Layer and Duct Flows," *Phys. Fluids A* **1**, pp. 318–325.

Shah, R.K., and Joshi, S.D. (1987) "Convective Heat Transfer in Curved Ducts," in *Handbook of Single-Phase Convective Heat Transfer*, eds. S. Kakaç, R.K. Shah, and W. Aung, pp. 5.1–5.46, Wiley–Interscience, New York.

Sharatchandra, M.C., Sen, M., and Gad-el-Hak, M. (1997) "Navier–Stokes Simulations of a Novel Viscous Pump," *J. Fluids Eng.* **119**, pp. 372–382.

Sharatchandra, M.C., Sen, M., and Gad-el-Hak, M. (1998a) "Thermal Aspects of a Novel Micro-pumping Device," *J. Heat Transfer* **120**, pp. 99–107.

Sharatchandra, M.C., Sen, M., and Gad-el-Hak, M. (1998b) "A New Approach to Constrained Shape Optimization Using Genetic Algorithms," *AIAA J.* **36**, pp. 51–61.

Shercliff, J.A. (1965) *A Textbook of Magnetohydrodynamics*, Pergamon Press, Oxford, Great Britain.

Sherman, F.S. (1990) *Viscous Flow*, McGraw–Hill, New York.

Shih, J.C., Ho, C.-M., Liu, J., and Tai, Y.-C. (1995) "Non-Linear Pressure Distribution in Uniform Microchannels," ASME AMD-MD-Vol. 238, New York.

Shih, J.C., Ho, C.-M., Liu, J., and Tai, Y.-C. (1996) "Monatomic and Polyatomic Gas Flow through Uniform Microchannels," in *Applications of Microfabrication to Fluid Mechanics*, eds. K. Breuer, P. Bandyopadhyay, and M. Gad-el-Hak, ASME DSC-Vol. 59, pp. 197–203, New York.

Shiloh, K., Shivaprasad, B.G., and Simpson, R.L. (1981) "The Structure of a Separating Turbulent Boundary Layer. Part 3: Transverse Velocity Measurements," *J. Fluid Mech.* **113**, pp. 75–90.

Shinbrot, T. (1993) "Chaos: Unpredictable Yet Controllable?" *Nonlinear Science Today* **3**, pp. 1–8.

Shinbrot, T. (1995) "Progress in the Control of Chaos," *Adv. Physics* **44**, pp. 73–111.

Shinbrot, T. (1998) "Chaos, Coherence and Control," in *Flow Control: Fundamentals and Practices*, eds. M. Gad-el-Hak, A. Pollard, and J.-P. Bonnet, pp. 501–527, Springer–Verlag, Berlin.

Shinbrot, T., Bresler, L., and Ottino, J.M. (1998) "Manipulation of Isolated Structures in Experimental Chaotic Fluid Flows," *Exp. Thermal & Fluid Sci.* **16**, pp. 76–83.

Shinbrot, T., Ditto, W., Grebogi, C., Ott, E., Spano, M., and Yorke, J.A. (1992) "Using the Sensitive Dependence of Chaos (the "Butterfly Effect") to Direct Trajectories in an Experimental Chaotic System," *Phys. Rev. Lett.* **68**, pp. 2863–2866.

Shinbrot, T., Grebogi, C., Ott, E., and Yorke, J.A. (1992a) "Using Chaos to Target Stationary States of Flows," *Phys. Lett. A* **169**, pp. 349–354.

Shinbrot, T., Grebogi, C., Ott, E., and Yorke, J.A. (1993) "Using Small Perturbations to Control Chaos," *Nature* **363**, pp. 411–417.

Shinbrot, T., Ott, E., Grebogi, C., and Yorke, J.A. (1990) "Using Chaos to Direct Trajectories to Targets," *Phys. Rev. Lett.* **65**, pp. 3215–3218.

Shinbrot, T., Ott, E., Grebogi, C., and Yorke, J.A. (1992b) "Using Chaos to Direct Orbits to Targets in Systems Describable by a One-Dimensional Map," *Phys. Rev. A* **45**, pp. 4165–4168.

Shinbrot, T., and Ottino, J.M. (1993a) "Geometric Method to Create Coherent Structures in Chaotic Flows," *Phys. Rev. Lett.* **71**, pp. 843–846.

Shinbrot, T., and Ottino, J.M. (1993b) "Using Horseshoes to Create Coherent Structures in Chaotic Fluid Flows," *Bul. Am. Phys. Soc.* **38**, p. 2194.

Sigal, A. (1971) "An Experimental Investigation of the Turbulent Boundary Layer over a Wavy Wall," Ph.D. Thesis, California Institute of Technology, Pasadena, California.

Sigurdson, L.W., and Roshko, A. (1985) "Controlled Unsteady Excitation of a Reattaching Flow," AIAA Paper No. 85-0552, New York.

Silhanek, V. (1969) "On Aircraft Longitudinal Motion after Boundary Layer Control System Failure During Take-off and Landing," Summary Report No. Z-13, Aeronautical Research and Test Institute, Prague, Czechoslovakia.

Simpson, R.L. (1970) "Characteristics of Turbulent Boundary Layers at Low Reynolds Numbers With and Without Transpiration," *J. Fluid Mech.* **42**, pp. 769–802.

Simpson, R.L. (1976) "Comment on 'Prediction of Turbulent Boundary Layers at Low Reynolds Numbers'," *AIAA J.* **14**, pp. 1662–1663.

Simpson, R.L. (1989) "Turbulent Boundary-Layer Separation," *Ann. Rev. Fluid Mech.* **21**, pp. 205–234.

Simpson, R.L., and Shivaprasad, B.G. (1983) "The Structure of a Separating Turbulent Boundary Layer. Part 5: Frequency Effects on Periodic Unsteady Freestream Flows," *J. Fluid Mech.* **131**, pp. 319–339.

Simpson, R.L., Chew, Y.-T., and Shivaprasad, B.G. (1981a) "The Structure of a Separating Turbulent Boundary Layer. Part 1: Mean Flow and Reynolds Stresses," *J. Fluid Mech.* **113**, pp. 23–51.

Simpson, R.L., Chew, Y.-T., and Shivaprasad, B.G. (1981b) "The Structure of a Separating Turbulent Boundary Layer. Part 2: Higher-Order Turbulent Results," *J. Fluid Mech.* **113**, pp. 53–73.

Simpson, R.L., Shivaprasad, B.G., and Chew, Y.-T. (1983) "The Structure of a Separating Turbulent Boundary Layer. Part 4: Effects of Periodic Freestream Unsteadiness," *J. Fluid Mech.* **127**, pp. 219–261.

Singer, J., Wang, Y.-Z., and Bau, H.H. (1991) "Controlling a Chaotic System," *Phys. Rev. Lett.* **66**, pp. 1123–1125.

Sirovich, L., and Deane, A.E. (1991) "A Computational Study of Rayleigh–Bénard Convection. Part 2. Dimension Considerations," *J. Fluid Mech.* **222**, pp. 251–265.

Sirovich L., and Karlsson S. (1997) "Turbulent Drag Reduction by Passive Mechanisms," *Nature* **388**, 21 August, pp. 753–755.

Smagorinsky, J. (1963) "General Circulation Experiments with the Primitive Equations," *Mon. Weather Rev.* **91**, pp. 99–164.

Smith, A.M.O. (1957) "Transition, Pressure Gradient, and Stability Theory," in *Actes IX Congrès International de Mécanique Appliquée*, vol. 4, pp. 234–244, Université de Bruxelles, Bruxelles, Belgique.

Smith, A.M.O. (1974) "High Lift Aerodynamics," AIAA Paper No. 74-939, New York.

Smith, A.M.O. (1975) "High-Lift Aerodynamics," *J. Aircraft* **12**, pp. 501–530.

Smith, A.M.O. (1977) "Stratford's Turbulent Separation Criterion for Axially-Symmetric Flows," *J. Appl. Math. & Phys.* **28**, pp. 929–939.

Smith, A.M.O., and Gamberoni, N. (1956) "Transition, Pressure Gradient and Stability Theory," Douglas Aircraft Company Report No. ES-26388, El Segundo, California.

Smith, A.M.O., and Kaups, K. (1968) "Aerodynamics of Surface Roughness and Imperfections," Society of Automotive Engineers Paper No. SAE-680198, New York.

Smith, A.M.O., and Thelander, J.A. (1973) "The Power Profile—A New Type of Airfoil," McDonnell Douglas Corporation Report No. MDC-J6236, Long Beach, CA. (Also available from U.S. NTIS; Document Number AD-773 655/6.)

Smith, A.M.O., Stokes, T.R., Jr., and Lee, R.S. (1981) "Optimum Tail Shapes for Bodies of Revolution," *J. Hydronautics* **15**, pp. 67–73.

Smith, C.R., and Metzler, S.P. (1982) "A Visual Study of the Characteristics, Formation, and Regeneration of Turbulent Boundary Layer Streaks," in *Developments in Theoretical and Applied Mechanics*, vol. XI, eds. T.J. Chung and G.R. Karr, pp. 533–543, University of Alabama, Huntsville, Alabama.

Smith, C.R., and Metzler, S.P. (1983) "The Characteristics of Low-Speed Streaks in the Near-Wall Region of a Turbulent Boundary Layer," *J. Fluid Mech.* **129**, pp. 27–54.

Smith, C.R., and Schwartz, S.P. (1983) "Observation of Streamwise Rotation in the Near-Wall Region of a Turbulent Boundary Layer," *Phys. Fluids* **26**, pp. 641–652.

Smith, D.W., and Walker, J.H. (1959) "Skin-Friction Measurements in Incompressible Flow," NACA Report No. R-26, Washington, D.C.

Smith, R.L., and Blick, E.F. (1966) "Skin Friction of Compliant Surfaces with Foamed Material Substrate," *J. Hydronautics* **3**, pp. 100–102.

Smith, R.W. (1994) "Effect of Reynolds Number on the Structure of Turbulent Boundary Layers," Ph.D. Thesis, Princeton University, Princeton, New Jersey.

Smits, A.J. (1990) "New Developments in Understanding Supersonic Turbulent Boundary Layers," in *Proc. Twelfth Symp. on Turbulence*, eds. X.B. Reed, Jr., et al., pp. IL4.1–IL4.19, University of Missouri, Rolla, Missouri.

Smits, A.J., and Dussauge, J.P. (1996) *Turbulent Shear Layers in Supersonic Flows*, American Institute of Physics, Woodbury, New York.

Smits, A.J., and Wood, D.H. (1985) "The Response of Turbulent Boundary Layers to Sudden Perturbations," *Ann. Rev. Fluid Mech.* **17**, pp. 321–358.

Smits, A.J., Spina, E.F., Alving, A.E., Smith, R.W., Fernando, E.M., and Donovan, J.F. (1989) "A Comparison of the Turbulence Structure of Subsonic and Supersonic Boundary Layers," *Phys. Fluids A* **1**, pp. 1865–1875.

Smits, J.G. (1990) "Piezoelectric Micropump with Three Valves Working Peristaltically," *Sensors & Actuators A* **21–23**, pp. 203–206.

Sniegowski, J.J., and Garcia, E.J. (1996) "Surface Micromachined Gear Trains Driven by an On-Chip Electrostatic Microengine," *IEEE Electron Device Letters* **17**, July, p. 366.

So, R.M.C., and Mellor, G.L. (1973) "Experiment on Convex Curvature Effects in Turbulent Boundary Layers," *J. Fluid Mech.* **60**, pp. 43–62.

Soderman, P.T. (1972) "Aerodynamic Effects of Leading Edge Separation on a Two-Dimensional Airfoil," NASA TMX-2643, Washington, D.C.

Soo, S.L., and Trezek, G.J. (1966) "Turbulent Pipe Flow of Magnesia Particles in Air," *I&EC Fundamentals* **5**, pp. 388–392.

Sovran, G., Morel, T., and Mason, W.T., Jr. (1978) *Aerodynamic Drag Mechanisms of Bluff Bodies and Road Vehicles*, Plenum Press, New York.

Spaid, F.W. (1972) "Cooled Supersonic Turbulent Boundary Layer Separated by a Forward Facing Step," *AIAA J.* **19**, pp. 1117–1119.

Spalart, P.R. (1986) "Direct Simulation of a Turbulent Boundary Layer up to $R_\theta = 1410$," NASA Technical Memorandum No. TM-89407, Washington, D.C.

Spalart, P.R. (1988) "Direct Simulation of a Turbulent Boundary Layer up to $R_\theta = 1410$," *J. Fluid Mech.* **187**, pp. 61–98.

Speziale, C.G. (1991) "Analytical Methods for the Development of Reynolds-Stress Closures in Turbulence," *Ann. Rev. Fluid Mech.* **23**, pp. 107–157.

Spina, E.F., Donovan, J.F., and Smits, A.J. (1991) "On the Structure of High-Reynolds-Number Supersonic Turbulent Boundary Layers," *J. Fluid Mech.* **222**, pp. 293–327.

Squire, H.B. (1933) "On the Stability for Three-Dimensional Disturbances of Viscous Fluid Flow Between Parallel Walls," *Proc. R. Soc. London A* **142**, pp. 621–628.

Sreenivasan, K.R. (1988) "A Unified View of the Origin and Morphology of the Turbulent Boundary Layer Structure," in *Turbulence Management and Relaminarisation*, eds. H.W. Liepmann and R. Narasimha, pp. 37–61, Springer–Verlag, Berlin.

Sreenivasan, K.R. (1989) "The Turbulent Boundary Layer," in *Frontiers in Experimental Fluid Mechanics*, ed. M. Gad-el-Hak, pp. 159–209, Springer–Verlag, New York.

Sreenivasan, K.R., and Antonia, R.A. (1977) "Properties of Wall Shear Stress Fluctuations in a Turbulent Duct," *J. Appl. Mech.* **44**, pp. 389–395.

Sreenivasan, K.R., and White, C.M. (1999) "The Onset of Drag Reduction by Dilute Polymer Additives, and the Maximum Drag Reduction Asymptote," *J. Fluid Mech.*, to appear.

Sreenivasan, K.R., Ramshankar, R., and Meneveau, C. (1989) "Mixing, Entrainment and Fractal Dimensions of Surfaces in Turbulent Flows," *Proc. R. Soc. London A* **421**, pp. 79–108.

Sritharan, S.S., editor (1998) *Optimal Control of Viscous Flow*, SIAM, Philadelphia, Pennsylvania.

Stanewsky, E., and Krogmann, P. (1985) "Transonic Drag Rise and Drag Reduction by Active/Passive Boundary Layer Control," in *Aircraft Drag Prediction and Reduction*, AGARD R-723, pp. 11.1–11.41, AGARD, NATO, Rhode-Saint-Génèse, Belgium.

Staniforth, R. (1958) "Some Tests on Cascades of Compressor Blades Fitted with Vortex Generators," NGTE Memorandum No. M.314, National Gas Turbine Establishment, Farnborough, Great Britain. (Also Aeronautical Research Council, CP-487, London, Great Britain.)

Steinheil, E., Scherber, W., Seidl, M., and Rieger, H. (1977) "Investigations on the Interaction of Gases and Well-Defined Solid Surfaces with Respect to Possibilities for Reduction of Aerodynamic Friction and Aerothermal Heating," in *Rarefied Gas Dynamics*, ed. J.L. Potter, pp. 589–602, AIAA, New York.

Stollery, J.L., and Dyer, D.J. (1989) Wing-Section Effects on the Flight Performance of a Remotely Piloted Vehicle, *J. Aircraft* **26**, pp. 932–938.

Stratford, B.S. (1959a) "The Prediction of Separation of the Turbulent Boundary Layer," *J. Fluid Mech.* **5**, pp. 1–16.

Stratford, B.S. (1959b) "An Experimental Flow with Zero Skin Friction Throughout Its Region of Pressure Rise," *J. Fluid Mech.* **5**, pp. 17–35.

Strazisar, A.J., Reshotko, E., and Prahl, J.M. (1977) "Experimental Study of the Stability of Heated Laminar Boundary Layers in Water," *J. Fluid Mech.* **83**, pp. 225–247.

Strykowski, P.J., and Niccum, D.L. (1991) "The Stability of Countercurrent Mixing Layers in Circular Jets," *J. Fluid Mech.* **227**, pp. 309–343.

Strykowski, P.J., and Sreenivasan, K.R. (1990) "On the Formation and Suppression of Vortex 'Shedding' at Low Reynolds Numbers," *J. Fluid Mech.* **218**, pp. 71–107.

Stuart, J.T. (1963) "Hydrodynamic Stability," in *Laminar Boundary Layer Theory*, ed. L. Rosenhead, pp. 492–579, Clarendon Press, Oxford, Great Britain.

Stull, F.D., and Velkoff, H.R. (1975) "Flow Regimes in Two-Dimensional Ribbed Diffusers," *J. Fluids Eng.* **97**, pp. 87–96.

Swanson, W.M. (1961) "The Magnus Effect: A Summary of Investigations to Date," *J. Basic Eng.* **83**, pp. 461–470.

Swearingen, J.D., and Blackwelder, R.F. (1984) "Instantaneous Streamwise Velocity Gradients in the Wall Region," *Bull. Am. Phys. Soc.* **29**, p. 1528.

Swearingen, J.D., and Blackwelder, R.F. (1987) "The Growth and Breakdown of Streamwise Vortices in the Presence of a Wall," *J. Fluid Mech.* **182**, pp. 255–290.

Swift, J. (1726) *Gulliver's Travels*, 1906 reprinting, J.M. Dent and Company, London, Great Britain.

Szodruch, J., and Schneider, H. (1989) "High Lift Aerodynamics for Transport Aircraft by Interactive Experimental and Theoretical Tool Development," AIAA Paper No. 89-0267, New York.

Tai, Y.-C. (1993) "Silicon Micromachining and Micromechanics," invited oral presentation at *AIAA Third Flow Control Conference*, 6–9 July, Orlando, Florida.

Tai, Y.-C., and Muller, R.S. (1989) "IC-Processed Electrostatic Synchronous Micromotors," *Sensors & Actuators* **20**, pp. 49–55.

Tam, C.K.W. (1995) "Supersonic Jet Noise," *Ann. Rev. Fluid Mech.* **271**, pp. 17–43.

Tang, J., and Bau, H.H. (1993a) "Stabilization of the No-Motion State in Rayleigh-Bénard Convection Through the Use of Feedback Control," *Phys. Rev. Lett.* **70**, pp. 1795–1798.

Tang, J., and Bau, H.H. (1993b) "Feedback Control Stabilization of the No-Motion State of a Fluid Confined in a Horizontal Porous Layer Heated from Below," *J. Fluid Mech.* **257**, pp. 485–505.

Tang, W.C., Nguyen, T.-C., and Howe, R.T. (1989) "Laterally Driven Polysilicon Resonant Microstructures," *Sensors & Actuators* **20**, pp. 25–32.

Tani, I. (1964) "Low-Speed Flows Involving Bubble Separations," in *Prog. Aeronautical Sci.* **5**, eds. D. Küchemann and L.H.G. Sterne, pp. 70–103, Pergamon Press, New York.

Tani, I. (1969) "Boundary-Layer Transition," *Ann. Rev. Fluid Mech.* **1**, pp. 169–196.

Tani, I. (1986) "Some Equilibrium Turbulent Boundary Layers," *Fluid Dyn. Res.* **1**, pp. 49–58.

Tani, I. (1987) "Turbulent Boundary Layer Development over Rough Surfaces," in *Perspectives in Turbulent Studies*, eds. H.U. Meier and P. Bradshaw, pp. 223–249, Springer–Verlag, Berlin.

Tani, I. (1988) "Drag Reduction by Riblet Viewed as Roughness Problem," *Proc. Japan Acad. B* **64**, pp. 21–24.

Tani, I., and Motohashi, T. (1985a) "Non-Equilibrium Behavior of Turbulent Boundary Layer Flows. Part I. Method of Analysis," *Proc. Japan Acad. B* **61**, pp. 333–336.

Tani, I., and Motohashi, T. (1985b) "Non-Equilibrium Behavior of Turbulent Boundary Layer Flows. Part II. Results of Analysis," *Proc. Japan Acad. B* **61**, pp. 337–340.

Tavella, D.A., Lee, C.S., and Wood, N.J. (1986) "Influence of Wing Tip Configuration on Lateral Blowing Efficiency," AIAA Paper No. 86-0475, New York.

Tavella, D.A., Wood, N.J., Lee, C.S., and Roberts, L. (1986) "Two Blowing Concepts for Roll and Lateral Control of Aircraft," Department of Aeronautics and Astronautics Report No. TR-75, Stanford University, Stanford, California.

Tavella, D.A., Wood, N.J., Lee, C.S., and Roberts, L. (1988) "Lift Modulation with Lateral Wing-Tip Blowing," *J. Aircraft* **25**, pp. 311–316.

Taylor, G. (1972) "Low-Reynolds-Number Flows," in *Illustrated Experiments in Fluid Mechanics*, pp. 47–54, National Committee for Fluid Mechanics Films, MIT Press, Cambridge, Massachusetts.

Taylor, G.I. (1923) "Stability of a Viscous Liquid Contained between Two Rotating Cylinders," *Phil. Trans. R. Soc. London A* **223**, pp. 289–343.

Taylor, G.I. (1935) "Statistical Theory of Turbulence," *Proc. Roy. Soc. London A* **151**, pp. 421–478.

Taylor, H.D. (1948a) "Application of Vortex Generator Mixing Principles to Diffusers," Research Department Concluding Report No. R-15064-5, United Aircraft Corporation, East Hartford, Connecticut.

Taylor, H.D. (1948b) "Design Criteria for and Applications of the Vortex Generator Mixing Principle," Research Department Report No. M-15038-1, United Aircraft Corporation, East Hartford, Connecticut.

Telionis, D.P. (1979) "Review—Unsteady Boundary Layers, Separated and Attached, *J. Fluids Eng.* **101**, pp. 29–43.

Telionis, D.P., and Werle, M.J. (1973) "Boundary-Layer Separation from Downstream Moving Boundaries," *J. Appl. Mech.* **40**, pp. 369–374.

Tennant, J.S. (1973) "A Subsonic Diffuser with Moving Walls for Boundary Layer Control," *AIAA J.* **11**, pp. 240–242.

Tennant, J.S., Johnson, W.S., Keanton, D.D., and Krothapalli, A. (1975) "The Application of Moving Wall Boundary Layer Control to Submarine Control Surfaces," University of Tennessee Report No. MAE-75-01210-1, Knoxville, Tennessee. (Also available from U.S. NTIS; Document Number AD-AO23536.)

Tennekes, H., and Lumley, J.L. (1972) *A First Course in Turbulence*, MIT Press, Cambridge, Massachusetts.

Theodorsen, Th. (1952) "Mechanism of Turbulence," in *Proc. Second Midwestern Conf. on Fluid Mechanics*, pp. 1–18, Ohio State University, Columbus, Ohio.

Theodorson, Th. (1955) "The Structure of Turbulence," in *50 Jahre Grenzschichtsforschung (Ludwig Prandtl Anniversal Volume)*, eds. H. Görtler and W. Tollmien, pp. 55–62, Friedr. Vieweg und Sohn, Braunschweig, Germany.

Thibert, J.J., Reneaux, J., and Schmitt, V. (1990) "ONERA Activities on Drag Reduction," in *Proc. Seventeenth Cong. of the International Council of the Aeronautical Sciences*, vol. 1, paper no. 90-3.6.1, pp. 1053–1064, ICAS, Washington, D.C.

Thieme, H. (1997) "Lower Palaeolithic Hunting Spears from Germany," *Nature* **385**, 27 February, p. 807.

Thomas, A.S.W. (1983) "The Control of Boundary-Layer Transition Using a Wave Superposition Principle," *J. Fluid Mech.* **137**, pp. 233–250.

Thomas, A.S.W., and Bull, M.K. (1983) "On the Role of Wall-Pressure Fluctuations in Deterministic Motions in the Turbulent Boundary Layer," *J. Fluid Mech.* **128**, pp. 283–322.

Thomas, F.O., Liu, X., and Nelson, R.C. (1997) "Experimental Investigation of the Confluent Boundary Layer of a High-Lift System," AIAA Paper No. 97-1934, Washington, D.C.

Thomas, F.O., Nelson, R.C., and Liu, X. (1999) "The Confluent Boundary Layer of a High-Lift System: Experiments and Flow Control Strategies," *AIAA J.*, to appear.

Thomas, L.B., and Lord, R.G. (1974) "Comparative Measurements of Tangential Momentum and Thermal Accommodations on Polished and on Roughened Steel Spheres," in *Rarefied Gas Dynamics 8*, eds. K. Karamcheti, Academic Press, New York.

Thomas, M.D. (1992a) "On the Resonant Triad Interaction in Flows over Rigid and Flexible Boundaries," *J. Fluid Mech.* **234**, pp. 417–442.

Thomas, M.D. (1992b) "On the Nonlinear Stability of Flows over Compliant Walls," *J. Fluid Mech.* **239**, pp. 657–670.

Thompson, P.A., and Robbins, M.O. (1989) "Simulations of Contact Line Motion: Slip and the Dynamic Contact Line," *Nature* **389**, 25 September, pp. 360–362.

Thompson, P.A., and Troian, S.M. (1997) "A General Boundary Condition for Liquid Flow at Solid Surfaces," *Phys. Rev. Lett.* **63**, pp. 766–769.

Thwaites, B. (1949) "Approximate Calculations of the Laminar Boundary Layer," *Aero. Quart.* **1**, pp. 245–280.

Tichy, J., Warnaka, G.E., and Poole, L.A. (1984) "A Study of Active Control of Noise in Ducts," *J. Vibration, Acoustics, Stress, and Reliability in Design* **106**, pp. 399–404.

Tiederman, W.G., Luchik, T.S., and Bogard, D.G. (1985) "Wall-Layer Structure and Drag Reduction," *J. Fluid Mech.* **156**, pp. 419–437.

Tien, N.C. (1997) "Silicon Micromachined Thermal Sensors and Actuators," *Microscale Thermophys. Eng.* **1**, pp. 275–292.

Tison, S.A. (1993) "Experimental Data and Theoretical Modeling of Gas Flows Through Metal Capillary Leaks," *Vacuum* **44**, pp. 1171–1175.

Tobak, M., and Peake, D.J. (1982) "Topology of Three-Dimensional Separated Flows," *Ann. Rev. Fluid Mech.* **14**, pp. 61–85.

Tokhi, M.O., and Leitch, R.R. (1992) *Active Noise Control*, Clarendon Press, Oxford, Great Britain.

Tollmien, W. (1935) "Ein allgemeines Kriterium der Instabilität laminarer Geschwindigkeitsverteilungen," *Nachr. Wiss. Fachgruppe Göttingen, Math.-Phys. Klasse* **1**, pp. 79–114.

Toms, B.A. (1948) "Some Observations on the Flow of Linear Polymer Solutions through Straight Tubes at Large Reynolds Numbers," in *Proc. First Int. Cong. Rheol.*, vol. 2, pp. 135–141, North-Holland, Amsterdam.

Townsend, A.A. (1956) *The Structure of Turbulent Shear Flow*, Cambridge University Press, Cambridge, Great Britain.

Townsend, A.A. (1961) "Equilibrium Layers and Wall Turbulence," *J. Fluid Mech.* **11**, pp. 97–120.

Townsend, A.A. (1970) "Entrainment and the Structure of Turbulent Flow," *J. Fluid Mech.* **41**, pp. 13–46.

Townsend, A.A. (1976) *The Structure of Turbulent Shear Flow*, second edition, Cambridge University Press, Cambridge, Great Britain.

Tritton, D.J. (1988) *Physical Fluid Dynamics*, second edition, Clarendon Press, Oxford, Great Britain.

Truckenbrodt, E. (1956) "Ein einfaches Näherungsverfahren zum Berechnen der laminaren Reibungsschicht mit Absaugung," *Forschg. Ing.-Wes.* **22**, pp. 147–157.

Tsahalis, D. Th., and Telionis, D.P. (1974) "Oscillating Laminar Boundary Layers and Unsteady Separation," *AIAA J.* **12**, pp. 1469–1476.

Tsai, H.M., and Leslie, D.C. (1990) "Large Eddy Simulation of a Developing Turbulent Boundary Layer at Low Reynolds Number," *Int. J. Num. Methods in Fluids* **10**, pp. 519–555.

Tsao, T., Jiang, F., Miller, R.A., Tai, Y.-C., Gupta, B., Goodman, R., Tung, S., and Ho, C.-M. (1997) "An Integrated MEMS System for Turbulent Boundary Layer Control," in *Technical Digest (Transducers '97)*, vol. 1, pp. 315–318.

Tu, B.J., and Willmarth, W.W. (1966) "An Experimental Study of the Structure of Turbulence Near the Wall Through Correlation Measurements in a Thick Turbulent Boundary Layer," Department of Aerospace Engineering Report No. 02920-3-T, University of Michigan, Ann Arbor, Michigan.

Tuckermann, D.B. (1984) "Heat Transfer Microstructures for Integrated Circuits," Ph.D. Thesis, Stanford University, Stanford, California.

Tuckermann, D.B., and Pease, R.F.W. (1981) "High-Performance Heat Sinking for VLSI," *IEEE Electron Device Lett.* **EDL-2**, no. 5, May.

Tuckermann, D.B., and Pease, R.F.W. (1982) "Optimized Convective Cooling Using Micromachined Structures," *J. Electrochem. Soc.* **129**, no. 3, C98, March.

Ueda, H., and Hinze, J.O. (1975) "Fine Structure Turbulence in the Wall Region of a Turbulent Boundary Layer," *J. Fluid Mech.* **67**, pp. 125–143.

Ueda, H., and Mizushina, T. (1979) "Turbulence Structure in the Inner Part of the Wall Region in a Fully Developed Turbulent Flow," in *Proc. Fifth Biennial Symp. on Turbulence*, eds. G.K. Patterson and J.L. Zakin, pp. 357–366, Science Press, Princeton, New Jersey.

Ulrich, A. (1944) "Theoretische Untersuchungen über die Widerstandsersparnis durch Laminarhaltung mit Absaugung," *Schriften Dtsch. Akad. Luftfahrtforschung B* **8**, p. 53.

Vakili, A.D. (1990) "Review of Vortical Flow Utilization," AIAA Paper No. 90-1429, New York.

Vakili, A.D., Wu, J.M., and Bhat, M.K. (1988) "High Angle of Attack Aerodynamics of Excitation of the Locked Leeside Vortex," Society of Automotive Engineers Paper No. SAE-88-1424, New York.

Vakili, A.D., Wu, J.M., Liver, P., and Bhat, M.K. (1985) "Flow Control in a Diffusing S-Duct," AIAA Paper No. 85-0524, New York.

Van den Berg, H.R., Seldam, C.A., and Gulik, P.S. (1993) "Compressible Laminar Flow in a Capillary," *J. Fluid Mech.* **246**, pp. 1–20.

Van Dyke, M. (1964) *Perturbation Methods in Fluid Mechanics*, Academic Press, New York.

Van Ingen, J.L. (1956) "A Suggested Semiempirical Method for the Calculation of the Boundary-Layer Transition Region," Department of Aerospace Engineering Report No. V.T.H.74, Institute of Technology, Delft, The Netherlands.

Van Ingen, J.L., and Boermans, L.M.M. (1986) "Aerodynamics at Low Reynolds Numbers: A Review of Theoretical and Experimental Research at Delft University of Technology," in *Proc. Int. Conf. on Aerodynamics at Low Reynolds Numbers*, vol. I, pp. 1.1–1.40, Royal Aeronautical Society, London, Great Britain.

Van Laere, L., and Sas, P. (1988) "Principles and Applications of Active Noise Cancellation," in *Noise Control Design: Methods and Practice (Noise-Con 88 Proceedings)*, ed. J.S. Bolton, pp. 279–284, Noise Control Foundation, Poughkeepsie, NY.

Van Lintel, H.T.G., Van de Pol, F.C.M., and Bouwstra, S. (1988) "A Piezoelectric Micropump Based on Micromachining of Silicon," *Sensors & Actuators* **15**, pp. 153–167.

Vargo, S.E., and Muntz, E.P. (1996) "A Simple Micromechanical Compressor and Vacuum Pump for Flow Control and Other Distributed Applications," AIAA Paper No. 96-0310, AIAA, Washington, D.C.

Vasilyev, O.V., Yuen, D.A., and Paolucci, S. (1997) "Solving PDEs Using Wavelets," *Computers in Physics* **11**, no. 5, pp. 429–435.

Verollet, E., Fulachier, L., Dumas, R., and Favre, A. (1972) "Turbulent Boundary Layer with Suction and Heating to the Wall," in *Heat and Mass Transfer in Boundary Layers*, eds. N. Afgan, Z. Zaric, and P. Anastasijevec, vol. 1, pp. 157–168, Pergamon Presss, Oxford.

Vidal, R.J. (1959) "Research on Rotating Stall in Axial-Flow Compressors: Part III—Experiments on Laminar Separation from a Moving Wall," Wright Air Development Center Technical Report No. 59-75, Wright-Patterson Air Force Base, Ohio.

Viets, H. (1980) "Coherent Structures in Time Dependent Shear Flows," in *Turbulent Boundary Layers*, AGARD CPP-271, paper no. 5, AGARD, NATO, Neuilly-sur-Seine, France.

Viets, H., Ball, M., and Bougine, D. (1981) "Performance of Forced Unsteady Diffusers," AIAA Paper No. 81-0154, New York.

Viets, H., Palmer, G.M., and Bethke, R.J. (1984) "Potential Applications of Forced Unsteady Flows," in *Unsteady Separated Flows*, eds. M.S. Francis and M.W. Luttges, pp. 21–27, U.S. Air Force Academy, Colorado Springs, Colorado.

Viets, H., Piatt, M., and Ball, M. (1979) "Unsteady Wing Boundary Layer Energization," AIAA Paper No. 79-1631, New York.

Viets, H., Piatt, M., and Ball, M. (1981a) "Forced Vortex Near a Wall," AIAA Paper No. 81-0256, New York.

Viets, H., Piatt, M., and Ball, M. (1981b) "Boundary Layer Control by Unsteady Vortex Generation," *J. Wind Eng. & Industrial Aerodynamics* **7**, pp. 135–144.

Vijgen, P.M.H.W., van Dam, C.P., Holmes, B.J., and Howard, F.G. (1989) "Wind-Tunnel Investigations of Wings with Serrated Sharp Trailing Edges," in *Low Reynolds Number Aerodynamics*, ed. T.J. Mueller, pp. 295–313, Springer–Verlag, Berlin.

Vincenti, W.G., and Kruger, C.H., Jr. (1965) *Introduction to Physical Gas Dynamics*, Wiley, New York.

Virk, P.S. (1975) "Drag Reduction Fundamentals," *AIChE J.* **21**, pp. 625–656.

Virk, P.S., Merrill, E.W., Mickley, H.S., Smith, K.A., and Mollo-Christensen, E.L. (1967) "The Toms Phenomenon: Turbulent Pipe Flow of Dilute Polymer Solutions," *J. Fluid Mech.* **30**, pp. 305–328.

Viswanath, P.R. (1988) "Shockwave-Turbulent Boundary Layer Interaction and Its Control: A Survey of Recent Developments," *Sādhānā* **12**, pp. 45–104.

Viswanath, P.R. (1995) "Flow Management Techniques for Base and Afterbody Drag Reduction," *Prog. Aero. Sci.* **32**, pp. 79–129.

Von Doenhoff, A.E. (1938a) "A Method of Rapidly Estimating the Position of the Laminar Separation Point," NACA TN No. 671, Washington, D.C.

Von Doenhoff, A.E. (1938b) "A Preliminary Investigation of Boundary-Layer Transition Along a Flat Plate with Adverse Pressure Gradient," NACA TN No. 639, Washington, D.C.

Von Kármán, Th. (1930) "Mechanische Ähnlichkeit und Turbulenz," *Nachr. Ges. Wiss. Göttingen, Math.-Phys. Klasse*, pp. 58–76.

Von Smoluchowski, M. (1898) "Über Wärmeleitung in verdünnten Gasen," *Annalen der Physik und Chemie* **64**, pp. 101–130.

Von Winkle, W.A. (1961) "An Evaluation of a Boundary Layer Stabilization Coating," Naval Underwater Systems Center, Tech. Memo. No. 922-111-61, New London, Connecticut.

Wadsworth, D.C., Muntz, E.P., Blackwelder, R.F., and Shiflett, G.R. (1993) "Transient Energy Release Pressure Driven Microactuators for Control of Wall-Bounded Turbulent Flows," AIAA Paper No. 93-3271, AIAA, Washington, D.C.

Waggoner, E.G., and Allison, D.O. (1987) "EA-6B High Lift Wing Modifications," AIAA Paper No. 87-2360-CP, New York.

Wagner, R.D., and Fischer, M.C. (1984) "Fresh Attack on Laminar Flow," *Aerospace America* **22**, March, pp. 72–76.

Wagner, R.D., Bartlett, D.W., and Maddalon, D.V. (1988) "Laminar Flow Control is Maturing," *Aerospace America* **26**, January, pp. 20–24.

Wagner, R.D., Fischer, M.C., Collier, F.S., Jr., and Pfenninger, W. (1990) "Supersonic Laminar Flow Control on Commercial Transports," in *Proc. Seventeenth Congress of the International Council of the Aeronautical Sciences*, vol. 1, pp. 1073–1089, paper no. 90-3.6.3, ICAS, Washington, D.C.

Wagner, R.D., Maddalon, D.V., and Fischer, M.C. (1984) "Technology Development for Laminar Boundary Control on Subsonic Transport Aircraft," AGARD CP-365, paper no. 16, AGARD, NATO, Rhode-Saint-Génèse, Belgium.

Walker, D.T., and Tiederman, W.G. (1990) "Turbulent Structure in a Channel Flow with Polymer Injection at the Wall," *J. Fluid Mech.* **218**, pp. 377–403.

Walker, J.D.A. (1978) "The Boundary Layer due to Rectilinear Vortex," *Proc. R. Soc. London A* **359**, pp. 167–188.

Walker, J.D.A., and Herzog, S. (1988) "Eruption Mechanisms for Turbulent Flows Near Walls," in *Transport Phenomena in Turbulent Flows*, eds. M. Hirata and N. Kasagi, pp. 145–156, Hemisphere, New York.

Wallis, R.A., and Stuart, C.M. (1958) "On the Control of Shock Induced Boundary Layer Separation with Discrete Jets," Aeronautical Research Council Current Paper No. 494, London, Great Britain.

Walsh, M.J. (1980) "Drag Characteristics of V-Groove and Transverse Curvature Riblets," in *Viscous Flow Drag Reduction*, ed. G.R. Hough, pp. 168–184, AIAA, New York.

Walsh, M.J. (1982) "Turbulent Boundary Layer Drag Reduction Using Riblets," AIAA Paper No. 82-0169, New York.

Walsh, M.J. (1983) "Riblets as a Viscous Drag Reduction Technique," *AIAA J.* **21**, pp. 485–486.

Walsh, M.J. (1990) "Riblets," in *Viscous Drag Reduction in Boundary Layers*, eds. D.M. Bushnell and J.N. Hefner, pp. 203–261, AIAA, Washington, D.C.

Walsh, M.J., and Lindemann, A.M. (1984) "Optimization and Application of Riblets for Turbulent Drag Reduction," AIAA Paper No. 84-0347, New York.

Walsh, M.J., Sellers, W.L., III, and McGinley, C.B. (1989) "Riblet Drag at Flight Conditions," *J. Aircraft* **26**, pp. 570–575.

Walsh, M.J., and Weinstein, M. (1978) "Drag and Heat Transfer on Surfaces with Small Longitudinal Fins," AIAA Paper No. 78-1161, New York.

Wang, Y., Singer, J., and Bau, H.H. (1992) "Controlling Chaos in a Thermal Convection Loop," *J. Fluid Mech.* **237**, pp. 479–498.

Wannier, G.H. (1950) "A Contribution to the Hydrodynamics of Lubrication," *Quart. Appl. Math.* **8**, pp. 1–32.

Wark, C.E., and Nagib, H.M. (1991) "Experimental Investigation of Coherent Structures in Turbulent Boundary Layers," *J. Fluid Mech.* **230**, pp. 183–208.

Warkentin, D.J., Haritonidis, J.H., Mehregany, M., and Senturia, S.D. (1987) "A Micromachined Microphone with Optical Interference Readout," *Proc. Fourth Int. Conf. on Solid-State Sensors and Actuators (Transducers '87)*, June, Tokyo, Japan.

Warnaka, G.E. (1982) "Active Attenuation of Noise: The State of the Art," *Noise Control Eng.* **18**, pp. 100–110.

Warner, J.V., Waters, D.E., and Bernhard, R.J. (1988) "Adaptive Active Noise Control in Three Dimensional Enclosures," in *Noise Control Design: Methods and Practice (Noise-Con 88 Proceedings)*, ed. J.S. Bolton, pp. 285–290, Noise Control Foundation, Poughkeepsie, New York.

Wazzan, A.R., Okamura, T.T., and Smith, A.M.O. (1968) "The Stability of Water Flow over Heated and Cooled Flat Plates," *J. Heat Transfer* **90**, pp. 109–114.

Wazzan, A.R., Okamura, T.T., and Smith, A.M.O. (1970) "The Stability and Transition of Heated and Cooled Incompressible Boundary Layers," in *Proc. Fourth Int. Heat Transfer Conf.*, eds. U. Grigull and E. Hahne, vol. 2, FC 1.4, Elsevier, New York.

Webb, R.L. (1987) "Enhancement of Single-Phase Heat Transfer," in *Handbook of Single-Phase Convective Heat Transfer*, eds. S. Kakaç, R.K. Shah, and W. Aung, pp. 17.1–17.62, Wiley–Interscience, New York.

Webb, R.L. (1993) *Principles of Enhanced Heat Transfer*, Wiley, New York.

Webb, R.L., Camavos, T.C., Park, E.L., Jr., and Hostetler, K.M., editors (1981) *Advances in Enhanced Heat Transfer*, ASME HTD-18, New York.

Wei, T., and Willmarth, W.W. (1989) "Reynolds-Number Effects on the Structure of a Turbulent Channel Flow," *J. Fluid Mech.* **204**, pp. 57–95.

Weiberg, J.A., Giulianettij, D., Gambucci, B., and Innis, R.C. (1973) "Takeoff and Landing Performance and Noise Characteristics of a Deflected STOL Airplane with Interconnected Propellers and Rotating Cylinder Flaps," NASA TM X-62,320, Washington, D.C.

Weis-Fogh, T. (1973) "Quick Estimates of Flight Fitness in Hovering Animals, Including Novel Mechanisms for Lift Production," *J. Exp. Biol.* **59**, pp. 169–230.

Wells, C.S., editor (1969) *Viscous Drag Reduction*, Plenum Press, New York.

Wells, C.S., Jr., and Spangler, J.G. (1967) "Injection of a Drag-Reducing Fluid into Turbulent Pipe Flow of a Newtonian Fluid," *Phys. Fluids* **10**, pp. 1890–1894.

Went, F.W. (1968) "The Size of Man," *American Scientist* **56**, pp. 400–413.

Werle, M.J., Paterson, R.W., and Presz, W.M., Jr. (1987) "Trailing-Edge Separation/Stall Alleviation, *AIAA J.* **25**, pp. 624–626.

Wheeler, G.O. (1984) "Means for Maintaining Attached Flow of a Flow Medium," U.S. Patent No. 4,455,045.

Whitcomb, R.T. (1956) "A Study of the Zero-Lift Drag-Rise Characteristics of Wing-Body Combinations Near the Speed of Sound," NACA Report No. 1273, Washington, D.C.

White, A., and Hemmings, J.A.G. (1976) "Drag Reduction by Additives: Review and Bibliography," BHRA Fluid Engineering, Cranfield, Great Britain.

Whitehead, A.H., and Bertram, M.H. (1971) "Alleviation of Vortex-Induced Heating to the Lee Side of Slender Wings in Hypersonic Flow," *AIAA J.* **9**, pp. 1870–1872.

Whites, R.C., Sudderth, R.W., and Wheldon, W.G. (1966) "Laminar Flow Control on the X-21," *Astronautics & Aeronautics* **4**, pp. 38–43.

Wickerhauser, M.V. (1994) *Adapted Wavelet Analysis from Theory to Software*, A.K. Peters Ltd., Wellesley, Massachusetts.

Wieghardt, K. (1943) "Über die Wandschubspannung in turbulenten Reibungsschichten bei verä nderlichem Aussendruck," Kaiser Wilhelm Institut für Strömungsforschung, no. U&M-6603, Göttingen, Germany.

Wilcox, D.C. (1993) *Turbulence Modeling for CFD*, DCW Industries, Los Angeles, California.

Wilkinson, S.P. (1988) "Direct Drag Measurements on Thin-Element Riblets with Suction and Blowing," AIAA Paper No. 88-3670-CP, Washington, D.C.

Wilkinson, S.P. (1990) "Interactive Wall Turbulence Control," in *Viscous Drag Reduction in Boundary Layers*, eds. D.M. Bushnell and J.N. Hefner, pp. 479–509, AIAA, Washington, D.C.

Wilkinson, S.P., Anders, J.B., Lazos, B.S., and Bushnell, D.M. (1987) "Turbulent Drag Reduction Research at NASA Langley-Progress and Plans," in *Proc. Int. Conf. on Turbulent Drag Reduction by Passive Means*, vol. 1, pp. 1–32, Royal Aeronautical Society, London, Great Britain.

Wilkinson, S.P., Anders, J.B., Lazos, B.S., and Bushnell, D.M. (1988) "Turbulent Drag Reduction Research at NASA Langley: Progress and Plans," *Int. J. Heat and Fluid Flow* **9**, pp. 266–277.

Wilkinson, S.P., and Balasubramanian, R. (1985) "Turbulent Burst Control Through Phase-Locked Surface Depressions," AIAA Paper No. 85-0536, New York.

Wilkinson, S.P., and Lazos, B.S. (1987) "Direct Drag and Hot-Wire Measurements on Thin-Element Riblet Arrays," in *Turbulence Management and Relaminarization*, eds. H.W. Liepmann and R. Narasimha, pp. 121–131, Springer–Verlag, New York.

Williams, D.R., and Amato, C.W. (1989) "Unsteady Pulsing of Cylinder Wakes," in *Frontiers in Experimental Fluid Mechanics*, ed. M. Gad-el-Hak, pp. 337–364, Springer–Verlag, New York.

Williams, D.R., and Papazian, H. (1991) "Forebody Vortex Control with the Unsteady Bleed Technique," *AIAA J.* **29**, pp. 853–855.

Williams, J.C. (1985) "Singularities in Solutions of the Three-Dimensional Laminar-Boundary-Layer Equations," *J. Fluid Mech.* **160**, pp. 257–279.

Williams, J.C., III (1977) "Incompressible Boundary-Layer Separation," *Ann. Rev. Fluid Mech.* **9**, pp. 113–144.

Williams, J.C., III, and Johnson, W.D. (1974a) "Semisimilar Solutions to Unsteady Boundary-Layer Flows Including Separation," *AIAA J.* **12**, pp. 1388–1393.

Williams, J.C., III, and Johnson, W.D. (1974b) "Note on Unsteady Boundary-Layer Separation," *AIAA J.* **12**, pp. 1427–1429.

Williams, T.I. (1987) *The History of Invention*, Facts on File Publications, New York.

Willis, G.J.K. (1986) "Hydrodynamic Stability of Boundary Layers over Compliant Surfaces," Ph.D. Thesis, University of Exeter, Exeter, Great Britain.

Willmarth, W.W. (1959) "Space-Time Correlations and Spectra of Wall Pressure in a Turbulent Boundary Layer," NASA Memorandum No. 3-17-59W, Washington, D.C.

Willmarth, W.W. (1975a) "Structure of Turbulence in Boundary Layers," *Adv. Appl. Mech.* **15**, pp. 159–254.

Willmarth, W.W. (1975b) "Pressure Fluctuations Beneath Turbulent Boundary Layers," *Ann. Rev. Fluid Mech.* **7**, pp. 13–37.

Willmarth, W.W., and Bogar, T.J. (1977) "Survey and New Measurements of Turbulent Structure Near the Wall," *Phys. Fluids* **20**, no. 10, part II, pp. S9–S21.

Willmarth, W.W., and Sharma, L.K. (1984) "Study of Turbulent Structure with Hot Wires Smaller than the Viscous Length," *J. Fluid Mech.* **142**, pp. 121–149.

Willmarth, W.W., and Tu, B.J. (1967) "Structure of Turbulence in the Boundary Layer Near the Wall," *Phys. Fluids* **10**, pp. S134–S137.

Wilson, C.E. (1989) *Noise Control*, Harper & Row, New York.

Wiltse, J.M., and Glezer, A. (1993) "Manipulation of Free Shear Flows Using Piezoelectric Actuators," *J. Fluid Mech.* **249**, pp. 261–285.

Wimpenny, J.C. (1970) "Vortex Generators," U.S. Patent No. 3,525486.

Winant, C.D., and Browand, F.K. (1974) "Vortex Pairing: The Mechanism of Turbulent Mixing Layer Growth at Moderate Reynolds Numbers," *J. Fluid Mech.* **63**, pp. 237–255.

Wood, C.J. (1961) "A Study of Hypersonic Separated Flow," Ph.D. Thesis, University of London, London, Great Britain.

Wood, N.J., and Nielsen, J.N. (1985) "Circulation Control Airfoils—Past, Present, Future," AIAA Paper No. 85-0204, New York.

Wood, N.J., and Roberts, L. (1986) "Experimental Results of the Control of a Vortical Flow by Tangential Blowing," Joint Institute for Aeronautics and Acoustics Report No. JIAA TR-71, Stanford University, Stanford, CA.

Wood, N.J., and Roberts, L. (1988) "Control of Vortical Lift on Delta Wings by Tangential Leading-Edge Blowing," *J. Aircraft* **25**, pp. 236–243.

Wood, N.J., Roberts, L., and Celik, Z. (1990) "Control of Asymmetric Vortical Flows over Delta Wings at High Angles of Attack," *J. Aircraft* **27**, pp. 429–435.

Wooldridge, C.E., and Muzzy, R.J. (1966) "Boundary-Layer Turbulence Measurements with Mass Addition and Combustion," *AIAA J.* **4**, pp. 2009–2016.

Wortmann, A. (1987) "Alleviation of Fuselage from Drag Using Vortex Flows," Department of Energy Report No. DOE/CE/15277-T1, Washington, D.C.

Wu, J., and Tulin, M.P. (1972) "Drag Reduction by Ejecting Additive Solutions into a Pure-Water Boundary Layer," *ASME J. Basic Eng.* **94**, pp. 749–756.

Wu, J.M., Vakili, A.D., and Chen, Z.L. (1983) "Investigation on the Effects of Discrete Wingtip Jets," AIAA Paper No. 83-0546, New York.

Wu, J.M., Vakili, A.D., and Gilliam, F.T. (1984) "Aerodynamic Interactions of Wingtip Flow with Discrete Wingtip Jets," AIAA Paper No. 84-2206, New York.

Wuest, W. (1961) "Survey of Calculation Methods of Laminar Boundary Layers with Suction in Incompressible Flow," in *Boundary Layer and Flow Control*, ed. G.V. Lachmann, vol. 2, pp. 771–800, Pergamon Press, New York.

Wygnanski, I., Sokolov, M., and Friedman, D. (1976) "On a Turbulent 'Spot' in a Laminar Boundary Layer," *J. Fluid Mech.* **78**, pp. 785–819.

Yager, R.R., Zadeh, L.A., editors (1992) *An Intorduction to Fuzzy Logic Applications in Intelligent Systems*, Kluwer Academic, Boston, Massachusetts.

Yajnik, K.S. (1970) "Asymptotic Theory of Turbulent Shear Flows," *J. Fluid Mech.* **42**, pp. 411–427.

Yajnik, K.S., and Acharya, M. (1978) "Non-Equilibrium Effects in a Turbulent Boundary Layer Due to the Destruction of Large Eddies," in *Structure and Mechanisms of Turbulence*, ed. H. Fiedler, vol. 1, pp. 249–260, Springer–Verlag, New York.

Yakhot, V., and Orszag, S.A. (1986) "Renormalization Group Analysis of Turbulence. I. Basic Theory," *J. Scientif. Comput.* **1**, pp. 3–51.

Yang, T., Ntone, F., Jiang, T., and Pitts, D.R. (1984) "An Investigation of High Performance, Short Thrust Augmenting Ejectors," ASME Paper No. 84-WA/FE-10, New York.

Yeo, K.S. (1988) "The Stability of Boundary-Layer Flow over Single- and Multi-Layer Viscoelastic Walls," *J. Fluid Mech.* **196**, pp. 359–408.

Yeo, K.S. (1990) "The Hydrodynamic Stability of Boundary-Layer Flow over a Class of Anisotrpic Compliant Walls," *J. Fluid Mech.* **220**, pp. 125–160, 1990.

Yeo, K.S. (1992) "The Three-Dimensional Stability of Boundary-Layer Flow over Compliant Walls, *J. Fluid Mech.* **238**, pp. 537–577.

Yeo, K.S., and Dowling, A.P. (1987) "The Stability of Inviscid Flows over Passive Compliant Walls," *J. Fluid Mech.* **183**, pp. 265–292.

Young, A.D. (1953) "Boundary Layers," in *Modern Developments in Fluid Dynamics: High Speed Flow*, ed. L. Howarth, vol. 1, pp. 375–475, Clarendon Press, Oxford, Great Britain.

Young, A.M., Goldsberry, J.E., Haritonidis, J.H., Smith, R.I., and Senturia, S.D. (1988) "A Twin-Interferometer Fiber-Optic Readout for Diaphragm Pressure Transducers," in *IEEE Solid-State Sensor and Actuator Workshop*, 6–9 June, Hilton Head, South Carolina.

Yun, W. (1993) "System Considerations for Integration of Microsensors and Electronics," invited oral presentation at *AIAA Third Flow Control Conference*, 6–9 July, Orlando, Florida.

Zagarola, M.V. (1996) "Mean Flow Scaling in Turbulent Pipe Flow," Ph.D. Thesis, Princeton University, Princeton, New Jersey.

Zagarola, M.V., Smits, A.J., Orszag, S.A., and Yakhot, V. (1996) "Experiments in High Reynolds Number Turbulent Pipe Flow," AIAA Paper No. 96-0654, Washington, D.C.

Zakin, J.L., Poreh, M., Brosh, A., and Warshavsky, M. (1971) "Exploratory Study of Friction Reduction in Slurry Flows," in *Chem. Eng. Prog. Symp. Seri.*, no. 67, vol. 111, pp. 85–89, AIChE, New York.

Zakkay, V., Barra, V., and Hozumi, K. (1980) "Turbulent Boundary Layer Structure at Low and High Subsonic Speeds," in *Turbulent Boundary Layers—Experiments, Theory and Modelling*, AGARD Conference Proceedings No. 271, pp. 4.1–4.20, Rhode-Saint-Génèse, Belgium.

Zaman, K.B.M.Q., Bar-Sever, A., and Mangalam, S.M. (1987) "Effect of Acoustic Excitation on the Flow over a Low Re Airfoil," *J. Fluid Mech.* **182**, pp. 127–148.

Zaman, K.B.M.Q., and Hussain, A.K.M.F. (1980) "Vortex Pairing in a Circular Jet Under Controlled Excitation. Part 1. General Response," *J. Fluids Mech.* **101**, pp. 449–491.

Zaman, K.B.M.Q., and Hussain, A.K.M.F. (1981) "Turbulence Suppression in Free Shear Flows by Controlled Excitation," *J. Fluids Mech.* **103**, pp. 133–159.

Zaman, K.B.M.Q., and McKinzie, D.J. (1989) "Control of 'Laminar Separation' over Airfoils by Acoustic Excitation," AIAA Paper No. 89-0565, Washington, D.C. (Also NASA TM-101379, Washington, D.C.)

Zaman, K.B.M.Q., and Potapczuk, M.G. (1989) "The Low Frequency Oscillation in the Flow over a NACA0012 Airfoil with an "Iced" Leading Edge," in *Proc. Conf. Low Reynolds Number Aerodynamics*, ed. T.J. Mueller, pp. 117–128, University of Notre Dame, Notre Dame, Indiana.

Zang, T.A., Hussaini, M.Y., and Bushnell, D.M. (1984) "Numerical Computations of Turbulence Amplification in Shock Wave Interactions," *AIAA J.* **22**, pp. 13–22.

Zaric, Z. (1972) "Wall Turbulence Studies," *Adv. Heat Transfer* **8**, pp. 285–350.

Zhang, F., and Sheng, C. (1987) "A Prediction Method for Optimum Velocity Ratio of Air Jet Vortex Generator," *J. Aerospace Power* **2**, pp. 55–60.

Zhuk, V.I., and Ryzhov, O.S. (1980) "Formation of Recirculation Zones in the Boundary Layer on a Moving Surface," *Fluid Dyn.* **15**, pp. 637–644.

Zilberman, M., Wygnanski, I., and Kaplan, R.E. (1977) "Transitional Boundary Layer Spot in a Fully Turbulent Environment," *Phys. Fluids* **20**, no. 10, part II, pp. S258–S271.

Žukauskas, A.A. (1986) "Heat Transfer Augmentation in Single-Phase Flow," in *Proc. Eighth International Heat Transfer Conference*, vol. 1, pp. 47–57, Hemisphere, Washington, D.C.

Index